TECHNOMIC
PUBLISHING CO., INC.

HAZARDOUS MATERIALS
EMERGENCIES
RESPONSE AND CONTROL

John R. Cashman

©TECHNOMIC Publishing Co., Inc. 1983
851 New Holland Avenue
Box 3535
Lancaster, Pennsylvania 17604

Printed in U.S.A.
Library of Congress Card No. 82-74318
ISBN 87762-324-4

CONTENTS

1 BRINGING HAZARDOUS MATERIALS INCIDENTS UNDER CONTROL: ON-SITE WITH A HAZARDOUS MATERIALS TEAM

In the stillness just before dawn, Fountain, Colorado lay asleep and unsuspecting as 13 miles to the north in Colorado Springs five improperly secured railroad cars, including a tankcar of flammable naphtha and a boxcar of dynamite, started rolling backwards. Slowly at first, then with ever-increasing momentum, the runaway cars headed down the long incline into Fountain. Within minutes, they entered the Fountain depot with a rush and impacted with the Kansas City Express carrying 34 passengers. The tankcar of naphtha split open, and the highly flammable liquid ignited spreading flames throughout the depot. The dynamite car exploded with a resounding concussion, showering debris for hundreds of yards. Three persons died, and 28 were injured in the holocaust. By mid-morning, 2500 people were milling about the site of the hazardous materials incident asking each other and officials: "How could this happen? Can it happen again?"

Catastrophies involing an unreasonable risk to health, safety, and property when transported in commerce."

The National Solid Waste Management Association defines hazardous wastes as—"Any wastes or combination of wastes which because of its quantity, concentration, or chemical characteristics poses a substantial present or potential hazard to human or animal health or the environment because such wastes are bioconcentrative, highly flammable, extremely reactive, toxic, irritating, corrosive, or infectious."

The National Fire Protection Association has identified the fire hazard properties of 1300 Flammable liquids, gases, and volatile solids; and has defined the fire explosion, and toxicity hazards for 416 hazardous materials. The Department of Transportation requires that 1600 different materials be placarded as hazardous when transported in commerce.

The Association of American Railroads has estimated that there are 1600 different commodities classified as hazardous transported on the nation's railroads today compared to 30 or 40 different products classified as hazardous carried by rail at the turn of the century.

The National Institute of Occupational Safety and Health has listed 28,000 *toxic* chemicals and identified 2200 of that number as suspected carcinogens.

A rapidly expanding volume of hazardous materials being manufactured, processed, stored, transported, and disposed of; coupled with an increase in incidents and spills that invariably followed has resulted in the formation of highly trained, fully experienced, and specially equipped "Hazardous Materials Incident Response Teams." Although no two such teams are alike, they can be roughly divided into three basic types: industrial teams, commercial teams, and public safety agency teams.

Nearly every major chemical manufacturing and processing firm in this country; such as

1

Union Carbide, FMC, Monsanto, Air Products Company, Dupont, and many others have company personnel who are on call at all times to respond to incidents ivolving their company's products. Dick Heinze, for example, is Chief Chemist for FMC Corporation yet reponds to pesticide emergencies around the country as a member of the National Pesticide Emergency Team Network. Similarly, L.J. Hudson responds to pressurized tankcar emergencies throughout the southeast as a member of the Air Products Corporation Response Team.

Commercial response teams are a relatively recent phenomenon: private-for-profit firms that respond to hazardous materials incidents, or provide related services, for any employer. Safety Systems, Inc. of Jacksonville, Florida (ramrodded by Captain Ron Gore, commander of the Jacksonville Fire Department's Hazardous Materials Team) specializes in "hands-on" hazardous materials training for industry and government as well as incident prevention, control, and investigative consultation. Hazardous Materials Technology and Services, Inc. is the "Red Adair of Hazardous Materials Incidents and Spills" with 500 miles of Jacksonville. (Its incorporators are all members of the Jacksonville Fire Department Hazardous Materials Team) Ryckman's Emergency Action & Consulting Team (R.E.A.C.T.) based in St. Louis operates a nationwide network of hazardous materials response personnel. Other major firms in this field are O.H. Materials, Inc. of Ohio and Illinois; (oil and hazardous materials spill containment and clean-up) Hulcher's Emergency Services of Illinois; (nationwide track rebuilding and derailment clean-up for railroads) and Peabody Clean Industry. (spill control and clean-up, hazardous materials clean-up, and waste oil disposal) There are also a few individual freelancers in the field; probably the best known is Jerry Cook, a chemist in Biloxi, Mississippi who specializes in on-site chemical analysis and pollution consultation for various railroads.

In January, 1977 the Jacksonville (Florida) Fire Department initiated one of the first "hands-on" hazardous materials incident response teams of any public safety agency in the United States. The unit is composed of three teams staffed by one officer and three men. Each team is on duty 24 hours and off 48. All members are volunteers with a minimum of two years and a maximum of 18 years experience as combat firemen with the Jacksonville department. The team's stationhouse is located in a central area providing quick and easy access to the expressway system, the port area, industrial and storage areas, and extensive railyards.

In its first 60 months of operation, the Jacksonville team responded to more than 1000 hazardous materials incidents within Jacksonville city and surrounding Duval County. To develop and maintain the expertise necessary to successfully respond to a wide variety of incidents, team members stress three critical areas: training, continuous preplanning, and aquisition and maintenance of an adequate inventory of specialized equipment.

Training, both academic and "hands-on", is never-ending. On any given day, team personnel are either undergoing training or providing it. There is in-service training; out-service training at schools, seminars, and workshops throughout the southeast; the provision of hazardous materials training to combat firemen throughout the Jacksonville department; and nearly continuous interaction with "visiting firemen" representing industrial and public safety organizations throughout the United States and Canada who spend from a day to two weeks living and working with the team in order to learn the tools, techniques, and disciplines developed in Jacksonville.

Captain Ron Gore initiated, trained, and currently directs the three shift, 15-man Jacksonville team. "With an increasing number of incidents, and a proliferation of hazardous chemicals, compounds, and products being manufactured, transported, stored, and processed within the city," recounted Ron Gore, "Wesly Yarborough, Chief of the Jacksonville department at that time, gave his blessing to formation of a specially trained, highly knowledgeable, and well-equipped hazardous materials team. The team was recruited from volunteers within the combat fire companies and placed in a centrally located station with easy access to the expressway system, the port, the storage complex area, and the railyards. As soon as we had the men we wanted, we began to beg, borrow, and . . . well, not quite steal . . . the specialized tools and equipment needed to deal with hazardous materials."

Such tools and equipment are as basic as a 13 volume reference library built into the cab of the team's remodeled pumper and as esoteric as a pH meter for determining the acidic or alkaline quality of a product. There are extensive patch and leak kits and materials. These include chlorine kits, precut rubber patches of various sizes and shapes, pointed wooden dowels and plugs that can be hammered into ruptured pipe openings and other leaks, hydraulic jacks and chains for clamping patches to tanks of all dimensions, and enough assorted pipe fittings, sleeves, brackets, compression plates, valves, elbows, and joints to make a plumber turn green with envy. Team members often work under frighteningly dangerous conditions, but their basic stock in trade is the ability to patch any leak in any vessel, container, cylinder, or tank.

"We've become quite expert in stopping leaks," agreed 18 year veteran firefighter, Bob Masculine, "although we don't claim to be repair people, and we don't want to gain a reputation for being in the repair business. We are a first aid company for leaks involving hazardous materials as a Rescue company is to victims. Our primary objective is to control the situation."

Additional equipment includes explosimeters, dosimeters, sorbent materials for pickup and disposal of small spills, dispersants, an infrared viewer for identifying heat patterns through walls, decks, floors and bulkheads; as well as foam, foam, and more foam.

"Most fire companies carry 15 to 20 gallons of foam," noted Captain Gore.

"Our two companies together carry 300 gallons."

Asked about rumors that he recruits team members like a Big Ten Conference coach recruits football players, Ron Gore laughed, then added: "Firemen establish a reputation. It's really eagerness that we look for, a desire to learn, a motivation to go out and train. Our people are like racehorses; you have to hold them back. We just can't use a man who thinks a fireman's most important duties are to polish the checkerboard and feed the Dalmatian."

"Although we train all the time," explained Captain Ron Gore, "we are still barely to the top of our boots as far as knowledge of hazardous materials is concerned. We average one hazardous materials call every other day; we would probably have double that number had we not trained all firefighters in the city to handle the most common types of hazardous materials incidents. Jacksonville is blessed in having a caliber of firefighters who, once they are trained, can easily handle the more routine alarms. Some calls they can't handle, often because they don't have the special tools and equipment we carry."

"First we train ourselves," continued Captain Gore, "with as much help from industry and government agencies as we can solicit. By ten each day we are in the field visiting chemical companies, railyards, the port area, and transportation centers. We practice on different types of vessels, vehicles, tanks, and cylinders used to store and transport hazardous materials. We study chemicals at the different storage areas; when we see things that are wrong, we ask people in charge to correct the situation. The industrial people train us in their particular products, vehicles, and storage tanks. They freely provide us with a great deal of expertise on how to handle potential incidents involving their products. We already know the philosophies, the basics, but we constantly talk with industrial people about the specifics: 'if that stuff spilled, what would be the proper way to handle it' We spend a lot of time on top of railcars and inside oil storage tanks. There are close to 100 different railcar designs; we have to be able to shut off valves, apply patches, and plug broken pipelines by touch. When you are wearing an acid-gas-entry suit and there is a corrosive or poisonous product pouring over you, drenching you; you can't see or hear a thing. You have to be able to make the repair by touch alone. We spend a lot of time practicing: turning shut-off valves, playing with pipes and couplings, and applying patches to damaged tanks just for drill.

We spend over eight hours a day in the field on preplanning, practice and training. In some fire departments, you have to beat the people into submission to go out and train. Here, it's the other way around. They are always asking—'Where are we going today? Let's go out and look at something.' "

"The hazardous materials incidents we are coming in contact with are so serious that there is no time for learning after we arrive on the scene," added Lieutenant Mark Chambers.

"Either we know what to expect and how to respond to it *BEFORE* we get there, or the potential for disaster is very great."

"We have done preplanning with every facility in this area that handles hazardous materials," noted Senior Fireman, Bob Masculine. "On each subsequent visit, we pull out the existing preplan sheet for a particular facility and review it with the people there. We ask them to make suggestions, or point out considerations we might never identify on our own. You have to lead them. The average industrial person is more interested in what his operation does rather than in what will happen when things suddenly go wrong. Without guidance, they will talk to you all day about how many widgets a certain machine can make in a day. We need to know where the light switches are located, how we shut down the ventilating system, what is located in that vat over there? What was the last incident you had, and what product was involved? What problem do you have most often? How much hazardous product do you normally have stored in the plant? Where is it? What products in what amounts are we likely to find sitting at the rear of the plant? After awhile, they begin to think like we do and start pointing out the things we need to know. When we complete a preplanning visit, there isn't much we don't know about a facility. All critical information goes on a preplanning sheet and is filed in a book that remains on the apparatus.

Jacksonville team members repeatedly stress the importance of maintaining and updating an adequate inventory of specialized tools and equipment. "The one item we could not do without," mused Bob Masculine, is the selfcontained breathing apparatus. These are an absolute must. Since we end up working in fumes, toxic vapors, and corrosive liquids so often; the acid-gas-entry suits are critical. We also have a four-hour mask; this is not a positive pressure mask, but it allows the wearer time to stage his preparation without having to come back for a new air bottle. We also have extension lines for our regular 30-minute air masks that supply an unlimited amount of air without being weighed down with the bulkiness of a tank. These are expensive items, but we have important materials that didn't cost a dime," added Masculine rummaging through a sizable collection of precut rubber patching material of all shapes and sizes. "We apply a tremendous number of patches on both pressurized and nonpressurized containers. This patching material looks expensive, but we get all we carry from a local rubber dealer who lets us scrounge around in his scrap heap whenever we run short."

"We have two pumpers that we altered a bit to accept our specialized gear," explained lieutenant Dick Morphew. "There is a library in a cab of each unit for the technical manuals we can't do without: tankcar manuals and schematic diagrams, chemical abstracts, tank truck manuals, service bulletins, various emergency services guides for hazardous materials, emergency preparedness plans for radiological and poison incidents, a guide to chemical hazards and reactions—a total of 13 manuals that receive pretty heavy usage. Even though we study such materials all the time, we often end up reviewing specific sections as we drive to an incident.

"We have explosimeters for monitoring air, fumes, and vapors," continued Morphew, "dosimeters to measure radiation accumulation; pH meters to determine the alkaline or acidic level of run-off; and an infrared scanner to detect heat patterns through metal and wood. Our two companies have on hand 300 gallons of foam, and all our foam nozzles and eductors are color-coded. Oftentimes we are spread pretty thin at an incident; we sometimes have to rely on whatever manpower is available. If we need A-FFF foam, we can just holler: 'Get the yellow nozzle and yellow eductor.' For high-expansion foam, we call for the red nozzle and red eductor.' "

"We have A and B chlorine kits," added Morphew, "and a number of other tools and materials for patching leaks; such as, hydraulic jacks and chains, compression plates, plugs and dowels. A few of our tools are nonsparking aluminum, bronze, or beryllium copper. We have to think constantly of potential chemical reactions—will the product react to the equipment and material we are using? Is the frangible disc we are using to repair an L.P.G. tank correct for the pressure of the tank we are working on? Will this rubber patch react with the

Phone_____ Address_____ Co.Name_____

Emer.Phone(s)_____ Basic Hazard_Petroleum Liq.

Pertinent Information_Master Electrical Cut-off (including loading rack) at rear

of Main Office on outside of Bldg. E-3 has a key to open front gates (electric).

3 nearest water sources_2 city hydst. in complex, 2 city hyds. just outside gates

to complex. Also, Engines can draft at Port Authority 2,000 ft. away.

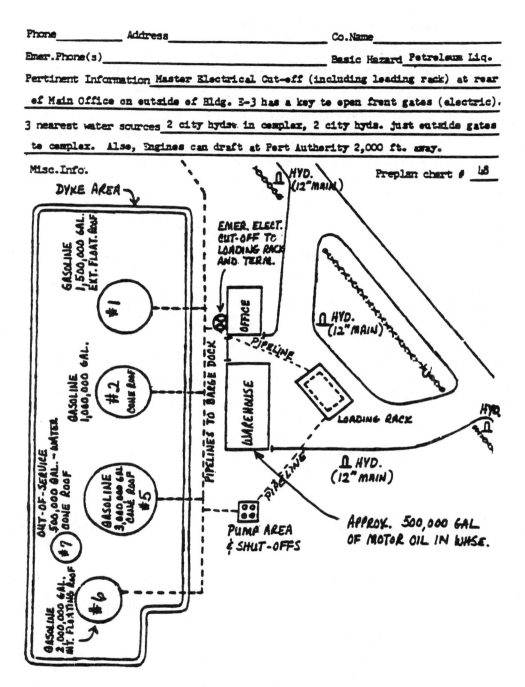

Basic Hazard—Petroleum Liq. **Pertinent Information**—Master Electrical Cut-off (including loading rack) at rear of Main Office on outside of Bldg. E-3 has a key to open front gates (electric). **3 nearest water sources**—2 city hyds. in complex, 2 city hyds. just ouside gates to complex. Also, Engines can draft at Port Authority 2,000 ft. away.

Figure 1. Preplan diagram of industrial facility.

product? We have rolls of sorbent materials for picking-up small spills, particularly on highways; plastic trashbags, funnels, mops, brooms and brushes. We have plastic drums for catching chemicals that react to metal, and siphon hoses for draining over-filled tankcars. A regular firehose filled with air from the air bottles makes a good floating boom for containing oil spills, and we can build a temporary dike to contain many fluids using ladders and canvas off the apparatus. You name it, we've got it,'' laughed Morphew coming up with a bottle of soap solution used for locating leaks.

To successfully answer any hazardous materials alarm, team members must identify the product, understand the product, know the environment surrounding the product, and know the proper response. Immediate and positive identification of the product involved in an accident can be critical. Many times the team must identify a product after they arrive on the scene. Oftentimes bystanders or facility personnel know only that "something" is causing a Hell of a problem. Team personnel may identify a product by placards, tankcar or truck identification numbers, or configuration of tanks or cylinders. They often call CHEMTREC or CHLOREP for guidance in understanding the reactive characteristics of a product and the proper response technique.

Team members must fully understand any product they are dealing with in order to plan their attack. By review of manuals, texts, guides; through experience and training; and by study of preplanning data they evaluate the threat. Is the product lighter or heavier than air? Is it water soluble or miscible? How does it react to other products? Is it an oxidizer? They consider flashpoints, boiling points, vapor expansion ratios, and tank pressures.

Team members must fully understand the mechanical environment in which hazardous materials are most often found. The valves, gauges, and piping of railcars, tank trucks, and loading racks must have been imprinted in their memory through experience and training so they can locate them while stumbling around in a vapor cloud or caustic spray wearing bulky acid-gas-entry suits. They must be able to look at an oil storage tank for one second and know the manner in which the tank operates, the type of petroleum product it is likely to contain, and be able to estimate the volume of product contained in the tank.

And lastly, team members must know the proper response or all their training, experience, specialized equipment, and preplanning will be wasted. They feel a fast response time is important, but they are careful to first evaluate the potential consequences of each action they plan to take.

They identify the type and degree of hazard(s) present. Is the product dangerous in terms of burns or frostbite? Is there a threat from inhalation, absorption, or ingestion? Is there a radiation hazard? Does the product pose a carcinogenic or biological danger? Such concerns may seem a bit far-out to regular fire service personnel, but the Jacksonville team deals with such products *all the time.* That's how they earn their pay.

They identify the product characteristics. Is it water reactive? Is it corrosive? Is it acidic or alkaline in case the product or run-off has to be neutralized? If vapor is present, they monitor constantly with exposimeters to determine danger of ignition of explosion, the boundaries of a vapor concentration, and to locate unsuspected pockets of vapor.

The team must relate products found at an incident to conditions on-site. They often erect windsocks as soon as they arrive at an incident, particularly if gases or vapors are present. They consider the temperature; does it pose a threat to pressurized tanks? Is fire or radiant heat endangering the product or its container? They check geographical conditions. Can the product, or run-off from the product be left uncontained; or is it flammable, poisonous, corrosive, acidic, alkaline and in danger of entering a water supply, sewer, or waterway? Can they boom it, dike it, absorb it, neutralize it, or disperse it?

While hundreds of product-related decisions are being made, team members must often simultaneously consider evacuation requirements and fire control techniques. Needless to say, the average team member doesn't become rattled too easily and loves to work under pressure.

After a hundred decisions, the team *may* be able to move in to patch, plug, or seal a leak or stabilize the situation in other ways. They may be able to dike or boom a simple spill, or use dispersants to "defuse" a spill of highly flammable liquids. They may use sorbents to pick up small spills, or call in a contractor with vacuum trucks or skimmers to handle larger spills.

Once the threat has been stabilized, the team is super careful to maintain incident vigilance and discipline. With hazardous materials, secondary incidents can be devastating. A flammable liquid washed down a sewer, or pesticide contaminated run-off entering a water supply, can pose a far more serious threat to public safety than the initial incident.

The Jacksonville Hazardous materials team has developed certain disciplines that guide their response to any hazardous materials incident.

1. Decrease Effective Response Time Through Knowledge and Preparedness: "Once you have a spill, it is too late to learn about the product, the vehicle it is transported in, or the container in which it is normally stored," explained Ron Gore. "You must have gathered the maximum possible amount of usable information beforehand. That's why we stress *both* academic *and* 'hands-on' experience. It doesn't help you much to know a particular product can kill you by inhalation, ingestion, and absorption if you don't know where the emergency shut-off valve is located on the type of vehicle normally used to transport that product."

2. Team Concept: A Highly Trained, Fully Experienced, Specially Equipped Group: "Any emergency service personnel can be trained to handle any type of emergency, including those related to hazardous materials," emphasized Captain Gore, "But it just stands to reason that people who have had extensive training and continuous expereience in hazardous materials are going to be better prepared to answer such alarms than people who respond just occasionally. All fire personnel should have a good grasp of what is required to respond successfully to common, everyday incidents; but a highly qualified team can be of invaluable assistance to an incident commander when there is a threat of significant life or property loss. A team will have been to so many incidents, dealt with so many chemicals and other hazardous substances, that they can relate better to the really hairy situations than the fellow who can merely say: 'We went to one incident a few weeks ago, and the product reacted this way.' That isn't good enough."

"When the adrenalin starts to flow, teamwork becomes super critical," agreed Bob Masculine. "When you are groping around under a tankcar in an acid-gas-entry suit with a toxic product cascading over you, you have to **know** that if you take off this part, without fail, without being assigned, your partner is going to pick up the piece you need next."

One inflexible discipline related to teamwork is that team members never, ever work alone. The old "Boy Scout buddy system" has proven to be effective time and time again. Although all team members have been in the emergency room at one time or another, no one has yet been seriously injured. Terry Daniels was wearing an acid-gas-entry suit while working atop a guishing tankcar when he got hit on the leg with a stream of cryogenic liquid that cracked the leg of his butyl suit like glass. Raising both arms in the designated signal of distress, he was carried off the tankcar and hosed-down by his partner. Bob Masculine ran out of air in a gas-acid-entry suit and had to be extricated by his partner. Another team member stepped into a hidden, three-foot deep sump filled with a corrosive product and had to be pulled out and hosed-down by other team personnel. "The product soaked my feet, legs, and thighs," admitted Ron Gore ruefully, "but it was my face that was red for three weeks afterwards."

3. Evaluate Before You Act: "Our primary objective is to control the situation," noted Lieutenant Dick Morphew. "We have learned from experience not to rush in. We lay back, analyze the situation, and attempt to evaluate the whole scene. We decide our tactics from a distance, sometimes with the aid of field glasses. We want to solve the problem; we don't want to become part of it.

"You can't take anything for granted," added Bob Masculine. "We arrived at one incident described as 'a chlorine leak' to find 11 different hazardous materials all jumbled together in layers. We first evaluate the situation, identify the product and its characteristics, consider evacuation requirements, and only then do we move ahead to stabilize the situation and achieve control."

4. Preplanning: Like any good salesman, team members know their territory. There is not a single major tank farm, railyard, storage area, port area, or warehouse containing hazardous materials that team members have not reconnoitered in depth. "We need to know what we are likely to find when we roll up to the main gate at three in the morning," stressed Ron Gore. On any preplanning visit, team members develop a diagram of each location visited that provides information that would be crucial in an emergency; such as, location of emergency shut-off electrical switches and mechanical valves; descriptions of tanks, cylinders, vessels, vehicles, and facilities encountered; location of nearest water supplies; and storage locations for various hazardous materials. All such preplanning data sheets are numbered and stored in a book on the apparatus. An alphabetical cross-index permits immediate location of the needed data sheet when a call is received. On each subsequent preplanning visit, the preplanning data sheet is updated as necessary.

"When we go to a facility to preplan, people there begin to view us as helpers, advisors, as a resource," noted Captain Gore. "They come to respect our knowledge and abilities. They also want to impress us that they are competent, knowledgable people; and perhaps with our assistance and guidance they become able to operate their facility in a safer fashion. Also, we learn the problems under which they operate. They learn our techniques and we learn theirs. When they do have a serious situation develop, they call us *at once*. We visit plants *every day*. We definitely feel we keep the number of incidents to a minimum by constant and continuous preplanning with all handlers of hazardous materials located in the city."

5. Establish A Command Post: "You cannot know everything about a product or a situation," explained Captain Gore, "but you must be able to surround yourself with people who have the necessary knowledge and expertise. There must be one location at a site where these people can interact with the incident commander. At any major incident, an incredible number of people can become involved and be quite necessary to successful control of the situation."

At one recent incident, it was estimated that 27 different knowledgable persons were involved in addition to the hazardous materials team. It is not unusual to have on hand representatives of the police, Civil Defense, E.P.A., the manufacturer of the chemical involved, the designer of the tankcar, the military, various federal-state-municipal agencies, private clean-up companies, earth moving concerns, the media, and disposal contractors. For an incident commander to maintain control and utilize these various skilled people to the the maximum degree possible, a command post is an absolute necessity.

6. Nonessential People Out of the Area: If a person does not need to be at the scene, if his experience is not necessary to stabilize the threat, get him out of there. This applies to response personnel every bit as much as it applies to curious bystanders. Once they have done their thing, see that they remove themselves to a safe distance. People want to be helpful. You have to control their access to the scene or you may find a guy in cut-offs and sunglasses looking over your shoulder as you crawl up the sice of a methyl bromide car in your acid-gas-entry suit and airpack. It happens all the time. Don't let it happen to you. Beware of the guy who thinks he is immune.

7. Have Knowledge of all Special Equipment and Expertise Available In Your Area: Few fire departments are likely to have on hand bulldozers or other earth moving equipment, empty tankcars and tank trucks necessary to transfer a product, specially trained and equipped track rebuilding crews, chemists, experts in pesticides and poisons, explosives and demolition people, vacuum trucks for scooping up spills, or similar equipment and expertise often needed during a major hazardous materials incident. You must know where you can obtain such assistance at three in the morning or on Sunday afternoon of a three day weekend. An itemized cardfile identifying all such equipment and services, the location where they can be obtained, and the name and number of the contact person is a must. For each item, an alternate source of supply should be identified if at all possible.

8. Hazardous Materials Incident Control Checklist: The Jacksonville team maintains a "Hazardous Materials Incident Control Checklist" for every incident they respond to. Such checklists are invaluable for subsequent training of response personnel, subsequent pre-planning visits to similar facilities, for providing continuity when one site commander and team goes off-duty and is replaced by another, to enable follow-up on contacts of a crucial nature, and to provide factual and timely basic data and information for any reports or testimony that may later have to be developed. This checklist is normally initiated as soon as an alarm is received, maintained and updated at the on-site command post throughout the duration of the incident, and ultimately filed at the hazardous materials team's stationhouse.

"What we have learned is quite transferrable," noted Captain Gore. "For any industrial organization, community, public safety agency, or consortium of various agencies concerned about potential hazardous materials incidents in their area; I recommend development of a "Hazardous Materials Incident Preparedness Plan' as a crucial first step."

The Jacksonville team developed the following 11-step guide to creation of such a plan.

1. Gather Data: Preplan community, shippers, and transporters as well as industrial and storage complexes; gather from them information concerning hazardous materials they handle, and discuss incident prevention and control procedures. Obtain hazardous materials reference material that will aid in determining correct handling procedures, and construct an emergency response and preplan book that will aid you at emergencies.

2. Develop an Action Plan: Develop an emergency plan for the firefighter and police officer who usually are first to arrive at the hazardous materials incident. Organize and develop an action plan for medical personnel, chemists, scientists, technical personnel, earth-moving and wrecking contractors, special hazardous materials incident contractors, as well as others, and have definite prior commitments from them.

3. Establish a Communications System: Assign incident leadership and/or command post resonsibilities prior to emergencies. Assign certain persons or groups to notify various special agencies; such as, Chemical Manufacturers Association (CHEMTREC) Environmental Protection Agency, State Civil Defense Headquarters, U.S. Coast Guard, or others that are equipped or needed. Designate a public relations or news media communicator.

4. Provide Training: Attend various hazardous materials seminars, conferences, and schools for training purposes as well as to obtain information to aid your own hazardous materials training program. Train all local groups that may be involved in responding to incidents. Develop for the public a hazardous materials awareness program. Provide special tools and equipment for hazardous materials incidents. Develop a special hazardous materials response team, or make prior arrangements to obtain a hazardous materials incident contractor. Have periodic mock hazardous materials incidents in your community.

5. Retrain and Update: *Never* become complacent concerning the possibility of a hazardous materials incident occuring in your community. *It can happen to you.* Continue retraining and updating hazardous materials incident prevention and control efforts in your community.

6. Size Up: Preplanning considerations? Incident location? What's involved? (type of hazardous material and its inherent nature) Incident conditions? (leaking, spilled, burning?) Time of day and weather conditions? Rescue and firefighting resources? Other considerations.

7. Call for Help: Additional rescue and firefighting forces. Police and Civil Defense assistance. Special agencies. (EPA, CMA, USCG,) Earth moving contractors. Product removal contractors. Hazardous materials incident control contractors. Other considerations.

8. Rescue: Immediate rescue. Intermediate evacuation. Major evacuation. Other considerations.

9. Exposures: Life—people. Priority property—product containers, explosives, etc. Secondary property—structures, woods, etc. Systems—water, communications, electric, highways, etc.

10. Extinguishment or Control: Water Control. Foam agent control. Inert gas or dry chemical control. Let it burn or leak down.

11. Overhaul: (Salvage and Restoration) Use precautions. Allow necessary personnel involvement only; keep other emergency personnel on remote standby. Allow no citizenry at scene until product transfer and clean-up is completed. *Again, use precaution.* Maintain incident vigilance and discipline.

Sooner or later, someone will always ask the Jacksonville Hazardous Materials Team the inevitable question: 'With all the training, preplanning, and specialized equipment; what *Really* happens when the world comes apart—when you are called in to stabilize an absolute catastrophe?

At approximately 1:50 A.M. on Sunday, February 26, 1978—90 years after naphtha and dynamite had obliterated much of Foutain, Colorado—Bay County Deputy Sheriff, Tom Loftin, pulled a U-turn on highway U.S. 231 in the darkened village of Fountain, Florida and headed south on 231 to continue his routine patrol. As he passed Brown's Grocery north of Youngstown, Loftin saw a flicker of light in his rearview mirror. Suspecting a burglary was in progress, he whipped his cruiser around to find a dozen persons sitting and lying about Brown's Grocery—retching and gasping for air, screaming of excrutiating pain in the groin, armpits, and lungs. Virgil Holman, a brakeman for the Atlanta and St. Andrews Bay Railroad, reported to Deputy Loftin. A train had derailed a mile away, rupturing a pressurized tank car of liquid chlorine; the chlorine had expanded to gas at a 460 to one ratio as it left the car causing a vapor cloud of greenish-yellow chlorine to drift over the nearby highway, killing eight and injuring 88 within minutes. Loftin simultaneously parked his cruiser across the southbound lane of highway U.S. 231 and used his radio to call for all possible assistance before he and Holman courageously set out, unprotected except for handkerchiefs held to the face, on a threequarter mile walk into the vapor cloud that would result in their locating and bringing out five desperately ill persons.

Jim Heisler, Bay County Civil Defense Director, arrived on the scene shortly after 2 A.M. and established a command post in his personal travel trailer 1.2 miles from the derailment site. For the next few hours, all attention was devoted to evacuating 2500 residents from the surrounding area and bringing out the dead and injured. As daylight approached, Heisler conferred with the Safety Director of International Paper and a chemist from Arizona Chemical regarding the possibility of patching the ruptured chlorine car. An unsuccessful attempt was made at daybreak just as evacuation was completed.

Heisler then put in a call to Captain Ron Gore 365 miles away in Jacksonville. Arriving at one Sunday afternoon, Ron Gore and other off-duty Jacksonville team personnel, immediately conferred with Heisler and other officials at the scene. After reviewing the train's manifest, they donned air packs and entered the derailment site to make a detailed analysis of the situation, which included diagramming the location, condition, and placement of each car in relation to others; in order to be able to develop a plan-of-action. On site they found approximately 30 cars, many of them stacked in layers—crushed, mangled, jumbled together within a haze of greenish-yellow chlorine gas that eddied back and forth in a light breeze. No birds, no crickets, no frogs, no grasshoppers could be heard or seen; there was absolute silence.

As the train had derailed, emergency brakes were applied automatically causing a "backing up effect" that stacked cars one atop another. The chlorine car was totally ruptured with jagged metal edges that made patching impossible. However, like a can of Coca-Cola that had been shaken then opened hours before, the chlorine "can" now sat docile, bubbling but releasing only small quantities of gas. The "fizz" was gone, and with it the danger of another deadly cloud, but the chlorine car was still dangerous. Next to it were three cars of caustic soda, one of which was leaking copious amounts into the sand. Although caustic soda is a neutralizer for chlorine, under any but controlled circumstances the two products can react violently. The chlorine, a combustible agent, was bubbling away next to two innocuous—looking boxcars. The Jacksonville team checked the car numbers against the train's manifest. *SUPER!*—A car of paraffin and a car of lubricating grease—all they needed to make an oxidizing rocket engine that could take them to the moon if they just had a detonator. No problem; a few yards away was a car of ammonium nitrate, an oxidizer capable of detonation from fire or severe shock. Under a four-deep stack of cars they found a liquid propane tankcar, upside down and imbed-

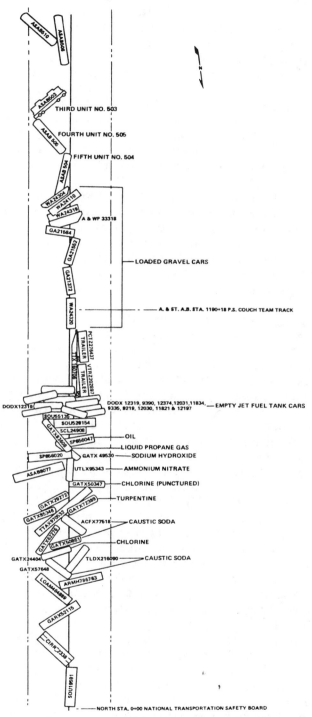

Youngstown, Florida Feb. 26, 1978. (From National
Transportation Safety Board Report #RAR–78–7).

Figure 2. Approximate car position after derailment.

ded in a carload of bricks, leaking flammable gas through a valve into the dome. Crawling through the sand on their backs, two Jacksonville men were able to tighten components in the upside-down dome and stop the leaks. A turpentine car lay ominously next to the chlorine car. Turpentine coming in contact with chlorine can cause automatic detonation. A number of empty JP-4 fuel cars spread about the area posed a problem due to fumes.

Because of their self-contained breathing apparatus, there was little communication among team members while on site. As they prepared to leave the area, Bob Masculine scribbled a note on his clipboard and showed it to Ron Gore. "Nice simple little chlorine leak we have here." Gore surveyed the incredible mass of highly flammable/explosive materials that surrounded them, then wrote on his clipboard: "Have you got a cigarette?"

The Jacksonville team assigned members on rotating shifts to monitor the site with explosimeters and then spent much of the night reviewing their diagrams and devising a step-by-step plan for stabilizing the multiple threats found on site.

At 7 A.M. Monday, the team initiated a practice that would be followed for the next five days. Before each phase of the operation, a meeting would be held with all persons having urgent responsibilities or duties for that particular phase to lay-out tasks, assign duties, and exchange information. Then they began work. A specialty crew from Hulcher Emergency Services in Illinois began to clear undamaged or nonhazardous cars from around the ruptured chlorine car as a Tyndall Air Force Base crash crew foamed down each car and the surrounding area, both to reduce the danger of ignition from sparks and to suppress the chlorine vapors under a blanket.

Earth moving equipment was used to construct an earthen dike around the ruptured chlorine and leaking caustic soda cars, to dam two small streams so caustic soda could not flow into Davenport Lake that provides part of the water supply for Panama City, and to provide a deep enough pool for drafting water to handle foam application needs.

A dozer also pushed dirt around each end of the apparently sound turpentine car as an added precaution; team members admitted they were most concerned at that time with turpentine and chlorine becoming mixed, resulting in detonation that would endanger all petroleum products in the area as well as the ammonium nitrate and L.P.G. Everytime a car was moved, the Tyndall A.F.B. crashtruck spread a blanket of foam.

By Tuesday morning the team was able to get at the severely ruptured chlorine car. An 80' by 80' pit was dug, and the chlorine car and a caustic soda car were dragged along a prepared trench to the edge of the pit. As with all operations, only absolutely necessary personnel having a specific job were allowed in the area. Standby-reserve personnel remained 2,000 feet away as Jim Heisler, the incident commander, watched each movement through fieldglasses and directed support operations. The caustic soda car was drained into the pit; and then, very cautiously, very slowly, the chlorine car was drained into the pit; and then, very cautiously, very slowly, the chlorine car was drained into the neutralizing pit.

Draining of the chlorine continued throughout Wednesday as less seriously damaged cars were righted and removed. Before each movement, a meeting was held to diagram each step, and a blanket of foam was applied. Explosimeters were used 24 hours a day to monitor the entire area for flammable gases and vapors.

By Thursday, the turpentine car had been rerailed and moved out and the contents of the chlorine car had been neutralized. Friday morning, after a final inspection and monitoring with explosimeters, and application of the ever-present foam "security blanket;" selected men began to remove a box car load of bricks tumbled about the L.P.G. car—brick-by-brick, one at a time—as all other personnel watched through fieldglasses from 2,000 yards away. After considering several options including total destruction of the car, pumping the car dry while it lay on its side, and leveling the car prior to off-loading; the last option was chosen. The car was foamed again, then righted, and the contents tansferred to an empty tankcar brought in on the rebuilt track. By ten Saturday morning, the damaged L.P.G. car was filled with water as an extra precaution, and evacuees were allowed to return to their homes.

The exhausted, subdued Jacksonville team held no celebration but merely climed into a truck for the long ride back to Jacksonville. There was little conversation, even when they eventually stopped at a crowded truckstop for dinner. Wearily seated around a large table in the crowded, noisy restaurant, their clothes yellowish-green and stinking of chlorine; the release finally came. First one man, then another, then the entire team spontaneously bowed their heads in silent prayer. For them, Youngstown was finally but a memory.

2 MAYDAY, MAYDAY, THERE'S BEEN AN EXPLOSION IN WAVERLY

Hazardous materials splattered upon the public consciousness of our nation with incredible force on February 24, 1978 when a single jumbo tank car carrying 27,871 gallons of liquefied petroleum gas (propane) ruptured with a Hiroshima-like fireball in downtown Waverly, Tennessee killing 16 persons and leaving scores more to stumble through the business district, skin dripping from their bodies like runny Saran-wrap. Two days before, a high-carbon wheel on a gondola car, its incompatible brakeshoes overheated by a handbrake negligently left in the applied position, broke seven miles outside of Waverly allowing the damaged wheeltruck to bounce along crossties through deserted countryside and finally derail 24 cars in downtown Waverly. Within its 25/32 inch thick steel envelope, tank car UTLX-83013 carried roughly twice the weight and three-times the volume of compressed, flammable gas permitted in a single tank car prior to the late 1950s. This benign-appearing yet massive bomb became an attraction, a curiosity luring townspeople into the area. When liquid propane leaked from that fragile container 40 hours after the derailment, it would instantly expand 270 times to highly flammable gas and the story of Waverly would be etched in 1700 degree heat.

From a simple derailment that initially caused no death or injury, and no property damage to other than railroad property, would come a hazardous materials disaster killing 16, scorching more than 80, decimating both the police force and volunteer fire department, and destroying a sizable portion of downtown. Millions of words have been written about Waverly by reporters, regulators, industry representatives, lawyers, and politicians. However, no one better understood the horror of Waverly than the survivors.

Initially, there was a degree of vigilance and discipline at the site, more than is normally the case in an isolated small town suddenly the scene of a hazardous materials derailment. Police and Fire Commissioner, Guy Barnett, immediately put into operation a preplanned disaster response operation. Within minutes, Waverly's small volunteer fire department joined city police officers at the scene as did emergency medical services crews from the Humphreys County Ambulance Service and deputies from the Humphreys County Sheriff's Department. Nearby residents were evacuated, and the area of the derailment was cordoned off to some extent. Fire personnel did exactly as they had been trained to do. They checked for fire; there was none, so they laid and attached hoses to nearby hydrants and set monitor nozzles to bracket the tank cars, ready to deluge the cars at the first sign of fire or leaking commodity.

A State Civil Defense team from Nashville arrived on the scene by 8:30 Thursday morning and recommended that an area within 440 yards of the derailed tank cars be evacuated and that gas and electric service to the area be shut off while the tank cars were lifted and moved.

Questions arose, and continue today, as to who had overall authority. It is difficult if not impossible to determine who really had the ultimate say in Waverly from the time train #584 derailed at 10:30 Wednesday night until tank car UTLX-83013 exploded at 2:53 P.M. on Friday, February 24, 1978. Once that fireball lit the sky, *nobody* wanted to appear to have made any decisions.

The initial evacuation perimeter crumbled bit by bit until Friday morning an order came down the chain-of-command to barricade only an area of three city blocks, basically the area that would be totally destroyed long before sunset. Businesses were allowed to operate even within that area. Although an order had been given initially to shut-off electrical and gas utilities within the derailment section of town, it appears that this was only partially done. If utilities had been shut-off, people might have been induced to leave the area until service was restored. They might have been angry, but they would have been safe.

The railroad directed its efforts toward clearing the main line so additional trains could roll through town. Cable slings were used to snatch and lift the two damaged tank cars, cars that have little structural strength from one end to the other except for that provided by the 25/32 inch thick walls of the tank itself. The track had been cleared by 2:15 P.M. Thursday and opened to rail traffic by 8:00 P.M. . . . 21½ hours after the derailment, yet by Friday noon little action had been taken to remove the tank cars from downtown or to transfer their cargo.

When a fireman refused permission for a tank truck to enter the area to discharge 9,000 gallons of gasoline at a bulk terminal within three blocks of tank car UTLX-83013, word came down the chain-of-command to let it in.

When firemen began to cascade water on the two tank cars as a protective measure to cool the cylinders, and the railroad became upset at having its crews work in muddy conditions, word came down the chain-of-command to shut-off the hoses.

Recommendations made, and actions taken, by the small Waverly Volunteer fire department were overruled or disregarded.

Although officials would later claim an evacuation was in effect at the time of the explosion, there was little evidence of this just prior to 2:53 P.M. Friday, February 24, 1978. A schoolbus moved slowly along Railroad Street within a few feet of tank car UTLX-83013, faces of children pressed against the windows, then turned right on Richland Avenue and slowly disappeared in the distance. Lloyd Florow, a retired fireman from Pasadena, Texas; had stopped by on his way to the store to observe the scheduled transfer of propane. Tommy Hornberger, age 19, was intrigued by the biggest excitement to hit Waverly in 20 years; although he had been run-off twice during the day, he was now back watching the action. The two biggest businesses in the area, Tate Lumber and Slayden Lumber Company, were both open and operating. Carman Oil Company was preparing to close, but employees had not yet left the premises. Two taverns in the basement level of the Spahn building were in the process of closing, while two blocks away bartender Wilbert Montgomery had a number of patrons at the Do Drop Inn. Seventeen employees were at work at the E&K Enterprises textile cutting operation in the old lock factory. Along nearby Richland Avenue, all dwellings appeared to be occupied.

At 2:53 P.M. an engineer and a brakeman in the cab of an L&N repair train unloading new crossties and rails at the Waverly derailment site, looked across the tops of three flatcars ahead of the engine and saw Waverly's world start to come apart.

"Well, looka there; that thing is fixin' to come unglued . . . there's smoke coming from around that tank," screamed the engineer. Workmen in the area had from five to seven seconds to recognize they were too close to the tank and try to save their lives.

"Lock 'er up," yelled the brakeman as he slammed the doors on the engine cab and crouched down behind a control panel.

Up ahead of the repair train, 21 year old Rex Gaut from Sterrett, Alabama was in the rigging of one of two Steel City Erection Service cranes. Just three hours before, Gaut has sat in front of UTLX-83013 smoking a cigarette and drinking a can of Dr. Pepper as a photographer from the Nashville *Banner,* Mike Coleman, snapped his picture. "Publish my phone number with it so the girls back home can call me," Gaut had joked. Girls mattered little to Gaut now. For a second he stared in awe at the greyish-white cloud of liquid propane now gushing from UTLX-83013, felt its wetness on his clothing, then shinnied down the derrick arm as fast as he could move. He never made it to the ground. Tank car UTLX-83013, that had sat as an attraction in the downtown area of Waverly for 40 hours, now came apart.

An indentation along the bottom-right side of the tank from the leading head to midway along the tank caused during the derailment now peeled open, and liquid petroleum gas (propane) rolled into the atmosphere, fighting to regain its normal gaseous state. As the fluid was exposed to the atmosphere, it expanded rapidly 270 times to gas causing a rushing of air filled with debris that terrified workmen now sprinting from the ruptured tank. In no more than seven seconds, propane gas reached an unknown source of ignition and flashed back to the gushing tank causing an immense explosion and fireball that covered the area.

Tank car UTLX-83013 separated into four major pieces. The short end portion of the tank was propelled 350 feet through the air in a northeasterly direction coming to rest in the yard of the Trolinger house on Richland Avenue. A second major portion, measuring 24 feet by 12 feet, traveled 150 feet through the air, coming to rest against Police and Fire Commissioner Guy Barnett's car parked on the North Church Street side of the Spahn building near the Starlight Lounge. A third portion, measuring 25 feet by 16 feet, was blasted high in the air and landed against the opposite side of the Spahn building 250 linear feet from its point of take-off. A limited number of smaller fragments were blown throughout the area. Then, for the first and only time during this sequence of events, the city of Waverly got lucky. The fourth and major portion of the tank which still contained the bulk of its cargo attempted to become an airborne, rocket-propelled projectile but rammed two boxcars containing heavy rolls of newsprint and sections of metal shelving and burned itself out.

Five people died immediately, 11 more lingered in agony from a few hours to a number of weeks before finally becoming victims; 54 others were burned seriously enough to be admitted to hospitals while an additional 42 received out-patient care. 1700 degree heat scorched the area, igniting any combustible materials including a great deal of human flesh. The wonder is not that so many people died, but that so many lived.

George Grundy of Waverly was having a cold one at the Do Drop Inn when he was suddenly blown out the front door onto a pile of snow. Another patron, his clothes blazing, tried to run past, but Grundy grabbed him and rolled him in the snow to extinguish the flames. Grundy then realized that he too was on fire, and with 20 or more patrons in hot pursuit headed for nearby Trace Creek. Standing in the creek, Grundy and the others could see they were on the perimeter of a blast area that covered three or more city blocks. They could only wonder about the poor devils inside that area.

George Crocker, an employee of Steel City Erection Service, was covered with liquid propane as he ran for his life. The fireball seconds later caught him in mid-stride and turned him into a running torch. "I made a dive under the wrecker," he says. "I had no idea there was any danger before the explosion." Crocker lived.

Howard Little, age 52, a claims representative with the L&N Railroad, was sitting in a stationwagon with Waverly's Fire Chief, Wilbur York. Police and Fire Commissioner, Guy Barnett, stood nearby. Suddenly finding himself on fire, Little struggled through the driver's side window that had been shattered by the explosion, ran a few paces, and rolled frantically in a snowpile. He lived.

Police Sergeant Elton "Tobe" Smith was on duty at the intersection of Willow and South Railroad Streets approximately 200 feet from UTLX-83013 when he heard a hissing sound. He sprinted down an alley but was overtaken by the fireball and severely burned. He lived.

Melvin Holcomb, supervisor of the railroad's clean-up crew, was incinerated sitting in the front seat of his car. His body would not be found until two hours or more after the explosion.

One workman's body was lifted high in the air by the blast and dropped on the roof of the Spahn building.

At Tate Lumber Company less than 170 feet from tank car UTLX-83013, James and Terry

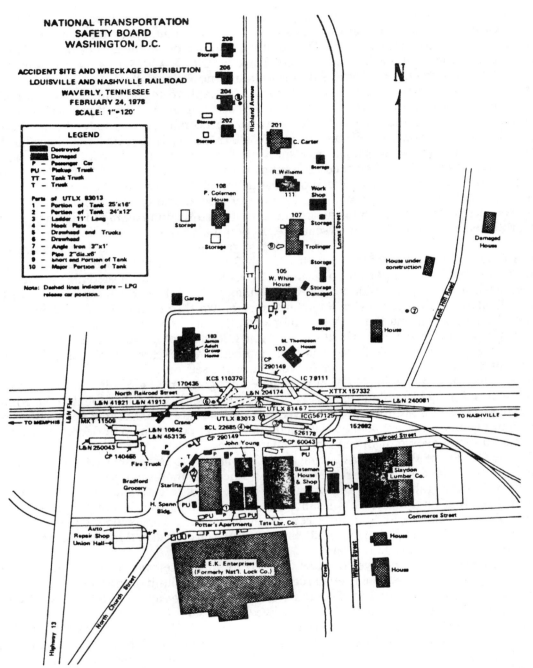

Figure 1. Plan of accident site and wreckage distribution (Courtesy National Transportation Safety Board.)

Hamm, father and son, were stacking and carrying boards with Van Johnson, age 26. At the sound of the explosion, the Hamms took off running in one direction while Johnson headed the other way. Johnson had taken only four or five strides when he was buried under a crumbling wall that protected him from the searing heat saving his life. Both the Hamms were fatally burned.

Bobby Story was returning to nearby Slayden Lumber Company after making a delivery. As he neared the railroad tracks, a truck stopped ahead blocking his way. Bobby glanced idly at the 33,000 gallon tank and watched in fascination as it ruptured before his eyes.

"I was looking directly at it when it exploded; that's what damaged my eyes. There was a puff of black smoke boiling around the car," he recalls. "Moments later, there was a dull thud, then fire shot straight up into the air with a tail of smoke and debris behind it, mushroomed out, and started falling all over the area. It was raining fire. The guy in front of me scrambled out of his truck and immediately burst into flames all over his body . . . he just cooked. He sprinted for the creek, but I never did find out who he was or what happened to him. I was slower to react, Thank God, and stayed in my truck. Within two seconds a tremendous heat wave, just scorching, rolled over my truck. I could see the heat coming, rolling at me; it was just like a dust storm . . . it was visible. It didn't last but a second of two; it just swept on over the truck and was gone. I didn't get out of the truck; that's all that saved me," concludes Bobby Story.

Jean Elvington is a trim, dark haired woman who started working parttime as a bookkeeper at Slayden Lumber Company and was eventually promoted to Manager. At 2:53 P.M. on February 24, 1978 she was waiting on customers at the front counter with UTLX-83013 framed in a large plate glass window at the front of the store. Jean Elvington was nervous. She was expecting anytime to be told to leave while they transferred the propane, but so far no one had come around to tell them to evacuate the area. Her daughter, Debby Wiles, was behind her in the office working on the books.

"I saw a puff of smoke by the tank car," Jean Elvington remembers, "It looked like a movie of an atomic bomb explosion; the flame and debris went straight up and then spread out like a mushroom. The front of the building crumbled; the cement blocks caved-in taking the window glass with them. The impact of the explosion knocked us all down. One customer got his knees and legs scraped, and I was thrown against the counter. My daughter started to run toward the back but was picked up and thrown against the doorframe. I started toward the backdoor when my mind registered that my daughter was still in the store. I turned around to go back for her, but she was nowhere in sight, and I just started screaming. Then the heat wave hit us, blistering my face. The heat was so very bad; everything was just scorched. My daughter came out of nowhere, grabbed my hand, and we bolted out the rear door."

At that moment, Bobby Story thrust his truck into reverse and gunned it around the back corner of Slayden Lumber Company, arriving just in time to pick up Jean Elvington and her daughter as they raced from the collapsing building. The three of them sat for a moment in the cab of Story's pick-up, mesmerized by the scene before them. There were people on fire, human torches, running in every direction but mainly toward nearby Trace Creek. A rain of flaming fuel was falling throughout the area, torching every building they could see.

"People were running toward the water," says Jean Elvington. "Their clothes were burning off. The worst thing was seeing all those people on fire and not being able to do anything."

Bobby Story put the truck into gear and began to move slowly through the carnage, hoping to get the women out of the area. He absentmindedly noted that the tape deck in the truck cab had melted into a gooey mass. He looked up to see Waverly police sergeant, Elton "Tobe" Smith, walking directly toward them. Smith was badly burned; his gunbelt and holster held a few charred strips of clothing on his body. In his enormously swollen right hand he clutched a melted walkie-talkie radio.

"The bones in Tobe Smith's hands could be seen clearly," notes Jean Elvington. "It looked like he was wearing surgical gloves that had ballooned out and become transparent. Yet he talked to us and remembers today everything that happened." Jean Elvington pauses a moment to remember the scene. "You can say that someone was terribly, terribly burned . . . but to look at him, talk to him, look at his charred body as he talks to you, realize the glob of tar in his hand is his radio . . . I can say it, but I can't really describe it. Certainly, I can never forget it. He was still functioning as a policeman, attempting to get people out of the area, trying to clear a path for vehicles carrying injured so they could get to the hospital. He agreed to go to the hospital with us. He talked to us all the way to the hospital, and when we arrived he walked in on his own power." Police sergeant Elton Smith returned to duty with the Waverly department some months later. Today, the skin on his face is as fresh and smooth as that of a young boy. Smith does not talk much about the agonies he went through while in the hospital, but will joke about the "free facelift" he received that makes him appear a decade younger than his 45 years.

During his relief break, 36 year old volunteer fireman, Riley Turner, had stolled over to a nearby store for a cold soda with no idea that he would soon become a living testimony to human courage for his indestructable will to survive against all odds. Now he headed back to his assigned position at a fire hydrant by the first house on Richland Avenue.

"I had my car blocking traffic and was in the process of checking the fire-plug," remembered Turner the following December as he lay in bed at home while his 14 year old son, Darold, continuously massaged the jumping nerve ends in his multilated back. "I guess I was like everyone else; we had been told again and again that there was no danger. I didn't have any direct dealings with the railroad company, but our chain-of-command assured us there wasn't any danger. When the train had derailed Wednesday night, I came on duty as soon as I got off work at the plant at 11:00 P.M. We immediately stretched our hose to wet down the tank cars. We had always been taught that if we found ourselves in a situation like this, we should keep water on any tank car carrying flammable gas. However, we got word later to shut-off the water, so we had applied the hoseclamps."

"Friday afternoon just as I got back from break," continued Riley Turner, "the word came down that they were going to start unloading one of the tank cars into a tank truck. I walked over to the hydrant on Richland Avenue and turned it on to charge my hose. I had my back to the tank; when I straightened up, it went off. Some people say it made a big 'Boom,' but all I can remember is a whoosing sound, and then I was on the ground."

The 1700 degree fireball rolled over Riley Turner, burning his left hand to a claw, melting his fireman's helmet down to a congealed glob the size of a small ashtray, and burning through the back of his heavy firecoat and coveralls down to the level of his rolled-down boots. "After I hit the ground," remembered Turner, "I instinctively started rolling over and over toward Trace Creek. I suppose I was on fire, but I really can't say. I held my breath, closed my eyes, and just rolled over and over until I felt myself go over the banking into the creek. In the course of our training, we had several films on that sort of stuff; it was something I just did naturally. I knew I had to make it to the creek if I was going to survive."

"As I was lying in the creek," continues Riley Turner, "there was a roaring sound for a second or two. I couldn't see anything but fire; all the air was blazing, just totally aflame. It was like looking into the fireplace at home with my face just a half inch from the flame.

It didn't scare me that much; I knew enough to lie still in the creek until I could think what to do. At the time, I didn't think I was burned all that bad. I knew my left hand was cooked; I could feel the pain, but I didn't feel much pain in my back. When I finally stood up, I could hear people hollering and a lot of sirens, but I was on the north side of the tracks and most of the injured were on the south. I couldn't see them through the flames and smoke. Just then Miss Thompson came out of a house next door, and I remember telling her to run. As she went on up the road, I made it out from between two houses to my car. It was a mess; the back window was blown out, and the paint was burned off. The engine started though, and I began driving through the flames with just one thought . . . to get to the hospital. There were two guys on the other side of the tracks, just totally burned. There was a pile of snow by the road, and one man was lying in the snow screaming. I just couldn't pass him. Both men were totally nude. Their hands were burned terribly; the fingers appeared to be burned off. The man on the snowpile had no ears; his body looked completely smooth, no hair, yet he was horribly charred. I went over and picked him up, and a piece of body fell off. I don't know if it was him or me. I put him in the back seat and went for the second man. He couldn't see and was walking around in circles. I got him by the hand, said 'Come on,' and led him to the car. He got in by himself, and I turned on the red light and headed for the hospital."

"I didn't use my left hand at all," he explains. "I knew it was badly burned, so I drove with the right. The two men in back never said a word until we got to the hospital although they were fully conscious. When a nurse pulled open the rear door, one of the men screamed: 'Don't touch me . . . Don't touch me . . .' but the other one just walked into the emergency room by himself. I don't have any idea of who they were, or even how old they were."

"At the Nautilus Hospital," continues Turner, "I got out of the car on my own; no one had to carry me or anything like that. My wife caught up with me there in the parking lot. I didn't realize that when I climbed out of the car she had seen that my back was just about gone. I was worried about our children who were visiting at my mother's near the explosion site and asked her to go check on them. I told her I wasn't hurt all that bad. When I turned to walk into the emergency room, she caught the attention of one of the nurses, Miss Baker, and pointed at my back. Miss Baker looked behind me and told me to come and sit down. She borrowed my pocketknife and cut off my clothes. She just snipped the collar of my coveralls and everything I had on fell off. The coveralls had crystallized from the heat and looked like a piece of plastic rather than cloth. I tried standing up but started getting dizzy, and one of the nurses told me to sit down. By then I had realized the back of my head and parts of my back were just char, so I lay down on a stretcher on the floor. Someone put a tag with my name on it around my ankle, and in just a few minutes they loaded me into an ambulance with Fire Chief York for transfer to Nashville. He was burned terribly and would die within a few hours, but all the 80 miles to Nashville we talked back and forth. Chief York, he and I were pretty tight before the explosion. He was an older fellow who had a lot of experience, and we used to talk a whole lot about fires. When the explosion went off and I held my breath, closed my eyes, and started rolling along the ground over and over until I got to the creek, I was just doing what Chief York had repeatedly told me to do." Waverly Fire Chief, Wilbur York, died shortly after arrival at a burn center in Nashville. Fireman Riley Turner is still a partial invalid.

Two men were sitting in a van in front of Bradford's store, approximately 80 yards from the exploding tank car. The windows on the van were tightly closed since the day had been cool although much warmer than the previous day or night. Dan Engle of the Public Service Commission was walking toward the vehicle when a white cloud suddenly blossomed from the tank car behind him.

"The cloud blocked my view; it was dense," recalls one man in the van, "just a greyish-white cloud. Rocks the size of fists were coming at us; instinctively I ducked . . . purely by reflex action from seeing something hurled at me. Two, maybe three seconds later . . . IGNITION. I didn't think of it at the time as a fireball, but merely as a fire. I had time only to turn toward the righthand door. I don't remember any concussion; I didn't hear any sound that I can remember. I remember trying to put out a fire in my hair. My clothes were burning; my synthetic britches welded together; the synthetic shell on my down-filled vest melted off. Suddenly there were feathers everywhere. My most distinct memory of that moment is the smell of burning hair. I remember thinking: 'God, someone's hair is burning up.' Actually, it was my mustache blazing away right under my nose."

Desperately, two men jumped from the van and headed for the nearby creek. By then the fireball was burning straight up from the tank car rather than outward. The survivor remained totally conscious. He walked over to the first ambulance he saw and climbed inside. He remembers telling a woman to close the door, but she was hysterical and paid no attention. Other survivors he saw had strips of flesh hanging from their faces as they tried to pull off what few pieces of clothing had not already burned off their bodies.

"When I turned away from the blast, flame may have rolled up and into my side of the van like a whirlpool," theorized the survivor many months later. "All doors and windows were completely closed; had they been opened I'd probably be dead. People were wallowing in the creek and in the snow trying to beat out the flames on their bodies. My swollen hands looked like gloves someone had filled with air . . . just raw meat . . . twice their normal size. People I knew didn't recognize me when I talked to them because my face was such a mess. My next door neighbor didn't know me until I spoke to her. My wife eventually brought me a mirror while I was in the hospital; my skin was charred black. I didn't realize for days how badly I was burned. It took a week for me to even be able to recognize time changes."

"At Waverly," adds the survivor, L&N seemed to have the attitude they had everything under control . . . and they got careless. Even now I always keep 50 to 70 feet back from the barrier at a railroad crossing until a train goes by. Others may pull right up to the barrier, but I am just not able to. I definitely want to see things change as far as the railroad is concerned. I will never forget the Waverly explosion. Trains are still derailing as regular as clockwork in this part of the country. I was upset by some things that were said at the time of the derailment and explosion. It was said . . . 'No one was in the area except those on official business' . . . **HELL**, a beer joint was open right there. The railroad was trying to repair tracks at the same time they were supposed to be unloading the tank cars. If they had taken some precautions, as was done at McEwen the following September, we probably would have had only about six people injured. There was a derailment of two polypropylene oxide cars at McEwen early the following September . . . far more dangerous than propane they say. That whole incident was handled in a vastly different manner. No one got close to it, and I mean **NO ONE**. They wouldn't even let helicopters fly over the area unless they were under direct orders from the railroad or civil defense.''

At 2:53 P.M. on Friday, February 24, 1978 Deputy Fire Chief, Bill McMurtry, was only seconds away from a place in history, perhaps the closest person to tank car UTLX-83013 to survive the blast unprotected by a building or vehicle.

''My back was toward the tank car,'' says McMurtry, a large man with bushy hair, ''and there had been so much noise I hadn't heard any leak others may have heard. All of a sudden, there was a giant 'Whoomp' far louder than the hanging gas heaters at the rear of the fire station make when the pilot light touches them off. I looked over my shoulder at the tank car just 125 feet away and thought . . . 'My God, that thing is on fire.' It looked like the whole world was on fire. I remember looking at the Clovis Chapel home where we had previously evacuated the retarded young women who lived there. Just every inch of that house was blazing. I was in the middle of the street near the long white building that had a couple of beerjoints in the basement, (the Spahn building) and my only thought was to release my hoseclamps located in a protected area there.''

''I took a step or two toward the hoseclamps, but it was like I was the Six Million Dollar Man on television; everything was happening in slow motion. Debris was flying through the air, but it was like I was in a dream,'' McMurtry remembers. ''I couldn't seem to make any headway no matter how hard I tried. A piece of telephone pole about four feet long sailed slowly by between me and my hoseclamps. It had an iron piece about 12 inches long bolted to it, and I could see the grain in the wood and the marks where a lineman's cleats had dug in. My mind told me I couldn't move through stuff like that, so I looked for anything to hide behind. There was only ten or 15 feet to cover to get the rearend of a car between me and the tank, but an unbelievable, searing heat caught me. I had seen a training film of a BLEVE (Boiling Liquid—Expanding Vapor Explosion) ten months before, and I knew I was too damn close to the tank to live. I made it to about even with the back wheel of the car and out of the corner of my eye saw a tremendous, rapidly expanding fireball. It took me a long time to realize what I actually saw was a reflection of the fireball in the car window. About then I just dove for the ground because I knew I wasn't going to make it to the car for protection. God, it was hot; nobody can describe how hot it got. It was horrible, and it ran through my mind that if I breathed I would sear my lungs. The force of the blast was pushing me along the ground, and I may have lost consciousness for a few seconds . . . I just don't know.''

''All of a sudden,'' continues McMurtry, ''it was no longer as incredibly hot as it had been. As I started to get up, there was a tingling on my neck, back, and head. My hands were huge blisters. A fellow ran by me screaming for help; flames were flaring out behind him. Everytime his foot hit the ground, he'd scream for help. That kind of transfixed me for a moment, then I looked down at myself. I wasn't like the running man; I wasn't on fire like everyone else I could see around me. They were black as the casing on a telephone, and I couldn't recognize any of them. I probably knew them all, but now they were just moving, burning, black forms. I looked at myself again to be sure; I just wasn't burned and blackened like the others. I was the only one at the scene I know of who was wearing the new fire retardant coveralls. My wife had bought them for me at Christmas, and this was the first day I had worn them. Even they had burned right through at the top of the back, but my cotton, dollar-store longjohns underneath had protected me somewhat even there. Wherever my skin had not been covered by the coveralls, however, I was a mess.''

"By then a firetruck was coming," remembers Bill McMurtry. As the driver jumped down from the firetruck, he just stared at me, his eyes as big as saucers . . . he just couldn't believe what he was seeing. I still had my radio in my hand and yelled at them at city hall to get us some help . . . that the tank had blown up. The dispatcher said they were unable to get out of the county because the power was knocked out. Then one of my firemen with a radio in his car came on over the radio: 'Mack, I am on the way,' and I could hear his siren going both over the radio and off in the distance. People were lying in puddles of melted snow, crawling in the street, and rolling over and over in little mud puddles trying to douse the flames that were eating them up. Vehicles were on fire all over the place, and a Civil Defense man came running up and asked for a fire extinguisher. He pointed at a van nearby; all the people inside were on fire even though the outside of the van appeared relatively unmarked. People had grabbed all the extinguishers off the truck, so the Civil Defense man tried to reach inside the rear door of the van for one he knew was there but burned his hands badly. The men inside the van were just torches, and someone screamed: 'What can we do?' I yelled: 'I can't help you. My hands are burned,' and held them up for him to see. I told him to use a hoseline off the truck without too much pressure and get those poor souls extinguished. He had water flowing in just a second and doused the inside of the van through its rear doors."

"The scene around us was horrible," continues McMurtry. "There were bodies burning in the street, yet people who should have been dead were up and moving about. There was no way some of those blackened bodies could be alive . . . but they were. I saw the two less seriously injured firemen were pumping water, so I started cussing into my radio to get some help. I asked the dispatcher to call the gas company and have them shut-off the natural gas lines in this part of town so we wouldn't have gas feeding the fires in houses that were ablaze all around us, but he reminded me that the phones were still out."

About that time, the fireman who had called McMurtry on his car radio, David Dillingham, showed up. "Get your gear on and cross the tracks," McMurtry told him. "Let me know what we have over there, particularly the condition of the other tank." Dillingham was back in a few minutes to report that just the one tank had blown while the other was still whole but had fire around it.

"I told him to do what he could to get some usable hoses into operation because all the hose we had laid and charged before the explosion was now split and useless," explained McMurtry. "It looked like someone had taken a giant razorblade and cut each section from one joint to the next. The tremendous heat in the fireball had melted the nylon in the hose, splitting it from end to end. I moved out then to the gas company and told them to shut-off the gas on this end of town. They said they had a man on the way to do it, but that turned out to be a lie. They told me later they couldn't do anything like that because it would shut-off the gas in McEwen and Dickson and everywhere else. I told the guy: 'Fellow, don't ever tell me that again; next time I tell you to shut it off I mean for you to shut if off. I don't care if anyone *ever* has gas again.'"

"We got very little cooperation at times," adds McMurtry. "Less than a block from where that tank blew there's an oil storage depot. Against my better judgement, and contrary to my orders, a tank truck had been allowed to come in that morning and unload 9,000 gallons of automobile gasoline. My orders were overridden because the L&N people told the political powers that be in this town that there was no danger. I had fought it all day long prior to the explosion. The on-site supervisor for the railroad had told me to gather up our hoses and get them out of there because they were getting in his way. Earlier, we had been ordered not to spray any more water on the tanks because it was making the ground too messy for the clean-up crews. I threatened to throw a wrench through the railroad supervisor's car windshield if he ran over my hose. I told him: 'Just like your job is with the railroad, our job is here trying to protect our town, and you're not helping one damn bit.' He said he was going to drive over the hoses, and I told him again that if he did I would put the wrench clean through the windshield and his face. I mean . . . it got that bad. I know he had his logistical problems. They were my problems too, because the longer those tanks stayed here the more problems I had."

"By this time, people were running down the street coming into the area," remembers Bill

McMurtry, "and I gave my coat and equipment to a New Johnsonville fireman I knew so he could help. We had only two or three of our own firemen still on their feet, and I figured the closest fire companies from out of town were 12 to 18 miles away. I headed toward the bridge to try to find someone with longrange communication equipment, but on the way there a guy hollered at me . . . 'I work for T.V.A.; what can I do to help?' It turned out he had a radio; he could only talk to the Tennessee Valley Power Authority office in New Johnsonville, but we told the operator there to send help. The operator said . . . "Where from?' . . . and I said . . . 'From any damn place man, wherever they can come from. We need manpower; there's another tank here that hasn't exploded yet, and if you think the first one was bad wait 'til you see the second.' "

"A Highway Patrolman came driving by," adds McMurtry, "and I tried to flag him down, but I guess I looked so darn awful he didn't pay any attention and drove on by to where an L&N man was directing him across the tracks to stop traffic. Within a couple of minutes . . . I don't know how long it was because time had no meaning at all . . . another Highway Patrolman came driving up and I stepped into the street to block his way. I don't know where he came from; I know most of the local Highway Patrolmen, but this guy was unfamiliar to me so I assumed he had come in off the Interstate. I told him we needed his communications equipment to get some help. He called Nashville, but they said they were aware of the situation and help was on the way. Come to find out, the Humphreys County Ambulance Service Director, Tom Doss, had broadcast a Mayday call as he ran out the door to come to the scene. His call had been picked-up and rebroadcast for miles around, and there was already help on the way from just about everywhere. I told the Highway Patrolman about the second tank, and he immediately began to block access to the site to keep out the people who by now were running up from every direction."

McMurtry received serious burns to his hands, head, and back but was far less seriously injured than a number of people who had been further from the tank car when it exploded. His hands are badly scarred and partially webbed, and one ear is nearly gone. When he finally left the hospital, his wife mentioned a mystery that had been puzzling her for some time. When she had retrieved his car some ten days after the explosion, it had been covered with duckfeathers. The feathers were all over the car, imbedded in the scorched paint. Only after McMurtry heard other survivors tell of duckfeathers all over the place right after the blast did he realize where they had come from. A number of men in the area had been wearing down-filled vests and coveralls to ward off the chill. When the fireball hit, it melted the synthetic fibers of the outer covering releasing the down insulation within. To this day, the clearest memory of more than one survivor of the Waverly explosion is of duckfeathers fluttering through the air, coating everything they touched. Perhaps their minds fastened on the duckfeathers to screen out the horror that was all around them.

Allman Adams, a volunteer engineering officer for the Waverly Fire Department, was sanding a fender at his brother's autobody repair shop a few blocks from the derailment site when a rumbling blast echoed through Waverly. Adams stepped outside in time to see a large section of tank car hang suspended for a second high above town then fall rapidly out of his field of vision. While his brother warned him that the second tank car had probably not exploded, Adams pulled his fireman's turnout gear from the rear of his 1971 Ford stationwagon and dressed quickly. He entered the site from the east by following East Railroad Street along the south side of the tracks until he reached Slayden Lumber Company less than 150 feet from the blazing tank.

"I couldn't move any further because of a wall of flames," he recalls. "A man was lying on his back just as nude as he could be, screaming: 'Help me.' I just pulled open the door, and he crawled on his hands and knees into the back seat; he was hotter than hell. I made a left turn, then a right, and there was a man with no clothes on crawling along on all fours. He held up his hand for me to stop, so I kicked open the front door and he pulled himself up and in. Both men were badly burned; they had been quite close to the tank. They had no skin at all as far as I could see. One of them didn't have but one shoe on; the other had both shoes but nothing else. They looked like freshly skinned animals; reddish fluid was pouring out of them, and their skin appeared to have been peeled right off. The next day when my son cleaned the car he had to get rid of a couple of double handfuls of rolled skin that had sloughed off them.''

"One of them never said a word; the other said: 'Call my wife and tell her I'm okay.' He gave me a New Johnsonville number, but I couldn't remember the last four digits. I never learned who he was or where he came from. I knew the man in the front seat beside me only because a small part of his clothing with a trademark or insignia on it had burned itself into his skin, and I could read the lettering on it. He came to the shop nine months later after the explosion and said: 'I want to thank you for saving my life.' I barely recognized him even then because he had no hair, no ears, just part of a nose, and no natural skin on his face.''

Chuck Webb, Assistant Fire Chief of the New Johnsonville Fire Department, had taken his family to Nashville early Friday morning to keep a doctor's appointment. On the trip home, he stopped in at Waverly about 2:30 P.M. to look over some material at the Sonic Drive-In where he had been asked to do some maintenance work.

As Webb pulled out of the Sonic Drive-In to head for home, he watched in fascinated horror as an awesome fireball mushroomed above downtown Waverly. "I just hit the red light and floorboarded the accelerator,'' he remembered later. "By the old lock factory a man who had been knocked down by the concussion but not burned was just getting to his feet as I drove by. Fifty yards or so further on, everything seemed to be burning. I slammed on the brakes, jumped out of the car, and told my wife to take the kids out of there. They could see destruction all around them and would later have nightmares about it. I also instructed my wife to call the New Johnsonville Fire Department and dispatch them under my orders.''

"A city patrol car had pulled in just ahead of me, and they were loading Police and Fire Commissioner, Guy Barnett,'' Webbs recalls. "I have been friends with Guy for a number of years, but I could barely recognize him . . . he was burned that bad. His clothes were shredded, and there was no hair left on his head. Guy was on his feet, and two of his officers were helping him into the back seat of the patrol car. From where I stood, I could see that he was in a bad way. I looked at his car the next day; although the windows had all been closed tightly, radiant heat had burned the interior dashboard clean away.''

"It's hard to remember now just what ran through my mind at the time; there was so much confusion. The fireball had disappeared in the few seconds it took to burn off the accumulated gas, but everything burnable in the area was now on fire . . . including a number of people. Some were rolling in mud puddles and patches of snow trying to extinguish themselves; others were running about like torches; a number had the clothes burned right off them, and their bodies had a charred look. A couple of the victims moving about looked like freshly skinned animals with fluids leaking out at different points of their bodies. Their clothing was either gone or burned into their skin. One fatality had been wearing a hat; the hat was gone, but the hatband was burned right into his forehead. A naked, badly charred man came running by; there were vehicles on fire, and every building I could see to the north and east was burning. A little girl was bleeding about the face; what little hair she had left was smoldering, and her clothes were in shreds. Jess Bowen III pulled up in one of the Luff-Bowen Funeral Home hearses, so I grabbed the little girl and took her over there. A severely burned man walked up to the hearse and said: 'Please, may I go?' I thought he was a colored man because his skin was black, but he was white. We put him in, along with a few others, when I spotted Fire Chief Wilbur York on fire about 50 feet away. He was lying on the ground trying to put himself out. I ran over, took off the white cotton shirt I was wearing and started patting the flames on his body.''

Chuck Webb pauses for a moment, his fullbodied voice breaking slightly as he continues. "Chief York was severely burned; his clothes were completely gone, and he didn't have any hair left on his body. He looked up at me . . . he knew who I was. I was very surprised at that. In his condition, I didn't expect him to know anything at all. He said: 'Chuck, it hurts,' and I told him . . . 'Just be quiet, Wilbur; I'll take care of you.' I looked around and spotted Rich McCoy, Editor of the local *News-Democrat,* and I just yelled . . . "Butch, come here; I've got to have some help.' Together we got Chief York extinguished and called to Jess Bowen to bring a stretcher. Tom Doss and a Humphreys County ambulance had arrived, however, and they took responsibility for the Chief. I said to Butch: 'Let's see if anyone else is alive,' and we started moving around.''

''There were two bodies lying on the other side of Chief York's stationwagon, but each was just a drawn-up, black mass. I couldn't even get to them as the heat was just too intense at that time. A man came up wearing one of those insulated jackets, and it looked like someone had shot through it with a shotgun; the insulation was sticking out all over. He was burned very badly around the face and hands. People were still running and screaming, and others were still smoldering, but a number of rescue workers had by now arrived on this side of the tracks. All the injured we could see appeared to have someone attempting to help them, so Richard and I crossed the tracks to the north side of the derailment scene to give what aid we could there. There were fewer people on the north side. The fire hoses hitched to a hydrant there were charged, but they were holed in numerous places, and the water was just running out of them onto the street. On Richland Avenue we found three bodies burned very severely with no clothes left on them, each drawn up into a knot. I wet a sheet we had been given by Tom Doss of the ambulance service and used this to smother the last of the flames on the first body and to cover it. You could tell they had been men, but that's about all. They were on their back or side with their knees drawn up to their chests. They were black, they were charred . . . that's all there is to it. Even had I known them, I could not have recognized them; they were burned beyond recognition. We found a fireman's helmet nearby, and one whole side was completely melted.''

''I suppose by now I was in a state of shock or close to it,'' adds Chuck Webb. That first ten minutes was just too awful to think about even now. Until this time, I had never thought about the second tank. It was my assumption when I first arrived on the scene that both tank cars had blown. Then we heard a screeching noise and looking through the flames and smoke saw the second tank car was still in one piece and venting. I guess Richard and I just turned white; I didn't look at him, and he didn't look at me, but I guess you could say that each of us independently became much concerned right then and there.

The safety valve on the second tank car was venting nearly continuously in an attempt to relieve pressure. There appeared to be flame impingement all over the bottom portion of the tank. Every time the safety valve would lift, we'd hear a rush of gas, and then the expelled gas would ignite about six feet above the valve. Had I realized earlier this car had not blown, I doubt I would have gone in there.''

Chuck Webb picked up two or three billfolds lying in the street and stuffed them into his pockets without opening them. ''I guess I really didn't want to know who they belonged to,'' he confesses, ''so I just passed them to a policeman later that evening. Recognizing that the second tank was still very much a threat, my mind forgot all about seeking out more victims and told me to start operating as a fireman. I had one of the county radios with me, but there were so many calls for help on the channel that I gave up trying to attract attention.''

''Rich McCoy and I became separated about this time as we each made our way back to the south side of the tracks where most of the people were. Deputy Fire Chief Bill McMurtry was there, burned pretty badly but trying to get some help and direct operations. One of his firemen, David Dillingham, and I managed to get their old firetruck started even though it had been badly fire-damaged. We drove the old truck back over to the north side of the tracks where by now the flames seemed even worse than they had been a few minutes before. It turned out that all the hose had been taken from the truck, so we had a firetruck that was no use at all. Dillingham saw the three bodies McCoy and I had extinguished before and recognized one as a Waverly Fireman he knew well. It affected him terribly.''

Chuck Webb borrowed a radio and managed to get through for a second to his own department's truck as it was enroute from New Johnsonville to Waverly. ''The driver acknowledged my call, but the other traffic blotted him out, and I was still trying to reestablish contact when they pulled in a few minutes later. Our goal then was to get some water on the second tanker as quickly as possible, but first we hooked to a dry hydrant. One large section of the blown tank car had fallen on top of a car and gone right through so deeply into the ground that it had ruptured a water line. We unhooked from that first hydrant and came back around by the bridge. By this time, Assistant Fire Chief Dutch Geissenhoffer of the Waverly department had arrived on the scene from his regular job over in New Johnsonville and he was doing a pretty good job. He knew about Chief Wilbur York and Commissioner Guy Barnett, and you could see the news had affected him deeply, but he was able to start getting things under control and bring some sense of organization to the total confusion that had existed up to now. He sent us around to the north side again to get some water on the second tank car. We hooked to one of the hydrants the Waverly department had been using before the blast and started throwing water on the tanker with a 2½ inch line. We tied the nozzle off so it would keep itself trained on the tanker because we weren't about to stay there and hold it. We then started fighting our way through the Slayden Lumber Company with a 1½ inch line until we met another department coming through from the other side. We were able to extinguish what fire remained there, and I believe we saved two of the Slayden buildings. By now, fire departments were arriving from throughout Tennessee . . . literally from all over the state. It was just unbelievable the number of fire-fighters that came into action. Several other groups put additional hoselines on the tanker. One department set-up a deluge gun at the south end of the tracks but didn't have quite enough pressure to reach the car from a safe distance. We unhooked our 2½ inch line and ran it through their deluge gun in tandem and were able to hit the tanker from 60-70 yards out. That's a more comfortable distance. Once you set a deluge gun you can leave it, so we were free to make a systematic search of the nearby dwellings on the south side of the tracks, some of which had burned while others hadn't. The windows were all blown out, and we were afraid there might be injured inside although we didn't find anyone. We continued to search for injured and extinguish outbuildings until ten or 11 that night when we were relieved by a truck from Jackson and I was able to send my personnel home until the morning.''

Chuck Webb was silent for a few moments, implying that he had come to the end of his story, but then he realized there was more he wanted to say. ''I was not at all pleased with the attitude of the State Civil Defense team that arrived after the derailment and directed things. I believe we would have been better off if they had stayed away, yet they came in and took over. Actually, they didn't have the authority to take over unless asked to do so. It appeared later that the Mayor of Waverly may have asked them to take over. I heard several comments from other firefighters about the attitude of the Civil Defense people not being what it should have been. They were not concerned. They didn't ask the advice of, or even consider, the Waverly Fire Department; they more or less just pushed them to one side and took over.''

''Also, I rode with some of the Nashville Metro Police that were working the area after the blast, and I was not at all impressed with their attitude toward the situation. Some of them were very smart-mouthed and sarcastic about the whole business. On the other hand, the National Guard came in and were a great help. There were a lot of actions that were political that I was not aware of or concerned with at the time, but later on the articles were written about who did what. The Mayor of Waverly seemed to be quoted as saying something to the effect that . . . 'If it hadn't been for the Civil Defense, Nashville Metro Police, and National Guard; they would have lost the whole city of Waverly.' I know I'm prejudiced, but if it hadn't been for the various fire departments they definitely *would* have lost Waverly. The various fire department personnel, the ones I talked to, did the best they could. The other people sat on their behinds in an air-conditioned office up on the hill and shouted orders about this, that, and the other thing while other men went into the site next to the second tanker and did the work. And up in the air-conditioned office on the hill they were sometimes sarcastic of the work that was done at the site after the blast. I've heard comments from several different officers and departments about the way they were treated and the attitude of the personnel on the hill.''

Richard McCoy, Editor of the weekly Waverly *News-Democrat,* was sitting in his office three blocks from the tank car pulling together what would be his last edition as editor. McCoy had recently decided to go into business for himself as the operator of a combination grocery store and service station over in New Johnsonville. He was editing copy dealing with Wednesday night's derailment when there was a loud rumble and the frontdoor of the newspaper building swung open on its own, then closed as if an unseen visitor had entered. "I didn't realize immediately that there had been an explosion," McCoy remarked later. "Everybody had been so darn casual about the handling of the propane that it never really had dawned on me it might explode."

Finally realizing what had happened, McCoy grabbed his camera and bolted out the door, heading down Church Street at a run, arriving at the scene in a dead-heat with the first ambulance. "I had no grasp of the scope of the situation at that time" he remembers. "My journalistic instinct was controlling my actions; I was taking pictures as fast as I could, but I was doing it in the Third Person. I didn't seem to be involved in the scene . . . terrible incidents were occurring all around me, but I was standing outside it all clicking the shutter and looking for the best picture."

In his first few minutes on the scene, Richard McCoy rapidly shot 40 frames of film, film that would eventually help to earn awards from the Tennesee Press Association to the Waverly *News-Democrat* for "First Place In Class For Medium Size Weeklies" and "Best Spot News Photos" for the year, and form part of a remarkable photographic record of the Waverly disaster. Disasters often go unrecorded in their first few moments, but at Waverly two professional photographers, Richard McCoy and T.O. Perkins, were on the scene within minutes and had the composure necessary to function as photographers in the midst of absolute chaos. McCoy, for example, didn't realize he had taken a photo of a horribly burned, completely nude man running past an ambulance until late that evening when he printed the film. Everything he saw for the first four minutes at the explosion site was viewed through the lens of his camera. Then . . . he was drawn into the action in a much more personal way.

"A guy went running by and called my name," he recalls. "It was Chuck Webb, Assistant Fire Chief over in New Johnsonville; he and I have been friends since grammar school. As he was moving away from me at a full run, he hollered over his shoulder: 'Butch, he's burning, come on.' I followed Chuck to where he had located Fire Chief Wilbur York's burning body. I felt there was virtually no one else in the area at that moment. Tom Doss's ambulance crew had loaded-up and headed for the hosital, and there was nobody else around. I dropped the camera on the hood of a car and joined Chuck. Chief York's body looked like a black 'Hefty' trashbag, except that his face was white as though at the moment of the fireball he had covered his face with his hands. He was burned so badly we couldn't distinguish between his turnout gear and his flesh. He had somehow managed to kick open the door of the stationwagon in which he had been sitting and roll ten feet. One of his shoes was lying nearby. His body had the texture of black canvas, like a log that had been left smoldering in the fireplace all night. Yet he was talking, and he seemed to be rational.

He was saying: 'My God, it hurts. It hurts.' Chuck was saying: 'Well, just take it easy chief; stay calm.' And Chief York said: 'Okay, I'm calm, but it hurts like Hell.' That really got to us. The man was in a ghastly condition, yet he had the composure to reassure us that he would remain calm because we told him to. He talked to us as we beat out the flames on his body. Chuck removed his white cotton shirt and put it around Chief York. It seemed like just a moment or two before Tom Doss was back with the ambulance after having made an initial run to the hospital about three-quarters of a mile away. Doss spread a blanket on the ground, and we started to roll the Chief onto it. He put his arms out and helped us get him onto the blanket. It was amazing . . . like a fireplace log had come to life. He just rolled over onto the blanket, and they took him away in the ambulance. He died later that night."

"By this time I wasn't thinking as a journalist anymore," admits Richard McCoy, "so we started to look for additional bodies. There were still very few people in the area as far as I can remember. I'm not sure what caused my tunnel vision; I didn't see other people or bodies. After talking to others later, I believe everybody went through this same experience. People seemed to have focused on one or two small things that were happening around them, blotting out things on the periphery, not really focusing on the magnitude of what was happening nearby. I can remember Chuck and I and Chief York, but when I developed my photos later I was stunned to see so many people in the different pictures. I had rather believed the three of us were the only ones there for five minutes or so.''

McCoy and Webb, carrying a blanket given them by an ambulance crew, then made their way across the railroad tracks through smoke, flame, and intense heat. They checked out two cranes, one engulfed in flames, but could locate no additional survivors. As they reached Richland Avenue, they were nearly surrounded by fire.

"At the foot of Richland Avenue," continues McCoy, "there were three bodies lying in a row. The first two were smoldering, and the third was completely on fire. They were obviously dead; there was no hope for these three, but I had a strange reaction. I just didn't want anyone to see them lying there smoking and burning. They looked very undignified with their clothes off. I guess it is natural; I wanted to cover them up, so I went and placed a blanket over the first body. Although it wasn't actually burning by now, the body was so hot the blanket ignited. I took it over to a nearby hydrant, where hoselines had been slit by the fireball as though someone had run a razorblade along the top of each section, and soaked it. I covered the first man, then did the same for the second, but the third body was too far into the flames and I couldn't reach it. All I could do was look through the flames at the poor soul.''

"Right about then, I looked through the flames and saw the second tank car hadn't exploded . . . I'm just not that much of a hero . . . I headed west toward the viaduct to cross the tracks away from the flames and the second tanker. I became separated from Chuck Webb and didn't see him again until some time later. Along the way I picked up a couple of billfolds, a few pieces of clothing, and some shoes. I found my camera on the hood of a car where I had left it and was on the verge of getting very, very sick. I just about lost it at that time.''

McCoy had expended the small supply of film he had brought with him, so he walked three blocks to the newspaper office and stuffed all the film he could find into his pockets. Back at the explosion site, firemen and policemen were now attempting to close off the area and wouldn't let McCoy re-enter. He cut east to the edge of the blast zone and was working his way back in when he met Lynn White of the Dickson County *Herald.* Together they waded Trace Creek, keeping out of sight of firemen and sheriff's deputies, and began to work their way through the blast area taking photographs as they went.

"Coming in from the east, it was much easier to assess the magnitude of the devastation," concludes McCoy. "All the buildings we could see were burning, and I had my first opportunity to really observe the area. Cars were popping, gas tanks were exploding, buildings would suddenly flare-up with a roar. There were little bands of firefighters all over the area by now, putting water on anything that looked like it might be saved. We walked around taking all the pictures we could and stayed well away from the second tank car. Eventually, a couple of buddies of mine on the Waverly fire department spotted us and knew we weren't supposed to be there. They started screaming at us to get the Hell out . . . so we did. This is the only incident that has ever affected me to the point where I almost became ill and stopped taking photographs. The disaster still weighs heavily on people's minds. Before the explosion, there was nobody willing to say they had the authority to make decisions. The original derailment was taken lightly; there was a carnival atmosphere, and suddenly the carnival turned bad. After the explosion, there was a feeling of rightous indignation but no one to direct it against. People were angry but could not decide where to direct their anger. Some of the people who might have been targets of that anger died in the blast, so how could you be angry with them? The psychological effects on some people were profound, yet people here have been very careful not to place blame.''

Police Investigator, Ted Tarpley, was a mile from UTLX-83013 on Highway 70 West at

2:53 P.M. that day. "I was following a police cruiser in my personal car when I saw a tremendous mushroom of fire over the town . . . just gigantic. I'd never before seen anything like it. The officer up ahead, Wallace Frazier, turned on his blue light and siren, and I was able to follow him through traffic to the scene. It was terrible . . . people rolling in the snow, standing in the creek, walking down the street with their clothes entirely gone. One of our officers, Tootsie Bell, was being helped into a cruiser by Officer Frazier, and I recognized a local man, Jess Turner, even though he was in a very bad way. Most of the victims, however, were just not recognizable even though I had probably known each and everyone for years. One man was sitting on his rearend with his knees drawn up to his chin; he was burned just terrible and his eyes were vacant . . . but he was alive. A bunch of burned men with no clothes on were walking around. If you've seen football players playing in the mud . . . just crusted with dark goo . . . that's what they looked like, but the 'mud' was skin. Flame was still falling from the air; buildings and vehicles were burning . . . some of the people looked like pieces of fried chicken. I'll always remember the people standing or lying in Trace Creek trying to put out their flames or cool their burns."

"I took a few people to the hospital," continues Tarpley, "but I have no idea who they were. I returned to the station and used the dispatch unit to call Fort Campbell in Kentucky; they sent helicopters immediately. I talked with the Chief of Police in Nashville and told him what we had. He said they would respond immediately, which they did. By this time the phones were ringing off the wall, and the radios were going crazy. We had a direct line to Civil Defense in Nashville, so I was able to tell them the situation. At that time, I assumed both tanks had blown. We hadn't been too concerned about the tank cars because the people with Civil Defense and the railroad had told us not to worry about them, that they handle them every day."

"I've seen people with their heads cut off, shot, cut-up every which way," adds Tarpley, "but nothing ever got to me as badly as when I walked into the hospital emergency room with my load of injured. All 12 or 13 chairs in that little room were filled with burned people, and the smell was terrible, just horrible. They were sitting there without a sound waiting to be treated, and the nurses were in there trying to help them. The burned people were reaching out to the nurses, yet they didn't make any noise. Nobody screamed, nobody moaned . . . just those hands reaching out gently in the silent room."

Immediately after the explosion, Ted Tarpley was appointed Chief of the Waverly Police Department.

At 2:53 P.M. on Friday, February 24, 1978 Sheriff E.C. Hall was sitting at his desk at the county jail in downtown Waverly. He had just put down the phone after talking to his wife, a beautician at one of the local shops, when he heard a tremendous explosion and turned to see through his office window a giant fireball 300-400 feet above town. He leaped for the door, instructed the dispatcher to call for help, and he and a deputy headed for the scene in separate cars.

"When I got there," says Hall, "people were running in every direction with their clothes on fire, hollering and screaming. The fire was everywhere. It had been just an hour and a half since I had last visited the scene to check on what was going on. I had talked with the Chief of Police while they were getting ready to transfer the propane, and everything had seemed fine. My men had been assigned to the area while they were uprighting the tank cars the day before. Once they got the cars upright, they said the danger was over and I could relax, so I pulled out my men."

"Now, people were running and screaming, trying to put out the fire on their clothes. I loaded two into my car; one was a local man, Jess Turner, and I believe the other fellow was a representative of the railroad. The railroad man was stumbling around with his shirt on fire, so I put snow on him to douse the flames. Jess Turner got in on the right-rear side of the car and sat quietly. He said he had lost his eyeglasses; hide was falling off his face and hands . . . just coming loose and falling down as I looked at him. I told the two of them I would be back as soon as I could, then went to confer with two of my deputies who had arrived on the scene. Some people were screaming. I think they could see the condition of other victims, and it made them wonder how badly they themselves were injured. They appeared to be reacting to the physical appearance of others, some who had their clothes completely burned off except for their beltloops or other small parts of apparel that had somehow been the only thing on their bodies not to burn. A number were rolling in the snow trying to extinguish themselves. Some of those walking around looked like they were covered with soot, and a number had hide sagging and falling from their bodies.''

"I saw the victims as they arrived at the hospital," continues Hall. "The scene there was just totally impossible to describe adequately. The hospital staff were doing a really good job. They were taking the victims in as fast as they arrived, separating the critically burned from the less severely injured, putting IVs into victims to counteract loss of fluids caused by the very serious burns. The injured were arriving in a continuous stream, many of them in private vehicles, others by means of a nearly continuous shuttle run by Tom Doss and the Humphreys County Ambulance Service.''

"It was amazing to see the condition of some who walked in without assistance. Roy Douglas, a city policeman, had just been driving through the area at the time of the explosion, but it got him. About all he had on his body was his gunbelt and pistol. I removed these from him. One of the Public Service Commission men, Engle, was in about the same condition, and I removed his gunbelt and pistol. (Engle would die some weeks later; Roy Douglas, although severely burned, lived.) Most of the injured were coherent although many looked like they should be dead. There was no room for any additional bodies; every inch of chair and floor space in the emergency room and nearby hallways appeared full as nurses applied IVs and cut charred clothing from people.''

"After a few minutes, when I had my first chance to look around, I began to feel that one terribly burned person sitting in one of the chairs was following me with his eyes. I was sure I had never seen this person before, but his body would shift position a bit as I moved about the room so his eyes could follow me. And then in one horrible second I knew . . .'Terry, is that you?' I went weak as that horribly burned head slowly moved up and down. Terry Hamm had been one of my deputies for a number of months until he went to work for Tate Lumber Company . . . and I hadn't even recognized him. He died later.''

"As Police and Fire Commissioner, Guy Barnett, was being wheeled out for transport to Nashville, he called out to me: 'Sheriff . . . how bad am I? Now don't lie to me . . . how bad am I?' 'Well Guy, you are burned quite a bit,' I told him, 'but you don't look all that bad' . . . and he didn't. I thought he would make it, but he died within two days.''

"We later established a temporary morgue at Humphreys County Utilities," concludes Sheriff E.C. Hall, "and began identifying bodies and notifying families. The next morning, we began a more systematic search through the rubble, but did not find additional bodies. We found a lot of personal belongings; such as, billfolds, keys, and that sort of thing. We had a refrigerated truck come in from Nashville Saturday morning, loaded the dead into it, and moved it up behind the hospital until local funeral homes could take responsibility. Dr. Francisco, the State Medical Examiner, and Dr. Bell plus some others worked on this identification and processing. They had to identify a number of bones brought in by search parties. They identified some as those of a dog and others as beef bones. Apparently, someone had brought a bunch of bones home from the supermarket for their dog, or perhaps someone had recently slaughtered a beef. The doctors had to be sure, and they identified everything.''

Five minutes before the blast, volunteer fireman David Dillingham was relieved of his duties at the hydrant on Richland Avenue so he could get some sleep before reporting for work as night dispatcher with the Waverly Police Department. He used his walkie-talkie to tell Fire Chief, Wilbur York, that he was leaving. "See you later," said York over the radio. On the way to his car, Dillingham stopped and chatted for a moment with Police and Fire Commissioner, Guy Barnett, and then headed for home.

"When I got to Highway 70 through town," recalls Dillingham, "I had a call on my radio from Bill McMurtry, who was Deputy Fire Chief at the time, about notifying a person to come and help them as they were a man short. While I talking with McMurtry, I traveled about a quarter of a mile and was parallel with the Waverly Shopping Center when the explosion occurred. The communication ceased. I looked over my right shoulder and could see smoke, particles and large pieces of debris, and feel intense heat although I was probably a third of a mile from the tank. I turned on the emergency lights and siren and headed around on back roads since the main road instantly filled with people trying to see what had happened. I parked my car in front of the Volkswagen repair shop about 100 yards from the blown tank car and was walking down the side of the long, white Spahn building pulling on my gear when a man coming toward me called my name. I took off my firecoat and put it around the man, assuming from the color of his skin that he was Black . . . I had no idea whatsoever that I knew this man. It was only as I was helping him into the car that I realized from his conversation that this was Frank Carver, my Fire Captain. I was terribly shocked. On the short drive to the hospital he talked to me off and on, and I kept looking over to where he was seated as he tried to use his hands but couldn't. He was in terrible pain; parts of his body looked greyish-black, but mostly his body resembled liquid red paint with ugly black scars on top."

"Back at the scene," continues Dillingham, "I located Deputy Chief, Bill McMurtry. A tremendously loud, steady roar in the area, like the sound of a burning kitchen match multiplied 1,000 times, made it impossible to talk normally. I put my hand on McMurtry's shoulder and yelled: 'Mack, what do you want me to do?' 'My God, don't put your hand on my back,' he yelled. He turned around, and I went numb. The fireball had burned away the back portion of his heavy firecoat and coveralls right down to the skin. I yelled: 'I'm sorry,' and he yelled . . . 'It hurts,' and that was it. He was badly burned yet physically in command. I was uninjured yet scared out of my wits, probably in shock, and barely able to think."

"McMurtry yelled something about me checking on the other tank and pointed through the fire. I started to move off, but he grabbed me and motioned for me to put on an airtank. I got an airtank off Engine 51 and headed for the tank cars. The fire was mostly to the ground with little or no smoke within five feet of the ground, so I was able to see well enough to locate the tank cars. I probably didn't get within 100 yards of them because the heat was so intense, but I could see a cave-like entrance through the flames that I took to be one half the exploded car. Just beyond was another large shape that appeared to have a burning fountain on top, the second tank car. It had fire all around it but was still in one piece. I didn't stay there and write a book about it, that's for sure. I saw what I needed to see, then turned and walked away. I went back to McMurtry and told him: 'Chief, it's whole but on fire. We've got to get out of here.'"

Dillingham managed to get the Waverly fire department's old Dodge firetruck started, and with some volunteers aboard, crossed the tracks via the Route 13 viaduct and entered Richland Avenue from the north end, stopping approximately 300 yards from the second tank car. "There was a man with a hearse there helping to evacuate people, and he said: 'There are some people down over there; do you want me to go and pick them up?' I told him . . . 'No, you better get this unit out of here in case that second tank comes apart.' He said: 'That's ALL I want to know,' and took off. I ran down to a hydrant about 50 feet from the tank and found our equipment ruined and useless. The hoses looked like someone had taken a knife and ripped them inside out. I picked up a 2½ inch nozzle made of chrome and it felt like a burned match, just limp. When I turned around and started walking back to the truck, I stumbled over what I though was a dog. The degree of intense burns made me think it was an animal, not human, but I looked around and saw a foreman's helmet . . ." Dillingham then realized the form at his feet was a volunteer fireman who he had seen alive and well just 30 minutes before.

Tom Doss is Director of the Humphreys County Ambulance Service; a fulltime, paid department with four vehicles and nine employees. In the first 18 minutes after the blast, Doss and his crews transported 39 seriously injured burn victims to nearby Nautilus Hospital. Doss feels the Waverly disaster did not have to happen.

"I sent a crew down to the site Wednesday night after the derailment as a precaution," he says. "By Thursday afternoon, I started seeing people smoking cigarettes down there, getting much too careless, so I brought our equipment and personnel back to our base station behind Nautilus Hospital. We remained here on full-alert ready to go at any time. I was sure there would be an explosion because they were just too careless handling those tanks. I went to the authorities and asked them to move all those people away from the area, but they said I worried too much. I told them they were being awfully careless letting visitors, sightseers, and everybody else come in there. Hell, even the lumber company workers went back to work right beside the tank cars, and a factory was operating nearby. A beerhall in the area was operating; beer drinkers were blown right out into the street when the thing went off. I went down there twice on Friday and said my piece. Nothing was done . . . nothing. It aggravated the Hell out of me, I'll tell you. I could never find anybody who was in charge down there. It was supposed to be Civil Defense, but I never could find anyone who would admit he was in charge. Everyone I talked to said: 'I don't know; see him.' I couldn't find anyone in charge; there was no one concerned enough to be safe."

"I went down there 20 minutes before the explosion and told an official what I thought needed to be done. He said . . . 'Tom, you worry too much.' I told him why I was worried. I've cleaned up many a mess after something like this, and that's why I worry. I worry about it until it's over with because I'm the one who is going to have to come in here and clean it up if something does go wrong. I know the agonizing shock you go through when you see people suffering and dead. That's why I worry."

Tom Doss is trained to worry, to be super cautious. "I've been in this business 30 years, and I always expect the worst. I never take anything for granted . . . never. When I arrive at an accident on the highway, I'm *expecting* live wires that will electrocute me; I'm *expecting* a gasoline tank to blow up; I'm *expecting* some idiot driving down the road to plow into us and kill us all. I'm always looking for a disaster because I've seen so many. I've been wrong lots of time times, but I'm still alive and my personnel are still alive. If I'd had my way, there'd have been a danger zone established down there, and only those people working on the train would have been allowed in . . . and *nobody* else; no sightseers, no bystanders, no policemen, no firemen, no nothin.' L&N would have been on their own."

"Everyone I talked to at the scene had the attitude: 'Nothing to it; nothin's going to happen.' The situation was new to them; they had never seen anything quite like it in their lifetime where I have seen a number. That's why I worry. You would think people would listen to a man who had been through this so many times before. You would think they might say: 'Well, he *might* know what he is talking about; he says he has seen it before. He went through it; he's seen it happen.' You'd have thought they would listen, but they paid no attention to me."

Frustrated, Tom Doss drove back to the ambulance headquarters behind Nautilus Hospital, arriving there shortly after 2:30 P.M. He had time only to inspect his crews to make sure they were ready to roll and decide which it would be this time . . . a chew of tobacco or a cigarette. Tom Doss was convinced there would be an explosion. "This thing is going to go anytime," he told his E.M.T.s sitting in the ambulance service office, "there is no doubt in my mind."

A tremendous explosion punctuated Doss's worry. The E.M.T.s were out the door and gone in two ambulances before Doss had a chance to spit. He swiveled in his chair and punched the key on the dispatcher's microphone: "MAYDAY, MAYDAY, THERE'S BEEN AN EXPLOSION IN WAVERLY," he said calmly into the microphone, then headed for the site in a third vehicle.

"As I pulled in, my two ambulances and a couple of hearses were loading patients," he remembers. "People on fire were running about the area or rolling in the snow. Fire and smoke were boiling up as far as I could see, and the second tank car had not exploded. I was literally terrified because the fire at the end of the tank looked so hot. I felt the second tank could not help but blow. Fire Chief York was lying on his stomach, and he was blazing. I grabbed snow and threw it on him, patted him with it, and loaded him into an ambulance. He had no clothes, no hair; he was burned from top to bottom. We put two more men in, one on the tank and one seated in a chair. They were walking, they were rational, but they were burned all over. I put sterile water on them to douse their smoldering bodies, and they just sat there. They were in shock, but not enough to put them out. Our main concern was transporting the patients as safely as we could; we were so close to the hospital we didn't worry about treatment. Load 'em up and go. We were at the hospital within a minute. I knew Chief York, but I never did learn who the other two were. I believe all three died. There was no place on them that wasn't burned. It was terrifying."

"I unloaded at the hospital and immediately went back to the scene, but my other ambulances had made it back ahead of me and had cleaned out the area. I didn't see anybody but the dead ones. By then they had established a danger zone, what they should have done when the train first derailed, but nobody did it. The Civil Defense didn't do it; the Mayor of Waverly didn't do it; the Chief of Police didn't do it . . . nobody . . ."

"No matter how well-trained people might be, when they come up against something like the Waverly situation, when they are expecting to be blown off the face of the earth at any moment, it is hard to face. But people went right in there and got the victims out. There was no hesitation at all, no thought of their own souls. There was no backing up; even while searching for additional dead and injured when everyone knew the second tank hadn't blown, there was no backing up."

"What aggravates me," Doss concludes with a sigh, "is that two or three incidents have happened since the disaster that indicated people haven't learned anything from it. We had a tanker truck overturn on Highway 70 East, and people swarmed in on that just like bees on honey. The Sheriff had to threaten to kill some of them to get them out of there. People just have that curiosity; if something happens, they gotta see it . . ."

Immediately after the explosion at Waverly, state and local officials said repeatedly that the area around the derailment site had been evacuated at the time of the blast.

Mr. and Mrs. Rockford Bateman lived in a small frame house nearby the derailed tank cars. "I didn't have any warning," says Olive Bateman, a small, elderly woman whose voice and manner are those of a younger person. "I had made soup in the kitchen, washed the dishes, and just entered the livingroom when the explosion threw me into the air and dropped me back down. The pictures were coming off the wall as I tried to run out the front door. I reached down to get my little dog, but he kept running away from me. Something hit me on the head, the doorframe was fixing to fall, and I knew I had to get out at once. My husband was close behind me, but he ran and got in the car while I ran as fast as I could to the corner. One of my nephews saw Mr. Bateman sitting in the car, so he ran through the flames and drove the car to safety. At the corner of Slayden Lumber, I looked back and saw my house going down . . . just going under the flames . . . everything we had was in that house and the bait shop on the side. It looked like the whole universe was on fire for a few seconds. It's all gone now."

As Olive Bateman scurried down the street between two walls of flame, a man lying in the street called out to her: "Get help, get help."

"Honey, help is on the way," Mrs. Bateman remembers calling. She ran through her daughter's house located at the corner, then crossed Trace Creek where her daughter met her on the far side and drove her to safety.

Shorty and Ella May White were at home in the second house from the exploding tank on Richland Avenue watching television with a neighbor, Margaret Thompson, who lived in the first house north of the tank cars on Richland Avenue. White had just returned from talking to Fire Chief Wilbur York. As he settled back into his chair before the T.V., the explosion occurred. The Whites and Mrs. Thompson ran out through the back door where they became separated. Mrs. Thompson was told to run for her life by a badly burned fireman (Riley Turner) making his way across the yard.

Further north on Richland Avenue, Mamie Yarbrough suffered a heart attack as she ran from her house, attempting to gather neighborhood children as she ran. Her daughter stated the family had never been told to evacuate prior to the explosion.

Martha Reed stated no one had ever come to her house to tell her to leave although she had heard a loudspeaker far away asking people to clear the area near the derailment.

Elizabeth Dean Carlew at 218 Richland Avenue said she called the Waverly city hall before the explosion and was told there was no need to leave her home. "Stay where you are until someone comes to tell you to leave," she says she was told. Within a few minutes, the tank exploded.

At 216 Richland Avenue, Linda Hooper had just returned home from work. "We were sitting down to eat when I felt a tremor like an earthquake. The windows were rattling; the whole house shook. I went to the front door and looked out. I saw flames and pieces of the train flying through the air. People were running, screaming, and hollering. The heat was so great it seemed to burn the skin right off my face. People were running in panic, trying to get away. People gathered in the churchyard and watched the explosion. All the way for 15 houses, people were at home when the tank exploded. Afterwards, the police came back and evacuated everyone from their houses."

A loss of such magnitude is especially tragic, and lingering, in a small town like Waverly where everyone knows everyone else. Men like Police and Fire Commissioner Guy Barnett, and Fire Chief, Wilbur York, were more than just town officials; they were friends and neighbors. Barnett had been Chief of Police for eight years and had been stationed in Waverly as a member of the Tennessee Highway Patrol for six years prior to taking the Chief's job. Wilbur York had been with the Waverly Fire Department for as long as there had been a department.

Walking reminders of the tragedy are a fact of life in Waverly; the living victims of the fireball will be on display for years to come. Some such victims are truly unique; by all past experience they should have died yet they are going about their lives as best they can. In doing so, they painfully remind townspeople on a daily basis of a simple derailment that ultimately was allowed to kill 16 and seriously injure more than 50.

As might be expected, the people of Waverly are generally uneasy about assigning responsibility or blame for the incident, particularly if any such censure might cause additional suffering to people in Waverly. There is a terribly sad acceptance of what happened, as though the delayed explosion that rent Waverly was an act of nature rather than an act of man. There was, initially, a strong feeling of anger but little or no common agreement toward whom or what such anger should be directed. There is still questioning as to who was in overall charge in Waverly from 10:30 P.M. on Wednesday, February 22, 1978 until 2:53 P.M. on Friday, February 24, 1978 when Waverly's world came apart. There is near-universal skepticism with claims that an evacuation was in effect at the time of the explosion; people know better.

Townspeople prefer to stress the uniqueness of the Waverly incident. Often, the longterm reaction is . . . "This had never happened before; what happened her was totally unexpected, completely without precedent." There *is* little precedent for the *exact* circumstances at Waverly where there was no fire before the rupture, there had never been any flame impingement on the tank was not punctured during the derailment, and there was no identified leakage from the tank prior to its coming apart.

Fault or responsibility for the Waverly tragedy will not be determined in Waverly but rather in the U.S. District Court in Nashville before Federal Judge, L. Clure Warden. Unlike the people in Waverly, the lawyers in Nashville have shown little hesitancy in attempting to assign responsibility. Approximately 50 personal injury suits seek a total dollar amount in excess of $362 million in compensatory and punitive damages. From a damage standpoint, the unique quality of this case is the type of injury and the magnitude of injury suffered by victims. There are injured parties who were never expected to live who not only survived but will appear in court as accusers. Immediate medical treatment, unusual burn-treatment techniques, and extensive cosmetic surgery will, in a number of cases, allow victims rather than their widows to appear in court. The personal injury cases are being heard in U.S. District Court rather than in Waverly because of a diversity of citizenship among both plaintiffs and defendants, a number of whom, both corporations and individuals, are from outside the State of Tennessee.

To date, there have been many accusations resulting from Waverly but few findings of fact. The process of determining and assigning legal responsibility is expected to continue for years.

THE FIRST RESPONDERS: PUBLIC SAFETY AGENCY RESPONSE ORGANIZATIONS

Denver Hazardous Materials Team Members Drop Suit-and-Tie Assignments to Respond 24 Hours a Day

Personnel of the Denver Fire Department Hazardous Materials Emergency Response Team don't sit around the stationhouse waiting for calls; as a matter of fact, they don't have a stationhouse to sit around. Each team member holds down a regular assignment with the department 40 hours a week, but responds instantly 24 hours a day for hazardous materials incidents within 116 square miles of territory. Assistant Chief Donald Penn is Drillmaster for the Denver Fire Department, while Captain Eldon Bullard is Assistant Drillmaster. Lt. John Marshall is the Hazardous Materials Coordinator for the Fire Prevention Bureau within the department, and Technician Miles Slocum is Fire Advisor to the Denver City and County Office of Emergency Preparedness.

In December of 1978, Assistant Chief Don Penn and Miles Slocum traveled to Tennessee to observe the hazardous materials control program operated by the Tennessee Office of Civil Defense, then returned to Denver to report to Fire Chief Merle Wise. Denver is blessed with longstanding cooperation between and among local emergency response agencies represented on the Colorado Committee for Hazardous Materials Safety. They continuously update and rewrite necessary response plans specifically as they relate to hazardous materials. Agency assignments and responsibilities are clear-cut and well maintained, but there is something extra in Denver . . . a feeling of mutual respect and support among agencies that is readily apparent to an outside observer.

Although the need for hazardous materials response equipment was well recognized by the City of Denver, the fire department operates on a fixed budget. The expense of a hazardous materials response vehicle would have placed it in a capital equipment category where it would have to compete for priority with new pumpers and other fire suppression apparatus.

Early in 1979 the Denver Office of Emergency Preparedness, a coordinating agency designed to initiate and bring together the efforts of various response agencies for a coordinated thrust, provided funding from a contingency fund reserved for purchase of pressing emergency-need items to obtain a ¾ ton Dodge van that was delivered to the Denver Fire Department in April of 1979.

As soon as team members returned from two weeks of intensive hazardous materials response training in Nashville funded by the Tennessee Office of Civil Defense/Emergency Preparedness, they began outfitting the van. Denver Fire Department repair and carpentry shop personnel turned the "shell" vehicle into a response van by installing shelving, applying insulation, and performing a great deal of straightening, cutting and altering. The vehicle was placed into service on July 20, 1979.

The Denver hazardous materials team operates off the "911" central dispatch system; when a call is received by the central dispatcher, he immediately activates the proper initial fire companies. As soon as the central dispatcher has positive information from the regular companies that hazardous materials are involved, he dispatches the team from their regular assignments by Pageboy. Team members receive verbal instructions over their pages, as much positive identification information as the dispatcher then has available as to the type of incident and the commodity.

"The main idea for the team is CONTAINMENT," notes Lt. John Marshall. "Our training has stressed containment. Our primary concern is to contain the product and stabilize the scene. Once that is done, we have time to contact the shipper or manufacturer and work on decontamination, clean-up, and removal. Denver is very heavily populated. A commodity flushed down the sewer system could create a secondary emergency far more serious than the initial situation. Also in Denver, we are relatively limited as to availability of water compared to many locations around the country."

"Denver does not have any great amount of heavy manufacturing," continues John Marshall, "but there are many facilities that store, handle, and process hazardous materials. Denver is a 'stopping point' for many such materials; much is sent here or travels through here but is not necessarily manufactured or used here. In addition, there is a fair volume of hazardous materials traveling through Colorado that is not scheduled to stop here. As for major users of hazardous materials, we have Rocky Mountain Arsenal at one corner of the city and Rocky Flats nuclear facility at another with a number of space and chemical companies in between. There is a limited number and amount of esoteric products in Denver, most notably car loads of hydrazine used as a rocket fuel, and various radioactive materials used at the Rocky Flats facility."

During the five month period of August through December, 1979 the team responded to 32 incidents of which they actually worked 15 in some way, shape or form. Of the 15, five were considered major incidents. Three were major rail tank car spills, including a 9,000 gallon tank car spill of 90 percent nitric acid solution that proved to be a three-day operation. At that particular incident, it was necessary to neutralize two pounds of nitric acid by bringing it to the proper pH using soda ash, then haul away 240 tons of dirt for disposal.

"We had two 'minor' derailments," adds Miles Slocum," 'minor' to the railroad but major to us. One of these involved five cars of 33,000 gallons each of nonodorized L.P.G., of which three cars were lying on their sides. In other incidents, we have handled two leaker tank cars that were discovered by railroad security. One was a plate-C (jumbo) propane car leaking in the valving system around the metering rod. An additional car that ultimately proved to be sound was stenciled 'Anhydrous Ammonia' but actually contained L.P.G.. Our toughest incident was the nitric acid car; our hairiest incident involved a damaged L.P.G. car that was moaning and groaning. Anytime you are dealing with a flammable gas vapor hazard, you have potential for serious problems. When we use a 'catch-all' arrangement to burn off a flammable gas, it is always a last resort. Under such circumstances, we monitor constantly and use water-fog repeatedly to knock down excess vapor."

In addition to initial hazardous materials response training in Nashville, the Denver team has participated in both long and short courses at the Colorado Training Institute and attended a National Fire Protection Association seminar in Denver. "We train as a team all the time," explains Miles Slocum, "using equipment off the rig, putting on the acid suits, and familiarizing ourselves with various materials. We borrowed a number of videotapes from the hazardous materials team of the Jacksonville, Florida fire department that we have used extensively in our training. The Jacksonville guys were really helpful; they couldn't do enough for us."

"One advantage of our present set-up," continues Slocum, "is that as Training Officer/ Drillmaster for the entire department, Chief Penn is as up-to-date as anyone can be on new types of foam and other materials and techniques that relate directly to hazardous materials control and containment. In addition, Chief Penn's people have put together a pretty extensive hazardous materials familiarization program for all personnel in the Denver Fire Department. This program is conducted each day through different pumper and truck companies around the city and is administered through the training division."

"Training is critical," agreed Lt. John Marshall, "but until you actually respond to a few incidents your knowledge will be limited. The more incidents you respond to, the more efficient and knowledgeable you become. There is is no learning experience quite like working at an incident and being responsible for achieving containment and control. The hands-on, actual incident situation is just so forceful a learning experience that you almost hope for more so you can hone your skills. The more your learn, the more you become aware of possible corrective measures that can be used in a specific situation. Training is crucial, but actual incident experience broadens your usable knowledge so much and so forcefully that its importance cannot be underestimated. Knowledge can be equated with respect for the product you are dealing with. Once you have worked with a specific commodity and developed a healthy respect for its characteristics and destructive capacity, you are far better qualified to deal with that commodity a second time. The more closely hazardous materials schools are able to replicate actual situations found on-site, the more successful such schools will be in preparing trainees for the real world. We just completed two weeks at the Colorado Training Institute where they had mock-ups and cutaways; plus actual components of tank car gauging devices, valves, and metering rods we could manipulate and play with in a safe atmosphere. This was extremely helpful. You need both the hands-on training to the extent it can be provided in a classroom, and you need to respond to a few incidents. An incident helps you to realize how little you know and how much there is to learn. An incident can be the greatest motivator a hazardous materials response team will ever have."

"Our van is as fully loaded as it can be," notes Assistant Chief Don Penn. "We now have to be very selective as to what goes into the van because nothing more will fit in unless we are willing to take something out. We rely heavily on the regular pumper companies for any additional equipment we may need at a particular incident. For example, we carry high expansion foam and a foam generator in the van, but we rely on the pumper companies to supply us with hoselines and water. We always have a regular company on-scene to provide back-up."

"We have some unique communications in the van that other fire departments might be interested in," added Miles Slocum. "Our walkie-talkie radios have a patching capability that allows us to punch a code number to obtain access to any telephone number. The van is equipped with a limited-access U.H.F. 'direction and control' channel that we use at the scene of an incident. When we don full Acid/Gas/Entry suits equipped with bone-conducting microphones at the back of the mask, all we have to do is drop one arm to activate a pressure switch and we are in communication with each other and can relay continuous information to our Incident Commander, Don Penn, throughout an attack and containment effort. We can be standing in a shower of hazardous commodity from a ruptured tank car here in Denver, Colorado and be talking directly with the CHEMTREC communicator in Washington, D.C., or with the tank car designer in another part of the country. You can't ask for more complete communications than that."

"There are four portables in the van," continues Slocum. "The Mayor, the police chief, and Denver Chief Merle Wise each have one; as do the Directors of Health and Hospitals, Public Works, and Safety. There are also eight of these units with the Special Services Unit of the Denver Police Department for operational situations. The frequency is not an everyday use frequency; it is always available for special operations."

In addition, the van is equipped with a 110 volt/550 watt inverter to permit operation of a Heathkit weather information kit. Denver's location within a natural half-bowl seems to play havoc with wind direction. At one recent spill, the wind shifted 360 degrees with sizable variations in velocity during a period of 30 minutes. The weather information unit sits atop a 14 foot mast that is mounted to a crossbeam as needed and provides moment-to-moment readouts on a digital meter in the van for wind speed, wind direction, barometric pressure, and temperature.

"We also have a tape recorder that operates off the same 110 volt power supply," adds Miles Slocum. "This is very helpful when we are receiving information from CHEMTREC, the car designer, or shipper. Any time you relay messages through a third party, there is always a chance the message will be garbled or misunderstood. One incorrect letter in the spelling of a

Hazardous Material Response Vehicle
City & County of Denver

VEHICLE:
    1979 Dodge ¾ ton Dodge Van

COMMUNICATIONS:
    Four Channel Fire Mobile
    Twelve Channel Police, Direction & Control Mobile
    Telephone Patch on Direction & Control Channel
    Direction & Control Portable Radios
    Fire Portable

PERSONAL PROTECTIVE EQUIPMENT:
    Pressure-Demand Breathing Apparatus
    Acidmaster Complete Enclosure Chemical Suits
    Two-Piece Chemical Suits
    Neoprene Gloves, Rubber Boots, Helmets
    Fire/Rescue Entry Suits
    Fire Bunking Gear
    Pressurized Eye Wash Unit
    Chemical Splash Goggles

CONTAINMENT EQUIPMENT:
    Mini X Foam Generator
    High Expansion Foam
    Metal X 30# Extinguisher
    AFFF Lightwater Extinguisher
    Chlorine Repair Kits
    Non-Sparking Special Alloy Tools
    Wooden Plugs (Various Sizes)
    Parafilm Wrap
    Plastic Sheet
    Absorbent Material

SUPPORT EQUIPMENT:
    Various Hazardous Material Reference Manuals
    Binoculars
    Cassette Recorder
    35mm Camera
    Non-Sparking Lanterns & Flashlights
    Five Function Weather Station
    550 Watt Power Inverter
    Lightweight Chemical Suits For Resource Personnel
    Disposable Suits & Gloves

TESTING EQUIPMENT:
    Gascope Combustible Gas Indicator
    Vapor Sampling Test Kit
    PH Strips
    Radiation Meters

Figure 1. Equipment List for Denver Fire Department Hazardous Materials Response Team's Vehicle.

product can make a world of difference. With the tape recorder, we can be sure what was said. We don't have to rely on memory, or be worried about misunderstandings.''

The Denver Fire Department maintains a resource list of heavy construction equipment available from all sources within city government. Whenever the needs of a particular situation exceed such resources available from city agencies, the hazardous materials team can fall back on a contract recently renewed between the Denver Office of Emergency Preparedness and the Associated General Contractors that is recognized in the city disaster plan. A simple telephone call can obtain any type of needed heavy equipment from a crane to a dump truck in a matter of minutes. The only charge to the city is for the operator's time; use of the actual equipment is provided as an emergency public service by the Associated General Contractors.

The hazardous materials Emergency response team of the Denver Fire Department forms an important component of a multi-agency response plan within the Mile High City. In Denver, it appears, interagency coordination and cooperation play a strong role in hazardous materials containment and control efforts.

Nearly One A Day for the Houston Fire Department Hazardous Materials Response Team

The Houston Fire Department's Hazardous Materials Response Team composed of 18 firefighters divided among three shifts went on-line October 5, 1979. As might be expected in a city that has often been called "The Chemical Capital of the World," Houston team members are a busy crew. They roll as a two-piece, combination Hazardous Materials/Rescue company with four or five men in a Rescue 1 vehicle and one man in a vehicle designated as H.M.-1. During calendar year 1980, the HMRT responded citywide to 335 reported hazardous materials calls of which 297 proved to actually involve hazardous materials. HMRT personnel worked 157 (53 percent) of the total 297 incidents while first-in companies were able to control 140 incidents (47 percent) without assistance from HMRT. In addition, the team averaged in excess of 60 Rescue calls a month as well as a number of fire suppression alarms within a response area that covers approximately one-seventh of the city.

With regard to type of hazardous materials incidents responded to by both first-in companies and the HMRT, leaks accounted for the most frequent problem with 140 separate incidents or 41.8 percent of the total. In addition, there were 55 spills, (16.4 percent of all incidents) 49 incidents involving fire, (14.6 percent of all incidents) 30 hazardous materials incidents not involving spill/leak/fire (nine percent of the total) and 23 accidents involving hazardous materials that did not result in an actual hazardous materials incident. An example of the last category would be a tanker-truck accident in which there was no release of hazardous commodity.

With reference to the type of container involved, fixed systems were the most frequently encountered problem with 100 separate incidents or 33.7 percent of all incidents. Tank trucks were the second most frequently encountered container problem with 60 incidents or 20.2 percent of the total. A general category of container including ships, barges, bottles, and storage tanks accounted for 58 separate incidents or 19.5 percent of all incidents. Drums accounted for 10.8 percent of all incidents, (32 incidents) portable tanks for seven percent, (21 incidents) cylinders for 6.4 percent, (19 incidents) and railroad tank cars for a mere 2.4 percent of all incidents. (seven incidents)

Roughly two-thirds of all hazardous materials incidents occurred within the eastern half of the city where loading docks, railroad yards, and petrochemical industries are concentrated. During 1980 HMRT personnel dealt with 63 different hazardous materials ranging from crude oil and diesel fuel to highly toxic phenol and dimethyl sulfide, highly flammable ethyl ether and acetylene, and extremely unstable methyl ethyl ketone peroxide and hydrogen peroxide. Although toxic, flammable, and unstable chemicals are a daily fact of life for the Houston Fire Department Hazardous Materials response Team, not a single member has been seriously injured.

Houston's HMRT operates two vehicles. H.M.-1 is a step-van similar to those used by bread companies to make deliveries to supermarkets and carries equipment and materials necessary for hazardous materials response including pressure demand SCBA, (Scott 4.5 and Type Two plus air lines) acid suits, proximity suits, and safety harnesses. H.M.-1 also carries 25 gallons of high expansion foam, 25 gallons of protein foam, 30 gallons of hydrocarbon emulsifier, 600 pounds of sand, 300 pounds of soda ash, plus high expansion foam generators, AFFF eductors, and protein foam eductors and nozzles. The unit is also equipped with a

Purple, K, Met-L-X, and CO-2 extinguishers. Detection equipment carried on H.M.-1 includes explosimeters, a pH meter, a CO meter, as well as a pyrometer and radiation monitoring kits.

An on-board reference library includes the *N.F.P.A. Hazardous Materials Guide, Firefighters Handbook of Hazardous Materials,* a complete *C.H.R.I.S.,* (Coast Guard Hazardous Materials Manual) D.O.T.'s *Emergency Action Guide, The Farm Chemicals Handbook,* a *GATX Tank Car Manual, Dangerous Properties of Industrial Materials* by Sax, *Recognition and Management of Pesticide Poisonings,* and *Safety Data Sheets.*

H.M.-1 carries all three chlorine kits; an "A" kit for small cylinders, a "B" kit for 1,000 pound cylinders, and a "C" kit for railroad tank cars. For patching and plugging materials the team carries railroad tank car sump covers, a railroad tank car "wild car" plug, (for use when product is pouring from a tank car after the belly cap has been removed with the bottom outlet valve open) lead wool, T-bolt patches, toggle bolts, sheet metal screws, pressure sensitive tape, wooden plugs, rubber stoppers, stainless pipe clamps, and duct tape. Additional patching and plugging materials include quick spray adhesive, foaming patch kits, red rubber gasket compound, neoprene gasket compound, sheet lead, epoxy kits, and fibreglass cloth.

To insure the team is completely armed for the wide variety of leaks they have encountered, (leaks accounted for nearly 42 percent of all Houston team hazardous materials responses during calendar year 1980) H.M.-1 also carries bungs and plugs, epoxy putty, duct sealing compound, pipe plugs from four inches to 10 inches, C-clamps from two inches to eight inches, silicone seal, "Form-A-Gasket", "PB-35 Sealant," stainless hose clamps, and assorted gas line clamps. Rounding out a very complete assortment of hazardous materials response equipment are "Dike-Pak" kits, plastic bags, chemical gloves, rain suits, goggles, ear protectors, ground rods, and grounding and bonding cables.

A second vehicle, Rescue-1, is a crew-cab unit equipped with a front-mounted winch, and an air cascade system used for supplying face masks, filling lifting bags and SCBA tanks, and cooling acid suits. As one team member noted: "The acid suits can be murder during our hot summers." Additional equipment on Rescue-1 includes a variety of forced entry tools and equipment as well as a hydraulic system and lines to power hydraulic exhaust fans, hydraulic sump pumps, (used occasionally for product transfer) and hydraulic power tools around flammable vapor atmospheres.

"The team is a combination Hazardous Materials/Rescue unit that responds citywide for hazardous materials emergencies and within a prescribed area for Rescue and box alarm fire suppression calls," explains District Chief Max McRae, Coordinator for HMRT. "This practice came about because we wanted to keep the men active. They do NOT want to get completely away from firefighting. As long as we can provide a high level of continuous activity for the men, we do not have a problem with them wanting to get back to an engine company."

Houston fire officials are quick to point out that much of the success enjoyed in initiating and developing HMRT can be traced directly to close and continuing interaction with major industrial concerns in the area. Industry continues to respond quite generously to many HMRT needs in the area of training and supplies. When the team ran out of chlorine cell putty recently, Dow Chemical in Freeport not only sent over an emergency supply of the patching material but followed up with a 100 pound supply. Dow also keeps the team supplied with a special foam patching material that its laboratory developed for Dow's own emergency response team. Shell Chemical's emergency response team and HMRT members enjoy a close working relationship, and Shell provided the team with a railroad tank car and some recovery drums for training and covered the costs for sending two HMRT members to the Safety Systems seminar in Palatka, Florida. Shell also hosted a four-hour seminar on pressurized and atmospheric tank cars at its Deer Park chemical plant for each of the three HMRT shifts, and has developed a special school on handling leaks and product transfer from overturned tank trucks for its own personnel that will include HMRT personnel in each future class. As District Chief McRae noted recently: "The success we have enjoyed in developing HMRT can be attributed in large part to industry, and industry continues to aid the team. HMRT will always call an industrial team for help when one is available. There is always a need for industrial teams who have a special expertise in handling a particular hazardous product."

Pressure Demand SCBA (Scott 4.5 and Type II, and air line)
Acid Suits
Proximity Suits
Safety Harnesses

Purple K Extinguishers
AFFF Extinguishers
Met-L-X Extinguishers
CO₂ Extinguishers

High Expansion Foam generators
AFFF Eductors
Protein Foam Eductors & Nozzles

Explosimeters
pH meter
CO meter
Pyrometer
Radiation Monitoring Kits

Portable Eye Wash
Drum Truck
Air Bags
Wrenches & other small tools
Vice and Pipe Cutter
Jacks and Chains
Brace and Bits
Entrenching Tools
Chain Hoists

NFPA HM Guide
Firefighters Handbook of HMs
C.H.R.I.S. Manual
Emergency Action Guide
Farm Chemicals Handbook
GATX Tank Car Manual
Dangerous Properties of Industrial Materials
Recognizing & Management of Pesticide
    Poisonings
Safety Data Sheets

Chlorine Kits (A, B and C Kits)

Plastic Bags
Chemical Gloves
Rain Suits
Goggles
Ear Protectors

RR Tank Car Sump Covers
RR "Wild Car" Plug

Ground Rods
Grounding & Bonding Cables

Supplies:
Lead Wool
T-Bolt Patches
Toggle Bolts
Sheet Metal Screws
Pressure Sensitive Tape
Wooden Plugs
Rubber Stoppers
Stainless Pipe Clamps
Duct Tape
Quick Spray Adhesive
Foaming Patch Kits
Red Rubber Gasket Material
Neoprene Gasket Material
Sheet lead
Epoxy Kits
Fiberglass Cloth
Bungs and Plugs
Epoxy Putty
Duct Sealing Compound
Pipe Plugs (4" to 10")
C-Clamps (2" to 8")
Silicone Seal
Form-A-Gasket
PB-35 Sealant
Stainless Hose Clamps
Gas Line Clamps

High Expansion Foam (25 gallons)
Protein Foam (25 gallons)
Hydrocarbon Emulsifier (30 gallons)
Sand (600 pounds)
Soda Ash (300 pounds)

Dike Pak Kits

Figure 2. Equipment and Supplies Carried On The Houston Fire Department Hazardous Materials Response Vehicle, "H.M.-1."

Although HMRT covers an area of roughly one-seventh of the city for Rescue and fire suppression alarms, the team is dispatched to any point in the city when the alarm involves a known or suspected hazardous material. The team will go up to five miles beyond city limits on the dispatcher's say-so if and when a request is received from another fire department. Beyond five miles outside of city limits, permission must be obtained from Houston Fire Chief, Vernon Rogers. "There is no problem," one fire official reported, "we will go beyond five miles outside the city. Chief Rogers was badly burned at the 1971 Mykawa Road incident when a railroad tank car of vinyl chloride BLEVE'd. He has a strong, personal commitment to hazardous materials preparedness and assistance to other fire service organizations."

Within the greater Houston industrial district that extends for a half-mile along each side of the 35-mile long Houston Ship Channel are located many industrial facilities which process petroleum and petrochemical products. Public and private interests in the area participate in a mutual aid organization known as CIMA. (Channel Industries Mutual Aid) Although the area served by CIMA is wholly outside the city limits, the Houston Fire Department is a member of CIMA and takes a very active part in its planning and operations. A Chief Officer, reporting directly to the Fire Chief, is assigned nearly fulltime to CIMA liaison, planning, and operational activities.

Houston team personnel make it clear that they rely heavily on the expert advice and information provided by CHEMTREC, the Chemical Transportation Emergency Center operated by the Chemical Manufacturers Association. HMRT contacted CHEMTREC during many hazardous materials incidents last year for information about the chemical involed, to verify the team's interpretation of its reference materials, or to locate a manufacturer or shipper. Representatives of most industrial complexes in the Houston area have agreed that HMRT call CHEMTREC when a need arises rather than the individual company, thus nearly eliminating the extremely difficult task of maintaining an updated list of local telephone numbers. Also, CHEMTREC now has computer contact with major chemical companies in Houston in order to speed industrial emergency response teams to a hazardous materials incident scene.

Captains in charge of each HMRT shift are responsible for continuous in-service training while District Chief, Max McRae, is responsible for outside training. Original training for the team consisted of 130 hours hazardous materials related instruction and 80 hours of Rescue related training. A number of major industrial concerns participated actively in this training including Texas Alkyls, Inc (alkyls) Diamond Shamrock Corporation, (chlorine kits) Stauffer Chemical Company, (corrosives) Coastal States Company, (L.P.G.) Charter Oil Company, (refinery operations) Richmond Tank Car Company (tank car construction) and Dow Chemical Company (patching/plugging of tank cars, tank trukcs, and drums).

Members of a 12-man back-up team formed recently received 32 hours original instruction that included hands-on training in drum patching, use of salvage drums, tank car bottom outlet leaks, "wild car" control, diking and booming, static electricity and grounding, chlorine kits, detection meters, foams, and air bags.

"We rely heavily on industry in the area for outside training," emphasizes District Chief McRae. "For the first six months of last year the chemical workers were out on strike, so our outside training program did have difficulty in securing people from the industrial response teams because they were all pulled into work on production during the strike. Normally, however, we can just about pick-and-choose from among some of the best and most knowledgeable people in the industry. Without fail, they are eager to make themselves available to assist us with training for the team." Even with the chemical workers strike, HMRT personnel attended four special training programs during the first six months of 1980: M.S.A. "Dike Pak" training, care and use of detection and monitoring meters, use of protein and high expansion foams, and a half-day program for each shift on stopping leaks in pressure and atmospheric tank cars hosted by Shell Chemical at its Deer Park facility.

Recently, the team completed its first annual seminar/workshop, a three-day session involving representatives from the Southern Pacific Railroad, Charter Oil Company, GATX, Union Carbide, the Houston Police Department bomb squad, DuPont, U.S. Coast Guard, Big 3 Industries, U.S.-D.O.T., and Shell Chemical. Subjects on the agenda were ones the team had expressed concern over during the year: railroad tank car construction, pressure cylinders, pesticide emergencies with emphasis on methyl isocyanate which is shipped through Houston in carload lots, and cryogenic tank truck emergency operations.

Team members have also traveled outside the Houston area to attend training at Texas A&M University, the National Fire Academy in Maryland, and the Safety Systems school at Palatka, Florida.

The Houston Fire Department is in the process of increasing training aids specifically related to hazardous materials control and containment available at the department's Fire Academy. A triple compartment, general commodity railroad tank car recently obtained from Shell Chemical will be used to create various types of leak situations on the ends, sides, and bottom outlet valves. The existing domes on the car are being replaced with a chlorine dome, an L.P.G. dome, and an acid dome. Freon gas will be piped to the various parts of the domes to provide a realistic simulation of pressure leaks. This triple compartment car will eventually replace a training aide consisting of a bottom outlet leg obtained from an overturned tank car at a derailment site. This unit is mounted on a portable stand to simulate bottom outlet valve leaks or a "wild car." Department training officers are currently seeking a chlorine one-ton container and a tank truck to complete the equipment they presently lack for a well rounded hazardous materials training facility within the Fire Academy.

HMRT shares its knowledge and experience with other fire departments. The team recently put on hazardous materials training programs for the Gulf Coast Firefighters Association, the Harris County Firefighters Association, and individual fire departments in Channelview, Cypress Creek, Jersey Village, and Champions.

HMRT also recently completed a hazardous materials program to familiarize 1670 Houston firefighters with the proper actions to take when suddenly confronted with an incident. Missouri Pacific's program on Recognizing & Identifying Hazardous Materials was used, and the functions and capabilities of HMRT were emphasized during the program.

District Chief McRae was asked recently if there have been any major changes in team operations overtime. "The original intent of HMRT was to stabilize a hazardous materials incident with the aid and advice of industry, then turn over the scene to industry for product transfer and clean-up," he notes. "BUT, it doesn't always work that way in practice. We have had a number of carriers we couldn't locate, and others that didn't have the means to handle a given situation. More and more, we have found ourselves getting into product transfer and clean-up; equipment and supplies on H.M.-1 have been expanded to handle limited product transfer and clean-up, and our training plan has been changed to focus more on this area."

The Houston Fire Department Hazardous Materials Response Team has been in operation less than 36 months. Yet the department is generally recognized as a leader in hazardous materials preparedness.

"No single factor can be credited completely for the success of HMRT in handling, almost daily, the many hazardous materials emergencies that occurred in Houston last year," reflects District Chief Max McRae. "Industry generously provided many hours of valuable training and furnished costly equipment and supplies. Fire department management was receptive to many of the team's recommendations and pushed through requests for new equipment and supplies that had not been budgeted. Dedicated team members, some of whom spent their own money and off-duty hours to improve the capabilities of HMRT, put in a lot of effort to get HMRT going and to keep it moving. It was this combination that provided the rapid development and successful experience of the Houston Fire Department's Hazardous Materials Response Team."

Hazardous Incidents Team One, Hillsborough County, Florida

HIT ONE (Hazardous Incidents Team One) was formally placed into service by the Hillsborough County, Florida Department of Fire Control And Emergency

Operations in December of 1979 after a year of training. From January 1, 1980 through April 30, 1981 HIT ONE responded to 74 hazardous materials incidents for which first arriving companies had requested assistance. The nine-man team composed of three Captains, three Driver/Engineers, and three firefighters; was formed under the training auspices of Chief of Training, Robert Mertens, and is commanded by James L. Brady, Chief of the Fourth Battalion. Each platoon of one Captain, one Driver/Engineer, and one firefighter is on duty 24 hours and off 48. Each off-duty member is subject to recall for major incidents through the use of Motorola paging devices.

Hillsborough County is the third largest county in Florida, encompassing 1037 square miles of land and approximately 24 square miles of inland water. The Department of Fire Control and Emergency Operations is headed by a Director, Chet Tharpe, who is, in effect, the overall Fire Chief for Hillsborough County.

"During the many months the team was in training it was formally placed into service, we purposely maintained a very low-key approach," remembers Battalion Chief Jim Brady. "We did not want to startle the community, or suggest that a special task force was in training because of ultra-hazardous chemicals in the area. As a result, we have received a great deal of cooperation from the industries we serve, including many contributions of equipment for the team."

HIT ONE's vehicle is a squad-type truck with a typical "Breadbox" component on the rear which features exterior and interior compartmentalization. The cab area is set-up both to transport members and to function as an inside planning area. There is a desk built-in to the cab area, and all reference materials are maintained there. "The reason we went with the crew cab is that it provides the entire team a place to discuss the situation while enroute," notes Chief Brady. "Team discussion will in transit to a particular incident, with reference materials and equipment readily available, has been quite valuable. The cab area can also function as a command post if necessary."

The vehicle has four different radio systems that can tie-in to the four-channel system of the Hillsborough County Fire Department, or to the mutual aid frequency as necessary. The vehicle and on-board equipment such as acid suits, chlorine kits, SCBA, chaining and patching gear, etc.; represents a cost of approximately $30,000. Included is a complete set of nonsparking brass tools—about 100 different items ranging from a small hammer to a large scoop shovel—that has proven invaluable in dealing with flammable and combustible liquids. Additional equipment includes a selection of explosimeters/gas detectors/"sniffers," assorted patches that can be used ready-made or hand-fashioned on-scene to meet particular requirements, synthetic plugging and diking materials, and airbags to assist in applying an instant patch to larger tanks. The vehicle is equipped with its own electrical generating system for power and lights.

In order to readily access chemical data, HIT ONE uses a small business computer (a Radio Shack model with "floppy disc" storage and 32,000 character memory that costs less than $2,000 including mobile installation) completely self-contained within their response vehicle that team members programmed themselves. That is, rather than use an available database, they created their own using information taken from a number of sources. HIT ONE personnel developed their own coded indexing system to further condense the mounds of information they wanted to store in the computer. The operator accesses the computer memory with a code key telling the computer the type of information being sought. He then types in, for example, the name of a chemical; the computer then displays a full screen of information HIT ONE personnel have previously selected, coded, and condensed when creating their "personalized" database. The team is currently experimenting with a telephone interface by which their on-board computer can access other computers throughout the country. If this application proves practical, HIT ONE will have nearly unlimited access to technical data and reference materials.

"Our most common incident has been gasoline tank trucks overturned, ruptured, and spilling," reports Chief Brady who has held positions in county fire service ranging from

PROTECTIVE CLOTHING:
  —Complete Nomex firefighting apparel for all members, including hoods.
  5-Acid/Vapor suits, Self Contained
  5-Aluminized Proximity Suits
  5-Self Contained MSA Breathing Apparatus with Composite Cylinders
  2-Chemical Splash Aprons
  3-Chemical Splash Goggles
  3-Bullard Shock Helmets with Ear Protectors
  —Assorted Gloves for special substances
INSTRUMENTS:
  1-Micronta Leak Detector
  2-MSA Explosimeters
  1-Hydro Carbon Detectors
  1-Leak Tracer
  1-Anemometer (Wind Speed & Direction)
  1-Continuity Meter
  2-Radiation Detector, Model 1-A
  2-Radiological Monitor, Model 6-A
  1-Aerial Radiological Monitor
TOOLS & EQUIPMENT: (Includes basic tools such as axes, pike poles, first aid)
  1-Complete Set Brass Hand Tools, includes small tools, sledge hammers, shovel, picks, chisels, etc.
  2-Chlorine Kit Type A
  2-Chlorine Kit Type B
  1-Chlorine Kit Type C
  —Chain, Assorted Size & Length
  6-50 lb Containers Lime
  1-Five Ton Com-a-Long
  3-Hydraulic Jacks, 20 Ton
  1-Large Steel Tong
  —Assorted Black Carbon Steel Patches
  —Assorted Wooden & Rubber Plugs
  1000' Barricade Tape
  2-10 × 50 Binoculars
  1-50x Spotting Scope w/roof mount tripod
  1-50lb Dry Sealant
  1-Heavy Duty Wheel Puller
  —Dry Chemical Fire Extinguishers, 60 lbs
  3-40' Nylon Straps, 2000lb Capacity w/lock connectors
REFERENCE MATERIALS:
  1-Chemical Data Sheets
  12-Assorted Reference Books on Hazardous Materials
  —Assorted Street, Sewer and Utility Maps
  —Assorted Topographic & Terrain maps
SPECIALTY APPARATUS;
  Light Plant and AC Generator with Accessories
  Maxiforce Air Bag System (For Patching & Lifting)
  Large Diameter Air Lines for Diking and Spill Containment
  Computer, on board w/complete mobile application & Disc drive for immediate reference on hazardous materials.

Figure 3. Tools/Equipment/Materials Carried by Hazardous Incident Team One of the Hillsborough County, Florida Department of Fire Control And Emergency Operations.

firefighter to Battalion Chief. "We have also responded to industrial chemical leaks. There is a highly industrialized area within the county, including an extensive phosphate industry and some ammonia pipelines, for which we provide protection. Overall, however, the most common types of incident has involved gasoline tankers. The HIT ONE team does work a regular fire suppression shift, but their primary purpose is to handle hazardous materials incidents. The team is what we consider a 'nonstrategic fire crew.' "

HIT ONE is always accomplished by a first alarm response whenever it is dispatched to a hazardous materials incident. A regular fire company responding in support of HIT ONE will supply the original complement of foam. If larger quantities of foam are required for a major incident, there is a completely self-contained, tractor-trailer foam unit with its own deck gun that can operate independently of other pieces of apparatus and carries 6,000 gallons of water and 200 gallons of concentrate. The unit is prepiped for eduction into a top-mounted stang gun with a Master Task Force 1,000 GPM nozzle. The foam unit was developed in-house in conjunction with MacDill Air Force Base fire officials after county firefighters had successfully used a similar MacDill unit on a number of occasions.

"Continuous training is a fact-of-life for hazardous materials response teams," stresses Jim Brady. "The first step of a long training program was familarizing team personnel with hazardous materials in general. This was accomplished through several seminars, special training using outside sources and instructors, and a detailed study of the National Fire Protection Association's program on hazardous materials. Our specialized training included working closely with MacDill Air Force Base, Chief Emmet Foil, in handling flammable liquid fires and related emergencies, including extensive training in the use and application of foam agents. Another important part of the training program has included working with local industrial complexes that store, use, transport, and manufacture hazardous materials. A great deal of 'hands-on' experience was gained through working with local industry and utilizing their knowledge of specific products/uses/precautions. In addition, transportation carriers and L.P. gas installations provided access to their facilities to better acquaint team members with specific situations."

"The team does a form of 'training exercise,' a term we prefer over inspection, at facilities that handle hazardous materials," adds Chief Brady. "We tour the facility, make our own preincident plan, then keep such plans on file with the team. In addition, the team follows a schedule in providing our individual fire stations what we call 'Fire Response Priorities For Handling Hazardous Materials Incidents.' They tell the engine companies what to look for, what they can expect, when to call for help, and what type of help to call for.

In addition, we make up a brochure or portfolio to distribute to industry, law enforcement agencies, and other organizations in the area so they know what we have in the way of tools/ equipment/materials, what is available for assistance, and what they can expect from us—as well as to obtain their reactions and feedback."

"We have been very fortunate; we have not had any injuries or fatalities at any of our incidents," notes Battalion Chief Brady. "Through hands-on, on-scene training under actual incident conditions we can tell over the months that each incident is being handled a bit better, brought under control a bit faster, handled a bit more efficiently. The more you actually 'do it,' the better you get.

"Hazardous materials present a serious but not impossible problem for the fire service. With our ever-expanding transportation networks, the potential for local authorities having to deal with hazardous materials incidents is ever-increasing. It is initially, and oftentimes ultimately, the fire department's responsibility to handle these situations when they arise. Every fire department need not have sophisticated equipment and specially designated personnel available 24 hours; however, every fire department should know where to obtain specialized tools/equipment/materials, what local resource persons can respond, and the time factors involved in acquiring such assistance when needed."

"A most important consideration for other organizations considering initiation of a hazardous materials response team is the personnel to be assigned. Members must *want* to be part of a team, people who are capable of performing 'above and beyond the call' because it is quite a responsibility when you assign someone this type of task. It requires a lot more than just being a firefighter, or just being a Captain. There is a lot more training, a lot more studying; these guys sometimes put in a 10 or 12 hour day just studying—aside from their regular station duties. The team effort must be real,'' concludes Jim Brady. ''You have to have the right people, the desire to do a good job, and the right equipment. Without any of these three components, there is no hazardous incidents team. The old story of . . . 'Let's take a truck, paint a sign on the side, and let some people ride in it' . . . does not work. Departments that are going to say they have a team, and thus commit themselves to a hazardous incident response cability, must be prepared to do a thorough and complete job of it. A team reflects on the entire fire service.''

Louisville Fire Department Hazardous Materials Team

More than a quarter-million individual shipments of hazardous materials move through the nation's transportation network each day by all modes: rail, highway, air, pipeline, and water. A study of the Department of Transportation predicts that total traffic in hazardous materials will increase by 100 percent within 12 years. It is estimated that hazardous materials currently constitute 20 percent of all goods shipped by all modes within the United States. More than 200,000 *bulk* shipments, of acids, corrosives, gases, pesticides, and other hazardous materials and wastes are in transit within this country each day. In the U.S. and Canada there are approximately 200,000 companies that in one way or another are involved in the manufacturing, packaging, shipping, and transportation of hazardous materials. A massive increase in the number and variety of hazardous materials being manufactured, transported, stored, processed and disposed of in this country; coupled with a parallel increase in the number of spills, leaks, derailments, ruptures, accidental releases, and uncontrolled discharges resulting from increased handling of such materials has brought a change of image for the fire service and a difficult challenge for the fire officer.

Like a number of progressive fire departments throughout the country, the Louisville organization is currently experimenting to find the most efficient and effective hazardous materials response and control organization from among a number of alternatives available. Overall, the Louisville Fire Department has 552 personnel engaged in fire suppression, and operates ten truck companies and 23 engine companies from 23 stations to cover an area of approximately 74 square miles. The Louisville department initially operated a special two-man chemical unit for a year but felt it was underutilized. On September 16, 1979 Louisville combined the chemical unit apparatus with an additional engine to form Engine Company 15 that operates as a regular engine company for a limited area of the city and as a hazardous materials team citywide.

Engine Company 15 is staffed by three platoons with a Captain, two Sergeants, and three Firefighters in each platoon. Personnel, who are all volunteers from among experienced combat firemen within the department, are on duty 24 hours and off 48.

''We went for volunteers because we wanted people who had specialized experience and a particular desire for this type of assignment,'' notes Major Robert Bailey, Public Information Officer for the Louisville Fire Department. ''We knew from the beginning the effort would require a lot of training, a great deal of time, rather significant involvement, and a high level of interest; and we tried to explain this right up front to candidates. Those who were selected jumped right in and did a Hell of a job.''

''Like a number of departments around the country, we in Louisville found ourselves faced with an increasing number of hazardous material/chemical incidents,'' explains Major Bailey. ''We realized that under the system we had been using, a one-vehicle chemical company, anytime it was dispatched we had to designate an engine company to support it. We were looking for a way to modernize our hazardous materials incident response.''

''We respond throughout the city as a hazardous materials team, but cover just our stationhouse area as a regular engine company," adds Captain Tom Sheehan, a Platoon Commander for Engine Company 15. "If we are on a structural fire call and receive a hazardous material alarm, we leave two men on the pumper supported by other engine companies and four of us respond to the hazardous materials incident with the chemical unit. Probably 98 percent of the time the engine and the chemical unit respond together to a hazardous materials call. The few times the engine has been tied-up at a structural fire, the chemical unit has responded by itself while another company was designated to support it at the incident scene."

During the past 12 months, in addition to normal fire suppression calls, the team has handled approximately 25 incidents plus a number of runs that turned out to be false alarms or did not involve a hazardous material. There have been a half dozen train derailments featuring various commodities, although acrylonitrile has been involved in at least two derailments. A wrecked and ruptured gasoline tank truck, a horizontal storage tank of #6 fuel oil on fire at the dome area, and a paint fire in a metal finishing plant were among recent incidents. The team uses a command post style of operation, establishes perimeter tapes to seal off a work area, and allows access to an incident site only to persons who have an absolute need to be there.

The team's chemical apparatus carries a variety of response equipment normally available from commercial suppliers such as proximity suits, acid-gas-entry suits, self-contained breathing apparatus, and explosimeters; as well as three 150 pound, wheel-mounted extinguishers of Ansul dry chemical, Metal-X, and CO-2 that can be accessed from dropramps at the rear of the apparatus. In addition, the team did a lot of scrounging. "We have a collection of plugs and clamps, plus pipeline caps from 1½ inch to 3½ inch that we put together ourselves," recalls Captain Sheehan. "We also carry duct seal and large blocks of putty-like material that will patch a variety of containers. For awhile there we were pretty regular visitors at plumbing supply houses and hardware stores."

"Much of our equipment selection resulted from plain, old common sense," agrees Major Bailey. "Out on inspections, we often obtained what we saw might be needed by procuring it from the industry we were visiting. An example would be a pipe that requires a different thread than those normally available. Local industry has been very good about helping us stock up with what we need."

Louisville team members say they have never yet seen a list of items that might be needed by the "average" hazardous materials response team because each locale is going to have different commodities, different containers, unique to that locale. They feel something used regularly in another town might not be needed in Louisville. A tool or device they use all the time might sit on the truck in another locale. The team uses experience gained through preplanning and at actual incidents to guide their selection of response equipment and materials to insure that limited space on the apparatus contains only equipment for which there is a proven need. In addition, the team periodically re-evaluates materials stored on the apparatus to determine if it has earned the space it occupies.

"The department has had a 'Target Hazard Program' in effect for about eighteen years now where we identify and preplan any building or complex we feel needs special attention," adds Major Bailey. "Target hazard booklets for facilities handling hazardous materials are routinely made available to Engine Company 15. Also, the team is encouraged to inspect citywide any industry and any facility they deem necessary, and all files developed by other companies that might be of interest to the team are made available."

Training is a continuous routine for the Louisville team. In addition to training in hazardous materials control and response provided by the department's training bureau, individual team members regularly attend seminars, training programs, and classes held throughout the country—as much as the budget will allow. As an example, a number of team personnel have journeyed to Florida to work and study with the Hazardous Materials Team of the Jacksonville Fire Department.

"Money is a problem in any fire department," emphasizes Tom Sheehan. "By utilizing a regular engine company for hazardous materials response, we feel we are getting maximum usage of personnel and equipment. We don't have these men sitting around the stationhouse waiting for a hazardous materials incident that might occur twice a month. They cover a regular fire suppression response area, but are free at a moment's notice to break away and respond to a hazardous materials incident. They are a regular, on-line fire company, but trained and teamed to handle hazardous materials incidents citywide as needed."

Captain Sheehan, when asked his opinion as to the most critical aspect of incident response judging from his personal experience, replied without a moment's hesitation—"Positive identification of the product. Until you get that waybill or shipping paper in your hand, you can hear all kinds of stories from people on the scene as to what the product in question might be. People want to be helpful in providing information, but they can unintentionally mislead you terribly. When everybody at an incident location is suddenly placed under great stress and strain, they may volunteer information that is far from correct. We make every possible effort to get that waybill, get the bill-of-lading, get ahold of that piece of paper to be *sure* of what we are dealing with. Oftentimes, industrial or transportation people will have only a rough idea what is in a container, or may know but want to keep the situation very lowkey. We have learned to get the waybill at all costs to allow a positive identification, then use our manuals and reference materials to come-up with the proper response and control techniques. We use *all* the well-known response manuals—N.F.P.A., Coast Guard, D.O.T.—possibly a half dozen. We often reference two or three different manuals to obtain the information we need to properly control a specific commodity. So far, we have been able to identify proper response control procedures for each commodity we have been faced with, but we are quite cautious in making our initial identification of products."

"Everyday new chemicals are being developed and placed into the transportation stream," concludes Major Bailey. "In some cases, the fire service has not kept abreast of the new chemicals and other materials added over the years. More and more incidents are occurring in cities like Louisville, and we have to be prepared to respond as efficiently as possible. I think it behooves all fire departments to take a hard look at the programs they have presently. Are they prepared to deal with hazardous materials? What specific preparations have they made? Could they handle a major incident tomorrow? We may not be able to handle any situation that arises, but we are a heck of a lot better prepared than we were a year ago. There are definitely areas we can improve upon, but at least we have realized that we are faced with a problem area that is growing every day. A year from now, the problem will be worse than it is today. In Louisville, we have managed to achieve a bit of a headstart, but three years from now we had better be even further down the road to hazardous materials preparedness."

County Fire Service Oil Spill Containment Effort Evolves into Hazardous Materials Response Team

Guilford County in north-central North Carolina covers 641 square miles around the city of Greensboro. A County Manager administers 36 county departments employing a total of 1800 personnel. The Guilford County Department of Emergency Services includes separate divisions for Fire, Ambulance/EMS, and Emergency Management. The Fire Service Division has been involved in oil spill containment and hazardous materials response since 1975 and has responded to approximately 200 incidents.

"We first got into a team-response to hazardous materials incidents when we formed an oil spill containment team in 1975," remembers Charles W. Porter, Director of Guilford County's Department of Emergency Services, and formerly a longtime firefighter with the Washington, D.C. department. "Within Guilford County, we have one of the largest inland bulk storage and handling networks in the world. Over one-half billion gallons of flammable liquids are stored in 200 tanks within the county. Tank trucks move out of here to points all over North Carolina, Virginia, South Carolina, and Tennessee."

"Guilford County is somewhat unique in its transportation systems," agrees D. Jerold Stack, Deputy Chief of Operations for the Guilford County Fire Service Division, and a charter member of the Guilford County Hazardous Materials Team. "We have the main line of the Southern Railroad on its way north to Washington, D.C.; and I-40 and I-85 junction here at Greensboro. Our extensive bulk storage facilities are served by two major pipelines: Plantation Pipeline System, and Colonial Pipeline. There is also a transcontinental natural gas pipeline that feeds northward through the western edge of Guilford County, and the county and city are becoming heavily industrialized. The area has the potential for a wide variety of hazardous materials emergencies; we have been aware of the situation for a number of years, and through planning have attempted to cope with it."

Although Guilford County's 37 fire stations are part of a basically volunteer department, Fire Marshal's Office and communications staff are paid personnel. All eight members of the dual-purpose oil spill containment/hazardous materials team are paid personnel of at least Fire Inspector grade, (comparable to Captain) higher echelon people who also get down in the ditches.

By utilizing every possible method available to them, team members have gathered an impressive array of response equipment; but they are quick to point out that such equipment would be of little value without the knowledge and expertise gained through long experience and weekly training. They stress that expert technical advice and continuing assistance from industrial and academic organizations and individuals has allowed them to accomplish a job they could not otherwise have done.

The truth is, however, that they have also done a heck of a lot on their own. "We were able to get chemical suits through the normal budget process, but a lot of our equipment is homemade," explains Chief Jerold Stack. "We can do a great deal with a half dozen 'S' hooks and a little blacksmithing. We have our own maintenance shop that so far has been able to fashion what we have needed. Every vehicle assigned to team members carries homemade dome-lid clamps that have proven extremely useful. These are hooked over the dome-lid of a tank truck, then screwed down to apply pressure and force the lid tightly against the dome-lid gasket to stop leakage through the lid."

Both Porter and Stack know how to stretch a dollar to achieve public service objectives. The team's oil spill containment trailer, originally donated by the Greensboro Oil Jobbers Association, has been stocked with sorbent materials, fenceposts, chickenwire, piping for inverted siphons, recovery drums, dome-lid clamps, and assorted containment equipment. Their "F.I.S.H." truck, (Fire Investigation Special Hazards) a combination hazardous materials response vehicle/command post, is a 1962 International with 3871 actual miles that was purchased as surplus for one dollar; it is loaded with response tools/equipment/materials ranging from homemade clamps and plates to purchased positive-pressure SCBA.

A large, industrial-type air compressor used to operate air-driven pumps (skimmers) in flammable vapor atmospheres was bought in Raleigh for $100. Maintenance shop personnel rebuilt and rewired it, installed it on an old ambulance chassis, and added much larger tires so it can be taken off-road. A self-contained "Light Wagon" unit was built using another old ambulance chassis and a generator obtained from surplus. Various sized metal compression plates and heavy rubber patches, as well as chains, used for sealing tank leaks are all cut in the maintenance shop. The metal compression plates are maleable steel; if they are not of the correct contour to fit a given tank, a sledgehammer is used to bend them to the desired contour. Chains are precut in five, ten, and twenty foot lengths and then color-coded by length and diameter. A color-code chart is posted on the side of the "F.I.S.H." truck so that an incident scene where manpower may be spread pretty thin, a team member can send a person not familiar with the equipment and have him select the proper length/diameter of chain by color alone.

Another homemade device is a Scott Air Pak tank with an air hose and 2½ inch connector; attached to an old 2½ inch firehose plugged at one end, it makes a quickly inflated containment boom. It's great for still water but will not work on choppy water. In addition to making their own windsocks, "S" clamps, dome-lid clamps, and assorted plugging devices such as a rubber ball/togglebolt plug; (used often for leaks in truck saddletanks; make the hole big enough to accept the togglebolt, then screw the ball down over the hold for a temporary but effective patch.) there is "The Thing." "That's Mr. Porter's invention," says Chief Stack, "a 500 GPM nozzle we operate off a 2½ inch hose when large volumes of water are required. It is just a portable, high-volume nozzle made in our shop, but it really works."

"We have a storage depot where we keep our bales of straw, additional sorbent materials, and sections of pipe for inverted-siphon dams," says Charley Porter. "Also, Colonial Pipeline has a trailer truck loaded with sorbent material that we can use. Shell Oil Company has their own team, yet all of their equipment is available to use on an emergency basis simply by hitching up to their trailer and towing it where we need it. They carry things like O.P.W. connectors for oil trucks. When we have to pump down an oil truck and need a special adaptor, we can be pretty sure Shell will have it on their trailer."

"Industry in this area has been fantastic to work with. Sometimes they are so darn generous that you are almost embarassed to go back and ask them for more . . . almost, but not quite," adds Porter with a grin. "We worked up a list of all kinds of heavy equipment— backhoes, frontend loaders, bulldozers—available to us from private industry 24 hours a day, seven days a week. We have never yet needed backhoes or frontend loaders that they were not on the scene within an hour. For truckloads of dirt or sand, the Department of Transportation will respond at our request. Also, there is a mutual aid agreement between and among Guilford County/Greensboro/High Point and all the surrounding counties; any equipment belonging to one is available to all the others. Assistance is near-automatic; we don't have to obtain permission to provide equipment on an emergency basis."

"Every vehicle we have carries a copy of the new D.O.T. manual," explains Chief Stack, "and I carry the condensed CHRIS manual (of the U.S. Coast Guard) in the back of my car. Our alarm room has a very extensive hazardous materials library: chemical handbooks and dictionaries, the complete CHRIS manual, D.O.T. standards, Chem-Cards, etc.. The vehicles of all members, as well as the 'F.I.S.H.' truck, have the capability to play 'Hawaii Five-O' and patch by radio-telephone to CHEMTREC so we do not have to relay through a dispatcher and risk the misspelling of a chemical, or have to rely on someone who might not fully understand what we are dealing with. We have the capability to talk directly to the manufacturer, carrier, tank builder, or CHEMTREC as needed and provide them information directly from the incident scene."

"Since the Greensboro/Guilford County Radiological Assistance Team operates out of our 'F.I.S.H.' truck during a radiological emergency, we maintain expensive Eberline radiation detection gear that responds to alpha, beta, and gamma radiation," explains Chief Stack who along with Captain Dan Shumate of the Greensboro Fire Department serves as Assistant Team Leader of the multi-agency radiological team. "The gear is too sensitive to be stored on the truck, so we kept it in our office. The same situation applies to our own Beta gas detectors, Draeger chemical detectors, and our more expensive gas/oxygen/vapor indicators. That way we can insure the batteries are fully charged and the sensitive equipment does not get damaged on the truck."

The Guilford County Fire Department Hazardous Materials Team uses certain operational methods that may or may not be familiar to readers. Because Chief Stack is also Safety Officer for the county's 1800 employees, and has an extensive background in Safety, he goes on-scene with the hazardous materials team *strictly as a Safety Officer* to insure that actions taken are performed in a safe and positive manner. Both Director Porter and Chief Stack feel the concept of a Safety Officer—a knowledgeable person whose over-riding responsibility on-scene is personal and public safety related to placement of apparatus, evacuation distances and areas, adequacy of protective equipment, elimination of potential ignition sources, decontamination procedures, maintenance of incident vigilance and discipline, and similar considerations— should be considered by other hazardous materials response teams.

The team is currently experimenting with a response strategy which has one Fire Inspector on-duty at all times whose first duty is to move immediately to an incident scene to assess the situation and determine necessary corrective action, manpower, tools, equipment, and materials. He then radios the central station where other team members have been checking in and provides his assessment of exactly what is needed.

The team also utilizes mechanics assigned to the Department of Emergency Services' maintenance pool. Each mechanic has a vehicle and will report to the garage to act as back-up, or assist in transporting equipment to the incident scene *and stay there* to insure the equipment operates properly. "Like any other operation," says Jerold Stack, "if you have equipment stockpiled for hazardous materials emergencies, a lot of it sits for long periods of time. Even though you take it out periodically and go through your preventive maintenance routine, there will always be some kind of specialized equipment failure whether you are dealing with firefighting equipment, hazardous materials response gear, or road graders here in Guilford County that may pull snow once a year. It is good practice to have maintenance people on hand so that if any of your equipment fails to start of does not operate properly, they are available to set it right."

"We preplan all oil terminal facilities," continues Stack, "capacity and location of each tank, the type of tank, where the drainage is, the railheads and sidings within the terminal, electrical and mechanical shutoffs, fire protection equipment, roadways, office buildings, fences/gates/entry points, hydrant locations, size of mains; all such special preplans are maintained within a book on the apparatus."

The diagrams for such preplans are works-of-art, all professionally drawn to scale with extensive detail. "A lot of work went into the preplan diagrams," admits Stack. "If you don't know what you are getting into, that is when you get hurt. Fire officers in the field gathered the necessary information and brought it back to three professional draftsmen employed temporarily under the Comprehensive Employment & Training Act who did the finished drawings."

"A notable aspect of response in the county is that we are *often* faced with potential entry of chemicals into water supplies and sources," emphasizes Jerold Stack. "Basically, all the drinking water used in the county originates within the county. In a large percentage of our incidents we get involved with damming and installation of an inverted-siphon. (used to separate from water insoluble chemicals having a specific gravity of less than one.) Many of our response actions are guided by the need to quickly remove a threat to the water supply. We carry a book on the oil spill containment trailer that shows every creek, stream, and waterway within the county. Also, we maintain soil surveys and topographic maps that delineate all roads and drainage patterns."

"We have roll upon roll of duct tape for suiting-up," notes Chief Stack. "Our chemical suits are completely bonded so vapors or drainage cannot enter, but we also tape over every seal to be absolutely certain team members have maximum protection. Also, for patching and plugging we carry a lot of duct-seal compound because we use it all the time. (Duct-seal compound is made by Johns-Manville and other companies for use by electricians in filling holes after running wires or pipes in buildings.) It is just a good grade of oil-based putty. You can work up a good glob and put it into a crack or hole—*as long as you don't knead it.* Minimum handling is the secret for successful use of duct-seal compound; hit it one time and one time only—then leave it alone. It works well on both gasoline and fuel oil containers."

"Our most complicated incident was a load of vinyl toluene that we played with for a week," remembers Stack. "This was a unique situation where CHEMTREC told us the truck had never left the plant yet we had it here in Guilford County. When the truck had left its home plant, the driver had put the steam to it thinking the product had to be heated. On the way down the road the commodity started generating its own heat and solidifying, a process that our advisors tell us can lead to an explosion. The consignee wouldn't accept the load, so the driver took it to a truck-stop. It sat there for a day getting hotter and hotter before they called us."

"We hooked inch-and-a-half and 2½ inch hoselines to the heating system of the tank; that is, we ran cool water through the piping on the tank designed to heat the cargo," recalls Stack.

"It was an insulated tank, so putting water on the outside would have done no good at all. The next step was to get rid of the insulation, so we literally peeled the tank.

We are not blasting anyone; CHEMTREC has always been very helpful to us, but in this instance they gave us incomplete information on the truck. The trailer had been manufactured by one organization and later modified by another, so the manufacturer's information we received on this particular truck was no longer correct. Neither the valving nor the piping were the same as the information provided. I was back at our office trying to relay information to people at the scene as to how and where to put the water into that thing. When I'd give them the information the trailer manufacturer would give me, it didn't work. Eventually, we had to learn from the *previous* owner exactly what he had done for modifications in order to get the systems to work. We ended up moving it to an isolated location and flooding it inside and out for a week."

Chief Jerold Stack reflected recently that of the team's approximately 200 incidents over seven years, by far the most common has been overturned tank trucks carrying a flammable liquid, usually gasoline. Surprisingly, the team has not yet had a serious fire at such an incident even though they have had as many as three separate gasoline trailers dump their loads in a single week.

"In the three most recent incidents, the largest commodity loss has been 200 gallons," reports Stack. "Currently, we use a skirt around the domelid cover that allows us to pump directly out of the truck while it is lying on its side. Before the dome-skirt method became available, we used canvas drop-tanks. (Portable, canvas catchbasins fire departments use when drafting water.) We positioned a droptank under the dome, allowed the product to flow into the droptank, closed the dome-lid cover with a homemade clamp, then pumped from the droptank. We would repeat this operation until the product in the truck was below the level of the dome, then pump directly out of the truck tank."

"We are talking about simple tank truck rollovers where there has been no rupture," cautions Stack. "The tank truck is lying on its side full of fuel, but there has been no major rupture and no fire. Of course, if there has been a serious rupture the fuel is going to be gone before we get there. We have had many tank rollovers occurring on the cloverleafs of the Interstate highway. A truck enters the access ramp too fast, and the liquid itself rolls the truck."

"During an average winter, we will have 15 to 20 incidents involving overturned tank trucks," estimates Charles Porter. "Several years ago a count we did over a 24 hour period at the intersection of I-40 and I-85 showed an average of 75 tankers per hour. With that kind of volume, you are going to have some incidents."

"We have been very lucky," says Porter, remembering another incident that could have been a disaster. "I was leaving the airport about 30 seconds away from the scene and was the first to arrive. The tank truck was over on its side, ripped open, and a train engine sat nearby. At first, there wasn't another soul around, then here come the engineer and the firemen walking down the road. They had known two blocks away that they were going to hit the tank truck, so they had locked the brakes and bailed out. The truck dropped 8,000 gallons of gasoline, but it never did ignite—no fire. We got backhoes into the area and went down the drainage ditch ahead of the gasoline and dug some sumps to catch it. We kept a layer of A-FFF foam on top of the sumps and used septic tank trucks to pump out the gas and carry it away."

"We don't clean 'em up," responded Jerold Stack recently when asked what the team had learned from its seven years experience. "We contain it, then let the reponsible party come in and clean it up."

"That was the biggest thing we learned;" agrees Charley Porter, "stay away from clean-up. However, we have changed that outlook a bit; now, if we do get stuck with clean-up, we charge $10.00 per hour for each man we have on the scene. I will get on the phone to industry if they don't have anyone on the scene and advise them of just that. Admittedly, we have been surprised at the response; no one has ever refused to pay . . . no one has ever complained. On *all* our incidents where a responsible party can be identified, we have billed—and always collected. We also charge for materials; if tools are broken, we bill for replacement costs. If we use 25 bags of sorbent material, two rolls of chickenwire, 12 fenceposts, and 100 bales of straw; we bill for replacement costs. With a local company, if they would rather go out and buy the material on their own and provide it to us as replacement, that is fine; but we do have the material and equipment replaced.''

We have learned from building a hundred or more inverted-siphon dams to keep the volume of water backed-up upstream as low as possible because the less water you back-up the less contaminated bank area you will have,'' notes Jerold Stack. "Also, many industrial people use a lot of absorbent pads. We, on the other hand, use them very little. We refrain from using sorbent materials if at all possible because they take forever to pick-up the product. We prefer to catch the product with inverted-siphon dams and skim it off with pumps. Our goal is to get it up as quickly and as safely as possible without playing with it for two weeks.''

"An essential component of response for us is the *availability of expertise,*" stresses Charley Porter. "The most important thing we have working for us is the expertise we can call on within the community, from industry, and the university . . . experts in pesticides, hydrocarbons, radiation. We absolutely depend on people like Dr. Stone of Ciba-Geigy; Dr. Burski from Burlington Industries; Dr. Knight with the University of North Carolina at Greensboro; Dr. Gladys Van Pelt, a Health Psychiatrist; Dayne Brown of the Radiation Protection Section/North Carolina Department of Human Resources; and a number of others. They are the experts; we are the doers. They report to the command post and work with us, or we will send a car or have the Sheriff's Department expedite their transport to the scene. We cannot overstate the crucial need for expert guidance. For example, we have large pesticide manufacturing and storage facilities in the area. What do you do when you have pesticides burning? What do you do with the run-off? Such questions require that we have ready access to experts qualified to provide answers. The key for us is the availability of expertise.''

"Like people all over the country, we have a lot of different responsibilites;'' concludes Chief Jerold Stack, "hazardous materials is just one of them. It is not something you can turn your back on and say—"Well, someone else will take care of that.' Hazardous materials response and control is a fire service function; it is recognized by us as a fire service responsibility.''

Philadelphia Fire Department Mans Three Chemical Task Forces

In 1976 the Philadelphia Fire Department under Fire Commissioner Joseph R. Rizzo assigned 150 firefighters from a total force of 3,000 to receive special training and equipment in order to form three Hazardous Chemical Task Forces. (HCTF) The three separate Task Forces are strategically located throughout the city with one in the northeast, a second in the northwest, and a third in the south end. Each HCTF is staffed with 50 firefighters and officers to cover all shifts; a total of 150 men teamed, trained, and equipped to handle hazardous materials response within Philadelphia.

Each Task Force consists of a 1,000 gallon pumper plus a ladder truck normally used for firefighting, a 1976 International chassis foam pumper with National Foam equipment, and a "chemical unit" much like a soft drink delivery truck with roll-up sides. Each HCTF is housed together and manned with specially trained personnel, designed to respond as a single entity on all incidents involving hazardous chemicals or materials. At times when all required Task Force components are not available from the same station, the Fire Communications Center is charged with insuring that the four components necessary to complete a Hazardous Chemical Task Force are dispatched.

Battalion Chief Harry Cusick has been active for a number of years in hazardous materials training and response. He teaches hazardous materials at the National Fire Academy in Emmitsburg, Maryland and recently helped to develop "Hazardous Materials Incident Analysis," a two-day course to be offered by the Academy on weekends.

"With the Hazardous Chemical Task Forces, we tried to take a different approach than we would with the typical third-floor-rear mattress fire," says Cusick. "We had experienced problems, and some injuries, with the 'go in and get it' type of approach when applied to hazardous materials incidents. Overzealousness is both our biggest hindrance and our greatest plus. We need the hard-chargers, the strivers, but we have to insure they fully understand what they are dealing with. Aggressiveness is what makes us good, but with hazardous materials it can create problems. The 'moth and candle syndrome' once appeared to be our biggest problem—the fact that we are constantly drawn to the flame. An overaggressive, zealous attitude can put you in a predicament where the immediate responder becomes part of the problem rather than part of the solution. With hazardous materials, you can easily find yourself in a situation where one of your first actions has to be evacuation of half your own personnel. Our Hazardous Chemical Task Forces represent an attempt to manage hazardous materials response rather than allowing the situation to manage us."

"An initial step was to significantly upgrade our hazardous materials informational data," continues Battalion Chief Cusick. "We isolated and identified potential problems by doing a citywide hazard analysis that covered both fixed sites and transportation routes. We approached industry, particularly those organizations that had experienced site-specific problems, and asked: 'What can we do to help one another?' As we became more aware of some of industry's inherent problems, and the auxiliary systems they had on hand that we could use, such as high-back and subinjected foam systems; we adapted such information to our tactics and moved ahead to complete detailed preincident analysis of specific installations. Concurrently, we selected the most advantageous locations for our Task Force personnel, reviewed equipment and apparatus we already had on hand, and identified additional necessary items. We brought in our three Division Chiefs and 13 Battalion Chiefs and reviewed with them the Hazardous Chemical Task Force concept. Using information developed during the hazard analysis phase, we assigned Task Forces to three specific locations, provided special equipment, and undertook a concerted training effort."

"If you are going to establish a Hazardous Chemical Task Force, you need a lot of money, extensive training, and continuous reinforcement of your efforts," adds Cusick. "There has to be a continuing commitment. The cost of equipment can be overwhelming. Our research and planning unit evaluated the various types of hazardous chemical suits available. The suits we purchased cost approximately $1600 *per suit*. With four suits for each of three Task Forces plus one for backup, a total of 13 suits, we had an expenditure of nearly $21,000 just for necessary chemical suits. Nonsparking tools are also very expensive. Buy the very best hammer available at the local hardware store and it might set you back $20 or so; a nonsparking hammer can easily run $100. Other departments should be aware that there is a significant cost factor involved in maintaining a high level of hazardous materials preparedness."

"In addition," says Cusick, "hazardous materials response is a special type of situation requiring continuous reinforcement training. We bring each unit in at least twice a year for concentrated instruction on flammable liquid fires, changes in hazardous materials processing techniques, container construction, and other identified training needs. Also, we send training officers to the Task Force stations during the year to provide repeated reinforcement of previous training. In addition, our Commissioner believes in the policy of rotation of officers. Every three years, officers who have worked on the Hazardous Chemical Task Forces are then assigned to regular engine and ladder companies. They have a background in hazardous materials, an awareness of proper procedures for hazardous materials response, that they carry with them to the regular line companies. We bring new officers in to work with the Task Forces, and they begin to build an appreciation for hazardous materials that will stay with them throughout their careers. We are constantly building a cadre of personnel within the department who have worked with and studied hazardous materials."

"We get a lot of assistance from private industry with our semi-annual training and other efforts," notes Fire Commissioner Joseph Rizzo who has been with the department for 34 years. "Allied Chemical and Rohm & Haas are both located within the city and have been very helpful, as have the trucking associations and the railroads. Chemical engineers, people who have extensive experience with specific chemicals, experts from the Chlorine Institute; we depend on a sizable group of people.''

"The specialized training our Hazardous Chemical Task Force personnel receive is not merely parochial education,'' agrees Chief Roger Ulschaefer. ''We use outsiders extensively. Representatives of the Philadelphia Gas Works have been particularly helpful. Basically, such persons provide us with insights on what we can do to help them, and what they can do to help us.''

An outsider observing hazardous materials preparedness within the Philadelphia Fire Department comes away with a feeling that seven factors are crucial to initiating and maintaining the high-level preparedness that is readily apparent; continuous training; provision of specialized tools, equipment, and materials; identifiable support for the program from the top on down; extensive prior planning; extremely effective organization; detailed written guidelines and procedures; and placement of a high priority on communications.

"If you are going to establish a Hazardous Chemical Task Force, you need a lot of money, extensive training, and continuous reinforcement of your efforts,'' adds Cusick. ''There has to be a continuing commitment. The cost of equipment can be overwhelming. Our research and planning unit evaluated the various types of hazardous chemical suits available. The suits we purchased cost approximately $1600 *per suit*. With four suits for each of three Task Forces plus one for backup, a total of 13 suits, we had an expenditure of nearly $21,000 just for necessary chemical suits. Nonsparking tools are also very expensive. Buy the very best hammer available at the local hardware store and it might set you back $20 or so; a nonsparking hammer can easily run $100. Other departments should be aware that there is a significant cost factor involved in maintaining a high level of hazardous materials preparedness.''

"In addition,'' says Cusick, ''hazardous materials response is a special type of situation requiring continuous reinforcement training. We bring each unit in at least twice a year for concentrated instruction on flammable liquid fires, changes in hazardous materials processing techniques, container construction, and other identified training needs. Also, we send training officers to the Task Force stations during the year to provide repeated reinforcement of previous training. In addition, our Commissioner believes in the policy of rotation of officers. Every three years, officers who have worked on the Hazardous Chemical Task Forces are then assigned to regular engine and ladder companies. They have a background in hazardous materials, an awareness of proper procedures for hazardous materials response, that they carry with them to the regular line companies. We bring new officers in to work with the Task Forces, and they begin to build an appreciation for hazardous materials that will stay with them throughout their careers. We are constantly building a cadre of personnel within the department who have worked with and studied hazardous materials.''

"We get a lot of assistance from private industry with our semi-annual training and other efforts,'' notes Fire Commissioner Joseph Rizzo who has been with the department for 34 years. ''Allied Chemical and Rohm & Haas are both located within the city and have been very helpful, as have the trucking associations and the railroads. Chemical engineers, people who have extensive experience with specific chemicals, experts from the Chlorine Institute; we depend on a sizable group of people.''

"The specialized training our Hazardous Chemical Task Force personnel receive is not merely parochial education,'' agrees Chief Roger Ulschaefer. ''We use outsiders extensively. Representatives of the Philadelphia Gas Works have been particularly helpful. Basically, such persons provide us with insights on what we can do to help them, and what they can do to help us.''

An outsider observing hazardous materials preparedness within the Philadelphia Fire Department comes away with a feeling that seven factors are crucial to initiating and maintaining the high-level preparedness that is readily apparent; continuous training; provision of specialized tools, equipment, and materials; identifiable support for the program from the top on down; extensive prior planning; extremely effective organization; detailed written guidelines and procedures; and placement of a high priority on communications.

With Philadelphia a major port, rail center, and industrial complex; fire officials felt early-on that they had to have specially equipped and trained hazardous materials response teams. "The Task Forces were necessary both for the protection of citizens and for the safety of our personnel," reflects Chief Ulschaefer. "Special knowledge and equipment was absolutely necessary. Many times we respond to an industrial location to find the industrial people who have the expertise for that location and that product are not available. We are dealing with the foreman. We find we must have the ability as first responders to handle a wide variety of chemicals and containers. Often the only advice you are going to get for a few hours is that from within your own command. Your men must have the knowledge, equipment, and expertise to handle a wide variety of situations. There is a constant upgrading of equipment and techniques because as the state-of-the-art changes 'out there,' we have to adapt and learn how to handle new hazards. The vast growth in products and processes within the entire petro-chemical industry has really impacted upon the fire service. The whole situation has changed; we are no longer just fighting Class A fires. We have to be ready to combat nearly any challenge the mind can devise."

"It is absolutely crucial for a major city fire department to have a highly trained, fully equipped nucleus of specialists to respond to hazardous materials incidents," adds Commissioner Rizzo. "The variety of chemicals we have been faced with is extremely broad, but so far we have had the ability to cope." Support from the top certainly doesn't hurt the operations of the Hazardous Chemical Task Forces. Commissioner Rizzo initiated the effort and even selected the name for the specialized units. The HCTFs are funded through the regular department budget, equipment has been obtained as a need was identified, and a high priority is placed on HCTF readiness and performance.

Quite recently, Philadelphia passed a "Right To Know Law" under which anyone having minimum quantities of certain chemicals on their property must report this fact to the Department of Licenses and Inspections. Anything over 500 pounds of certain dry chemicals, or a 55 gallon drum of liquid, must be reported. In this was any citizen has the right to acquire information on certain hazardous materials being manufactured/stored/used/transported or generated within the city. "We could previously go into industrial occupancies and identify chemicals on the premises," reports Chief Ulschaefer, "but with the new ordinance the responsibility for reporting is on the handler. There are 475 types of chemicals involved."

Prior planning, effective organization, and detailed written guidelines and procedures go hand-in-hand in creating high-level hazardous materials preparedness within the City of Philadelphia. At least once a year, fire department personnel stage an orientation on every chemical plant, refinery, and other hazardous target within the city to learn what is taking place within such locations, observe changes that have been made since the last visit, and identify materials currently stored or handled there. These visits also foster good working relationships that will come in handy should there be an incident.

Five years ago the department implemented a program whereby refinery and chemical industry facilities were assisted to upgrade their plant fire protection services in order to mitigate identified potential hazards. Where once it was relatively common to find unmarked tanks in such facilities, now every tank is marked. Fire officials cite exchange of information between the fire service and refiners as particularly rewarding, noting that some large refineries spent millions of dollars upgrading their plant fire protection and identifying specific hazards. An interchange of information was born that has been both constant and continuous.

Philadelphia participates in a tri-state agreement among Pennsylvania/Delaware/New Jersey by which all chemical industries and the fire services identify and agree to share specialized tools/equipment/materials when one is faced with a major incident.

Separate, written Standard-Operating-Procedures are in force for Hazardous Chemicals And Materials, Petroleum Properties And Chemical Plants, LNG/PPG Emergencies, Radiological Incidents, Railroads, and Disposal of Hazardous Chemicals. With regard to general hazardous chemical emergencies, Task Force units always use a command post, rely on CHEMTREC as needed, and utilize a staging area—a designated area outside the perimeter of the incident where first aid equipment/standby manpower/logistical support can be marshalled. Hazardous materials locations are preplanned by local fire companies, updated on an annual basis, and station exercises are conducted by all platoons to familiarize members with conditions and to discuss specific firefighting operations that may be encountered. Preplans are forwarded to all three Task Forces where they are maintained in a separate book. In such plans, particular attention is paid to identification of key plant personnel; the plant's water system; pumping equipment; fire protection systems; utilities; drainage; fire fighting capabilities; foam and other fire suppressant materials available; and identifications of hazardous materials.

In plants having their own fire brigades, a company employee is identified as the Plant Emergency Coordinator and equipped with a portable radio capable of receiving and transmitting on the "F-3" fire band. The Plant Emergency Coordinator, who uses the radio call designation "PC-1," must have the ability to supply or have at hand the following basic information: (in addition to a status report on what is burning and what has been done prior to the arrival of fire department personnel.) plant maps; fire protection system diagrams; product pipeline diagrams; electrical system diagrams; fire suppressant materials available; guides for fire department command personnel; and diagrams of the plant drainage system.

Initial responding units are normally notified by the Fire Communications Center (FCC) of the gate number, the street location within the plant, and the number of the operating or storage unit involved—information normally obtained by the FCC from the plant telephone operator during a follow-up call after the plant's city fire alarm box has been activated. First arriving units establish a command post and initiate contact with the Plant Emergency Coordinator to develop certain priority information: life hazard; product or products involved and their chemical characteristics; exposure hazard; fire protection systems available; and plant fire brigade activities thus far.

With regard to bulk shipments of liquefied natural gas or liquefied petroleum gas by marine vessel, whenever such a shipment is destined to arrive at a petroleum property or chemical plant in the City of Philadelphia, the plant is to notify the Fire Communications Center no less than 96 hours in advance and provide the following information: the location of the unloading site and the estimated time of arrival of the vessel; the intended length of time the vessel is expected to be in port; the estimated time of departure; any changes in the itinerary; the name and registry of the vessel; and the size and capacity of the vessel. Such information is then transmitted to the Fire Marshal's Office, and to the Assistant Chief/Fire Fighting Forces. The Fire Marshal's Office contacts the U.S. Coast Guard Dangerous Cargo Officer at Gloucester City, New Jersey to ascertain if the vessel has been properly inspected prior to its arrival; and arranges with the Coast Guard for a joint inspection of the receiving facility to insure compliance with applicable regulations. A fire officer from the first alarm response assignment accompanies the facility inspection team. Fire companies in the docking area are provided with all pertinent information about the shipment.

There are also written guidelines for LNG/LPG emergencies not involving bulk shipment by marine vessel. Along with general information for handling gas emergencies; such as, definitions and characteristics, requirement of a 2,000 foot evacuation perimeter, tactical instructions, directions for stopping the flow if possible, and a warning not to extinguish the fire if the gas flow cannot be stopped—instructions are provided on how to request assistance available 24-hours a day from the Gas Supply Department of the Philadelphia Gas Works regarding valve shut-down, fuel transfer, etc..

A Railroad Operational Procedure stresses the need to maintain communications at all times with the various railroads that serve Philadelphia, and provides telephone numbers to be used when requesting information regarding incidents or emergencies. Major railyards for each railroad are listed individually with a telephone number for each. Procedures are also given for stopping rail traffic, and for having high voltage power lines de-energized. Instructions are provided on consultation with freight train conductors or engineers to obtain information listing cargos and the location of specific hazardous items within the train so as to enable firefighters to isolate hazardous materials not involved in a fire. Additional guidelines are spelled out for incidents involving loaded freight or tank cars parked in railyards or on railroad rights-of-way or sidings, as well as procedures for having the Fire Communications Center secure pertinent information from the yardmasters of each of the four major railroads in the city. For incidents involving loaded freight or tank cars on private sidings, instructions are given for a series of possible methods to identify the cargo. Included in these methods are contact by the FCC with the various railroads, and a call to CHEMTREC. The fire dispatcher may also contact the Operation Control Center or Movement Director and have a train crew dispatched to move rail cars at the request of the on-scene commander. Additional instructions cover, among a number, construction of earthen dams to contain spills, consultation with railroad personnel when dealing with electrically-powered railroad engines or cars, methods employed to have a catenary system (system of wires supported between poles and bridges supporting overhead contact wires normally energized at 11,000 volts) shut-down, and emergencies in railroad tunnels.

With incidents involving a suspected hazardous material on the Conrail system, the FCC can call a Conrail emergency number with the number of the tank car or freight car and receive the following information: the material's commodity code; (STCC) proper name of the material; physical properties; description of its characteristics, hazards, and reactions; procedure if the material is on fire or involved in fire; personal protection; and evacuation procedures, if applicable.

For a hazardous materials inquiry on the B&O Railroad, Conrail's computer system is available for emergency use if a copy of the waybill, and thus the STCC number, can be obtained for the rail car involved. (The STCC number is a series of seven digits assigned to any commercial product shipped by rail. Any known hazardous material will have an STCC number beginning with 49: e.g., 4905410—hydrogen sulfide.)

All command vehicles, Hazardous Chemical Task Forces, support units, and the FCC have been issued *Emergency Handling Of Hazardous Materials In Surface Transportation* (in addition to other reference manuals) published by the Bureau of Explosives/Association of American Railroads.

In addition, the Philadelphia Fire Department has a written Standard-Operating-Procedure for Radiological Incidents, and one for procedures to follow when disposing of hazardous or unidentified chemicals that may be delivered to a fire station by a citizen.

All procedures mentioned above provide detailed guidance for *both* the fire-ground commander and the Fire Communications Center. The FCC, located in the basement of fire headquarters, is definitely a nerve center during response to hazardous materials incidents. During a serious incident, senior officers receive progress reports through the FCC every ten minutes, and the Center makes automatic notifications as dictated by procedure: other agencies that need to be informed, the U.S. Coast Guard if a situation occurs near the waterfront, clean-up and earthmoving contractors when needed, the city water department if entry into any sewer system becomes necessary, etc.. Also available if needed are a large mobile communications unit and a roomy emergency command post vehicle.

Greensboro Fire Department Utilizes Extensive Preplans for Highway & Railroad HAZ MAT Incidents

"The basic purpose of the Greensboro Fire Department Hazardous Materials Team is to respond to fire suppression calls and to handle hazardous materials incidents as they occur," says Captain Dan Shumate, a shift commander with the team. "There was no way we could have *just* a hazardous materials team with incident response as its only function; the team is a regular fire company within its stationhouse area, but specially teamed/trained/equipped to respond citywide to hazardous materials emergencies. If we are on a fire call, there is a procedure for getting us relieved from the fire scene so we can respond to the hazardous materials incident."

The progressive Greensboro, North Carolina Fire Department under Fire Chief R.L.

Powell, Jr. operates 15 stations with 200 total personnel. Station 8 on South Chapman Avenue is home to the department's all-volunteer Hazardous Materials Team. As with all engine companies in Greensboro, the hazardous materials company is staffed by one Captain and three firefighters for each of three shifts. Each shift is on duty for 24 hours and off 48, although team members are subject to 24-hour recall for hazardous materials incidents.

"We took a reserve engine, a 1958 American LaFrance 1,000 gallon pumper, and stripped certain equipment from it to provide room for our hazardous materials response equipment," continues Captain Shumate. "We have 800 feet of 2½-inch hose loaded on one side and 150 feet of 1½-inch on the other side that is preconnected to an inch-and-a-half foam eductor. This apparatus, designated 'Haz Mat—99,' carries 100 gallons of A-FFF foam with 15 gallons in storage. The main tools, expertise, and training for hazardous materials response is available from the team, but normally there will be two other pumpers responding with 1800 feet of 2½-inch hose. In any major line-laying situation, we will use their hose; so we have kept to a minimum the amount of hose we carry on Haz Mat—99. We have two 2½-inch foam eductors plus two 60-395 GPM fog nozzles which we recently purchased. Within the overall department, we have a total of 690 gallons of foam onhand, and we have made arrangements with industries and organizations in the area to supply us with additional foam on an emergency basis."

The team also carries chemical suits, nonsparking tools, combustible gas/oxygen indicators, and a chemical detector unit with a variety of detector tubes. For radiological incidents within Greensboro or surrounding Guilford County, the Greensboro Fire Department Hazardous Materials Team supports and works with the Greensboro-Guilford County Radiological Assistance Team, (R.A.T.) a multi-agency radiological response team staffed by personnel from six agencies of local government. In addition to his duties with the Greensboro Fire Department team, Captain Shumate functions as Assistant Team Leader of the multi-agency R.A.T. team.

For additional equipment that might be needed at hazardous materials incidents, the Greensboro Fire Department Hazardous Materials Team may obtain support from three fire department squads that run within the city. These units are basically minipumpers, actually Rescue vehicles with firefighting capability. They carry rescue tools and forcible-entry equipment and run both medical and rescue calls. Within their stationhouse areas, they also cover small fire calls such as car fires. Each squad truck mounts a 300 gallon water tank with a 250 Hale pump and 30 gallons of foam. Included among the special equipment carried by each of the three squad trucks are air bags, a Hurst tool, assorted handtools, com-a-longs, cutting tools, air chisels and blades, floodlights, chains, "Super-Scissors," and complete first aid equipment.

"We have arranged with businesses in the area to obtain scrap rubber for use in our patch kits," adds Dan Shumate, "and some of the woodworking shops, particularly a business that makes wooden lamps, supply us with wooden plugs and dowels. Also, the Hazardous materials company has established a reference library within Station 8. We contacted all the engine companies and stations within the city and had them send us a copy of their preplans for each of the known chemical companies within their respective areas. We made two copies of each of these preplans; one went on the Haz Mat—99 apparatus, while a second copy is maintained within the Station 8 reference library. To supply a weatherproof compartment on our open-cab LaFrance, we had to have the garage build a special compartment. After the expense, time, and effort spent on preplans; we wanted a guaranteed dry, secure place on the apparatus in which to keep the plans and other reference materials so the first rain or snowstorm would not ruin our efforts."

"Until now, a great deal of hazardous materials training has been provided the entire department," explains Captain Shumate, "although now we will be putting our hazardous materials training emphasis on the team—along with the other companies. For example, *every* apparatus within the department carries hazardous materials manuals: N.F.P.A.'s *Fire Protection Guide On Hazardous Materials,* D.O.T.'s *Selected Hazardous Materials,* and the *Field*

Protection Guide On Hazardous Materials.'' We are presently evaluating the U.S. Coast Guard 'CHRIS' manual.''

Besides a great deal of training in hazardous materials provided by the training division of the Greensboro Fire Department, team personnel arrange with local agencies and industrial organizations—such as Piedmont Natural Gas and a major trucking company—for familiarization visits to facilities, review of equipment and processes, and hands-on training with various types of containers. The department has sent men to Florida for training with Safety Systems, Inc.; and Safety Systems has provided seminars in Greensboro. Fire officials trace initial interest within the department for the development of a hazardous materials response team to the National Fire Protection Association 24-hour course which has been broadly used.

S.O.P. for Highway Emergencies

In recent years, Greensboro has become a highly industrialized area. Located within the city and surrounding Guilford County is an extensive petroleum bulk storage network fed by two major pipelines. An estimated 200 storage tanks hold up to one-half billion gallons of flammable liquids for distribution by tank truck throughout North Carolina, South Carolina, Virginia, and Tennessee. Several years ago, a traffic count taken at the junction of I-40 and I-85 over a 24-hour period showed an average of 75 tank trucks an hour. The Greensboro Fire Department has had a highway disaster control plan in effect for two years. The experience with the highway preplan was so positive that last year the department completed a similar Standard Operating Procedure For Railway Emergencies.

''We experienced a number of incidents on the highway where we did not have an adequate water supply,'' explains Dan Shumate. ''Recognize that the Interstates are federal highways, so we could not put signs along the roadway giving directions to the nearest hydrants. We had each company go out and locate distinctive landmarks along I-85, I-40, and U.S.-29 that we featured in zonal maps. Basically, we created a matrix of individual maps detailing the routes of major highways through the city. (See Figure 4) Each individual section (map) was assigned its own identifying number. At an incident on either of the two Interstates or U.S.-29, say a tanker truck on fire or a chemical spill, the first arriving officer reports the nature of the emergency, its location, and identifies the zone where the emergency has occurred. If he describes his location as . . . 'Zone 851-North,' . . . we know where he is on Interstate-85 and what lane he is in.''

When the first arriving officer activates the highway emergency plan, the Communications Center dispatches two additional engine companies; one is designated as the Responding Company, the other the Assisting Company, by the officer on the scene who controls the show and tells later arriving units where to go. The Responding Company always goes to a predesignated location on the highway; the Assisting Company always goes to the proper hydrant. Once dispatched companies are assigned a role as either Responding Company or Assisting Company, they refer to their copy of the *Highway Disaster Plan Manual,* a collection of detailed zonal maps and explanatory narrative that identifies hydrant locations, various types of occupancies in the area, routes and approximate distances for laying lines, etc.. The Assisting Company responds to the most suitable hydrant and lays line(s) to the designated location of the Responding Company.

''Without the zoned guide, dispatched units would have little idea where the nearest hydrant was located, and might lose valuable time trying to find their way off the Interstate and through nearby residential or industrial neighborhoods,'' notes Captain Shumate. ''With the zonal guide, a pumper coming down the Interstate can tell where to stop to be closest to the nearest available hydrant. Merely driving down the Interstate, you just do not see these hydrants. They are by no means on the highway; they are 400-600 feet or more off the highway in abutting residential or industrial areas. There are no hydrants whatsoever on the

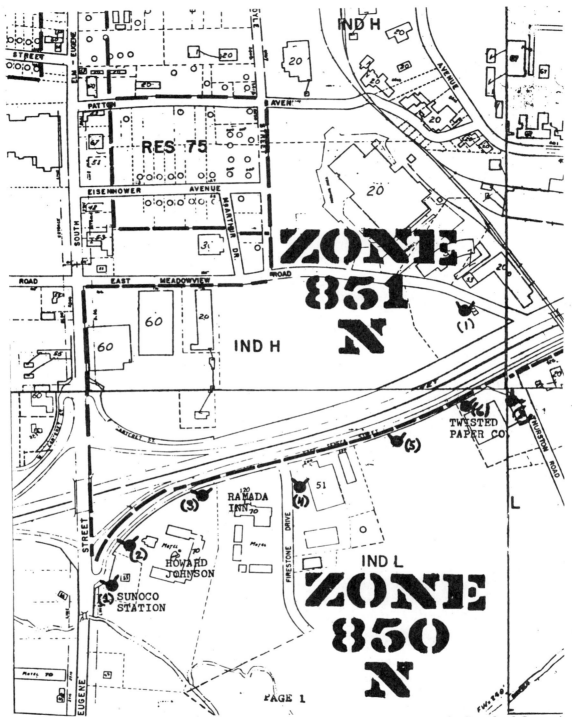

Figure 4. "Sample Zonal Map Taken From Greensboro Fire Department's Standard-Operating-Procedure For Highway Emergencies."

There are 21 separate maps covering 42 zones in the highway emergency preplan that covers Interstate Highways I-40 and I-85, and U.S.-29. Narrative provided for each map tells the Responding Company pumper exactly where to stop on the highway (by referring to prominent landmarks) so the Assisting Company can lay hose to it from a specific hydrant identified on the map. **Example:** If the Assisting Company attaches to Hydrant #4 in Zone 850-North, the Responding Company pumper is instructed to . . . ". . . stop in the northbound lane of I-85, 400 feet east of Ramada Inn at Firestone Drive."

Notice that major industrial and commercial occupancies are identified as to type (chemical storage, motel, petroleum storage, etc..) by use of code numbers.

Interstates. In the example given, the Assisting Company goes to 'Zone 851-North' and pulls its lines through fences and across ditches (the distance to each hydrant is given on the preplan document) to the Responding Company pumper located at a predetermined spot on the highway, and then runs lines to the first pumper already on the scene. The preplan stops a lot of radio traffic and a lot of confusion. A copy of the 'manual' is carried by every company, even the ladder companies and the squads.''

1.   Get information from authorized personnel if possible
     a.   Inform dispatcher of nature of spill & notify proper authorities
     b.   Life involved
     c.   Types of chemicals (correct spelling) Call Chemtrec if necessary
     d.   How much chemical involved
     e.   Leakage, fire, how long?

2.   Personnel
     a.   Approach scene carefully
     b.   Look for life involved
     c.   Check placard for types of chemicals (Check book)
     d.   Check for leakage
     e.   Check for fire
     f.   Inform Team commander of situations

3.   Plan of Action
     a.   Evacuation Distance
     b.   Placement of Apparatus and men
     c.   Water Supply
     d.   Inform men of type of chemical
     e.   Protective clothing required at all spills
     f.   Other chemicals close by
     g.   Breathing apparatus needed

4.   Fighting Hazardous Chemical Fires
     a.   Know exactly what the chemical is
     b.   Know what it's reaction is to water
     c.   Know what to do if two or more chemicals mix
     d.   Know how long chemicals have been burning
     e.   Know what to expect of sudden change in color or brightness of flame
     f.   Know what tank to cool if two or more tanks are involved
     g.   Know safe distance to fight fire

5.   Spill but no fire
     a.   Get as much information possible from authorized personnel
     b.   Don't go by smell or guess
     c.   Look for frostline on tanks
     d.   Allow No Smoking in area
     e.   Get an explosive meter to measure the amount or density of chemicals
     f.   Make sure area is secured

6.   Spill with Fire out of Control
     a.   No closer than 2000 feet
     b.   Use deludge gun
     c.   Never advance, let it burn
     d.   Evacuate immediately

Figure 5. Hazardous Materials Incident Checklist Issued To All Units of the Greensboro Fire Department.

S.O.P. For Railroad Emergencies

"A similar plan for railroad emergencies that was completed this year required considerable work, probably more than with the highway document," adds Shumate. "Each company within their own district went out and gathered information on which facilities or businesses along the right-of-way might create a particular hazard, and identified entry-points to the railroad track. They traveled the entire trackage through the city, canvassed the nearby neighborhoods to locate exactly the various hydrants, identified and planned routes for the laying of lines to every point on the railroad, and took measurements of all possible lays to all points along the trackage. We used zone numbers already on the city map we were working with to identify specific maps within the overall matrix. You tell a man you are in 'Zone 10' and he needs to hook to 'Hydrant 3', and he knows exactly where to go and how much hose he will have to lay. That is, on the back of this particular zonal map it explains that Hydrant 3 is at Cypress and 14th Streets and that the hydrant is located 200-feet from the track."

The preplan of the railway system through the city includes identification of predetermined routes into each particular zone, hydrant locations, areas where evacuation would present special problems in the event of a major emergency, (such as that section of the main railyard adjacent to the Greensboro Coliseum complex) and buildings that could present additional problems due to their close proximity to the railroad; i.e., hospitals, schools, nursing homes, chemical companies, and various types of manufacturing or processing occupancies. Also delineated on each zonal map is a minimum evacuation boundary (1,000 feet) from the rail line so response personnel can determine from a street location whether they are within or outside a minimum evacuation distance from the incident.

Because the highway and railroad preplans have proven themselves in use, the department is currently developing a similar preplan for floodplain areas. "We expect two major benefits from the floodplain preplan," notes Chaptain Shumate. "When flooding is expected, we will have a good idea of which areas are likely to cause problems, how best to get the people out of such areas, how we can get equipment and manpower into the area, and how best to dispatch and route equipment around these areas to respond to other alarms. Another consideration in developing the floodplain preplan was that in some chemical spills the commodity has entered creeks and other waterways. We needed to know where all these creeks and waterways flowed, and what kinds of facilities along their banks might be causing a spill of unknown source. Now, if a chemical is located in a waterway, we have a better idea of what facilities might be causing it, the specific products they handle, and we are able to trace it back to its source in a more knowledgeable search. Also, we know immediately the best route to use for bringing in bulldozers or other heavy equipment to do our diking, as well as the best locations for installing dikes, booms, or sumps. In creating these various preplans, it was not one person or one company doing the legwork; the assignments were handed out to all line companies so that they intimately learned their respective areas. They are going to have the 'Manual,' but they also did the preplan work for their area; and they will remember the specific situations they encountered when working their way through the neighborhood. Thus, the 'ownership' of the plan is with the line companies."

North Charleston District Uses Written Hazardous Materials Response Standards

At 10:45 A.M. on Monday, July 2, 1979 a forklift truck struck a valve on a 6,000 gallon tank of dimethylamine at a chemical storage facility in Charleston, South Carolina opening a 1½ inch hole in the bottom of the tank. Dimethylamine is a flammable liquid/corrosive/environmentally hazardous substance with an NFPA 704M system rating of "4" for flammability and a "3" for health. Firefighters from the North Charleston District Fire Department sealed off the area, evacuated a trucking company immediately downwind, set-up a master stream to dilute the commodity and control the flammability hazard, and drove a wooden plug into the tank to stop the leak. Sand trucks from the state highway department were called in and dikes were constructed to contain run-off. Acetic acid and a carbon-based compound were used to neutralize the chemical. As a direct result of this incident, a hazardous materials response team was formed within the North Charleston District Fire Department.

"The dimethylamine incident indicated to us that the potential for a major hazardous materials incident involving life exposures is quite real," noted Assistant Fire Chief Gerald Mishoe recently. "The first-in company had acted immediately to control the vapor problem; and Charleston County Police, the U.S. Coast Guard, the South Carolina Department of Health and Environmental Control, (DHEC) and the Charleston County Emergency Medical Service, as well as industrial personnel, combined efforts quite effectively to control the situation fairly quickly. However, the experience made us realize there was a great need for more training, more specialized equipment, and more coordination with other agencies in handling this type of incident. In September of 1979 we organized a volunteer, 16-man hazardous materials response team from among the 70-personnel in the department."

A basic training program in hazardous materials response was initiated that created a lot of interest among firefighters who had not volunteered for the special team; eventually, North Charleston District went to department-wide hazardous materials training. Currently, every firefighter has been trained in hazardous materials response, and two trucks carry special response gear as well as handle heavy rescue and provide truck company support at fires. The designation "Team" refers to the two trucks and the equipment on them, but the "Team" in terms of personnel is the entire department.

Fifteen separate fire departments are required to adequately serve Charleston County. The North Charleston District is a 45-square mile area roughly in the middle, and encompasses the bulk of heavy industrial development within the county. Major railyards and Charleston International Airport are both located within the district, and Interstate Highway 26 runs right down the middle. Nearby is the busy Port of Charleston. Industrial development has brought obvious benefits to Charleston, but has contributed to a situation where some industrial areas are located immediately adjacent to residential areas.

One LPG storage farm, for example, contains 45 60,000-gallon storage tanks separated from a major residential neighborhood by a fenceline. Within Charleston, there is a recognized potential both for industrial/fixed-location incidents and for incidents related to any of the five modes of transportation.

Although the North Charleston District Fire Department obtained hazardous materials training from commercial trainers, and had representatives of local companies and the trucking industry give classes, a sizable portion of the training effort involved team members canvassing their response area to take 300-400 color slides of specific, *local* situations and facilities.

"We used the slides because we wanted to show personnel not only that there were major hazards in the area such as those attached to the railyards and the Interstate, but also to show there were potential problems not quite so obvious: a 500 gallon propane tank behind a local forklift company, a flammable liquid storage area at a sugar bag company, a loading dock at a local warehouse, and stored flammable gases at a welding supply company," explains Assistant Chief Mishoe. "We tried to show potential situations all the way from heavy industry to far less obvious locations such as one of the new 'U-STORE-IT' miniwarehouses that are being built all over the country. We spent a tremendous amount of time on railroad incidents and different types of railcars because trains come back and forth in front of our station all the time."

Although acquisition and use of the color slides was initially a training-related function, it had obvious benefits for preplanning as well. "In our initial training program, we pretty much scoured the area to find our major hazardous materials problem areas," says Mishoe. "Now, we feel we pretty much have our eyes on those locations where problems are most likely to occur. We hope we will not find a lot of surprises, although I expect we will find some things we may need to do a little differently when we do additional preplans."

North Charleston District's "Squad 10" truck is a 1948 chassis that previously mounted a generator and was used as a light plant before being converted to a hazardous materials response truck. It is stocked with rubber patches, tapered plugs of various sizes and dimensions, hoseclamps, vice-grips with welded extensions on the jaws, assorted rags and wooden wedges/dowels for sticking in holes, sliced innertubes, and assorted handtools. The "Squad 10" truck carries enough gloves, both butyl rubber for use with a variety of chemicals and nitrolatex for strong corrosives, supplied by a local industry to equip first-response personnel from the department rather than just the personnel attached to the "squad 10" truck. "Our cascade system is probably the most useful piece of equipment on the truck," says Captain Hampton Shuping, Jr., "a 1400 cubic foot, four-cylinder system. We use it all the time. We carry a 55-gallon recovery drum; when we use sorbent material, we pick it up loaded with chemical, seal it in the recovery drum, and either ship it out or dispose of it ourselves depending upon the specific product recovered. We use two varieties of sorbent material . . . the pad-type, and clay-like material such as they use in automobile garages. We carry 200 pounds of the clay-like sorbent."

"In 30-gallon drums we carry 90-gallons of A-FFF concentrate and 60-gallons of ATC/alcohol-resistant foam, plus two one-gallon containers of wetting agent," adds Captain Shuping. "We carry both acid-gas suits and proximity suits, as well as four extra airpacks. There is a small handpump, originally used to pump diesel fuel into our own trucks at prolonged fire operations, we use to transfer product from a leaking drum to a sound one; it comes in handy when a trailer truck carrying a mixed load of drums has one or two that are defective or holed. It will handle diesel and fuel oils with no problem. We also carry a small siphon hose—a 'Georgia creditcard';—we get called to motels quite often where an automobile gas tank is running over into the parking lot. We merely siphon out enough to get the situation under control. We carry four-foot sections of PVC pipe. When responding to overturned tankers or major flammable liquid spills, we first attempt to contain the spill. We throw a section of PVC pipe down in the bottom of a ditch, build a dike over the top of it, and temporarily close the pipe with a piece of plywood or whatever is handy. Water being used to divert or flood the spill carries the fuel into the ditch and is caught behind the dike. The mixture settles; the fuel comes to the top of the water; the pipe at the bottom of the dike is opened to let the water underneath flow out; and then the flammable liquid is picked up."

"Before we used this system, we had a tanker truck turn over in a heavily populated area where we had to use a tremendous amount of water to divert the fuel into a storm drain that led to a big ditch where we could dike it," adds Gerald Mishoe. "We were concerned with the flammability hazard, so a lot of water was used. The gasoline dumped amounted to 8,900 gallons, but we wound up in a 24-hour operation with tankers and pumps picking up 24,000 gallons of gasoline/water. We learned from that experience that we had tripled our spill problem by combining all that water with the gasoline. Now, we use this simple piece of sewer pipe, drain the water out from under the fuel, and then pick up the fuel. Most flammable liquids are lighter than water and insoluble, and will separate from water and come to the top." (the technique will work on any product that is insoluble and has a specific gravity of less-than-one.)

"Working in acid/gas suits, wearing SCBA, with a relief valve blowing off next to you, it is very difficult to communicate; so we use a three-foot-by-five-foot chalkboard and simply write visual messages back and forth," notes Captain Shuping. "We carry a number of manuals including NFPA, Bureau of Explosives, the new D.O.T. manual, the *O.S.H.A. Pocket Guide To Chemical Hazards,* and some others. We have a combustible gas detector as well as an O-2 deficiency meter for checking manholes and other confined spaces. We have enough equipment to be a 'band-aid' team; get in there and contain it, stop it, and pick up small spills—our usual rule-of-thumb is 50 to 100 gallons. If the incident is more sizable, say a tanker truck dropping several thousand gallons, then we closely coordinate our efforts with the U.S. Coast Guard and the South Carolina Department of Health and Environmental Control. (DHEC) With the Coast Guard, DHEC, and fire department working together, there has not been anything we've come up against that we couldn't handle."

Squad 12

2 - 150' Coils ¾'' Rope
1 - 15 lb. CO$_2$ Extinguisher
1 - 20 lb. Dry Chemical Extinguisher
1 - Oxygen/Acetylene Cutting Torch
1 - Stokes Basket and Harness
1 - 14' Extension Ladder
2 - Safety Belts
1 - Broom
1 - Fire Rake
1 - Flap
1 - 3800 Watt Dynamote Power Unit
2 - 50' Extension cords & 500 Watt Floodlights
Assorted Cribbing Blocks
2 - Long Backboards
1 - Rescue Saw (Homelite DM50)
1 - Bolt Cutters
1 - Halligan Tool
3 - Rolls of Plastic
1 - Roll Felt Paper
1 - Shovel
1 - Sledge Hammer
2 - Buckets
1 - Smoke Ejector
1 - Ajax Air Chisel
1 - Water Vac
1 - 10 Ton Port-A-Power
1 - 4 Ton Port-A-Power
2 - 20 Ton Floor Jacks
2 - Flashlights
1 - 25' Extension Cord
1 - Combustible Gas Detector
1 - Sprinkler Kit
3 - Wooden Plugs

2 - 6 ft. Pike Poles
1 - 10 ft. Attic Ladder
1 - Chain Saw
2 - Pressure Demand Breathing Apparatus
4 - Extra Air Cylinders
1 - 36'' Pry Bar
1 - Oxygen Inhalator
2 - Extra O$_2$ Cylinders
3 - Cervical Collars
4 - Spanner Wrenches
1 - Trauma Box
1 - OB Kit
1 - Blanket
1 - Sheet
1 - Hydrant Wrench
2 - Crescent Wrenches
1 - Claw Hammer
2 - Meter Covers
1 - Channel Lock Pliers
1 - Aircraft Snips
3 - Screwdrivers
2 - 2½'' Double Males
2 - 2½'' Double Females
10 Gallons AFFF Foam
1 - 60 GPM Foam Eductor & Nozzle
1 - 95 GPM Foam Eductor & Nozzle
1 - 240 GPM Foam Eductor & Nozzle
1 - Akromatic 60-400 GPM 2½'' Nozzle
4 - Chemical Gloves (prs.)
1 - Oyltite Stik
1 - Epoxy Ribbon
1 - Hurst Rescue Tool
1 - 2 Ton Come-A-Long

Squad 10

1 - Compressed Air Cascade System
2 - Salvage Covers
175 Gallons AFFF Foam
2 - Acid/Gas Suits
4 - Demand Style Breathing Apparatus
1 - 55 Gallon Recovery Drum
2 - 50 Lb. Bags Hydrated Lime
2 - Traffic Cones
2 - Proximity Suits
6 - Assorted Wooden Plugs
20 - Pairs Chemical Gloves
1 - Canvas Cover
1 - Bucket
1 - Extension Cord
1 - Rubber Mallet
1 - Pry Bar
3 - 50' Sections 1½'' Chemical Hose
3' Piece 4'' PVC Pipe
3 Gallons G.V.C. Wetting Agent
1 - 6' Pike Pole

4 - 50 Lb. Bags Scatter Sorbent
1 - Roll Conweb Sorbent
3 - Pair Vise Grips
1 - Set Assorted Wrenches
1 - Hose Clamp
1 - Bull Horn
2 - Spanner Wrenches
2 - Screwdrivers
7 - Assorted Rubber Plugs
10 - Band Clamps
1 - Channel Lock Pliers
1 - Chalkboard
1 - Roll Felt Paper
Assorted Rubber Patches
2 - Shovels
1 - Epoxy Ribbon
1 - Oyltite Stik

Figure 6. North Charleston District Fire Department Equipment List.

"During any radiological incident, we have the capability of calling in a U.S. Navy team attached to the nuclear submarine base that is located here," says Assistant Chief Mishoe. "We will take a quick reading with the field survey meter (Geiger counter) we have, then bail out if there is a leak and let them handle it. We also use the Explosive Ordinance Disposal Team from the naval base if bombs or explosives are involved. With regard to acids/alkalis, we do not have a pH meter but do use pH papers. In possibly a half-dozen incidents this year, we took quick readings with pH papers that were later checked by meter. The meter readings proved to be very close to the readings we had obtained with the paper, so we feel pH papers are sufficient for our needs. The paper is reasonably quick; you merely dip it into the fluid you are testing, get a reaction on the paper, then match the paper against a chart. We had an incident recently where a commodity was running off the highway as if someone had been dumping it as they drove along. A quick test with the pH paper indicated a pH down around 2½ to 3. I followed it out and found a tanker truck with a busted valve had poured 3,000 to 4,000 gallons of sulfuric acid over a three or four mile stretch of highway. It was a pretty bad situation; we had to use a vacuum truck to pick up the big puddles, then lay down soda ash to neutralizer the remainder."

For quite some time now the North Charleston District Fire Department has used a formal, written *Hazardous Materials Response Standard,* much of which was borrowed from the Phoenix, Arizona Fire Department; that covers Objectives, Definitions, Dispatching, Actions of the First Arriving Unit, Size-Up, Formation of an Action Plan, Control of the Hazardous Area, Use of a Hazard Zone and an Evacuation Zone, Use of Non-Fire Department Personnel, and General Guidelines. "We use a written *Standard* because we found some of our major problems related to response to hazardous materials incidents included exposure of personnel and equipment," explains Gerald Mishoe. "Under the *Standard* the first-in company will ease-in and size it up. They have written instructions—whether it be two pieces or six pieces responding—on how to approach the scene and designate a safe staging area for additional arriving units. The *Standard* stresses that hazardous materials incidents require a more cautious and deliberate size-up than most fire situations, and that care must be taken to avoid committing personnel prematurely. Additional units stage around the perimeter and allow the first-in unit to evaluate the situation. The first-in unit then calls in other units as they are needed. The *Standard* has kept us out of trouble; sometimes personnel forget to follow it at minor spills, but it has definitely kept us out of trouble at larger incidents."

"In addition, because we use Coast Guard, DHEC, and industrial personnel; there is a section in the *Standard* on use of non-fire department personnel and our responsibilities toward them," continues Mishoe. "At least a half-dozen times, the *Standard* emphasizes the need for a command post and a person-in-charge. The command person is responsible for the safety of *all* parties involved, not just firefighters but assisting personnel from industry and various agencies as well as citizens who might be exposed. We felt it was necessary to put the load on somebody's shoulders; to insure that all requirements are met, someone has to know they are being held responsible. The *Standard* starts off by giving us an *objective* we are trying to accomplish; it *defines* what we consider to be a hazardous materials incident; and it covers everybody's *responsibility* starting with the dispatcher. It lays out the first arriving unit's responsibility regarding size-up and development of an action plan, control of the hazardous area, and use of non-fire department personnel. All firefighters have been trained in the *Standard*; if the first-in guy forgets to cover some of the points, hopefully there will be somebody coming in to back him up who will check to make sure all points have been covered."

Hazardous materials incidents responded to by North Charleston District firefighters range from a 3500 gallon spill of dimethylamine through hydrochloric and nitric acid spills down to what may be the "smallest" hazardous materials spill on record. The local branch of a major, national chainstore called one night to say that something was in the air and people on the premises couldn't breath. Engine company personnel wearing breathing apparatus found two one-ounce bottles of nitric acid in the safe of the jewelry department were broken and the acid was reacting to the metal of the safe. "We respond to a great number of gasoline spills and tend to treat gasoline with a tremendous amount of respect," says Gerald Mishoe. "We believe gasoline is one of the worst hazards we deal with simply because we have developed a lackadaisical attitude toward it because we see it and use it so much in our everyday lives. People tend to think of 'hazardous materials' as BLEVEs and railroad explosions, but of the ten incidents a month that we average most involve small spills."

Regarding 'lessons learned,' Assistant Chief Mishoe reflected recently that—"In our particular area, there is the potential for having quite a few different agencies involved in a hazardous materials incident because we have a local office for the Coast Guard, DHEC, Civil Defense & Disaster Preparedness; as well as the naval base, several police/fire/EMS agencies, and industry. We found there was a definite need for on-site coordination. If someone does not establish command early in the incident, you may wind-up losing control and having six or seven people saying six or seven different things. Also, we found that the situation of responding to incidents and not knowing what was involved as a problem in the past. We learned from our dimethylamine incident that we needed to be a little slower in sizing-up and not commit too much equipment or too many personnel prematurely. Overall, we have found that although you have all sorts of reference books and many varied agencies available to supply help; if a fire department will equip itself, have a halfway decent training program, and establish a good command procedure—fire department personnel will be able to handle the majority of incidents by themselves. We have found it takes some very simple procedures, some rather common moves, and simple equipment to control most incidents. Our experience has been that once you have airpacks, which most fire departments have, and acquire some sorbent materials and acid/gas suits; much of the additional equipment required is already available to most fire departments. When fire departments think about forming a hazardous materials team, they first think—'Gosh, we have to buy all that equipment.' We already had all our rescue equipment. We had the necessary protective clothing with the exception of acid/gas suits. The foam equipment was available. It was mainly a matter of coordinating equipment, isolating it so we could get to it when needed. You *could* get very elaborate, but there is really no need. Beat around, beg, borrow, and steal whatever you can get your hands on. We are not above rummaging around in a dumpster; the little rubber plugs we use came out of a dumpster."

"We learned there are a lot of people around who have a tremendous amount of expertise; DHEC, Coast Guard, industry, other agencies," adds Mishoe. "Once we started getting together on some of these incidents, we learned we have a tremendous amount of resources right here locally. Everything that is here, everything we use with the exception of a few pieces of equipment, has been here all along. The expertise, the people, and the agencies have been here all along; it was just a matter of putting it all together in a reasonable, usable form."

"We really don't have anything fancy," concludes Mishoe. "Mainly, we have regular fire department equipment that we managed to muster in one place. The key to the whole thing is people; people who have the training, experience, and common sense to know what to do. We feel we have some good people in our fire department who meet those requirements. There is a lot of publicity about large incidents. If people will begin to prepare for the smaller incidents, learn how to respond well to the smaller incidents; they will find the big incidents don't occur all that often and when they do the preparation, experience, and coordination they have achieved in working the small incidents will be helpful to them. When we first got into EMS, people said: 'It is not your job to do that; you are a firefighter.' Who else is going to be in there at a highway spill or industrial accident? Who will be there first? It is our nature and our position to handle hazardous materials incidents, so we might as well be prepared."

Small Volunteer Department Makes Extensive Commitment to Hazardous Materials Preparedness

Fire department personnel are normally the first responders to hazardous materials incidents. Since a majority of the estimated 29,000 fire departments in the United

States are manned by volunteers; the knowledge, training, and equipment possessed by volunteer fire departments is often a critical factor in incident control. At a time when severe budget constraints are a fact-of-life for all departments, the level of hazardous materials preparedness achieved by some volunteer fire departments is nothing short of amazing. Down on the east coast of Florida at Stuart, the 25-man Martin County District 2 Volunteer Fire Department has demonstrated how much can be done with minimum dollars and maximum dedication.

In July of 1979, the department under then-chief Ed Smith responded to a truck fire on the highway to find the fully-involved cargo consisted of chlorine, peroxides, and an epoxy-type product. Ed Smith, who had retired as a Colonel after 30 years in the Army and then served as Chief of the Martin County District 2 department from late 1976 through late August of 1981; noted recently: "We saw the need and recognized that somebody better do something." The "Something" became a very knowledgeable, highly trained, well equipped hazardous materials response team.

"The big push came after the truck incident out on the highway," says Fire Chief Fred Monks, a charter member of the hazardous materials response team who was recently promoted from Captain. "Our department has always been very dedicated to any job we take on. We try to stay up on the latest thing, emerging fields such as hazardous materials. Chief Smith began acquiring informational materials; since he and Captain Ron Gore of the experienced Jacksonville Fire Department Hazardous Materials Team are both circus buffs, he knew where to go to obtain good advice regarding the tools/equipment/materials we would need. Also, the Grumman Aerospace facility at Martin County Airport ordered a new crashtruck and donated their old one to us; we cover the airport during their off-duty hours, and assist during their normal working hours."

"We reserved the Grumman truck as a hazardous materials response vehicle," remembers Chief Monks, "painted it, reworked it, and stocked it piece-by-piece. As we began gathering equipment, we started attending class after class after class. Four of us attended the course at Palatka, Florida; we trained with industry, commercial trainers, and the railroad. We train and drill on hazardous materials all the time. Two classes are scheduled this month and one next. We even repeat courses," grins Monks. "We don't handle hazardous chemicals every day, so even if training is repetitious, we take it. We have our own hazardous materials training field with a tank truck and flammable liquid pits installed on a piece of county property just down the road. Currently, a pipe scenario is being built that will allow us to apply water or pressure to make patching and plugging drills more realistic."

There are 12 fire departments in Martin County. District 2 covers a 12 square mile area for firefighting and the entire county for hazardous materials response. Of the 15 firefighters on the Martin County District 2 Hazardous Materials Response Team, nine are from District 2 while six are members of the Palm City Fire Department. Any qualified firefighter within the county may apply to join the team, but the training requirement is very stiff: 85 hours of formal hazardous materials response training. Any person who has made the team has demonstrated an undeniable commitment to the volunteer fire service.

"Through what we have learned in classes, read in fire service publications, and advice obtained from experienced response teams; we have gathered our equipment over-time," explains Chief Monks. "We use the truck cab as a small, satellite command post where we keep our bookwork, list of emergency phone numbers, information on Chemtrec, (the Chemical Transportation Emergency Center in Washington, D.C. operated by the Chemical Manufacturers Association) and various reference materials such as Chem-Cards, the Coast Guard CHRIS Manual and the N.F.P.A. and D.O.T. manuals. The chief's car, which we use as the main command post, carries a duplicate set of manuals and reference materials. We hope soon to have a telephone system in the truck cab so we can talk directly with Chemtrec/carriers/shippers rather than having to patch through a dispatcher or other intermediary and risk losing critical information. That is a problem we have faced—secondhand conversations. The intermediary may not understand what we say, or he may obtain information on a completely different chemical from the one we are dealing with. One wrong letter in a chemical name can make a great difference.

Team members set a small, homemade windsock on a pole as soon as they arrive at an incident. They have boxes of assorted handtools ranging from paint scrapers to vice-grips; many times it is necessary to scrape off paint or rust to get a good seal for a patch. They modify a number of vice-grips by welding metal extensions to the jaws; these allow them to clamp-off hoselines such as on an L.P.G. delivery truck. A local rubber company that makes boots and hoses provides them with scrap odds-and-ends of rubber that are cut into assorted sizes and shapes for use as patches. They carry C-clamps, S-hooks, pipe plugs/caps/clamps/sleeves, hoseclamps, assorted tape, mops/buckets/shovels/brooms, commercial sealant and adhesive products, a chlorine "A" kit, and a device for measuring the depth of oil film on top of water that permits an estimation of the total volume of a spill. For applying patches to larger containers, they carry chains, compression plates, hydraulic jacks, com-a-longs, loadbinders, and cable. There are traffic cones, meters and "sniffers," bung/chime tools for use on metal drums, assorted gloves, pipe and crescent wrenches, pipe threading tools, hacksaws and blades, boltcutters, binoculars, and a Polaroid camera. They carry totally-encapsulating chemical/acid suits, SCBA, foam, 'Jan-Solve," sorbent materials, and soda ash. A canon on the truck roof can be operated by one man from inside the cab, and can throw either foam or water. The truck carries 750 gallons of water and 50 gallons of foam.

"We are getting a bit 'maxed-out,' " laughs Chief Monks, "but we could get additional equipment on the truck if we had to. There is definitely a system to it all. We probably loaded, unloaded, reloaded, and rearranged the truck 40 times. We have different categories, locations, and compartments for stowage; everything definitely has its own location. When we have an incident, we try to have a man with us who knows the truck, but if we are short of manpower we can tell a man exactly which compartment to go to for a specific item. Also, our 'Equipment Storage List' is broken down by compartments and other *specific* locations. It is *not* thrown on there helter-skelter by any means; it would not be of use to anyone if it were. We have to know what we have and where it is."

"At last count, we had expended $15,000 to $18,000 on hazardous materials response equipment, not including the truck," adds Fred Monks. "That amount is required. You can't just say—'We will use this particular piece of equipment every time.' Some of the money came from the county budget, and we run fund drives as do many volunteer departments. We are not bashful about how much we spend, because we feel we are not abusing the money, that it is being used to good purpose. It's the same as you hear in the military—'Take care of your weapon, and it will save your life.' We do take care of our equipment because it is what protects both ourselves and the public."

Types of Incidents:

"The Martin County area is attempting to attract industry," notes Chief Monks, "but there is no real industry here at this time. We have a lot of water plants to supply the great number of condominiums in the area, so we have experienced a number of chlorine incidents where we had to suit-up and cap-off leaking cylinders using a chlorine "A" kit. Because we do have so many water plants, there is apathy. People don't realize the cylinders are supposed to be chained. Cylinders are not being checked; they are just sitting there rusting. There is bulk fuel storage in the county, but there have been only two incidents connected to it. A 7,000 gallon semitrailer loaded with #2 diesel fuel overturned and ruptured two of the risers. We removed the risers, patched them, sealed the dome, foamed it down, transferred the load, and cleaned up the area. A gasoline truck caught fire as it was being off-loaded; we extinguished the fire and patched a ruptured valve. Other than that, it has been an acid spill here, a chlorine leak there and a bit of gasoline here and there—nothing of major proportions. Our major industry is senior citizens who have come here to retire, and a great number of condominiums to house them. Building is our major industry.

Special tools and equipment carried by Martin County District 2 Volunteer Fire Department Hazardous Materials Response Team 1962 Dodge 4WD American LaFrance high pressure booster with a midship mounted 100 GPM Hale pump, a 750 gallon booster tank and a 50 gallon foam (6%AFFF) tank. A turret nozzle is mounted on the cab with two 1'' 200' booster lines and two higher pressure water/foam nozzles.

Tools and Equipement carried include:

Cab Compartment
2 Koehler hand lights and charger
2 Flashlights
20 × 50 Binoculars
Wind Socks
Reference Library
CHEMTREC input and return data forms
Martin County Maps

Left Front Compartment
Blackboard and chalk
4 Spare Scott bottles in rack
Hooligan bar
Pry bar
Spare hose gaskets

Left Side of Apparatus
10' Section of 2½'' hard suction
8' Wood pike pole
7' Fiberglas pike pole
Pick axe
Flat head axe
2 Scott Air-Paks in protective bags
20 lb Dry chemical extinguisher
2½ gal Loaded stream extinguisher

Right Front Compartment
4 ton Port-a-power
2 Bio Pak 45's with spare bottles
Tool box (see that section for contents)
8' Piercing nozzle with 1'' to 1½'' adapter
Emerson resuscitator

Right Side of Apparatus
24' Extension ladder with Skull Saver
15 lb CO₂ Extinguisher
20 lb class ''D'' extinguisher
Hose spanners and bracket

Left Rear Compartment
3 Sets asbestos gear (pants, coats, helmets, and gloves)
2 Acid Suits

Center Rear Compartment
Smoke ejector

Right Rear Compartment
35' 5/8'' Manila rope
9' × 12' Salvage cover
50 lbs Metal-X for Class ''D'' portable extinguisher
Smoke ejector door bar
Radiological monitoring instruments
Large rubber patches

Top Rack
50 gal 6% AFFF foam
10 gal Jansolv-60
100 lbs Soda ash
100 lbs Plug-N-Dike
3-50' Sections 2½'' hose for floating boom
Push broom
Kitchen broom
Coal shovel
Round point shovel
Flat shovel
2 - #2 Wash tubs
2 - #10 Buckets
3 Traffic cones
2 Bundles wood shims
2 Wheel chocks
12 - 18'' × 2'' × 4'' & 18'' × 4'' × 4'' Wood cribbing
15' ¼'' Chain
20' 5/8'' Chain
2 - ¼'' Chain binders
2 ton Cable hoist
2 Booster crank handles
2½ gal. Gasoline can

Chlorine ''A'' Kit
22 Adams pipe clamps ¾'' to 8''

Tool Box on Top Rack
Hazardous materials kit with plugs and patches
1 - 2½'' Double male hose coupling
1 - 2½'' Double female hose coupling
1 - MAC #120 Chain saw
Federal voice gun
1½ ton Hydraulic jack and handle
24'' Bolt cutters
Rubber mallet
DART Surface film measuring instrument
4 Ziamatic protective head covers
24'' Pipe wrench
30'' Pry bar
26'' Hand saw
10' 3/8'' Vinyl tubing
1 - 2½'' male hose plug
1 - 2½'' female hose plug with adapter for air application for floating boom
Tray #1 1/8'' to 2'' pipe caps
Tray #2 1/8'' to 2'' pipe caps
Tray #3 #20 to #44 hose clamps
Tray #4 ''S'' hooks
    Pipe clamps, Tape, Metal screws (mixed)
Top Tray 2 — pair safety goggles
    Chain saw files
    Reflective tape
    2 - rolls ribbon epoxy
    2 - hose spanners
    Pliers
    Paint scraper
    Garden trowel
    Garden claw
    Chain saw bar wrench
    2 — recharge cylinders for Ansul ''D'' portable extinguisher
On Order 2 Acid suits, Chlorine ''A'' Kit

Engine 201
1979 Emergency One 1,000 GPM Pumper with 1,000 gal booster tank.
Angus 120 GPM Low expansion foam nozzle 1½''
Angus 120 GPM Eductor 1½''
2 - Akron 95 GPM Eductor 1½''
Akron 6' Piercing nozzle 1½''
30 gal Angus 3% Fluoroprotein Foam

Chief's Station Wagon
MSA Explosimeter with 8' and 20' hose and 4' sniffer tube
10 × 50 Binoculars
Hazardous materials repair kit with patches and plugs
Hand tools

Station
70 gal Angus 3% Fluoroprotein Foam
50 gal 6% AFFF Foam

Tool Box in Right Front Compartment of Engine 202 (Hazardous Materials Response Team)
14'' Pipe wrench
12 Screw drivers (various sizes, flat and phillips)
Welders vice grip
Duct tape
2 - 3'' ''C'' clamps
Can CRC 5-56 Lubricant
2 - 1'' Hose spanners
8 piece Combination wrench set
7 piece Metric combination wrench set
5'' Punch
9'' Punch
18'' Punch and pry bar
8'' Pipe wrench
3 Blade pocket knife   Putty knife
10'' Crescent wrench   2 - 2'' ''C'' clamps
7'' Vise grip   Water faucet handles
10'' Vise grip   Scissors

Figure 7. Special Tools and Equipment Carried by Martin County District 2 Volunteer Fire Department Hazardous Materials Response Team.

Just then a call came in on an incident that may sound strange to firefighters in other parts of the country, but is farily common along Florida's east coast. A cylinder and a 55-gallon drum came floating in on the tide and wedged in rocks in front of the House of Refuge, an old lifeboat station. In one recent month, approximately a dozen different containers (L.P.G., compressed air, ether, and diesel fuel) washed ashore. Usually, the containers have been in the water for extended periods and container markings are illegible. "Today's incident was stricly manual labor," added Chief Monks that evening. "The two containers lodged under a limestone formation, and we were racing the tide to get them out. Normally, they float ashore on a nice sandy beach which makes it easier and safer. Still, there is a technique for handling unidentified containers; you have to take precautions; maintain incident vigalance and discipline, or you could just as easily be dead. By no means have all the containers and drums we have taken off the beach been 'empty.' "

What Have They Learned:

"It's a continuous learning process," reflects Fred Monks. "We have made many mistakes, but none serious. We try to go by the book, and have procedures by which to operate, but there is no guarantee we will not make mistakes. We take polaroid photos, both for future training and to use during a critique that is held after every incident. We sit down and talk about our handling of the incident and critically review the photos. 'What were *you* doing? Who did not have their gear rolled-up? Who was not in breathing apparatus? Who was not doing this or that?' By reviewing the photos we find things we did not notice while on-site. The critiques and film are later used for training. It is a continuous, circular process by which we attempt to learn from our mistakes."

"We approach any incident as if it were the most serious incident ever," stresses Fred Monks, "until we know exactly what we have, then we may back down a bit depending on the degree of hazard that appears to exist, eliminating the unnecessary equipment and gear. We establish a command post, cordon off the area, perform a size-up from a safe area using binoculars if warranted, and identify both the commodity and the tools/equipment/materials we will need to handle it.

At a reported chlorine leak, for example, a two-man team (never just one man) will don acid suits and breathing apparatus and make *two* trips into the danger area. On the first trip the two-man team will establish the type of container, identify exactly where it is located, determine where it is leaking, and assess what tools and equipment will be necessary to correct the situation. If a very basic action, such as turning a valve, will not correct the problem; they return to the Incident Commander to be supplied.

"Some of the best advice we have received is pretty basic," says Fred Monks, "like when you go in on a propane cylinder leak, don't forget they have valves to turn them on and off. We estimate that one-third of the time we go in and turn a valve, and the problem is remedied except for clean-up. Another third of the time we can handle it with a pair of vice-grips—don't worry about crushing the line—it is already damaged. Clamp the vice-grips on and the leak is stopped. So much of it is common sense if you don't let your adrenalin run away with you. It is a serious situation but getting excited is not going to help one bit. Take your time, think about it, discuss the situation; you are a team and no man is going to do it all alone, so work together as a team. By working together team members can see a lot more than one person alone; by discussing the situation, you come up with more ideas. Seldom is there any one good method for controlling a leak."

"We try not to get into clean-up, but rather attempt to render the situation safe, neutralize the commodity; we are not in the clean-up business," adds Chief Monks. "We *will* assist with clean-up, but we feel our primary mission is to attack the problem, neutralize the product, and render the situation harmless. That is the way we have been taught and trained."

Chief Monks worries that many departments may say—"Hazardous materials, why it would cost us $100,000 to put a team together," when actually many departments already have the bulk of the necessary tools and equipment in their toolbox or on their apparatus. They often can centralize tools/equipment/materials that are available, modify existing equipment, or fabricate certain items. "Some departments have their own shop," he notes. "We don't, but the county mosquito control agency within the Department of Environmental Services has made a lot of equipment for us. They have been outstanding. We go down there and say—'I need such-and-so; I need a big patch like this so I can fit it onto a railroad tank car.' They rolled a piece of steel with eyes and a screw on it, and there it was. Every fireman has a contact for something like this, a person he can go to and say—'I need . . .' It may look primitive, but the key question is—'Does it work?' You have to come up with a workable item. An item made by the best welding shop in the world may not work if they don't understand your needs. Volunteer departments are particularly aware of the tremendous value of individuals and agencies within the community who respond when you say—'I need . . .' "

4 INDUSTRIAL RESPONSE TEAMS

"This is CHEMTREC. Do you Have a Transportation Emergency Involving Chemicals?"

When a transportation-related hazardous materials incident occurs anywhere in the United States, information and assistance is truly just a phone call away. By dialing 800-424-9300, first responders and carrier personnel can obtain immediate information about the product involved to aid them in proper control and containment efforts. If the situation warrants, the shipper or other experts will be contacted for additional advice and/or on-site assistance. Operated by the Chemical Manufacturers Association, the Chemical Transportation Emergency Center (CHEMTREC) located at 2501 M Street, N.W. in Washington, D.C. is both a source of chemical control information for carriers and first responders *and* a notification and alerting system for industrial response teams. All major chemical manufacturing companies participate in CHEMTREC. For first responders, CHEMTREC is the fastest way to get an expert at their elbow.

John C. Zercher has been Director of the Chemical Transportation Emergency Center since it began operation on September 5, 1971. He came to CHEMTREC after 25 years in production and transportation with Celanese Corporation. "To man the Center we wanted reliable, longterm shiftwork employees," says Zercher. "My wife suggested we go to the military to obtain retired noncommissioned officers. We started with a staff of seven military retirees: three marine sergeants, three navy chiefs, and one air force sergeant. All were master sergeants or above with 20 years military experience. They were dedicated, disciplined, and all had stood watch in the military. They do not become unglued when the pressure is on. Of the seven who started here in 1971, six are still here after 11 years. Our communicators have a tremendous amount of experience in their particular function."

From its opening day on September 5, 1971 through July of 1981 CHEMTREC received calls relating to 17,370 separate transportation emergencies. The total number of calls received is far greater and includes multiple calls pertaining to a single incident, calls related to non-transportation emergencies, nonemergency requests for general information, calls related to delivery problems and disposal requests, and miscellaneous calls. CHEMTREC has even received 3500 calls for Howard Johnson's motel reservations because the Center was assigned a toll-free number that two years previously had been released by Howard Johnson's. Only three out of every ten calls received by CHEMTREC's emergency communicators have dealt with true emergencies. During 1981, the percentage of true emergency calls was even lower, about 23 percent. This situation exists even though CHEMTREC's emergency telephone number, although widely distributed to emergency response organizations, shippers, carriers, and manufacturers; has never been promoted to the public-at-large. The CHEMTREC number will not be found in any telephone directory. Unjustified use of the CHEMTREC emergency number plagues efforts by the sponsoring chemical industry to maximize effectiveness of the CHEMTREC system. Nonemergency calls may prevent real emergency calls from getting through. Thus, it is important to reiterate that the Chemical Transportation Emergency Center was designed to provide information and assistance in the event of transportation-related chemical emergencies—spill, leak, fire, exposure, or accident. CHEMTREC is *not* intended and is *not* equipped to function as a general information source.

When a transportation-related chemical emergency is reported to a CHEMTREC communicator, he will attempt to determine the essential information: name of the caller and a call-back number, location of the problem, shipper or manufacturer of the product involved, type of container involved, rail car or truck number, carrier name, consignee, local conditions on-site, etc.. The communicator will then provide the best available information on the chemical(s) reported to be involved, indicate specific hazards represented by the chemical, and advise on what to do—as well as what not to do—for crucial first steps if the chemical is involved in spill, leak, fire, or exposure. Having advised the caller of initial emergency actions to be taken, the communicator will then immediately notify the shipper. Responsibility for any additional action, which may include dispatching expert assistance to the scene, then rests with the shipper.

It should be noted the speed and completeness of response that CHEMTREC can provide is directly related to the quantity and quality of information made available by the caller. For this reason, emergency response personnel should make every possible effort to obtain reasonably complete information before placing a call to CHEMTREC. (For the types of information CHEMTREC communicators will be interested in, refer to "INFORMATION NEEDED BY CHEMTREC" carried as Illustration 1.)

The completeness of information provided CHEMTREC by emergency response personnel can vary widely. In some cases the caller knows the product(s) involved and is calling for reassurance and confirmation as to product characteristics. In other cases the caller knows the product(s) but does not know specific characteristics and is calling to obtain additional information. Sometimes the caller does not know the product(s) but has some information that may or may not indicate the characteristics of the product(s) involved; such as, container shape, size, and color; placard or label; or other observable indicators. Often the caller may not know the product(s) involved but may be able to name the carrier, manufacturer, or user. CHEMTREC would then use all its available resources and experience to contact the carrier, manufacturer, or user: determine what the product(s) are; determine their inherent characteristics; and insure a call-back to the emergency response organization. In some cases the caller may not know anything about the product(s) except that "something" is causing a problem. This is perhaps the most difficult situation for the caller and the communicator to deal with. In such cases, the communicator will attempt to use available bits of information to identify the actual product. For example, railroad tank cars carry their own individual reporting marks, a combination of letters and numerals on the left end of every car, which identify the owner and the car number. With this information, the CHEMTREC communicator can often immediately identify the shipper if the shipper owns the car; if the car is owned by a leasing company, identification of the product will take a bit longer. CHEMTREC maintains an extensive file of railroad tank car reporting marks in order to speed the process of company-product identification when no other identifying information is available.

Approximately 200 companies that are members of the Chemical Manufacturers Association (to be a member of C.M.A., a company must make a chemical conversion. C.M.A. member companies represent over 90 percent of the production capacity of basic industrial chemicals in the U.S. and Canada.) plus 650 nonmember companies are tied-in to the CHEMTREC system. Such companies are obliged by the fact of their participation to provide information on their products for reference by CHEMTREC communicators when an emergency occurs. Information on chemical names is stored in a series of tubfiles; information pertaining to tradename chemicals is maintained in looseleaf notebooks. Approximately 75 percent of

"CHEMTREC"
(800) 424-9300

"CHEMTREC" stands for "Chemical Transportation Emergency Center" and is a public service of the Chemical Manufacturers Association. It provides immediate advice for those at the scene of emergencies and then promptly contacts the shipper of the chemicals involved for more detailed assistance and appropriate follow-up.

INFORMATION NEEDED BY CHEMTREC

1. WHAT HAS HAPPENED _____

2. WHERE _____

3. WHEN _____

4. CHEMICAL(S) INVOLVED _____

5. TYPE & CONDITION OF CONTAINERS _____

6. SHIPPER & SHIPPING POINT _____

7. CARRIER _____

8. CONSIGNEE & DESTINATION _____

9. NATURE & EXTENT OF INJURIES TO PEOPLE _____

10. NAUTRE & EXTENT OF PROPERTY DAMAGE _____

11. PREVAILING WEATHER _____

12. COMPOSITION OF SURROUNDING AREA _____

13. WHO CALLER IS AND WHERE HE/SHE IS LOCATED _____

14. HOW AND WHERE TELEPHONE CONTACT CAN BE RE-ESTABLISHED WITH CALLER OR ANOTHER RESPONSIBLE PARTY AT THE SCENE

_____

Illustration 1.

emergency alls to CHEMTREC pertain to chemical names while 25 percent of calls deal with tradename products.

In addition to providing immediate control and containment information to local emergency response personnel suddenly confronted with a hazardous materials emergency, CHEMTREC also serves as a communications hub to initiate provision of information and/or on-scene assistance by technically expert industrial response organizations. For certain products, manufacturers combine efforts through mutual aid networks to provide immediate, maximum levels of assistance to local response personnel regardless of the shipper or the location where an incident occurs. Mutual aid networks that can be activated by CHEMTREC have been established by shippers of chlorine, pesticides, hydrogen cyanide, vinyl chloride, phosphorus, and hydrofluoric acid. When a transportation emergency involves one of these groups of products, the CHEMTREC communicator's first call may be to someone other than the shipper.

For example, in 1970 the National Agricultural Chemicals Association formed the "Pesticide Safety Team Network," a cooperative, voluntary program operated as a public service by the Association and participating companies; designed to minimize the risk of injury arising from the accidental spillage or leakege of pesticides. Fourteen Association members—Chevron, Shell, Stauffer, DuPont, Dow, Diamond Shamrock, Hopkins, Mobay, Velsicol, FMC, Union Carbide, Monsanto, Wilbur-Ellis, and Helana—currently cooperate in furnishing personnel, equipment, and expertise for prompt and efficient clean-up and decontamination of pesticides involved in major accidents. More than 45 individual safety teams currently make up the Pesticide Safety Team Network. Ten Area Coordinators have each been assigned a specific area of the United States in which to receive from CHEMTREC reports of any incident involving pesticides occurring in that area and insure response in one of several ways to make certain potential hazard to the public has been reduced or eliminated.

Immediately following receipt of an emergency message from CHEMTREC, the Area Coordinator will phone the caller and advise him that the problem is being handled by a pesticide expert and receive any necessary information regarding the problem. The Area Coordinator will communicate with the manufacturer or producer of the involved product and agree upon an emergency procedure to follow. The person reporting the incident to CHEMTREC is then recontacted and advised of immediate steps to take and that a safety team, if needed, is on the way, either from the manufacturer or dispatched by the Area Coordinator from a roster of teams in the area.

The "Chlorine Emergency Plan" (CHLOREP) developed by the Chlorine Institute, a trade association whose member companies represent over 99 percent of the total product moved in the United States and Canada; includes 62 trained industrial response teams ready to respond quickly to chlorine emergencies anywhere in the United States or Canada. Each team is assigned a geographic area of coverage and will react to any chlorine emergency in that area regardless of the type of shipment or its origin.

Currently, 28 Chlorine Institute member companies sponsor one or more teams consisting of an emergency contact person, team leader, assistants, and a home coordinator. The CHLOREP team leader and the individual in charge at the emergency scene jointly decide whether the services of a team will be required. If a team is required, one is dispatched immediately. CHLOREP teams work with teams that may be sent from a greater distance by the shipper; if the emergency extends over a long period of time, the shipper's team will relieve the CHLOREP team.

CHLOREP teams are trained and equipped to use emergency kits developed by the Chlorine Institute. The "A" kit is specifically designed to handle and control leaks in chlorine cylinders; the "B" kit is for ton containers; and the "C" kit is for tank cars or cargo tanks. In the United States, the Chlorine Emergency Plan (CHLOREP) is activated through CHEMTREC. In Canada, CHLOREP cooperates with both the Transportation Emergency Assistance Plan (TEAP) administered by the Canadian Chemical Producers Association, and the Canadian Transport Emergency Centre (CANUTEC) administered by Transport Canada. When a chlorine emergency occurs anywhere in the U.S. or Canada; CHEMTREC, TEAP, or CANUTEC immediately notifies CHLOREP.

In like manner, the Phosphorus Emergency Response Team (PERT) program was established by the manufacturers of elemental phosphorus (ERCO Industries, Ltd., FMC, Hooker, Mobil Chemical, Monsanto, and Stauffer) to assist in providing response to transportation emergen-

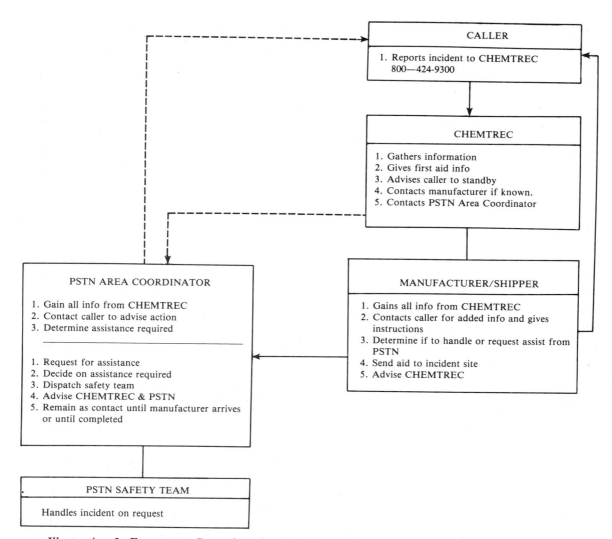

CALLER

1. Reports incident to CHEMTREC
   800—424-9300

CHEMTREC

1. Gathers information
2. Gives first aid info
3. Advises caller to standby
4. Contacts manufacturer if known.
5. Contacts PSTN Area Coordinator

PSTN AREA COORDINATOR

1. Gain all info from CHEMTREC
2. Contact caller to advise action
3. Determine assistance required

1. Request for assistance
2. Decide on assistance required
3. Dispatch safety team
4. Advise CHEMTREC & PSTN
5. Remain as contact until manufacturer arrives
   or until completed

MANUFACTURER/SHIPPER

1. Gains all info from CHEMTREC
2. Contacts caller for added info and gives
   instructions
3. Determine if to handle or request assist from
   PSTN
4. Send aid to incident site
5. Advise CHEMTREC

PSTN SAFETY TEAM

Handles incident on request

Illustration 2. Emergency Procedure for Handling Accidental Spills of Pesticide Chemicals.

cies involving phosphorus. The purpose of PERT is to provide technical assistance to any public or private agency requesting help during a phosphorus emergency; assistance provided may range from a telephone response to on-site aid depending upon the specifics of a particular incident. In the United States, activation of PERT is handled by CHEMTREC while within Canada activation is handled by CANUTEC.

Local responders faced with a transportation emergency involving chemicals can contact CHEMTREC by dialing 1-800-424-9300 day or night. (from Washington, D.C., call 483-7616. Marine radiotelephones, and callers outside the continental United States, can reach CHEMTREC by calling 202-483-7616 Collect.)

Distribution Emergency Response by Shell Chemical

No treatment of industrial hazardous materials response teams would be complete without mention of Shell. Among hazardous materials shippers, carriers, manufacturers, and regulators; Shell has a strong reputation for quality in responding to transportation incidents involving company products. Shell's Transportation Emergency Reporting Procedure (STERP) was established to facilitate prompt communication from the scene of any transportation emergency incident involving Shell products or transportation equipment to a competent Shell person. The intent is to assist operations at the scene by providing accurate technical information and, if necessary, additional Shell participation. Essential data about a transportation incident is communicated by designated Reporting Centers to the specific Shell Response Action Group having product responsibility.

The major financial responsibility for taking action at the scene of a transportation emergency incident rests with the carrier. Shell's efforts under STERP are aimed at offering assistance, furnishing technical and safety information, providing for alternative storage and handling procedures, and assuring that corrective action is being taken. Assistance may be provided by telephone; or if the situation warrants, Shell will send a Response Action Team to the scene to render additional help. Shell's plan is to over-respond with a team rather than hold back.

STERP utilizes two Reporting Centers, actually pipeline dispatching offices that are manned 24-hours a day seven days a week, as an initial communications link between the scene of an incident or the person reporting an incident and the appropriate Shell product expert. A specific Shell product expert receiving a call from a Reporting Center accepts full Shell responsibility for taking the appropriate response action. He makes contact directly with the person requiring assistance as well as other authorities at the scene of the incident, provides emergency action information, and activates a Response Action Team if he deems it necessary or desirable.

Depending upon the product involved, the STERP reporting procedure may initiate response action from either one of two groups within the overall Shell organization. If the material involved is an oil product, a Response Action Team will be alerted and dispatched from Shell Oil's product distribution section. If the incident involves a chemical product or propane, a Response Action Team is furnished by Shell Chemical. Although the reporting procedure is the same for both chemicals and oil products, the specifics of incident response will be different. Only response by Shell Chemical is covered in this section; response by Shell Oil is treated in another section of this chapter.

As Manager of Transportation Safety And Regulations, Richard L. Way is responsible for assisting all Shell locations to comply with federal and state regulations related to transportation, as well as the management of emergency response efforts for Shell Chemical. An easy-going, softspoken man, Way is widely known and highly respected throughout the industry. "The formal notification component of the STERP system was put into effect by Shell in 1970 and tied-in to the Chemical Transportation Emergency Center (CHEMTREC) in Washington, D.C. when that service was initiated in 1971 by the Chemical Manufacturers Association," explains Dick Way. "In late 1975 we formed a Shell Chemical Emergency Response Program that identified and formalized chemical emergency response teams within the company. Team members have fulltime, regular jobs within Shell facilities around the country; when there is an incident, they respond to provide technical assistance to the carrier. Shell Chemical uses common carriers for transporting chemicals over-the-road; we do not operate proprietary equipment except for four trucks at one location on the west coast. We respond primarily as technical advisors to protect the public and emergency services people on-scene, to prevent contamination of the product if it can be recovered, and to provide the safety advice to get the job done."

"We found there was a great deal of apprehension on the part of our refinery people to be involved in a chemical incident," continues Way. "Generally, this was due to lack of knowledge and experience with the chemical products. It was much easier to train the chemical people to handle the oil products than it was to train the oil people to handle the chemicals. This problem applied to agricultural chemicals as well; staff involved in pure chemical areas such as solvents are hesitant about handling pesticides because they don't know enough about them. For this reason, we told our manufacturing facilities—'We will not call you out to handle an incident involving a product you do not understand. If you are a producer of epichlorohydrin, you will get called for a spill of that product whereas someone who does not produce that specific product will not be called.' This philosophy complicated our response a bit, but it gave a great deal of confidence to team members who knew they were not going to be handling products they did not understand. We looked to our manufacturing facilities primarily to supply team members with expertise in containers. For example, if an incident involves tank cars, the responding team will include people who daily fill tank cars and understand what the machinery is all about. Also, because they handle the products daily and know without having to look at a lot of reference material how to do it safely, we depend on them to know the day-to-day requirements for protection of people. Manufacturing facilities were used to supply people who know the product, work with it routinely, and know the equipment. Additional team members with particular areas of expertise were then drawn from other areas of the company; such as our research laboratory, a quality control office, the public affairs group, and the head office. If we have a major chemical incident, the response team would consist of our or five Shell people but only two of them would be likely to come from one of our manufacturing locations where the product was actually made.''

Chemical emergency response personnel within Shell number approximately 100 people assigned to 21 teams located at manufacturing plants, three major distribution terminals, and two research laboratories throughout the United States. "We told the plants we did not want them to send less than two people on any incident—we thoroughly subscribe to the buddy system—and that we wouldn't ask for any more than four," adds Way. "Whenever there is an incident, my office becomes the control center here in Houston; we provide a 'war room' or communications center from which we attempt to support the team in the field with whatever company resources are necessary, and we follow the incident through to its conclusion. We may handle response a bit differently than some of the other companies that have programs. A number of companies are divisionalized, and they tend to receive the message that there is an incident and then turn responsibility over to whatever division manufactures the product without much additional head office involvement. To the contrary, we keep involved in an incident at the head office level, and have devised communications systems by which we maintain contact with our people in the field. Response team members go on detached duty from their regular assignments and report to the Response Coordinator during an incident. This does not mean they can't call back to their homebase for equipment or assistance, but merely that there is a person in head office who is responsible and has control of the total company response for each incident until it has been concluded.''

"We tried to sit down and synthesize some typical incident scenarios," recalls Dick Way, "and it became a very frustrating effort; we very quickly gave it up on the basis that hardly any two incidents were exactly alike. We do provide our people with general guidelines. Since they respond principally as technical advisors rather than hands-on clean-up workers; we suggest, Number One, that they learn who is in-charge of the incident, and that they work with all persons on-scene to insure safe and effective operations. Most of the things we do are based on past experience and an attempt to do things better the next time. For example, we operate a formal command center here in Houston because we learned that when a single coordinator attempted to handle a relatively major incident during office hours it was like a nonsectarian firedrill. People were willing to help, and they came from their offices to the coordinator's office but they had to go back to their own offices to use the telephone. We decided to put two or three people together in one room with telephones so they could keep current on what was going on and make telephone calls without losing touch. Our response control room has a four-station/call director phone system where everybody has a button and can tap into any of the various conversations. Also very helpful is a conference call setup for as many as five people at different locations around the country. We have a Texas Instruments Silent 700 computer terminal for running air dispersion and water dispersion models. Shell maintains a 24-hour computer timesharing system that allows us to use a phone line, dial in the correct code to gain access to the computer, and have the computer run various dispersion programs. We use the same type of portable computer terminal for receiving hardcopy notification from CHEMTREC that there has been an incident. There are nine Chemical Response Coordinators in all including myself, and each of us has a terminal at home so we can receive incident reports 24-hours a day.''

"During a major incident you have a lot of people both within the company and outside the company who want to know what is going on, so we divide the various chores among ourselves," notes Dick Way. "One man might be getting the team selected and on the road; another might be handling notifications to management; a third may be maintaining contact with the incident scene, the railroad, or a fire department. A fourth person is really the guy who coordinates everything to insure that it works. That is the theory behind the command center operation. We don't use it too often, and I thank management for allowing us to keep it available. Nowadays, an unused office is quite a luxury. We believe having the command center available at all times is a necessity, and it has definitely proven itself in the past.''

"On the west coast we have three refineries, two of which are called 'manufacturing complexes' where we produce *both* refined oil products and chemicals, so along the Pacific coast we have five response teams plus a pesticide team at a research laboratory in Modesto, California," adds Way. "Denver has a manufacturing location for pesticides, and the Wood River facility near St. Louis, Missouri is another 'complex' manufacturing both chemicals and oil products. Down on the Gulf coast we have four manufacturing locations and a research laboratory: Deer Park manufacturing complex and our Westhollow research center in the Houston area; Norco complex in the New Orleans area, Geismar near Baton Rouge, and a relatively small but growing pesticide manufacturing plant in Mobile, Alabama. In the Ohio area, we have a polymers plant near Marietta. El Paso, Illinois is the site of our pesticide formulation and packaging facility. There is a polymers plant at Woodbury, New Jersey; plus product distribution plants at Porland, Oregon; Chicago, Illinois; and Sewaren, New Jersey. We do not have any trouble deciding which team goes if there is an incident in, say, Kansas. Part of the Coordinator's job is deciding what he needs and how and where he obtains the people best qualified to respond to a specific incident. A company plane is almost always available during nighttime hours. If we have to get personnel to an incident in a hurry, a corporate aircraft will often be used for transport; although we use commercial airlines in some instances because the corporate planes are based in Houston and it is more expedient to get from Denver to Kansas City, say, by commercial airline. A response team will normally fly into a major airport that is fairly close to the scene of an incident. From there they may have an automobile trip, or rent a helicopter, although we find that we can normally get within driving distance quite easily.''

"The performance of manufacturing plants is generally determined by their ability to maintain their costs of production at the lowest level in cents-per-pound or cents-per-unit," explains Way. "We determined early-on that we would transfer all the charges for incident response to a head office account, my account. I budget somewhere on the order of $60,000 to $70,000 a year for this activity. I do not accumulate the salaries of the actual responders, however; they are all staff people rather than hourly employees. The extra time they spend on response is not an additional cost for Shell, so we do not count salaries against response but rather just the out-of-pocket costs. We consider the overall expense for response quite economical considering the size of our operation, the hazards associated with many of the products we ship, and the number of incidents we respond to.''

As for equipment carried by Shell Chemical Response Action Teams, Dick Way notes: "Soon after the formal response program was initiated, there was a derailment in upstate New York in February of 1976. The azaleas were blooming here in Houston when the response team boarded a plane; when they landed in New York state, it was 20 degrees below zero and they spent the first day looking for warm clothes. Later, we sat down with all of our people for a discussion as to what kinds of tools/equipment/materials our people should have. We identified three different types of equipment to keep on-hand ready to go. First, we devised a list of equipment to be supplied to each team member: personal equipment, summer and winter clothing, rain gear, boots, gloves, and other items to protect a responder from both the product and the elements. We also came up with a list of equipment that is packed in bags ready to go; team-type equipment such as breathing apparatus, an explosimeter that is a must tool for us, a tape recorder, barrier tapes, cameras, clipboards, and similar items. Finally, we asked each of the teams to provide themselves with the special equipment they would need to handle their own particular products; such as pressure guages, epoxy patching kits, gaskets and the like. If you manufacture pesticides that are packaged in 55-gallon drums, the type of equipment you carry is quite different than if you are a team member from a location that ships principally by tank car. Also, if your plant ships mainly by general purpose tank car, your equipment needs are quite different from the pressure gauges and pressure tubing required by a team from a facility that ships propylene and vinyl chloride in pressure tank cars. Each team brings together those things they feel they will need at an emergency that are not likely to be readily available at the site. Such equipment must be easily transportable; our chemical teams do not generally have response trailers or trucks that could be used to transport extensive gear since they respond principally as technical advisors. Our oil products response teams, on the other hand, do use equipment trailers since they are oriented more to hands-on containment and clean-up and often work on Shell-owned vehicles and containers.''

Shell Chemical response teams use Material Safety Data Sheets as their primary source of information on specific products. They also tend to use the National Fire Protection Association's *Fire Protection Guide On Hazardous Materials,* the Department of Transportation's *Emergency Response Guidebook,* and the Bureau of Explosives/Association of American Railroads *Emergency Handling of Hazardous Materials In Surface Transportation.* Individual teams also normally have at least one of a number of varied toxicology textbooks or handbooks so that particularly at railroad incidents they will be able to determine toxicity levels and reactivity potential of non-Shell chemicals that might be involved. A critical reference source for each team is the *STERP/STERN Handbook* (Shell Transportation Emergency Reporting Procedure/Shell Toxicology Emergency Response Network) with its names and telephone numbers of Shell experts and physicians available 24-hours a day to provide technical assistance to the teams.

Response personnel tend to be longtime Shell employees who are completely familiar with the tasks they are assigned as response team members since they perform these same tasks in company plants and offices on a normal, day-to-day basis. For pure response aspects of team operations. Shell stages a national meeting one year attended by one or two members from every team who are brought up-to-date on response techniques and progress of the overall program. Alternate years, as many as six regional meetings are held around the country so

EMERGENCY RESPONSE EQUIPMENT

Personal	Team	Special Team
Coveralls, Summer & Winter	Breathing Apparatus	Tool Kit
Hard Hat & Liner	Explosimeter	Epoxy Patching Kit
Face Shield	Cameras	Thermometers
Goggles, Splashproof	Tape Recorder	Sampler
Sweat Shirt	First Aid Kit	Sample Containers
Rain Suit	Binoculars	Gaskets
Gloves, Leather &	Flashlight (6 V.)	Pressure Guages
Impervious	Reflective Tape	
Flashlight	Pennant Tape (100 Yds)	
Rubber Boots or Galoshes	Clipboard, Lighted	
Ron Kit	Ear Plugs	
Travel Bag	Cyalume® Emerg. Lights	
	Travel Bag	

Illustration 3.

more team members can attend. Each team comes prepared to discuss a particular incident so other teams can learn from the experience. As an additional form of training, Incident Reports submitted by each team after every incident are circulated to all teams. Thus, all response team personnel are able to review incidents handled by other Shell responders. Incident Reports have a minimum, standard format, but teams are able to add additional pages and attachments in order to cover whatever they think was important in handling a specific incident. These reports are widely distributed and read within the company.

"We are a pesticide producer," says Way. "I'm not sure where we stand in the overall market, but we have three plants producing pesticides. Practically all our pesticides are shipped in containers of five gallons or less, and drums are the containers involved in nearly half the 150 to 160 incidents a year that we cover. The other major type of container involvement is tank cars. There are very few tank truck incidents involving Shell chemical products. Railroad emergency response people tell us that most of the tank car incidents they encounter are caused by the shipper; loading operations may not have been done as well as they should have been; the bottom outlet cap may not have been put on with enough leverage; or someone may have failed to replace a gasket. The best therapy we know of it this should happen at Shell is to send the guys who were responsible for filling the car out to fix the car."

According to Way, to some extent there is a common thread to the mechanical causes of incidents. "To give you an idea of a typical problem, on pressure tank car equipment more often than not the car is stopped in a railyard where someone hears a whistling of escaping gas. Sometimes, and this is more serious, there's an iceball formed around a fitting where liquefied gas is escaping, evaporating, and chilling causing ice to form. Most often, however, the first indication of concern is the hissing of gas escaping from a packing gland, or from a safety relief valve that has not reseated properly after it has relieved due to impacting of cars connecting or jolting at rolling switches or for other reasons. With general purpose tank cars in which we are shipping liquids, a bottom outlet cap may be found to be leaking and someone discovers it at a switchyard or when the car goes through a periodic inspection enroute. These leaks are generally quite minor. Another type of incident involves packages. It is quite common to have a truck pull into a loading dock where the carrier plans to consolidate freight and find that some piece of machinery has tipped over and knocked five-gallon containers off their pallet causing one or two to leak. These are not major incidents, but they can cause damage to people if carriers don't ask for help and advice. Then there are the incidents you see featured in the press where a truck trailer tips over and spills its contents, possibly causing personal injury or property damage. There is really no pattern of commonality to such incidents. They can occur out in the boondocks and not be a very significant public event, or in the worst possible places such as right downtown. They are difficult if not impossible to predict."

An "Oil & Chemical Spill Incident Report" is filed every time a Response Action Group member is notified of an incident and every time a Response Action Team is dispatched to an incident. These reports are used to evaluate the effectiveness of response actions, as "case histories" for training purposes, and to identify possible corrective methods and procedures. They provide a good indication of the variety of incidents responded to, and the problems Shell personnel may have to deal with on-site.

* * * * *

On January 18, 1978 at 12:10 P.M. CHEMTREC notified Shell of a derailment on the Western Railroad of Alabama at Milstead about 20 miles east of Montgomery that had occurred 90 minutes before and involved two Shell tank cars. Tank car SCMX-2513 and SCMX-2662 were reported as wrecked and having discharged their contents of 25,000 gallons of acetone (DMK) and 25,000 gallons of methyl ethyl ketone (MEK) respectively. A caustic soda car containing product of another manufacturer was also breached and its contents were mixing with the Shell solvents. A butyl acrylate car was also in the wreck but was not leaking. Because of the magnitude of the spill, and the potential for fire and/or environmental damage, the decision was made to dispatch a Shell Response Action Team from the company's Norco manufacturing complex in Louisiana where the Shell tank cars had originated.

By 1:25 P.M. R.J. Richard and R.C. Fernandez had assembled their personal and emergency gear at Norco and were standing by. Since no corporate aircraft was available, arrangements were made for them to leave by commuter airline at 3:50 P.M. Meanwhile, other Shell personnel arranged airline bookings and obtained tickets, reserved a rental car at the Montgomery airport, obtained detailed directions for driving from Montgomery to the derailment site, identified the railroad employee in charge of the scene, identified a supplier of vacuum trucks operating in the Mobile area, and arranged for a Shell public affairs representative to meet Richard and Fernandez at the incident scene.

Richard and Fernandez landed in Montgomery at 5:40 P.M. and arrived on-site at 6:45 P.M. As they drove up, the odor of methyl ethyl ketone was detected some 100 yards from the accident scene. The acetone car, SCMX-2513, was split open for two-thirds the circumference of the car and wedged into a "V" shape by a flatcar loaded with lumber. SCMX-2662 with a three-foot diameter hole in one end had lost its load of methyl ethyl ketone except for 200 gallons or so. UTLX-43194, a caustic soda car next to the MEK car, was leaking from around the saddle on the bottom side of the car. There appeared to be no flammable liquid south of the point of impact. A dam had been hastily constructed about one-half mile north of the road crossing/point of impact. It appeared most of the liquid had seeped into the ground just under the two solvent cars and the caustic car near where a deep hole had been made at the point of impact. Diesel fuel from the overturned locomotive and a truck that had been sitting astride the crossing acted as a blanket over the spilled solvents.

Richard and Fernandez then looked at the remainder of the derailed and overturned cars and found the lead two locomotives upright but derailed. The third locomotive was lying on its side west of the main line. Beyond the overturned locomotive were three overturned boxcars jammed together. Two piggyback truck trailers loaded with fireworks lay in the general area of tank car NATX-21138 loaded with butyl acrylate. The derailment had been caused when a truck towing a lowboy trailer carrying a large piece of earth-moving equipment became stuck on the rail crossing. The truck driver was attempting to jack the lowboy trailer off the tracks when the train appeared. Tractor, trailer, locomotives, and cars were thrown in all directions.

Although the solvents released by tank cars SCMX-2513 (acetone) and SCMX-2662 (MEK) are both highly flammable, there had been no fire and no injuries. Richard and Fernandez guessed the solvent/air mixture was too rich to ignite when sufficient heat had been present. They made their first report to headquarters about 9:30 P.M. and requested data on possible reaction of butyl acrylate with caustic should the acrylate car develop a leak. At this time, the team's primary concern was to remove liquid solvent from the area, or cover it with sand. Also, team members recognized the empty solvent cars would require safe explosive limits prior to moving. The railroad was unable to secure a vacuum truck, so plans were made to cover the area with sand. There was no evidence any product had reached the Tallapoosa River located one-and-a-half miles away. By 10:45 P.M. the team had received preliminary word that there was no apparent danger if butyl acrylate and caustic mixed, but that a contact would be made with the manufacturer and the preliminary judgement confirmed in the morning. They cautioned the railroad to do no pulling of damaged cars until they okayed it as they were concerned that interlocking and scrapping of metal might cause a spark that could set the whole area ablaze. They again suggested the railroad cover the area with sand as much as possible, and by 11:00 P.M. were able to leave for dinner.

When Richard and Fernandez returned at seven the next morning, they discovered that someone had moved the tractor and lowboy trailer. Railroad workers had attached a sling to the lumber car and were bringing up a caterpillar tractor to jerk it loose. They stopped this activity and again explained what could happen from just one spark. Then they rechecked the two solvent cars, confirming that the acetone (DMK) car was empty and the MEK car held approximately 200 gallons. They purchased a six foot length of ¾ inch rubber hose to siphon the liquid from the MEK car in order to proceed with decontamination of the car. They opened the domes on both solvent cars to create a draft. Siphoning of the MEK continued at a slow rate. When all liquid was removed by 3:00 P.M., both cars were flushed with 1,000 gallons of water provided by the Tallassee Volunteer Fire Department. A suction blower was obtained from the Montgomery Fire Department and installed on the dome of the MEK car. Prior to installation of the blower, an explosimeter check indicated 50 percent on the DMK car and 100 percent on the MEK car. Richard arranged with the railroad supervisor to have someone maintain the blower operation during the night. By 9:00 P.M. explosimeter readings showed 20 percent for the DMK car and 60 percent for the MEK car.

Richard and Fernandez returned to the site at 6:00 A.M. on the 20th and again used an explosimeter to check both solvent cars; readings for both were zero percent. Sand from a nearby gravel pit had been spread over all liquid in the area. The railroad was advised that pulling apart of the wreckage could now begin. There was a bad moment when in attempting to rerail the lumber car, the rear trucks were jammed shearing the pin and creating a large spark, but no ignition occurred. The butyl acrylate car was positioned parallel to the main line for offloading into tank trucks. The caustic soda car, completely empty by this time, was pulled to the west side of the tracks as were the two solvent cars. Richard and Fernandez departed the scene at 1:30 P.M. on January 20th, approximately 43 hours after their arrival.

*   *   *   *   *

Shortly after 2:00 A.M. on August 6, 1978 Dave Williams, a field inspector for the Bureau of Explosives of the Association of American Railroads, called Shell's Wood River refinery near St. Louis, Missouri to report that six Shell tank cars loaded with propane had derailed on the Illinois Terminal Railroad near Collinsville, Illinois. A Response Action Team of C.R. Woodford and A.H. Caldieraro was alerted at 2:15 A.M. and arrived at the derailment site by 3:30 A.M. They found six loaded 33,000 gallon propane tank cars off the rails. GATX-97275 was lying on its side burning from a 60 to 24 inch puncture. The area had been evacuated for a quarter mile radius, roadblocks installed, and fire hoses were being activated by Maryville and Collinsville Fire Departments. SOEX-3368, an insulated car lying immediately south of the burning car, had a 1½ inch fire hose attached to its undercarriage. WRNX-30107, just to the north of the burning car, had a two-inch monitor nozzle positioned to the rear. Both lines were activated to lay a water blanket between the fire and adjacent cars to maintain desired cooling and keep the tank cars from venting. Using a combustible gas indicator, Woodford and Caldieraro checked each remaining car for leakage in the dome areas, but all appeared to be sound. All personnel, with the exception of a fire watcher, were evacuated and the contents of GATX-97275 were allowed to burn off. There was some concern over two additional tank cars; both ends of SOEX-3177 were dented approximately eight inches. SOEX-3368, a DOT-105A jacketed pressure car, had its insulation jacket peeled back in one area, but the limited amount of flame that had impinged on the car hit the insulated area.

The derailment had occurred in a deep valley a few hundred yards from Interstates 55-70. Liquid propane escaping from the ruptured car turned to gas as it hit the night air. Heavier-than-air gas/vapor traveled approximately one-half mile along the right-of-way until it reached a source of ignition, then flashed back to the ruptured car which erupted in a large fireball. One man sitting in his pick-up truck at a rail crossing waiting for the freight train to pass got out of his vehicle and ran. He received 2nd and 3rd degree burns. A woman sitting in a passenger car on the opposite side of the tracks remained in her car and was later treated and released. In spite of the size of the fireball, there were no other injuries.

By Monday morning, August 7th, the entire cargo of GATX-97275 had burned off and the car would be allowed to cool overnight before starting wreckage clearing operations.

By Tuesday morning, there were no explosive vapors in evidence and Hulcher Emergency Services, a clean-up contractor brought in by the railroad, began to rerail undamaged cars. Two undamaged propane cars were rerailed, then the burned out car, GATX-97275, was moved to an isolated area in early afternoon and flooded with water. The railroad installed six sections of new track by 6:00 P.M. and three empty 33,000 gallon cars were spotted adjacent to wrecked and loaded cars. Hulcher crews transferred product from the first car between 7 and 11:00 P.M., then ceased operations until daylight. In the mornng, two more damaged cars were emptied of product between 7:00 A.M. and 3:15 P.M. The area evacuation was then lifted, and Hulcher personnel reentered the area to rerail the remaining cars. By 4:00 P.M., Woodford and Caldieraro were on their way back to Wood River.

*   *   *   *   *

W.D. Kollmeyer, an organic chemist at Shell Development Company's Biological Sciences Research Center in Modesto, California was awakened at 6:20 A.M. on Sunday, August 3, 1980 by a telephone call from Shell headquarters in Houston, Texas. Notification had been received through the Chemical Transportation Emergency Center (CHEMTREC) in Washington, D.C. that a fire had occurred in a trailer at a truck terminal in Fresno. Sixty gallons of Shell's VAPONITE 2 EC Insecticide packed four one-gallon jugs to the case were involved. Kollmeyer immediately called the Fresno Fire Department dispatcher to verify the basic information, obtain directions to the truck terminal, and inform fire officials that he would be there in three hours. Drawing his trousers on, he called his own plant in Modesto and arranged with a security guard and boiler attendant to obtain a company car. By 8:00 A.M. Kollmeyer had picked up the company car; loaded emergency gear, literature, and a self-contained breathing apparatus; and headed south for Fresno. Upon arrival at the trucking terminal, Kollmeyer was met by the terminal assistant manager and a Fresno Fire Department chief officer. A truck trailer had burned, possibly due to arson; the fire had been extinguished with water, and an undetermined number of one-gallon VAPONITE 2 EC insecticide containers had been broken. Large volumes of water and emulsion had run off onto the asphalt yard and then into the gutter on a nearby street. The fire department had blocked off the street, and Kollmeyer warned all personnel to avoid contact with the emulsion and not to breath the vapors. The Assistant Fire Chief reported that no one had suffered any ill effects, and Kollmeyer provided him with a copy of the Materials Safety Data Sheet for VAPONITE 2 EC.

VAPONITE remained on the rear of the trailer amid burned rubble, and a number of broken bottles were visible. The aluminum siding and top of the back half of the trailer were largely gone, having either melted or burned during the fire. It appeared that essentially all the freight on board had suffered some sort of damage either from direct fire, smoke and soot, or water. The carrier had already contacted IT Corporation to handle clean-up and decontamination. An IT representative from Bakersfield and a vacuum truck and work crew from Martinez were enroute to Fresno.

Going through the manifest, Kollmeyer noted a shipping order for 23 cans of insecticide manufactured by another company. A red five gallon can that had been pulled off the rear of the truck by the fire department was examined and found to contain a pyrethroid insecticide. Kollmeyer called Houston, and it was agreed that he should alert the second chemical company via CHEMTREC. CHEMTREC agreed to make the notification.

VAPONITE is an organic phosphate insecticide-acaricide effective as a fumigant, stomach and contact poison; it has an LD-50 (Lethal Dose-50 Percent) of 55 mg/kg, and can be absorbed through the skin. It is used on livestock for flea, tick, and mite control and on edible crops for spiders, wasps, and flies. It has an effective residual action of two to three weeks.

The non-Shell product is a synthetic pyrethroid compound used as a contact and stomach poison with an LD-50 of 1030 mg/kg. It is used to control a variety of insects such as weevils, aphids, and corn borers.

Early in the afternoon, a local health representative arrived at the terminal to obtain background information and assurances that everything was being handled properly. By 2:30 P.M. IT Corporation's crew had arrived from Martinez, as had a representative of the company that manufactured the synthetic pyrethroid compound. The trucking terminal manager arrived and asked what items might be salvaged. Because an undetermined number of VAPONITE bottles had broken and then discharged or burned in the fire, and owing to the extensive fire and smoke damage throughout the trailer, Kollmeyer estimated the entire cargo had suffered vapor contamination at the very least. Any item made of wood, fiber, cloth, leather or other absorptive material was contaminated and would have to be discarded. According to the shipping papers, diverse items such as clothing, shoes, plates, and stereo equipment were on board. The only items that might be salvaged were two large metal objects, an airplane engine and a large safety cab for a caterpillar tractor.

IT personnel agreed with this assessment that except for the airplane engine and safety cab, the entire cargo would be appropriately discarded when IT began clean-up and decontamination. The trailer would also have to be discarded since about half the aluminum siding and top was gone, the main structural frame had sagged drastically at the midpoint, and the rear tires were scorched. The carrier's terminal manager agreed that the trailer and its cargo appeared to be a near-total loss.

The IT foreman was able to secure locally a very large dumpster into which the trailer contents would be loaded, and made arrangements to take all the contaminated cargo to a privately owned Class A landfill in Western Fresno County that could handle toxic wastes.

After Kollmeyer conferred with all parties, the terminal was closed for the night with a private security guard and a fire department officer remaining all night. The terminal manager agreed to be present at four the next morning when two carrier employees would arrive for the morning shift in order to warn them to keep out of the contaminated zone. Kollmeyer then departed for his home in Modesto, arriving there at 9:00 P.M.

The next day, August 4th, a Monday, Kollmeyer was back at the terminal in Fresno by 9:00 A.M. IT personnel were working on clean-up and decontamination of the trailer, yard, and gutter. They erected small, earthen dams in the gutter so run-off water and chemicals could be sucked up by the vacuum truck, then used caustic to neutralize the pesticide.

A Deputy Chief of the Fresno Fire Department requested a letter from Kollmeyer on proper treatment or disposal of firehose which had been contaminated during the firefighting effort. This Kollmeyer later provided. The emergency services coordinator for Fresno County wanted to know if any extra hazard might arise from a mixture of Shell's VAPONITE and the pyrethroid compound compared to either product alone; that is, synergistic toxicity. Kollmeyer replied that he didn't know. By then, representatives from the California Department of Health Services, California Regional Water Quality Board, and the Fresno Metropolitan Flood Control District had arrived on-scene. Kollmeyer reviewed the situation with them. In his estimation, very little if any of the waste water from the fire had gotten into the public drain. He noted that small amounts of VAPONITE would eventually degrade harmlessly in water, and that he would appraise them of the maximum of pesticide lost by taking inventory of what was recovered from the fire debris. They agreed to determine where the public drain discharged.

At 3:00 P.M. the IT crew began removing the remaining VAPONITE from the rear of the trailer wearing protective clothing and air respirators, or full-face masks with organic vapor cannisters.

At 3:45 P.M., the Fresno Fire Chief called to say that a firefighter who had been on the scene yesterday but not directly involved in the firefighting had checked in at a hospital with symptoms of headache and nausea. Kollmeyer read the chief symptoms of overexposure as contained in the Materials Safety Data Sheet for VAPONITE: headache, constricted pupils, blurred vision, weakness or tightness in chest, nausea, abdominal cramps, diarrhea, excessive sweating and salivation. He provided office and home telephone numbers of a Shell Physician assigned to the Shell Toxicology Emergency Response Network, (STERN) and asked the chief to alert him immediately if the Shell physician could not be reached. Kollmeyer checked back with the fire chief two hours later, but there was no further information on whether the firefighter had been intoxicated by the pesticide or was suffering from an unrelated ailment.

IT workers recovered thirty one-gallon containers of VAPONITE from the burned out trailer that were either totally intact or broken on top but still contained two-thirds or more of their original contents. Thus, approximately 30 gallons of the original 60 gallons were recovered. The recovered VAPONITE was poured into a 5-10 percent sodium carbonate solution in a 55-gallon drum which was periodically checked with Ph paper and then pumped into IT Corporation's vacuum truck. Contaminated cartons, general cargo, broken glass, and empty containers were placed in a large metal dumpster for transfer to a Class A landfill where the vacuum truck would also unload its contents. By the end of Monday, Shell's VAPONITE had been removed along with five to ten percent of the trailer's contents.

On Tuesday, Kollmeyer again called the fire chief to check on the condition of the firefighter who had checked in to a hospital, but was unable to get an answer at either of the phone numbers he had been given. By phone, he discussed with various persons at the truck terminal his opinion that loose asphalt immediately adjacent to the trailer should be removed wherever it was obviously weakened by chemical contact. There was some concern over how much chemical product might have entered a public drain. Kollmeyer stated that from his observation of the spill and drain, he felt a negligible amount of run-off water containing some emulsified chemicals might have entered the drain. It had been determined that the drain entered into a "box" and when this box filled, as in the rainy season, the overflow went to the municipal sewer system. Kollmeyer noted that small amounts of VAPONITE in water would degrade harmlessly before the rainy season arrived in late fall.

By Tuesday morning, August 6th, all the trailer cargo had gone into the dumpster, and final decontamination of the trailer remains was scheduled to begin. Contaminated asphalt surfaces had been washed with sodium carbonate solution, but definite stains were still evident. The asphalt adjacent to the northeast corner of the trailer was very loose and crumbling, and it was arranged to suck this material into the vacuum truck waste tank. Kollmeyer informed the terminal manager that it was Shell's position that all contaminated asphalt should be removed, and offered to provide an official letter to this effect. Kollmeyer was then able to return to his normal duties with Shell Development Company's Biological Sciences Research Center.

Eventually an inspector for the State of California Regional Water Quality Control Board reported that the clean-up and decontamination procedures were sufficient to mitigate any short or longterm water quality problems, and found the small quantity of run-off that entered the city's storm drain system to be inconsequential insofar as water quality matters are concerned.

*   *   *   *   *

At 1:24 P.M. on July 27, 1981 Shell was notified by CHEMTREC that tank car GATX-16289 carrying phenol and shipped from Shell Chemical in Houston, Texas to a consignee in Louisiana was sitting in the Avondale, Louisiana yards of the Missouri Pacific Railroad. The trainmaster was concerned that product appeared to be sloshing from the dome area whenever the car was moved.

A Response Action Team from Shell's Norco manufacturing complex in New Orleans consisting of W.E. Torres and A.E. Naquin was alerted at 2:00 P.M. and departed the plant at 2:30 P.M. arriving on-site an hour and fifteen minutes later. Torres and Naquin met with the trainmaster who advised them that GATX-16289 had been isolated on a sliding one-quarter mile north of the station. Railroad personnel had reported that product was splashing out of the dome each time the tank car was moved and that the dome was not completely closed.

Torres and Naquin took several photographs of the dome area and noticed that vapors were escaping via a one-inch gap between the dome cover and the dome. The cover was cocked to one side due to a one-inch spacer on one of the front dome bolts under the dome handle. Of the six dome bolts, two were loose, and one was off completely. The two loose bolts were made of stainless steel and had gaulled threads that caused problems for Torres and Naquin as they worked to open the dome. They opened the dome, measured the outage of commodity within the tank, checked that the dome gasket was in good condition, reclosed the dome lining up the spacers so the cover would fit flush on the dome, and tightened all bolts. Two bolts had to be hammered in order to tighten them down, and the team made a note in their report that these bolts would have to be replaced before the car was shipped again. They sealed the dome with a numbered seal, and then checked the top unloading assembly. The plug in the two-inch ball valve was less than hand-tight, and the valve was cocked slightly open. They also noted in their report that the handle on the valve was broken and should be replaced. They installed the two-inch valve plug with a wrench and checked that the one-inch valve plug was tight, then sealed the valve assembly with another numbered seal.

In their report, Torres and Naquin noted that it had been raining in the area, and there was no evidence of product on the outside of the car. Dogs and nuts which had caused problems were made of stainless steel which increases chances of gaulling. The two inch valve on the deepline had been partially open, and the plug was only hand-tight. Also, closing the dome cover tended to tip the valve handle open. The car had a screwed valve which can be loosened when disengaging unloading lines. They also noted that car loaders should be more diligent in their inspection, and in securing and sealing of closures and valves. Returning to the railyard depot, Torres and Naquin advised the trainmaster that the car could be released for delivery to the consignee, and returned to the Norco plant by 6:30 P.M.

Shell has taken a number of steps over the years to respond to a recognition of the real-world environment that shipping containers have to endure. They have gone to heavier small-package containers than are required in an effort to provide greater puncture resistance. A number of products have been upgraded into tank cars with more steel in the jacket in an effort to reduce damage if the car should be derailed. Approximately half the tank cars used by Shell are owned by Shell; the remainder are leased. Shell applies the same requirements to leased cars as to proprietary equipment and will not lease a car that is not up to Shell standards.

Shell has a reputation for providing assistance and training to local emergency response organizations including fire departments. Although the company does not broadcast its policy in such matters, it is known to be quite liberal. "We do feel that it would be unreasonable for Shell to turn its back on an incident," admits Dick Way. "We will do whatever we can to assist. We have made a number of calls recently on products that were not ours simply because the local authorities knew we had a capability and asked us to help them. We have worked very closely with a number of fire departments, and have individuals on our staff who are themselves volunteer firefighters and have been very active in providing training and in setting up exercises to simulate incidents. Locally, we have been very close to the Hazardous Materials Response Team of the Houston Fire Department. They have a very good team, and working closely with them has been a mutually rewarding experience. We encourage any industrial organization in a specific locale to get into it with their local fire department. I used to feel there was a real hazard in having firefighters who knew little about chemicals and had very few incidents to train on do anything other than try to control an incident until experts could arrive. Nowadays, however, a number of paid departments in major cities that have sufficient manpower and equipment to handle a major hazardous materials incident receive a good amount of special training. Once that happens, they are probably as capable as most industrial response teams, and all they need is technical advice and assurances that they correctly understand a given situation. The Houston Fire Department folks don't pretend they know everything, and they do call us whenever there is an incident involving Shell products, and sometimes when the incident does not involve Shell products."

The final session of a recent three-day workshop put on by the Houston Fire Department's Hazardous Materials Response Team, (HMRT) for example, was a combined exercise for HMRT and Shell Chemical Company's Response Action Teams to determine future training needs. A simulated "emergency" featured a leaking tank car of epichlorohydrin on a railroad siding with a man down. An engine company, unaware of the exercise, was called to the Houston Fire Academy and told to handle the emergency, as would occur when the closest engine and HMRT are dispatched to an actual incident. HMRT and an ambulance arrived at time intervals approximating arrival times at actual emergencies. As arranged beforehand, HMRT called CHEMTREC; CHEMTREC contacted Shell, and a Shell team was brought to the scene. Shell activated its national command center and had its national hazardous materials team handle communications, had two chemists on the scene with field analytical equipment, and established communications between the paramedic on-site and Shell physicians in downtown Houston.

"Some time ago, the Houston Fire Department HMRT was called out to an incident where a refrigerated van had been sitting locked for two or three days with its refrigeration unit running," remembers Dick Way. "Someone had noticed the refrigeration unit was no longer running, got concerned, and called the fire department. HMRT personnel discovered some 60 to 70 full unlabeled drums inside the truck; there was not a label on any of the drums, yet they were in a refrigerated, reefer-type van. They cautiously opened one of the drums and found it loaded with a rather gooey material with a greenish cast to it. They took a very small sample and put it on the ground to see if they could ignite it, but couldn't get it to do much. They attempted to get hold of any organization that could provide some laboratory testing, but as it was late on a Friday night they were unable to reach anyone. Eventually, they called Shell to learn if we could help. We knew it was not our material from the description of the containers and contents, but I contacted one of our research people who might be able to provide some guidance, and he volunteered to go to the scene. It was his opinion that the product was not reactive, and he recommended just isolating the trailer and letting it sit until we could get more information. In the meantime, EPA appeared on the scene and took a sample. It was a rented truck and had been reported stolen by the rental agency. The next morning, while we were all trying to figure out what we had on our hands, someone called from New York City after apparently hearing a news story and said—'Hey, that's mine; it's concentrated lime juice.' Although the incident turned out to have an amusing ending, it did represent the seriousness involved with an unidentified, possibly dangerous material. You almost have to treat such a situation as the worst thing that could occur until you find out exactly what you are dealing with."

As to what has been learned through the operation of Shell's Response Action Teams, Dick Way stresses that incidents generally are not nearly as serious as they first appear to be. "First reports are almost always inaccurate," he says. "That fact alone is dangerous because long experience can lull you into a false sense of security. We make every effort to get someone on-scene as rapidly as possible who is capable of evaluating the seriousness of the situation. Also, we have found there has been a great improvement in the ability of carrier personnel, warehousemen, and emergency responders to handle minor problems. If we can convince ourselves over the phone that an incident is minor and people onscene have the qualifications to handle it; we will give them a go-ahead, get a report back, and then stand-by just in case. I guess that is the major change we have undergone; we probably don't respond to as high a percentage of incidents as we did in the past."

"The reason we got into emergency response is that we feel it is part of being a good corporate citizen, that you step forward and do what you can to mitigate an incident," adds Way. "There was a time within the industry when it was common to have a 'No Comment' for accidents that occurred; that time has long since passed. We feel the public has a right to know; the news media have a right to inquire. We have a very open attitude about letting people know what has happened, what we are doing to help correct the situation, and what we are going to do in the future to help prevent simular situations. No one likes to admit mistakes or the failure of materials that they recommended; but incidents do happen occasionally. Truck drivers are sometimes required to take evasive action in traffic that can cause a trailer to jackknife, possibly turning over the trailer and causing a hazardous material to leak. Then, of course, the public becomes incensed because they are exposed to these things. In reality, it may not have been the driver's fault. In general, our attitude is to be open, to face up to whatever responsibilities we may have, and not try to hide a thing."

Richard Way serves as Chairman of an industrial group formed some years ago to advise the Chemical Manufacturers Association on how to make CHEMTREC work better. "We did a study in 1976 of CHEMTREC's first five years of operation and tried to visualize what CHEMTREC should be doing for the next five years," he says. "We quickly discovered two very startling facts: first, not every fire department or emergency service group in the country knows about CHEMTREC . . . which is a disheartening fact in itself. We obtained estimates from the firefighters' trade association that indicated perhaps half of all fire departments don't know about CHEMTREC. The second thing we learned was that not all chemical companies had emergency response capabilities. Since then we have concentrated on these two areas within the CHEMTREC advisory group. We have conducted workshops for chemical company members on how to organize effective emergency response programs; we have conducted workshops on what response teams might expect, what equipment is available to them, and what techniques they should employ when responding to an incident. These efforts are continuing. We are publicizing CHEMTREC to the public the best way we know how. We have mailed brochures on CHEMTREC to every fire department and every police agency in the United States, and we are getting ready to do it again . . . urging them to use CHEMTREC as the fastest means to get an expert at their elbow. We feel strongly that every carrier, every shipper, needs the ability to have a contact on a 24-hour basis to provide assistance to anybody who is having trouble with a chemical product.

Also, I would like to see more recognition of the dedication of industrial responders who are willing to stake their spare time and expose themselves to a bit more danger than would be the case if they remained in their normal work environment. We have hardly any way to adequately compensate such people for the job they are doing. They definitely need recognition. I therefore would like to offer all kinds of accolades to the emergency responders who actually make it on the scene.''

T.E.A.M.: New England Nuclear

New England Nuclear, one of the world's leading manufacturers of radioactive chemicals for research and radiopharmaceuticals for medical diagnosis, has a reputation for effective organizational management and a record of committed involvement in efforts directed toward the safe transportation of hazardous materials, particularly radioactive materials.

New England Nuclear, acquired by Dupont in April of 1981, has an outstanding growth record, having increased its sales tenfold in the past ten years. NEN reported sales in the fiscal year ending February 28, 1981 of $94.2 million, 15 percent above fiscal 1980; net income of $11 million was 20 percent ahead of the amount earned the previous fiscal year. NEN and its subsidiaries employ about 1600 people at manufacturing and research facilities in Boston and Billerica, Massachusetts and at sales offices in the U.S., Canada and Europe.

The same attention to detail, the ability to work cooperatively with a wide variety of interests, that has keyed such a sales performance is also evident in New England Nuclear's emergency response program known as "TEAM." (Transportation Emergency Assistance Members.)

"TEAM is a very carefully selected acronym that conveys the theme of New England Nuclear's reponse effort," says Ermes DeMaria, Compliance And Operations Officer, who is known to many hazardous materials professionals around the country for his effective work with the Hazardous Materials Advisory Council and his continuous missionary work in organizing, promoting, and participating in training seminars directed toward shippers/carriers/response personnel. "We are just one organization among a number in a response to a hazardous materials incident involving radioactive materials. That is, we are working WITH OTHERS, and it is important that we recognize the others."

TEAM is a public service that provides immediate, expert advice for those in a need of assistance with a problem or incident involving a New England Nuclear product during distribution. TEAM operates around the clock to receive calls from any point in the U.S.. In the event of any problem involving a New England Nuclear shipment, call 617-667-9538 and request "TEAM." The action word, "TEAM," alerts the telephone operator to transfer the call without delay to a member of the response team.

TEAM is New England Nuclear's Transportation Emergency Assistance plan. It is a public service that provides immediate expert advice for those in need of assistance with a problem or incident involving a New England Nuclear product during distribution.

TEAM stands for Transportation Emergency Assistance Members. It is composed of representatives from the Corporation's Environmental Control and Corporate Distribution Departments.

TEAM operates around the clock—24 hours a day, seven days a week—to receive calls from any point in the United States. In the event of any problem involving a New England Nuclear shipment, *CALL "TEAM" AT 617-667-9538.*

How Does TEAM Work?

The following statement appears on all shipping papers and shipper's certification:

> "For assistance in a transportation emergency and/or problem with this shipment, *CALL (617)667-9538,* day or night, and request *"TEAM."*

The following Corporate Distribution Information Symbol appears on all packages next to the address label—

**IN THE EVENT OF
TRANSPORTATION EMERGENCY:**

DIAL

AND REQUEST

*Transportation Emergency Assistance Member—
a 24-hour service provided by*
NEN **New England Nuclear** ®

The "action word" **TEAM** alerts the telephone operator to transfer the call without delay to a member of the **INITIAL RESPONSE** *TEAM.*

The **INITIAL RESPONSE MEMBER** records the detailed nature of the problem on a *"TEAM Response Report"*.

If there is concern for the protection of personnel or property, the **INITIAL RESPONSE MEMBER** will contact the **SUPPORT** *TEAM* for direction.

Each team member involved summarizes his action on a *"TEAM Response Report"*. All *"Response Reports"* will be submitted to the Corporate Distribution Compliance Office by the next work day.

Illustration 4.

"Many emergency response programs accumulate data; 'there was a problem; we corrected it; very good, let's get back to work.' " says Ermes DeMaria. "We attempt to go further, to use the data to improve performance. We operate off the assumption that every safety problem can be corrected by an improvement in the packaging, in the documentation, etc.. This is not always true; a plane crash could destroy my theory, but our assumption is that from the data collected we can improve the safety system. We maintain the data on file and analyze it periodically. If there is a particular type of package that got involved with wetness or deterioration of the outside surface, then it becomes immediately visible, and we can take corrective action. The system also provides data as to how many releases we have stemming from normal conditions of transport. All our packages, 99.99999 percent, are Type A packages—performance oriented to meet normal conditions of transport. They are not accident tested. Over the past three years we have shipped in excess of one million packages; there were two instances where any Type A quantities of radioactive material were released—two releases out of over one million shipments, so I know these two incidents by heart. The first occurred at Stapleton Airport in Denver when a package dropped off a baggage cart and was run over twice by a truck. The second resulted when a small cargo aircraft crashed near Columbus, Ohio."

"There have been only two releases out of a million-plus shipments, but 350 other incidents where someone had a need-to-know," emphasizes DeMaria. "A package might have become wet in the rain, or might have been dropped in a puddle and the obviously wet package was noticed by someone who came along later. Perhaps condensation has appeared around a package causing people to become concerned. Each of these situations 'suggested' a bigger problem, and somebody had a need-to-know . . . we MUST respond to that need-to-know. The TEAM operation works extremely closely with state and local people responsible for radiological problems within their local areas. Every TEAM member has a copy of *Directory of Personnel Responsible For Radiological Health Programs* (HEW Publication, FDA-78-8027) that is updated every six months. If there is a problem, say in Colorado, we call the person in Colorado responsible for radiological health programs. We don't try to handle the situation ourselves. We KNOW state and local people have an interest, and we work with them. Normally, the states provide the on-scene technicians while we provide them information as to what they are dealing with. As a sidenote, TEAM is in the CHEMTREC tubfile, and we have responded to incidents involving products from other shippers; this doesn't bother us."

Most of the calls to TEAM come from emergency responders; police, firefighters, civil defense people and similar first responders, DeMaria notes. Carriers make the second largest number of calls. "We often have to start a dialogue with the caller to obtain useful information about the package. 'Can you see a label?' (the hazardous materials label, not the address label, that is a primary communication device to tell what is in the package.) Usually the caller has not seen the label; he normally has to go back and take another look. The first level of information is provided by the label. A label tells us the product is fully regulated—it is not a small quantity. The color of the label is important; if it is all white, or yellow, we know what kind of radiation might be involved. Three lines would indicate the highest level of radiation. We ask for the 'contents' information from the label; how many curies. Also, the 'Shipper's Identification Number' is a unique number. If given that number, we can tell from the computer record exactly what this particular shipment consists of. The first part of a normal conversation is totally involved with attempting to learn what we are dealing with."

Although nearly all training programs directed toward hazardous materials responders we are aware of stress a need to obtain the shipping papers in any transportation-related incident, Ermes DeMaria had a four-word answer to our question as to how response personnel making calls to TEAM could be more helpful to themselves and to TEAM. "GET THE SHIPPING PAPERS," he emphasizes. "Of all the incident calls I have been involved with over two-and-a-half years, there has NOT been one instance where the caller had obtained the shipping papers. Responders do NOT get the shipping papers even though the information contained on them is crucial if we are to provide effective assistance. Emergency responders have to know how to clearly define the problem; get the shipping papers—this cannot be over-emphasized. The responder must know how to read the shipping paper, be aware of the type of information contained in it, and be able to relate this information to the shipper.''

Ermes DeMaria believes that both shippers and response personnel need to be aware of the effect their actions or lack of action can have on media coverage of an incident. ''If a reporter on-scene sees things being done efficiently, questions being asked and answers being given, he will report that the situation is under control, that a good job was done. Conversely, if a reporter gets run-arounds, if he sees a lot of confusion, if his questions are not answered; he will report that too. In any emergency response program, there are questions being asked and answers being given. There needs to be a feeling of credibility, an assurance that the job is being done right. This is critically important.''

For hazardous materials reponse personnel, the following information on New England Nuclear shipments is provided. During a recent calendar year, 36 percent of special nuclear material went by air on passenger-carrying aircraft; approximately 15 percent went by cargo or freight aircraft. Whatever mode the package may be shipped by, it eventually goes by highway. Trucks carry many of the shipments as does the U.S. Postal Service. On the other hand, United Parcel Service carries only a small amount of shipments. U.P.S. will handle only exempt quantities; (Limited Quantities) they will not take Yellow II or Yellow III because of possible exposure of undeveloped film.

When you have a transportation emergency involving a New England Nuclear product during distribution, call TEAM at 617-667-9538.

Rockwell International's Radiological Monitoring Team

Rockwell International operates the Rocky Flats facility near Denver, Colorado as a contractor for the U.S. Dept. of Energy. The main activity at Rocky Flats is the manufacture of plutonium and uranium components for nuclear explosives in support of the nuclear national defense program. These materials are owned by the federal government and are not shipped in commercial transport but rather are moved by U.S. Dept. of Energy vehicles and personnel. However, because Rockwell personnel through day-to-day monitoring of such materials on the plant site have developed and maintain an expertise in the monitoring of many different types of radioactivity in nearly any environment; the company as part of its contract has available a radiological monitoring team to respond in assistance of federal, state, or local government should a radiological incident occur within Colorado. That is, the operation of Rockwell International's Radiological Monitoring Team is somewhat independent of the products handled at the Rocky Flats facility in that the team has the capability to monitor any type of radioactive material that could be used or transported in commerce within the U.S.

As Director of Health, Safety and Environment at the Rocky Flats plant; Dr. Robert Yoder, a Health Physicist, provides advice, support, direction to the plant in all matters dealing with environmental health and protection, worker health and protection, and facility adequacy and safety. A number of response personnel, and at least one writer, have benefitted greatly from Yoder's clearcut, extremely informative demystification of radiation monitoring instrumentation during training at the Colorado Training Institute. ''Our team consists of approximately 25 individuals and is primarily prepared to deal with transportation accidents but could respond to other types of radiological emergencies were that required,'' notes Dr. Yoder. ''We have appropriate equipment to do radiological monitoring in essentially any environment we might find ourselves in within the State of Colorado; and are prepared with a trailer for clothing, accommodations, instrumentation, and electronic support for instrument repair so we are able to send a cadre of trained personnel as well as support necessary to keep them in the field. The team is prepared to depart the plant with one hour's notice anytime day-or-night when activated by the state health department, the Dept. of Energy, or any other authorized

Action Flow Chart

A call for *TEAM*

During NEN work hours

NEN switchboard transfers call to **INITIAL RESPONSE *TEAM***

After NEN work hours

Answering Service
1. Completes **CALL REPORT**
2. Contacts **INITIAL RESPONSE *TEAM***
3. Submits **CALL REPORT** to Compliance

An **INITIAL RESPONSE *TEAM*** member or a **SUPPORT *TEAM*** member contacts the caller

Office of Corporate Compliance

1. Records information
2. Contacts caller if **SUPPORT *TEAM*** not required
3. If required, contacts **SUPPORT *TEAM***
4. Submits **RESPONSE REPORT** to Compliance

1. Directs Response
2. Each member involved submits **RESPONSE REPORT** to Compliance

TEAM Organization

INITIAL RESPONSE *TEAM*

SUPPORT *TEAM*

COMPLIANCE OFFICE

ENVIRONMENTAL CONTROL

CORPORATE DISTRIBUTION

MANAGEMENT

Illustration 5.

agency that may need assistance. In the case of Rockwell shipments for which we have primary concern, we are able to respond within about 15 minutes because we know when the trucks leave the plant and where they are located. However, our main objective for the team is to be ready and capable to respond in support of local government agencies that have overall responsibility for a radiological incident. We do not take charge of the scene, or tell people what they have to do, but offer what advice we can with regard to our best judgement.''

TEAM PROCEDURE

TELEPHONE ANSWERING SERVICE:
1. When a call for *TEAM* is received, fill out a *TEAM* CALL REPORT.
2. Tell the caller that a Transportation Emergency Assistance Member will return his call and to remain at the telephone until the call is returned.
3. Call one of the *TEAM* members on duty:
   A. Record the time he was reached.
   B. Relay the T.C.R. number as well as all of the information on the CALL REPORT.
4. Place the CALL REPORT in the envelope provided and mail immediately.

ANSWERING SERVICE *TEAM* CALL REPORT

Date of Call _____ Time Rec'd _____ T.C.R. No _____

Name of Caller _____
Telephone No. _____ (Area) _____ Extension _____

Location of Caller:
Street _____
City _____ State _____
Company or Agency _____

Alternate people to contact:
Name _____
Telephone No. _____ (Area) _____ Extension _____
Address _____ City _____ State _____
Company or Agency _____

Name _____
Telephone No. _____ (Area) _____ Extension _____
Address _____ City _____ State _____
Company or Agency _____

Urgent—return call immediately _____
Expedite—return during regular work hours _____
Information only _____

NEN Employee contacted _____ Time _____
Comment _____

Mail immediately to: **New England Nuclear Corporation**
Treble Cove Road
North Billerica, Mass. 01862
Attn. Compliance Office
Corporate Distribution

Signature _____ Date _____

TEAM ED/178-1

Answering Service TEAM Call Report

TELEPHONE ANSWERING SERVICE:
1. When a call for *TEAM* is received, fill out a *TEAM* **CALL REPORT.**
2. Tell the caller that a Transportation Emergency Assistance Member will return his call and to remain at the telephone until the call is returned.
3. Call one of the *TEAM* members on duty:
   A. Record the time he was reached.
   B. Relay the **T.C.R** number as well as of the information on the **CALL REPORT.**
4. Place the **CALL REPORT** in the envelope provided and mail immediately.

INITIAL RESPONSE AND SUPPORT *TEAM*:
1. Fill in the **T.C.R.** number and preliminary information.
2. Record detailed account of problem.
3. Determine the necessary action.
4. If assistance is required from the **SUPPORT TEAM**, record name of *TEAM* member and time contact was made.
5. If **SUPPORT** is required, the call should be returned as soon as possible to the person or alternate responsible person that requested assistance.
6. Summarize your evaluation and action.
7. Sign and send to Compliance Office, Corporate Distribution by next working day.

TEAM PROCEDURE

INITIAL RESPONSE AND SUPPORT TEAM
1. Fill in the T.C.R. number and preliminary information.
2. Record detailed account of problem.
3. Determine the necessary action.
4. If assistance is required from the SUPPORT *TEAM*, record name of *TEAM* member and time contact was made.
5. If Support is required, the call should be returned as soon as possible to the person or alternate responsible person that requested assistance.
6. Summarize your evaluation and action.
7. Sign and send to Compliance Office, Corporate Distribution by the next working day.

TEAM RESPONSE REPORT

Summary of Call Report: _____ Time Report Rec'd _____ T.C.R. No _____
Priority Status: Urgent—return call immediately _____
Expedite—return call during regular work hours _____
Information Only—no return call required _____

Caller _____ Telephone _____

Alternate Contact _____ Telephone _____

Detailed Account of Problem:
Person contacted _____ Time _____ Date _____
Nature of problem _____

Location of Shipment _____
Name of Carrier _____
Identifying marks: Bill No. _____ Package (NENC) No. _____
Label: W-I _____ Y-II _____ Y-III _____ others _____
No. of Boxes _____ No. of Boxes _____ No. of Boxes _____
Products & Quantities involved _____

Name of Consignee _____
Address _____
SUPPORT *TEAM* contacted Yes No
Name _____ Time _____
Name _____ Time _____
Name _____ Time _____

Detailed Report of Response _____
☐ cont'd. on rear

Signature _____
Date _____

TEAM ED/178-2

Illustration 6.

Team members are Health Physicists, Radiation Monitors, and Electronics Technicians who work at the Rockwell plant. "One of the key things in any emergency response program is having a knowledgeable cadre of people who can assess the situation and respond correctly," adds Dr. Yoder. "Part of our strength is the knowledge of people we have doing the normal day-to-day work at the plant. For example, the Electronics Technicians do the regular maintenance of instrumentation equipment here on the plant site all the time; should we be asked to respond to an incident off-site, they would merely transfer their talents to the field. There is very little special training required for such people except for making sure they know where the equipment is located at all times, that they know how to calibrate it, and through exercises insuring that they can be alerted and moved out within no more than one hour day-or-night. Team members do know a bit more about how to repair their equipment than the average worker might; there is always a possibility an Electronics Technician might not be available when you need one, so *all* team personnel are given a bit more information than normal on the maintenance and repair of instrumentation."

Like the Maytag repairman, team members must be continuously ready to respond, but the Rockwell team at Rocky Flats has been activated for an actual off-site emergency only once in the past ten years; a transportation-related incident for which the team provided support. To date, the few radiological incidents that have occurred within Colorado have been ones that local authorities have also been able to handle without outside support. With regard to the national scene, Dr. Yoder says, "I don't know of any response team under contract to the Dept. of Energy that has ever responded to an incident where there was a life-threatening situation, where one had to take some action to save a life because of radiation. The types of radioactivity seen in transportation accidents have been of such a low level that from a health standpoint it was almost insignificant. Certainly from a public standpoint an incident may have been alarming, and radiation is certainly not something that one would want to leave in the environment, but it has not been a situation where someone has had to make snap judgements in order to protect life."

Should the Rockwell International Radiological Monitoring Team be activated at any time, the minimum first response team is composed of three people: a Supervisor or Health Physicist, and two Radiation Monitors. The next group to leave the plant would include an Electronics Technician and additional Radiation Monitors to provide back-up support. "The main thrust of our activity is to assess the scene with regard to the radiological qualities that may be present," explains Robert Yoder. "Are there any radioactive materials outside of a container; if so, how much, how many, how far has it spread? If a call is received, we would first alert the Chief Health Physicist at the plant. He/she gathers as much information as possible about the accident scene and what is required, immediately dispatches a crew to prepare the response trailer and get it ready for departure, and dispatches three team members who pick-up their equipment at a central point and leave immediately in one of our four-wheel drive vehicles. Here at the plant, we continue to assess whatever additional information we can gather about the scene so that we can provide appropriate back-up support to the first team already enroute. We have two-way radio communication with all our vehicles, and have connections through the Colorado Emergency Radio Network so we are able to talk with our team and others at the scene to get the latest information."

"We are prepared with a number of instruments that can be dispatched with team members," continues Yoder. "What we actually decide to carry may be dictated in part by the kind of information available from those requesting assistance. We carry alpha radiation detectors; these would probably be scintillation detectors that detect alpha particles by the scintillation they make on a CRT screen somewhat like a television screen. Little blips of light can be counted with appropriate counting systems. We also carry Geiger-Mueller counters that are very good for gamma radiation and beta radiation that might be present. Had we any indication that neutron sources could be involved, we would carry neutron detectors in order to be able to measure the presence of neutrons separately from any alpha, beta, or gamma radiation that might also be present. We have portable power supplies that allow us to carry a spectrometer, a fairly elaborate piece of equipment that allows us to assess the kind of radioactive material involved. We can envision occasions where we might not know what we are dealing with; the spectrometer allows us to make a positive identification of the material. We also have scintillation counters that are sensitive to gamma radiation, and may even be sensitive to some of the plutonium materials that could be present in the environment should the incident involve smoke detectors or something like that. A normal assortment of instruments would include an alpha scintillation counter, a beta/gamma Geiger counter, a gamma counter, possibly one neutron counter, and probably a spectrometer. Scintillation counters are rather finicky, and therefore we would take one per person. Were it required, we would bring along additional instrumentation in the trailer when it was dispatched shortly after the initial response team.''

"We take an Electronics Technician along to make the instrumentation repairs necessary, to insure the instruments are properly calibrated, and to see to it there is no problem in having the proper equipment to do the work we need to do," adds Yoder. "Some of the equipment we use is delicate; there are also environmental factors that can come into play. The equipment is designed for use inside areas where the temperature is 70 degrees and the humidity is 60 per cent, a normal working environment. If you take that same equipment and all of a sudden respond where the temperature is ten degrees below zero, you sometimes have to do a little tinkering with the high voltage power supply and other things in order to have them work properly in such an unusual environment.''

If and when team members are requested to respond off-plant, in addition to equipment used to measure radioactivity they carry fullface gasmask-type respirators, shoe covers, gloves, and anticontamination clothing which is really just heavy cotton coveralls with hoods that cover the head. All openings on protective clothing are taped to provide a barrier between the individual and the environment. The team also carries equipment to set-up barriers so as to limit access to an area that could be contaminated, utilize procedures for logging people into and out of the area, and use dosimeters for measuring accumulated radiation to insure that no one spreads radiation beyond the immediate scene. A crucial consideration in the opinion of team members is to limit the size of the area they have to deal with.

"There are standard-operating-procedures used at the plant, but at an accident scene the normal, routine practices and values may have to be modified to deal with the situation at hand," says Dr. Yoder. "It is important to understand that all we do is measure the radioactivity. Team members are aware of D.O.T. regulations regarding radioactivity that could be transported on a vehicle, knowledgeable of Nuclear Regulatory Commission requirements with regard to radioactivity in an uncontrolled environment, and understand all international regulations that apply. We take the measurements and advise the people in charge as to what the measurements are and how they relate to the guidance that has been issued by regulations, and may even make a recommendation as to what should be done, if anything. Therefore, we do not have step-by-step guidlines for action on-scene. We don't say . . . 'If you see this number, do this; if you see that number, do that.' We have to rely on the judgement of the individuals at the scene: their knowledge of what radioactivity in the world is, the values they might run across, the health effects of such values, and their ability to make recommendations as to actions that need to be taken.''

Dr. Yoder stresses that the Rockwell team is a group of technically expert people who monitor and assess radiation, and advise officials on-site as to the significance of their findings. "We are not there to do patching, plugging, or clean-up," he emphasizes, "that is not our charter. We will help; we will advise; but such functions as clean-up are really the responsibility of the government entity responsible for control of the scene.''

"No matter who responds," concludes Robert Yoder, "whether it be Rockwell or someone else, what you are really relying on is the capability of your technical staff to understand the materials they are dealing with, and their ability to respond in real time to the kinds of problems they are likely to be presented with. The right man can use all kinds of different equipment to get the right answer: the wrong man may not be able to make the necessary assessments as quickly as he might. We rather pride ourselves on having an in-depth group of people who are well qualified to respond. Some are more expert in one area than another, but they are all expert in the general area of emergency response."

On Time and Intact—Hazardous Materials Response and Control at DuPont

For more than 175 years, the DuPont Company has been involved in the manufacture and shipping of hazardous materials. In the early 1800's, the company shipped black powder in specially sprung, clearly identifiable wagons pulled by six-horse teams hung with warning bells. Young boys were hired to run ahead of the wagons in small towns and villages yelling—"Powder Wagon Coming! Powder Wagon Coming!" Today, DuPont is a combined enterprise of 170,000 people with annual sales of $32 billion derived from four main areas of production: chemicals, plastics, specialty products, and fibers. Daily, 100 million pounds of DuPont materials, both finished products and intermediates, are in transit by truck, train, plane, and barge. During an average year, DuPont will ship in excess of 30 billion pounds; or roughly 140 pounds for every man, woman, and child in the United States. The company manufactures and sells 18,000 different chemicals. During 1981 DuPont acquired Conoco thus adding production and transportation of petroleum products to the actvitiy levels described here.

About 80 percent of DuPont shipments move by rail aboard 9,000 proprietary and leased rail cars. DuPont's own truck fleet, 200 tractors and 400 semitrailers, carries 1.8 billion pounds of material a year, while substantial additional amounts are transported by common carriers. Nearly a third of all shipments are classified as hazardous for transportation purposes. The company annually ships nearly 12 billion pounds (5.4 million metric tons) of hazardous materials throughout the world. While company-owned vehicles and vessels carry 25 percent of all DuPont shipments, for fully 75 percent of all shipments the company relies on the services of more than 3,000 truck, rail, marine, and air transportation companies. With a total of 1.2 million shipments a year, on any given day DuPont is likely to have 50,000 shipments in motion. Possibly no other company has comparable experience in shipping the volume and variety of hazardous materials handled by DuPont. Over the years, that experience has resulted in practices, procedures, methods and programs to insure that DuPont hazardous materials are delivered on time and intact. Both within the chemical industry, and for industry as a whole, DuPont's record for the safe transportation of hazardous materials is second to none.

The safe transportation of hazardous materials does not just happen; it is managed. In 1972 DuPont's Transportation and Distribution Department, a central traffic department established to coordinate distribution matters and lead the development of improved methods of product transportation, became concerned over a growing incidence and severity of transportation incidents involving company products. During the first quarter of 1972, there had been a sharp upturn in both the number and severity of transportation-related incidents. The accident rate alone was 2.5 times higher than the average of the preceding five years. Heavily dependent on carriers, particularly railroads, for the transport of its products, the company appointed an Interdepartmental Committee For The Safe Distribution of Hazardous Materials. The committee's task—define the extent of the problem facing the company and recommend specific actions to correct the situation.

Early in 1974, after 18 months of detailed study, the Interdepartmental Committee For The Safe Distribution of Hazardous Materials presented its findings and recommendations. Three recommendations within the overall report were of central importance in defining program directions for the following decade.

1. Safe distribution of hazardous materials is an extension of the manufacturing process and needs serious corporate attention in terms of handling procedures, documentation and manufacturing follow-through.

2. To maintain continuity and underline responsibility, the duty of insuring safe distribution should remain with the manufacturing department.

3. Plant sites and shipping points must be subject to regular visits and inspections by a central, corporate audit group.

Basically, the committee found that although the company had been doing well in terms of methods then available, such methods were no longer good enough and new methods had to be found. The central theme of the committee's report—that distribution must be considered an extension of the manufacturing process—meant that it was no longer acceptable for DuPont to merely do the best job possible in terms of manufacturing, packaging, shipment preparation and documentation; and then leave the delivery phase solely to the carrier. Soon, the physical distribution process became as important to each operating location as cost, safety, yield, and other factors in the production process by which such locations had previously been evaluated.

To turn recommendations into concrete action, DuPont initiated the RHYTHM program. (Remember How You Treat Hazardous Materials) C.R. Bigelow, currently Division Manager-Domestic for Transportation and Distribution, who was a central figure in the development of RHYTHM, explained at the time the feeling behind adoption of the RHYTHM acronym. "It signifies a smooth-running, repetitive system that can become second nature to people handling hazardous materials, a procedure that creates a rhythm so insistent that a missed beat becomes a signal of potential trouble."

As developed over the years, RHYTHM has eight basic components required for implementation.

1. Establishing commitment and organization by tying plant managers to the success of the program.

2. Determining and selecting committees, alternative organizational structures, and means of insuring commitment.

3. Developing a means of analyzing hazardous materials handled, including searching out hazards posed by products, intermediates, and wastes.

4. Defining the type and amount of training needed and establishing a training schedule.

5. Implementing and documenting the training program as a means of promoting the overall concept.

6. Deciding which policies and procedures need to be audited and implementing necessary reviews.

7. Designing and establishing follow-through procedures for a continuous audit system.

8. Defining methods of working with outside agents—shippers, warehouses, freight handlers, and others—involved in movement of the company's hazardous materials.

The popular concept of such a program among many shippers is that of a vehicle for meeting government regulations. DuPont, however, has always emphasized RHYTHM as a means of attaining DuPont standards for safety and by so doing automatically meeting, and frequently exceeding, governmental requirements, Regulatory compliance is just one component of the total effort.

To change RHYTHM from a mere acronym to a productive program with measurable, positive benefits in terms of hazardous materials safety required a lot of work. The first step was to build an awareness of RHYTHM among all levels of DuPont personnel.

The twin goals of RHYTHM are to make every possible effort to prevent accidents from happening by use of consistent preventive measures; if and when hazardous materials accidents do occur despite the best efforts of company personnel, knowledgeable company employees are trained, teamed, and equipped to respond quickly. The prevention phase of RHYTHM is designed to instill correct safety practices in the day-to-day operations of DuPont employees who come in contact with hazardous materials during manufacturing or shipping. Formal training sessions, programmed instruction courses, booklets, movies, slide/tape presentations, posters, a newsletter—a wide variety and large volume of information and training aids was created to heighten the awareness of DuPont employees and agents to the need for safe distribution practices, and to provide information on how to do the job correctly and safely.

RHYTHM consists of four basic elements: Physical Distribution Guidelines, Training Aids, Commodity Distribution Data, and Shipping Point Surveys. A *Physical Distribution Guidelines Manual,* known throughout DuPont as "the Orange Book," was created to serve both as a reference book and a training material source. Twenty-six indexed sections complement federal regulations on-hand at each shipping point and cover regulatory agencies, shipping containers, transportation emergency procedures, and similar topics. Shorter versions of the manual have been distributed to more than 1,400 company agents including public warehousemen and terminal operators to guide them in meeting with DuPont and government requirements.

Programmed instruction courses known as Rhythm Individualized Training (RIT) that have been prepared for "Cargo Tank Inspections," "Tank Car Inspections," and "Response To Transportation Emergencies" stress nuts-and-bolts oriented, hands-on, how-to instruction in various aspects of safe handling of hazardous materials. Additional RIT course training aids now being prepared include "Shipping Containers," "Material Classification," "Shipping Small Packages," "Hazard Labels And Markings," "Regulatory and Non-Regulatory Agencies," and "Shipping Papers." In developing "Response To Transportation Emergencies," for example, lessons learned by DuPont employees throughout the world in responding to countless transportation emergencies was synthesized in three volumes. Volume 1 covers the structure of national response organizations, including DuPont's own Transportation Emergency Reporting Procedure. (TERP) Volume 2 covering plant site preparation details non-emergency-scene functions such as the care and use of personal protective equipment, emergency response vehicles and equipment, and pre-emergency public affairs considerations. Volume 3 covers on-scene treatment of both tank car and cargo tank containers, small container treatment, and containment and clean-up procedures on both land and water.

DuPont provides "Commodity Distribution Data Sheets" (CDD's) for carriers and emergency services personnel on 400 commonly shipped bulk chemicals. Each CDD's contains information on various government and carrier rules and regulations for the commodity designated. Significant physical properties of the material and specifications of approved shipping containers are listed, along with emergency procedures to be followed in the event of spill, fire, or explosion. For each material, CDD's contain on one sheet the freight classification, hazard class, shipping information, generic name, common name, chemical name, chemical formula, materials of construction, and an identification of compatible materials. A copy of the CDD's is provided with each tank truck shipment. The Commodity Distribution Data Sheets are used primarily for transportation, and are in addition to Material Safety Data Sheets that are used for broader purposes.

The fourth element of the RHYTHM program was development of a shipping point survey, or site audit procedure, to measure program effectiveness. This internal audit, coupled with Bureau of Explosives plant surveys, provides feedback on areas that are successful and those that need strengthening. Site audit surveys are conducted by engineers from DuPont's Safety And Fire Protection Division at both company-operated and non-company-operated facilities which ship and receive hazardous materials. Both preventive measures and some aspects of emergency response are covered during site audits. Auditors will examine shipping records in general and pull specific shipping papers for a detailed review. A facility must be prepared to demonstrate that a particular shipment was properly prepared for loading, properly inspected, and shipped in accord with both government regulations and DuPont standards. DuPont standards are often substantially more demanding than comparable government regulations. "God helps plants if they do not perform well on these audits," noted one employee. "If a plant flunks an audit, there will be Hell to pay in Wilmington. A poor audit evaluation is a reflection on the operation of the plant, and therefore a reflection on the people who work there. An audit is treated as an extremely serious process."

While instituting some of the most extensive hazardous materials control programs of any industrial organization, DuPont has also built an emergency response capability second to none. As Emergency Response Supervisor within DuPont's Transportation And Distribution Department, Thomas L. Hamberger coordinates the company's entire program of emergency response, including the operation of TERP. (Transportation Emergency Reporting Procedure) "All our efforts are directed at being sure that DuPonters are as well equipped as humanly possible to advise and assist the responsible carrier in dealing with transportation emergencies involving DuPont products," says Hamberger.

DuPont receives rapid notification of such incidents from CHEMTREC, the Chemical Transportation Emergency Center in Washington, D.C. sponsored by the Chemical Manufacturers Association. "Last year we had 214 incidents resulting from a total of 1.2 million shipments," adds Hamberger. "The severity of many of these incidents was limited; 50 percent involved 55-gallon drums or smaller containers. Approximately 50 percent of the notifications forwarded to us by CHEMTREC originated with public safety agency responders such as police officers and firefighters. I would estimate another 25 percent originated with carrier personnel. Very few notifications are made by private citizens."

In March of 1980, DuPont went on-line with a data terminal link to CHEMTREC known as "HMER-1" (Hazardous Materials Emergency Response) developed by the Long-Lines Division of American Telephone & Telegraph for the chemical industry as a result of inquiries by the Inter-Industry Task Force On Rail Transportation of Hazardous Materials. When CHEMTREC is notified of a transportation incident involving hazardous materials shipped by DuPont, a CHEMTREC communicator first provides emergency response information to the caller, then transmits incident information in a standardized hard-copy format to DuPont's corporate headquarters in Wilmington, Delaware. The CHEMTREC communicator also telephones a DuPont 24-hour contact point to warn that a message has been sent and provides the message number. The DuPont contact point notifies Hamberger that a message has been received. During normal office hours, Hamberger merely types in the proper code at a terminal in the T&D section and obtains a hard-copy printout of the emergency message. During nonworking hours, Hamberger or one of five other response coordinators use portable terminals at their homes to receive the printout. The coordinator then alerts the industrial department involved which in turn alerts the shipping point that activates predetermined emergency response procedures. The shipper determines the specific response action to be taken.

"The specific data terminal we use is a 'Com-Stor' floppy-disc apparatus," notes Hamberger. "Presently, there are 23 chemical companies on-line plus the Coast Guard's National Response Center and the Association of American Railroads' Bureau of Explosives. The major benefits of the data terminal system are that it is accurate, rapid, reliable, and cuts down on internal communications. It eliminates verbal transmission of information; you don't transpose numbers; you don't misspell chemical names. Basically, we have a dedicated, direct line to CHEMTREC. I get a call, and a DuPont employee says—'CHEMTREC just sent you message 1234.' I go to my computer, look at the file number menu and confirm that the last message was Number 1234, type in an access code, and in 45 seconds have a hard copy of the message. I can see immediately the time the local responder called, the date, the name of the person calling, his title and location, a callback telephone number, the product, its point of origin, its destination, the type of container, the carrier—and I have a three or four sentence report of what happened. (See sample CHEMTREC message sent by HMER-1) Messages are numbered in chronological order as they are sent by CHEMTREC, so my next message may have a number widely removed from the previous message. The message can be accessed as both a display on a CRT screen and a hard-copy printout. A standard format is used for messages, so we do not have to search through each message to find the crucial information. I know where critical facts will be located within each message."

"Our Transportation Emergency Reporting Procedure (TERP) consists of a centralized communications network and a decentralized response network," notes Hamberger. "All incident notifications and communications pass through DuPont corporate headquarters here in

```
CHEMTREC REPORT  [2408] + [UKPAI-AR----]        SPECIAL       [       ]
DATE             [22SEP81]                       CALL RCVD          [0717]
TAPE             [44/81](NOTE:  INFORMATION BELOW) ENDED           [0721]
                        (SUBJECT TO VERIFICATION)
POINT OF CONTACT
NAME/TITLE       [DISPATCHER 70                                          ]
ORGANIZATION     [MONTGOMERY COUNTY FIRE DEPT.                           ]
LOCATION         [EAGLEVILLE, PA.          ] PH#  [215-539-8770          ]

PRODUCT NAME     [TEFLON TYPE 30                                         ]
SHPR/ORGIN       [DUPONT (TRADE NAME)                                    ]
CONSIGNEE/DEST   [------, N.WALES, PA.                                   ]
CONTR:  QTY/TYPE[IN AN OVEN (55 GALLONS)                                 ]
CARRIER NAME    [N/A                                                     ]

PROBLEM DESCRIPTION AND LOCATION                                         ]
[LOCA:  IN PLANT FOR ---------,N.WALES, PA.                              ]
[PBLM:  THEY HAVE AN OVEN FIRE INVOLVING THIS MATERIAL.  CALLER WANTS TO ]
[KNOW WHAT THE DECOMP PDCTS ARE.  TOLD HIM THAT DECOMP IS MORE THAN LIKELY]
[HYDROFLUORIC ACID AND OTHER FLUORINE COMPOUNDS AND TO MAKE SURE FIREMEN AR]
[E WEARING SCBA AND FULL PROTECTIVE CLOTHING.  STANDBY FOR CALL FROM MFGR ]
[FOR FURTHER INFO ON PDCT.                                               ]

PD/FD RESPOND [YES ]   POP AREA [PLANT ] WEA/TEMP  [UNK                  ]
```

Illustration 7. HMER-1 Data Terminal Printout-Sample Notification from CHEMTREC to DuPont Emergency Response Coordinator.

Wilmington, Delaware, but our response teams are located throughout the United States at company manufacturing facilities and shipping points. DuPont personnel respond in-person to approximately 50 percent of the 200 or so incidents a year that involve DuPont products. We attach responsibility for safe transportation and handling of hazardous materials to the departments that make specific products. The word to the departments is—'If you made it, it is yours.' This applies to emergency response as well as to preventative measures. DuPont views the safe transportation of hazardous materials as an extension of the manufacturing process.''

"There is no rigid rule as to the structure of a DuPont emergency response team," continues Tom Hamberger. "What is good for a plant in Memphis that manufactures one product and uses one type of shipping container may not be good for a plant in New Jersey that produces a totally different product and uses different shipping containers. Teams are structured to fit their particular product, shipping containers, and the transportation mode most commonly used. A single plant may have different teams for different products. However, every ER group has a team leader, usually a person out of the production group who knows the product intimately. We never have to teach a team leader about the products he may have to handle in an emergency situation because he had grown up working with such products. A number of team leaders will have advanced degrees; you have to understand that DuPont hires an inordinate number of Ph.Ds. However, there is no requirement that a team leader have a Ph.D. or that he be a Chemist; he could be a Mechanical Engineer. Even though our response team members will constantly consult manuals and guidelines for a particular incident, they always possess that all-important firsthand information about the products they respond to. Usually a team will also have one or two people from the plant's mechanical group because of their nuts-and-bolts ability. Also, someone from shipping is usually on a team because he knows the containers, regulations, and shipping papers. It is also customary to have a team member from the plant's technical group, possibly a Ph.D. who works in the laboratory; he goes along as the team's technical advisor. Depending upon the nature of a specific incident, the plant would activate from two to five people.''

"Response tools/equipment/materials/vehicles vary by individual plant depending upon the products shipped and the containers used," adds Hamburger. "Some teams may have only a 'flyaway kit' consisting of two heavy cases with padlocks and rope handles. If the incident is close by, they throw these in the back of a pick-up truck and take off. If the incident is further away, they transport their kit by aircraft. For those teams that have emergency response vehicles, the equipment on the vehicle is often dedicated to emergency response and is always available."

DuPont ER vehicles range from custom designed vans, to pick-up trucks, to large step-vans. As a contractor for the Department of Energy, DuPont's Savannah River Plant may assist D.O.E. in carrying out provisions of D.O.E. and Interagency Radiological Assistance Plans in the five-state area of Alabama, Florida, Georgia, North Carolina, and South Carolina. Savannah River uses a step-van equipped with sophisticated radiation counters and monitoring equipment that pulls a trailer with a 50-foot telescoping antenna and a ten Kw generator with a 60-gallon fuel tank. An additional truck carries five kits of radiation detection equipment and protective clothing plus eleven footlockers of supplies.

The Repauno Plant in Gibbstown, New Jersey that manufctures industrial chemicals uses a four-door crewcab unit. Actually a dual-purpose vehicle, the interior is equipped as an ambulance, and lab technicians who are trained first aid personnel operate the vehicle on the plant site for medical emergencies. The truck's side compartments contain emergency response equipment including breathing apparatus, protective clothing, assorted handtools, and items used for on-scene repairs and product transfer.

DuPont's chemicals, dyes, and pigments plant in East Chicago, Indiana uses a customized van loaded with ladders, acid suits, breathing apparatus, fire hoses and extinguishers, lights and a power source, and a drum pump used to transfer product from a leaking drum to a sound one. The van responds to both on-plant and off-plant emergencies. For off-plant emergencies, the van may be used by any one of several response teams that specialize in handling specific chemicals. In such cases, kits with fittings and gaskets to aid in on-the-spot repair of rail cars or trucks are added to the van's equipment.

The response vehicle at the Louisville Works where elastomer chemicals are produced is a step-van used by four different response teams depending upon the product and container involved in a specific incident. The checklist of tools/equipment/materials carried on the van and organized by numbered compartments covers five pages of typescript. The van is equipped with one fixed and two portable radios, a CB radio, and a mobile telephone. The fixed radio serves as a base station at an incident scene.

The Chambers Works in New Jersey utilizes two identically equipped one-ton Emergency & Disaster trucks which serve as repair trucks on the plant grounds until an off-plant emergency occurs. Response equipment is permanently stored in a locked and sealed shelter and loaded on the trucks for each incident. Equipment includes chemsuits, air packs, and a portable emergency shower; as well as a complete plumbing shop including tubing, plugs, and gaskets of various materials. Also available are general hardware tools, fire extinguishers, a multipurpose saw, com-a-longs, a blower, and emergency lighting.

"DuPont has both a written policy for emergency response and extensive written guidelines for the handling of transportation emergencies," continues Tom Hamberger. "DuPont will be involved; we will advise, we will assist because the safe transportation of hazardous materials is considered to be an extension of our manufacturing process. We will not turn around and look the other way, or say—'Yup, you have a problem.' There is no discussion of whether we will or will not respond; commitment to emergency response is not open to debate. It is corporate policy to bring total DuPont resources to bear on emergency response to hazardous materials incidents involving company products. We will help reduce the hazard, remove the potential danger, and mitigate the situation. The potential liability considerations, if they ever arise, will be sorted out later. We are not asking our people to be the Red Adairs of hazardous materials; but within the confines of their personal safety and areas of expertise, using good prudent judgement and their knowledge of equipment, our people are directed and trained to respond fully and effectively. They are also given the necessary authority to get the job done. Effectively, we view our response team captains as DuPont management at the scene of an incident. In essence, if the entire corporation including the Chairman of the Board gets involved in an incident, they report to the DuPont team captain at the scene. He is in-charge; nobody attempts to second-guess him. The man-in-charge is expected to make smart decisions. We happen to be a very decentralized organization; we are measured within that decentralized organization on the basis of results. A site manager has a responsibility and concern to insure that he assigns capable people to handle emergency response.''

In the event of a transportation emergency, all appropriate DuPont personnel are notified of the incident promptly, and action is taken to correct the situation and minimize the affects on people, property, and the environment. Even when called upon to handle transportation emergencies of other manufacturers or carriers which in no way involve company products, shipments or locations; DuPont personnel are instructed to provide requested assistance when a real emergency exists and DuPont is the nearest available source of needed expertise.

Most DuPont shipments of hazardous materials are handled by common carriers rather than by company personnel and proprietary equipment. Company procedures make it clear that primary responsibility for dealing with a transportation emergency rests with the carrier, and that DuPont's role should be to advise and assist the carrier and emergency services personnel at the scene by providing technical information on products and equipment. In cases where the carrier is unable or unwilling to fulfill its responsibility, and action is necessary to correct a dangerous situation; DuPont responders are instructed to provide direct assistance such as transloading or patching of containers to reduce hazards.

In general, DuPont personnel on-scene are to take whatever action is necessary that, in the opinion of the response team leader, will serve to mitigate dangerous circumstances. The final authority for taking action rests with the designated response team leader. If a carrier will not take responsible action at an incident scene to protect public safety, DuPont response teams are instructed to assume a leadership role with respect to DuPont products or equipment.

Each DuPont shipping location is required to develop a plan for responding to transportation emergencies involving their own materials and to provide assistance to other shipping points when requested. Although many factors affect the composition of a specific team, the "average" team is generally made-up of a team leader with supervisory authority and product knowledge plus team members with expertise in shipping/manufacturing, maintenance/equipment, and environmental/technical areas. Every DuPont response team responding on-site assigns one team member as a "Safety Lookout" responsible for observing overall activity at the scene in order to insure the safety of other DuPont personnel.

Emergency response plans also include procedures for . . .
• Obtaining nonscheduled air transportation to the emergency scene.
• Obtaining cash for emergency crew members.
• Training emergency crews.
• Keeping personnel familiar with response procedures, the hazards associated with chemicals that may be involved in an incident, and ways of dealing with the media.
• Periodically reviewing and updating the emergency response plan.
• Training persons to handle the public relations function at the scene.
• Arranging for manning of a central contact center to provide a plant point for contact with the response team at the scene. The center is intended to provide consultation and technical information from any company source, provide backup personnel support as necessary, and arrange for additional equipment or logistical support as requested from the incident scene.
• Chartering special means of approved transportation to insure a speedy arrival.
• Maintaining a running log of events on-scene, and issuing a written report at the conclusion of each incident.

A plant emergency response team leader is responsible for assembling and informing an emergency crew and dispatching it to the incident site. He coordinates the team's efforts on-site so as to obtain desired results without endangering personal safety of team members, and alerts the appropriate medical group for possible consultation from the emergency scene. The team leader must also keep appropriate management informed as necessary, issue a summary report of incident operations, and execute any required follow-up procedures.

When a transportation incident occurs involving either regulated or nonregulated company products, the DuPont facility that shipped the material is responsible for insuring that a timely, effective response is made. When an incident occurs at a great distance from the shipping point, DuPont plants located closer to the accident scene are prepared to respond upon request from the plant that shipped the material. In other words, the DuPont site closest to the incident is to provide an on-scene presence as soon as possible to advise and assist the carrier and emergency services personnel. However, this immediate response by a closer DuPont facility does not relieve the shipping point of its responsibility to dispatch its own response team to provide ongoing or longterm guidance at the scene. Ultimate responsibility to assure an effective response remains with the shipping point.

The Product Pros—Shell Oil Company's Response Actions Teams

Shell Oil Company operates 85 product distribution facilities throughout the United States from which 225 company-owned tank trucks safely deliver millions of gallons of gasoline every day. No company has done more to develop, advance, and shape start-of-the-art methods and procedures for responding to tank truck rollovers. Handled correctly, a rollover accident can be turned from a potential hazard into a safe clean-up operation; handled incorrectly, or by guesswork, a rollover can quickly change from a relatively safe clean-up operation into a conflagration. To minimize guesswork and maximize planned, deliberate emergency response to tank truck rollovers, Shell Oil Company has emphasized six basic components in developing a high quality response action program.

1. Welding of knowledgeable Shell personnel into Response Action Teams trained, teamed, and equipped to safely handle road emergencies involving Shell vehicles and products.

2. Development of detailed emergency response contingency plans for all Shell Oil Company product distribution facilities.

3. Identification and provision of proper tools and equipment for safe handling of tank truck rollovers.

4. Establishing a communications system able to quickly dispatch necessary manpower and equipment to the scene of a rollover any place in the country.

5. Identification and/or internal development of preferred pump-off methods and procedures.

6. Sharing of expertise developed by Shell personnel with industrial and public safety agency organizations.

Response Action Teams:

In April of 1979 Shell Oil Company's products distribution department identified a core group of 26 knowledgeable employees selected from Shell's eight distribution areas and brought them to San Jose, California for a week. Rather than a standard training session, the week in San Jose was designed to facilitate a consolidation of experience and a sharing of ideas between and among a group of product distribution professionals possessing a high degree of mechanical aptitude. Their goal—to identify the best methods of response to tank truck rollovers including preferred methods for off-loading gasoline from overturned trailers. For their "Final Examination," the 26 were divided into four teams, and each team was assigned a different rollover situation staged at the San Jose Shell plant. Each team had two hours to off-load the overturned tank, filled with water for safety purposes, and upright the trailer. Each tank truck had been rigged to pose a different, difficult mechanical challenge. To complicate matters, highway patrolmen and firefighters had been enlisted to both help and harangue the Shell teams. Observers were assigned to needle the teams and test their skills at handling irate bystanders.

In a typical exercise, one team was presented with a loaded trailer rolled over on its right side at approximately a 120 degree angle on a grassy area adjacent to a freeway exit. The dome cover appears to be holding, and there are no other leaks. A fire department is on the scene, and the fire chief is preparing to dispose of the gasoline in the tank by washing it down the sewer. The team must convince the chief there is a better way and then do it.

Experience and mechanical knowledge possessed by team members quickly became evident. Carl Frimodig, fleet maintenance superintendent, West Orange, New Jersey; and Bill Geis, automotive supervisor, Sewaren, New Jersey; came up with an air drill that exerted about 90 pounds of pressure on the one-eighth inch thick aluminum hull of the tank in cutting a hole through which the product was removed. The group using this method required 45 minutes to solve the problem; the other three groups required the full two hours. When the 26 participants at San Jose returned to their respective Shell facilities around the country, they formed the nucleus of the Shell Response Action Teams.

"The San Jose session was precipitated by a 1978 incident in Willow Grove, Pennsylvania," according to F.A. "Al" Ely, Manager of Safety and Health for Products Distributuion with Shell Oil Company. "A tank truck rolled over under a railroad overpass and leaked a quantity of gasoline. Our people handled the incident in accord with what we felt at the time were correct response measures and methods. There were no serious problems, but in reviewing the Pennsylvania incident later we had to recognize that we had been very lucky. We immediately began research and initiated efforts to develop formalized response methods for tank truck rollovers."

"Our main concern has always been to get the gasoline out of the trailer before we do anything to right the vehicle and haul it away," emphasizes Al Ely. "There are people in the industry who believe you can use a wrecker or crane to pick up a loaded trailer, set it on its wheels, and then drive or tow the vehicle away. Our contacts with various trailer manufacturers and experts in the field convinced us that righting a loaded trailer is not a safe procedure. An aluminum trailer, although it meets all design specifications, is designed to travel down the road on its wheels. When an aluminum trailer is not on its wheels, all design criteria are out the window. The reinforcing structure of a tank trailer is along the underside; when a trailer is on its back or side, the stresses on the shell are such that we do not feel it is safe to fool with it in any way while it contains product. Our thrust is to first remove the gasoline, render the trailer safe for handling, and only then right the unit and move it. The San Jose meeting was scheduled to insure that company personnel would be capable of responding to tank truck rollovers in a professional manner; in a manner that would be safe for our people, for the public in general, and for the environment."

"The people who came back from San Jose became the central core of our Response Action Teams," adds Ely. "They disseminated methods and techniques so that currently at all locations around the country where we have proprietary equipment we have people trained in proper response methods. We feel contingency plans developed by individual facilities to cover 'pre-incident' considerations are a critical aspect of the overall program. If a tank truck rolls over at two o'clock on a Sunday morning, that is no time to begin looking through the phonebook to learn who to contact; all that has to be done beforehand."

Continginency Plans:

In developing a contingency plan for a local Shell facility, responsible personnel travel a two-way street. Not only do they formalize arrangements, contacts, and communications within the company but within the community as well. Local authorities are contacted to acquaint them with the capabilities of Response Action Teams and the fact that teams will respond to incidents involving Shell products or vehicles. Prior contacts are established with, and a contingency phone list created for, contractors who can provide cranes and wreckers, spill clean-up, additional labor, vacuum trucks, sand and gravel, and foam. The contingency phone list includes all government agencies that could become involved—fire, police, environmental, transportation, Coast Guard, as well as other state and municipal officials.

Internally, local facility contingency plans identify all Shell personnel who may need to be contacted; and require initiation and use of an "Incident Log." Also, a checklist is developed to guide use of planned, appropriate actions. Contingency plans consider stabilization of an accident scene, establishment of relationships with authorities on-site, and initiation of communications from the scene to a Shell Response Action Group coordinator who coordinates the overall company response. Contingency plans also consider control of spectators, positioning of equipment, establishment of roadblocks, and possible need for evacuation.

Tools And Equipment:

"Each Response Action Team uses an equipment trailer that contains all the tools/equipment/materials that may be needed at an incident," says Al Ely. "When a call comes through—and thankfully we are currently averaging only three calls a year on Shell trucks throughout the entire country, although occasionally we respond to equipment not owned by Shell because local authorities know we have a special capability and say they desperately need help—all the team need do is assemble its members, hook a vehicle to the trailer, and be on their way. The trailers generally contain personal protective equipment, a variety of tools, blocking and bracing materials, and the ancillary equipment needed to perform the pump-off operation."

Once an accident scene has been stabilized to protect life safety, the first piece of equipment out of a response team's trailer is an explosimeter. *Nothing* happens until readings are well within the safe range, and constant operation of the explosimeter continues throughout the operation. While some Response Action Team members check for leaks and seal any that are located; others consider environmental concerns by stringing booms, using sorbent materials, or blocking off catchbasins and sewers as necessary and if safe to do so. The team will then establish a step-by-step procedure for removing gasoline from the trailer using one of five pump-off methods preferred by Shell personnel. Only after the trailer has been emptied do team members move ahead to right the trailer and clean up the scene.

Communications And Dispatch:

When a tank truck loaded up with up to 8,500 gallons of gasoline rolls over, it is essential that specially trained, highly experienced Response Action Teams be alerted and move to the scene as quickly as possible. Therefore, the oil products teams are tied-in to the overall STERP (Shell Transportation Emergency Reporting Procedure) system. The phone number for CHEMTREC (the Chemical Transportation Emergency Center in Washington, D.C., an emergency notification and response system established by the Chemical Manufacturers Association in 1971) is displayed on the sides and both ends of Shell tank trucks. When a rollover is reported to CHEMTREC, CHEMTREC notifies STERP reporting centers in either Wood River, Illinois or Pasadena, Texas that are manned 24-hours a day. The communications center calls a specific Response Action Group member located nearest the rollover. The RAG member, an area manager of Shell facilities, becomes the emergency response coordinator who dispatches and maintains communications with the nearest Response Action Team. In most cases, the RAG coordinator uses a Shell central dispatch office as his base of operations. Following a predetermined contingency plan, and using Shell dispatchers and telephones to handle communications, the coordinator activates or calls out necessary additional resources ranging from empty tank trucks to towtruck operators to clean-up contractors. The Response Action Group (RAG) member manages the incident.

Preferred Pump-Off Methods:

At the scene of an accident, the Shell truck driver, if physically able, is expected to shut off the engine, lights, and battery-disconnect switch before getting out of the cab. He is instructed to move himself and anyone else on the scene well away from the rollover; then call, or have someone else call, local police and fire agencies, his own dispatcher, and CHEMTREC. He is also to assure that no flares, flashbulbs, or cigarettes are allowed in the area. Because the further gasoline spreads the greater the risk of ignition, he visually examines the tanker to assess the leak/spill situation. He may be able to immediately stop or slow down any flow. If not, he will be able to provide a report to the first response agency arriving on-scene. He reminds initial responders of the importance of containing the product within a limited area. Flushing gasoline may move it away from the tanker, but it also endangers a much wider area. Also, the driver is instructed to inform authorities that a Response Action Team is on the way and will take responsibility for removal of the product and vehicle and clean-up the area.

Although a specific Response Action Team may have as many as six members, as few as two members are able to handle most situations. Once the team has established communications with the RAG coordinator by means of telephone or mobile communications equipment, first actions on-scene are designed to stabilize the situation. Working from a predetermined checklist, the team captain makes assignments. One team member checks the area for flammable vapors; use of an explosimeter on a continuing basis is the most basic of safety practices. Two members are assigned to check the tank compartments for leaks and to stop or contain any leaks that are located. Other team members take action to contain any product that has already spilled, and the tank is blocked and braced to prevent shifting as product is pumped out.

Once the situation has been stabilized, the team carefully plans the safest approach to gaining access to tank compartments and pumping-off the gasoline. Factors that effect the selection procedure include the position of the overturned tanker, damage to the tank, accessibility of tank fixtures used in certain pump-off methods, and the amount of product in the tank.

As a result of this assessment, one of five preferred pump-off methods is chosen. With any method, grounding and bonding is all-important. The rolled vehicle is grounded to a rod with the bonding cable connected first to the trailer and then to the rod. Additional grounding and bonding precedes any flow of product from one container to another. The pump-off vehicle, hose couplings, down-spouts, and recovery pans and tubs are all grounded.

If the trailer is lying on its right side, and the regular discharge lines are undamaged and facing downward, product can be removed through the discharge lines. A bottom-loaded trailer will have product in the discharge lines that must be drained into a grounded container before pump-off is initiated. Then, a 90-degree quick coupler is fitted onto the discharge outlet and a pump-off hose attached to the coupler. The internal valve must be opened in order to obtain product flow from the tank compartment. The vent must be blocked off or kept closed; otherwise, opening the internal valve will open the vent allowing product to enter the vapor recovery system or spill out the vent opening. Once the vent is closed by blocking off the airline that activates it, by repositioning the vent boots and plugging the vapor recovery piping, or by plugging the vent cover; the internal valve can be opened allowing product to flow into the discharge line.

On units with air-activated internal valves where the air line from the internal valve to the vent is outside the tank, the line is blocked off, disconnected, and a source of compressed air—about 50 pounds—is used to open the valve. Tire air, bottled air, or air from an emergency vehicle will do the job.

On units with the air line inside the tank, the internal valve can be opened mechanically.

The pump-off vehicle, or vacuum truck, is brought up to the side of the overturned tanker and positioned uphill and upwind since its engine and exhaust system could be a source of ignition. The pump-off vehicle is bonded to the rolled vehicle, one continuous bonded system to prevent static electrical sparks. With the units bonded together, the valve open and the vent blocked; product is pumped from the tank compartment until it falls below the level of the valve and suction is lost. The remainder of the product can then be recovered by opening the dome cover, or by removing the internal valve or the vent—whichever is above the product line and is accessible.

A grounded downspout is attached to the opening and the pump-off hose is attached to the downspout which has been grounded from the tank to the spout. In situations where the discharge lines are pointed upward, and the piping is either damaged or undamaged, product in the discharge line is drained off into a grounded container. The discharge piping is removed, a discharge adapter with victaulic coupling is attached to the internal valve, the pump-off hose attached to the coupler, and product pumped off to below the valve. The internal valve is removed, and product remaining in the tank pumped off through the grounded downspout. Any method involving removal of the internal valve requires reinstalling of the valve before uprighting the unit so that any product not pumped out doesn't drain out. Also, opening internals also opens vents, so the vents must be blocked off or prevented from opening in any procedure involving opening the internals.

Delivery vehicles with more than one compartment contain bulkheads separating compartments, and baffles within compartments for reinforcement, and to prevent shifting of load. All baffles have a half-circle drain hole at the bottom, a round vapor escape hole at the top, and an access hole at the middle. New Shell Oil Company delivery vehicles have additional holes at three o'clock and nine o'clock which allows more thorough compartment draining when the vehicle is on its side. However, even without these additional holes, the downspout can be positioned behind the baffles through the center access hole, and much of the product can be removed.

When a tanker is completely overturned, product can be recovered by removing the internal valve and pumping out through the valve opening. The Response Action Team disconnects the victaulic coupler, removes the valve bolts, pulls out the valve, and pumps the product out through the grounded downspout.

On a bottom-loaded trailer, the discharge lines or pipes will be full of product. Before removing the coupler, the product should be drained back into the tank by opening the internal valve. Once again, the vent will be opened unless first disconnected.

When a unit is on its side, removing the internal valve could also be used to recover product, providing product has been pumped down to or below the level of the valve. The discharge lines, whether facing upward or downward, contain product and must be drained as previously described. A special adaptor developed by Shell is then connected to the internal valve and the pump-off hose connected to the adaptor. The vent should be blocked off, and the internal valve opened either by compressed air or mechanically. Product is pumped off to below valve level, or until suction is lot. The hose and coupler are removed and the valve unbolted and removed; the grounded tub should remain in-place since there will be a small loss of product as the valve is unbolted and removed. The remainder of the product in the tank is pumped out through the grounded suction hose, or downspout; and the valve must be reinstalled before the unit is uprighted.

In cases where the rollover rail is not damaged, and the vapor recovery system is accessible, product can be safely removed through the vapor recovery system. The Response Action Team uses a connector to link the vapor recovery outlet to the pump-off hose since they are of different sizes. With the hose attached, the vent is opened allowing product to flow through the vapor recovery system. Product is then pumped off to the level of the vent; the remainder of the product can be pumped out the vent opening, through the dome, or through the internal valve opening using the grounded downspout or grounded suction hose. If the vapor recovery piping or the rollover rail are damaged in any way, then the method of pumping off through the vapor recovery system cannot be used. The vapor recovery system can be checked for leaks by connecting the vacuum truck to the vapor outlet. If the system holds ten inches of vacuum for one minute, it is intact and can be used for product recovery.

Entry to a tank compartment can also be made through an aluminum funnel placed over the hatch to recover product. A gasket or seal is placed around the hatch. The cover is placed over the gasket and clamped down. The pump-off hose is attached, and the pump is activated. The hatch is opened and product pumped out. When product reaches hatch level, the grounded pump-out tube is used to recover the rest of the product.

Another method of entering an aluminum tank filled with gasoline is by drilling a three-inch diameter hole into the tank and pumping out through a grounded pump-out tube. Equipment used in this method includes an air-powered drill with three-inch hole saw, airhose, pressure regulator, and a source of air that can generate 100 pounds of pressure. The drill requires four cubic feet per minute of air at 90 pounds pressure. Maximum revolutions are 500 per minute. Full protective gear, including a safety line manned by a team member on the ground, helps protect the man on top. A ladder is set in place and secured, and a compressed air bottle to provide air for the drill is secured near the ladder. A man goes up, selects a spot for drilling in-line with the dome cover, and takes the drill. Drilling in-line with the dome cover prevents hitting a baffle. Firefighters lay down a flushing stream of water directly onto the drill which cools, flushes chips away, and lubricates the drillbit. With light but steady control, and letting the drill do the work, the team member drills into the compartment. When he has drilled through, a process that takes about a minute, he inserts the grounded downspout, attaches the hose, and comes down off the tank. To minimize exposure, there is no one on top during the pump-off procedure. After the product is pumped off, a tapered wooden plug is inserted into the hole, and the unit can be righted. The advantages of the drill method are that each hole requires only a minute to drill and no additional product is spilled. Before adopting the drill method, which is the preferred technique, Shell research did extensive testing to insure that the drilling process would not generate sparks or heat enough to be a source of ignition. Research indicated the technique is absolutely safe on aluminum tanks. Experience supports this.

Sharing of Expertise:

The five pump-off methods and procedures detailed in the preceding paragraphs were taken from two slide-tape presentations, *More Than Just A Resource* and *Preferred Pump-Off Methods,* developed by Shell and made available without charge to major oil companies and common carriers, and to jurisdictional agencies to acquaint local responders with the capabilities of and the methods used by Shell Response Action Teams.

Shell has a reputation for working closely with local fire and police agencies. Personnel at local Shell oil products facilities throughout the country provide training, lend or demonstrate special equipment, and participate in simulated incidents with area firefighters and police officers. Response Action Team procedures are designed to have Shell specialists, firefighters, and police officers form a team at the scene of a tank truck rollover. Through initiation and annual updating of contingency plans, Shell employees maintain ongoing relationships with jurisdictional agencies.

Shell is more than liberal about sharing experience developed to respond safely to tank truck rollovers; the company has taken concrete actions to insure that pump-off methods preferred by Shell are generally available to the industry. In September of 1981, Shell conducted a two-day tank truck rollover school at Texas A&M University's Oil And Hazardous Material Control Training Division in College Station, Texas. Texas A&M, world famous for providing quality training in fire control/hazardous materials response/oil spill control, adopted the program as part of its regular oil and hazardous materials control curriculum and currently offers a two-and-a-half-day version four times a year that is open to industrial representatives, fire and public safety personnel, contractors, and others involved in operations surrounding volatile liquid tank truck rollovers. One day is devoted to tank truck design and appurtenances; the remaining day-and-a-half emphasizes hands-on exercises.

R.L. Osmon of Shell currently serves as chairman of a Task Group appointed by the American Petroleum Institute's Highway Transportation Committee to develop an API Model Emergency Response Plan to assist those companies currently without a formal response plan for highway emergencies involving tank trucks. The Task Group was formed on the premise that companies have responsibilities to react to emergencies in which they are involved. Because of the ever increasing number of products industry develops and transports, society's traditional emergency response agencies may have difficulty maintaining the technical expertise to safely deal with transportation incidents. Industry must be ready to step in and provide the expertise necessary to insure public safety.

"We have a reluctance about responding to other people's incidents," admits Al Ely. "We are not in the response and clean-up business. We have responded to incidents not involving Shell products or vehicles because as good corporate citizens we cannot in good conscience allow a situation to exist that is dangerous to the public. One of the reasons we make such efforts to share our expertise is because we think it is only fair that the industry as a whole, and each company within the industry, be able to handle its own problems. Although liability and cost must be considered, our main concern is for the safety of our own people. It is unfortunate enough if we have an incident of our own and our people have to go out and expose themselves to danger. We just do not think it is fair to ask them to respond to someone else's problem. Simply because of the nature of what we ask them to do, all Response Action Team members are volunteers who handle emergency response in addition to their regular duties. Although we have done it in the past, we are reluctant to respond to incidents not involving Shell products. We will do so only on the basis that there is no other way and there is real danger to the public if the incident is not handled quickly. We leave it up to the local Shell manager whether or not to employ Shell resources to handle somebody else's problem."

"Our Response Action Teams, along with detailed contingency plans, dedication of necessary tools/equipment/materials, and development of preferred pump-off methods; have provided us with confidence that we can properly handle a rollover situation," concludes Al Ely., "Of course, our primary thrust is to avoid emergency situations through the skill of our drivers and driver training in defensive driving techniques. If an incident does occur, however, the Response Action Team program has given us the ability to respond in such a manner as to protect our employees, the general public, and the environment."

5 STRIKE TEAMS

They Were The First: The Bureau of Explosives

Although hazardous materials incident response is often thought of as a relatively recent phenomenon, field inspectors with the Bureau of Explosives of the Association of American Railroads have been doing it for seventy-five years. Operations of the Bureau for the Safe Transportation of Explosives and Other Dangerous Articles began with the opening of the Office of the Chief Inspector at 24 Park Place, New York City on June 10, 1907. Although vast changes have occurred throughout the transportation industry during the past seventy-five years, the basic mission of the Bureau of Explosives has not changed at all. Then, as now, its goal was and is to promote the safe handling and transportation of hazardous materials; to collect, analyze, and disseminate information of these materials, and to provide emergency assistance for the benefit of member railroads.

In the late 1800s the railroads were the major transporters by land of raw materials. Trucks, airplanes and pipelines had yet to make their appearance. During the first few years of the 20th century, a series of rail incidents involving explosives led the Senate Commerce Committee to address the problem of the transportation of explosives by rail. The E.I. DuPont Company was at that time the primary manufacturer of explosives, and shipped large amounts by both rail and ship. DuPont and the Pennsylvania Railroad were able to convince the Senate Commerce Committee that, indeed, there was a problem, and Congress needed to act but that the solution to the problem should rest in private hands rather than with government. A bill was passed in 1904 creating within private industry an organization known as the Bureau for the Safe Transportation of Explosives and Other Dangerous Articles, and early in 1905 the Pennsylvania Railroad and most of the other major carriers, along with DuPont, established the Bureau with one Chief Inspector. Within two years the Bureau had established a laboratory in South Amboy, New Jersey, that devised a group of tests to determine the relative safety of different types of explosives and had initiated a field force of inspectors who periodically visited railroads and shipping points in the community to determine whether or not shippers and carriers were taking prudent action with respect to loading, storage, and transportation of explosives. These actions were the earliest efforts in the United States to control the transportation of hazardous materials.

Because the railroads at that time were really the only transporters of hazardous materials and had a lot at stake, railroads into the 1920s provided the focal point for early development of hazardous materials regulations. In 1921 the Interstate Commerce Commission required the Bureau of Explosives to publish its regulations as a tariff, and then adopted the Bureau's regulations as the regulations of the federal government. In the United States, the railroad industry wrote regulations for the transportation of hazardous materials, and the government adopted these same regulations. The Interstate Commerce Commission relied on the expertise of the Bureau of Explosives in formulating I.C.C. regulations.

Since 1922 there has not been a death caused by the transportation of explosives by rail. The Bureau was thus successful in its original goal, and over time other industries handling materials that fell within the definition of explosives-and-other-dangerous-articles were covered by regulations initiated by the Bureau. In over 500 specific and separate instances the Interstate Commerce Commission required that industries seeking approval of certain materials must go to the Bureau of Explosives for that approval.

The Bureau's relationship to control of hazardous materials actually transcended the railroad industry and impacted upon other modes of transportation as well. At various times the Bureau has had recommendation and approval responsibilities for truck shipments of hazardous materials, acted as an agent for the Federal Aviation Administration, and was one of the principal authors of the International Air Transport Commission's hazardous materials regulations.

The Bureau of Explosives acted as the agent of the Interstate Commerce Commission until the creation of the Department of Transportation in 1967 when the safety function formerly performed by the I.C.C. was transferred to D.O.T. The Department of Transportation then began to exercise its legal responsibility for discretionary control over the transportation of hazardous materials and slowly withdrew delegations of authority previously granted to the Bureau of Explosives.

"We recognize that the Bureau of Explosives means something to the Federal Railroad Administration, the Department of Transportation, the Association of American Railroads, and to that segment of the railroad industry involved with hazardous materials," says Michael R. Miller, current Manager of Field Operations for the Bureau. "The people who are not completely familiar with the Bureau are the emergency response agencies such as police and fire departments. Part of the reason is that our name, the Bureau for the Safe Transportation of Explosives and Other Dangerous Articles, may well be antique. We are no longer involved as much in the transportation of explosives. We still do the inspections and all the things we did before, but the transportation of chemicals and gases has progressed so far during the past 75 years that to be up-to-date we should probably change our name to 'the Bureau of Hazardous Materials.' However, we have a traditional name, and I guess we always will. Many people are surprised to learn who we are, what we do, and the fact that we operate in both the United States and Canada."

"Historically, the main mission of the Bureau was to write the regulations," notes Miller. "We also started out by saying—'You will have inspectors; not only will you write regulations, but you will bring in inspectors to insure that the regulations are complied with.' Our philosophy has always been that compliance with carefully drawn regulations means safety. Currently, we have one senior inspector and two inspectors in Canada, and two senior inspectors and 14 inspectors in the United States. As Manager of Field Operations, I have direct control of all the inspectors. Persons attached to our Washington headquarters staff include Roy Holden who is Manager of Technical Services; he handles anything to do with bulk containers. Thomas Phemister, as Director of the Bureau of Explosives, is responsible for our entire operation. Ann Mason is Manager of Environmental Services. There is also a Manager of Operations Analysis, James Perry, who handles the publication and updating of our chemicals, lists, tank car register, analysis of field functions, and anything dealing with computers; and Elizabeth Rabben, Supervisor of Military And Intermodal Services whose responsibility it is to act as liaison between the military and the railroad industry for the shipment of explosives and other dangerous articles, work with the military and domestic sectors in the development of safe blocking and bracing methods, and act as the Bureau's small package expert for trailers and containers/intermodal transportation/boxcars."

"The Bureau operates a laboratory in Edison, New Jersey," adds Roy Holden, the Bureau's tank car expert. "Primarily, the lab deals with the classification of materials with regard to hazard class. Say you have a bucket of something. What is it? How should it be classified? How hazardous is it? Is it explosive? The lab also does approvals of certain materials for the Department of Transportation.''

"Dispatch of field inspectors depends upon the type of incident being responded to," explains Mike Miller. "An inspector will respond whenever requested, and may respond when not requested if in our judgement we would be able to assist a railroad. Our view is that we are here to assist the railroads anytime they ask for help. If they ask for response, cooperation, assistance; we respond. All major railroads are members of the Association of American Railroads, but in a limited number of cases we might respond to smaller railroads that are not members. If a railroad needs our help, we are going to be there."

"We have a three-month training program for field inspectors during which they come to headquarters and are provided a presentation by each member of the staff on their particular area of expertise," adds Miller. "They are trained in the regulations by the senior inspectors, then participate in a three-month on-the-job training period during which they spend day and night with a senior inspector learning what to do for emergency response. Normally, field inspectors have five to six years experience with a railroad when employed by the Bureau. We want people who are able to work on their own and able to make decisions, yet realize there are several things out there they do not know. Our people are expected to know whether or not a mechanical problem can be fixed. If there is a union railroad employee on-scene qualified to handle a necessary rapair, he would have to do it. That is, if the railroad has a qualified person from their mechanical division on-scene, our inspector would first advise him what has to be done, and then probably help him do the actual job. If there is no qualified union employee available, our man will do the job. When they go out to a railroad, our people are considered officials of that railroad, able to perform the same duties and make the same decisions as an official of that railroad; however, they are there only as a guest of the railroad and are working for the senior railroad official in-charge."

Bureau of Explosives field inspectors with support from headquarters staff handle roughly 200 railroad/hazardous materials incidents in an average year of which approximately 150 will require an on-scene response while 50 or so will be handled by telephone. There are certain commonalities among incidents. A lot of times the problem is a ruptured disc, particularly with acids or corrosive materials. Also, a fitting not properly secured is quite common: a bottom outlet cap left loose and hanging on its chain; a valve not completely closed; a manway cover with its bolts off on a nonpressure car; or a plug left off on a pressure car. Such problems often stem from a mistake in unloading or loading of the car, a failure on the part of the shipper.

Last year, Bureau of Explosives personnel responded to 55 to 60 derailments, approximately 120 leaking tank cars, and 12 boxcar or T.O.F.C./C.O.F.C. incidents.'' (Trailer-On-Flatcar/Container-On-Flatcar) Boxcar and trailer/container incidents are small package problems; surprisingly, they are often the ones that cause the headaches. As one Bureau employee noted, "With a tank car, you fix or isolate a valve, or you off-load the car. Boxcars/trailers/containers, on the other hand, are often two or three day events because you have to sort out exactly where the leak is, determine what you have to do about the leak, and then proceed with decontamination/reloading/blocking/bracing.''

"The percentage of tank cars owned by railroads is very small," says Mike Miller, "the great majority are owned by shippers. The regulations say the shipper is responsible for securing a car, and we hold the shipper responsible for problems that can be clearly traced to improper loading or securing of the car. When a Bureau field inspector responds to a leaking tank car incident, he completes a detailed report as to what he found wrong with the cars and sends a copy to the inspector responsible for the area from which the car originated. Say an inspector in Cincinnati has a problem with a car that originated in Houston. He sends a copy of his report to our inspector in Houston who will take that report and visit the shipper. The best way to solve a problem with hazardous materials as far as tank cars are concerned is to go to the source—the shipper. Our man will make an inspection of that shipper's facility and do everything he can to help people there secure their cars correctly and comply with D.O.T. regulations.''

Roy Holden is Manager of Technical Services for the Bureau of Explosives, and one of the country's foremost authorities on tank cars. ''When we find a type of car that did not perform well, we have the capability of locating every similar car within a day and stopping it dead if necessary,'' he emphasizes. ''This year we had what appeared to be a very minor tank car incident in Massachusetts which in our opinion, when examined by a competent fracture analyst was actually very serious. This finding triggered the deadlining of about 2000 similar cars, and the railroads are spending $9,000 to $11,000 per car to correct the flaw. A problem can surface in something as simple as a stainless steel sampling line on an L.P.G. car. If you just bought a carload of gas, before you accept it you take a sample through a quarter-inch sampling tube. There is a little nipple with an ordinary pipe-thread that connects to a very heavy coverplate. The part was observed to be cracking and causing a leak in a very few instances. We were able to prove exactly why it did what it did, and ordered a change in the weight of the pipe and material used to make the nipple. Such improvements do make a difference in overall safety because with even a small leak you are going to have people getting very concerned. The problem with the steel sampling line was not recognized as a problem until one of our inspectors identified it as such. Any time we get a crack or a leak we do not understand, it triggers an in-depth investigation of why the container did not perform properly. We have to know why it was inadequate, what is the likelihood of the same situation occurring again, how many similar containers are in use, and whether or not we must stop them in their tracks. We also have to determine proper corrective action and establish a timeframe for getting it completed. Eventually, the computer will be able to perform the tracing function for us. Presently, however, the details of construction on the older tank cars—some of which may remain in service for 40 years—have not been added to the computer memory bank. Sometimes it is easier when we find a fault to identify the owners of all similar cars, get them on the phone, and have them make a detailed inspection of all the suspect cars they operate.''

With ever-increasing numbers of agencies and organizations represented at any hazardous materials incident, perhaps the most often asked question related to incident response is . . . 'Who's In Charge Here?'' ''We are there to advise the senior railroad official on-scene,'' responds Mike Miller.

''He is in-charge of cleaning up the wreck and taking care of his railroad as far as I am concerned. It is his property, and he knows best how to handle the situation. He may defer to us or to any other emergency response people that have been brought in. He wants to get it cleaned-up, get the fire out, mitigate the spill as quickly as possible because the railroad wants to get its business back in order and get their trains running again because that is what railroads do for a living. The Bureau of Explosives, on the other hand, handles hazardous materials; we tell them what to do with the tank car. They know how to run the cars down the track, but they want our expertise. In the last five years the process has been complicated by an influx of government involvement, and civilian intervention on the part of fire departments.''

''We recognize that fire departments are there to help us,'' adds Miller. ''We emphasize that our inspectors are to get out and talk to local emergency responders such as fire departments and provide training when possible. However what fire departments lack is training in hazardous materials. Railroads have done one heck of a lot of work and put a lot of money into training fire departments. Unfortunately, there have been articles written in fire service and related publications with themes that suggest the interests of railroads are incompatible with those of response agencies. We do feel that at a railroad incident the railroad is in-charge. Recently at a railroad incident in an eastern city the local fire department did a great job; conversely, at an incident in a midwestern city the fire department response was not so great. I received a call from a railroad superintendent who reported a pencil-sized stream of acrylonitrile leaking from a tank car with approximately 12 inches on the ground. He reported that the fire department was on the scene, and that he had provided firefighters with the emergency response data necessary to properly control the spill.

I cautioned him, 'Please tell the fire department they MUST NOT put water on the acrylonitrile.' He ran outside; five minutes later he was back to say . . . 'Too late.' It cost a million dollars to clean-up that acrylonitrile spill because the fire department responded, and the first thing they did was throw water on it. There was absolutely no reason for that to have happened. The only reason the fire department should have been there was to make sure there was no fire, or if there had been a fire to control it and get people out of the area. As to 'Who's In Charge Here?' . . . unfortunately a lot of times it is civilian people who are not sure of what they are doing. There is a film called 'BLEVE' that most response people are familiar with. In the original version of the film, which I understand has now been taken out, a fireman is on top of a ladder with a hose spraying a tank car. The fire chief says . . . 'We don't know what is in those tank cars' . . . yet he has a man up on a ladder spraying waterover a burning tank car not knowing what is on fire. That is the kind of thing the National Fire Protection Association, the railroads, and the Bureau of Explosives are trying to get rid of. We do not need that type of response; it can get somebody killed.''

"One of the first things any of our people do when they arrive at an incident scene is survey the area being very careful to *verify* the information they already have," explains Roy Holden. "They make *sure* they know what product is involved. They check the volume of any spill and the topography of the area, as soon as they have checked in with the senior railroad official on-scene. They also introduce themselves to people on-scene such as firefighters, environmental people, clean-up contractors, and government officials. Presumably, at any incident in any inspector's district he will already know the railroad officials involved; he will not be meeting them for the first time. An inspector's normal, day-to-day duties require that he inspect yards, stations, and procedures of any railroad that handles or ships hazardous materials. He covers his district making sure he knows the railroad officials whether he is responsible for a three or four state area or a one-state area like California. After a certain amount of time in his district, he will become known to all railroad personnel who are likely to be involved in response to an incident. Any emergency response organization tends to feel . . . 'I know that person; I know what I can expect from him.' If they trust a person, they just naturally tend to work closely with him. Our inspectors attempt to develop such a relationship with the railroads in their district; as well as with fire departments, federal agency representatives, and environmental people.''

"The railroads, of course, may bring in private contractors," adds Mike Miller. "Our people know who the private contractors are, and usually know who will be responding—the actual person. For example, Art Proefrock of Hulcher Emergency Services and I have worked a heck of a lot of derailments together. We each know how the other works. It gets to be a situation where you see the same people at incidents.''

"There are many decisions that have to be made on-scene," says Roy Holden. "You may have additional damaged cars that may or may not be leaking. You attempt to examine them to determine if they are in condition to be picked-up and rerailed, whether the commodity has to be transferred on the spot, or whether it would be better to move the car before you transfer product. Making these decisions is an experience thing. You must consider both the type of product in a particular car and what that specific container can withstand. A tank car is not 'just a tank car;' of the 200,000 or so that are out there, there may be 100 different types. Each type is designed for a specific function, and the degree of containment represented by different types of tank cars may vary widely. That is one of the reasons there are detailed specifications for tank cars. A lot of mistakes are made by inexperienced people on-scene. An emergency response organization may report they have a gas car leaking from a bottom outlet, yet that is impossible. We know gas will be in a 112, 114, or 105 car—a pressure car. We know there won't be a bottom outlet. Instantly, we know it is some other problem or other fitting they are talking about. This is a problem in communications, and a Bureau inspector must be constantly aware of the problem. In the early stages of rail incidents, we *often* receive misinformation.''

''You must first carefully identify what you have on scene for cars and products, the problems that can be identified for each car, and prepare recommendations as to corrective procedures,'' continues Holden. ''What needs to be taken care of first? If it is still night, is the material and equipment you need for corrective action something that can be brought up at night? Do you need to call in a specialist? Will it be necessasry to off-load materials, and can you acquire the empty containers and transfer equipment that may be needed? The initial need is always for correct information. *Verify* what you have, catalog the necessary actions, make a plan, then go ahead and carry it out. To identify what is in a particular car, we use a consist (a document that shows the position of each car in a train) and the waybills.

If you happen to catch something traveling in an 'N.O.S.'—category, (Not Otherwise Specified)—say you have a 'Flammable Liquid-N.O.S.'—call the shipper and get the *exact* properties of the material. Normally, we use radios to get such information. We have to depend on the carriers for radios because different railroads use different frequencies. We cannot carry a radio with enough crystals, nor would we know the various frequencies in use at a particular scene. Oftentimes, however, a telephone can be installed in just a couple of hours, or you might use a radio link to a microwave system. Generally, the railroad company will supply the communication signals that we need.''

''We have guidelines for response,'' adds Miller. ''You respond when you are told to respond; you respond when there is a leak; you respond when there is a request. Many of the government agencies have textbooks on the subject that say you will respond when certain things happen—you will make this decision when this, this, and this happens—that provide a model where you plug yourself in. We don't do that, and I doubt we ever will because our actions are based on and determined by past experience in similar situations. You don't go back to the basics when making emergency response decisions anymore than you do when you decide to go eat. What you are going to do when you get there depends on what you find in the refrigerator. A lot of considerations and decisions on-scene become second nature through experience; because of that fact, we double-check ourselves. For a major incident, we are going to be responding from headquarters; what one of us doesn't catch, another should be able to. We also double-check the inspector. Because of long experience, you know you are not as good as you might like to think you are. There is always somebody there to check you.''

Michael Miller was asked recently if fear is a consideration even among experienced personnel who have responded many times. ''Fear is always a factor,'' he says. ''If I showed you a derailment—where we had two L.P.G. cars leaking, and maybe an acid car—and you stood there in your rubber boots and protective gear and said—'I'm not afraid to go in there; let's do it'—then I would become afraid. I would have a very high level of anxiety about going in with a cowboy who might get me killed. Even some longtime railroad employees with extensive training will say at a derailment: 'I don't want to go in there; those tank cars scare me. I know exactly how that tank car is built. I can tell you it is 7/8th inch grade B-212 steel, has four inches of insulation, and a 1/8th inch jacket. I can tell you where every fitting is located on the car, but I don't like that stuff.' That is why our people are there. It is not that they don't have a fear of their own, but the fear is diminished by experience. Like anyone else, we too are afraid of doing anything that we are not familiar with. That is one of the problems of emergency response to hazardous materials by fire departments; they are scared to death of what they think might happen. They have seen the 'BLEVE' film, and they think they know what is going to happen here; so they move in and pour 100,000 gallons of water on an acid car that has a ruptured disc, and all we are getting is fumes. Or they go in and evacuate a half-mile area for an L.P.G. car that has a valve leaking from its seat—which means I can smell the product if I open up the dome and sniff around. Too often the man unloading a car recognizes there is a problem, he reports it to local officials, and all of a sudden the local officials are in there saying 'WORST CASE.' A lot of unnecessary problems are generated out of fear that causes people to overreact. Therein lies the reason for extensive educational work railroads and industrial organizations do with local response agencies. A good healthy fear is definitely not bad, but it must be balanced with knowledge, experience and training or some poor decisions can easily be made.''

Although Bureau of Explosives personnel have been responding to railroad hazardous materials incidents for 75 years, there have been a number of changes in the way they respond according to Miller. ''Things have changed drastically because of the new and exotic chemicals being produced. It is no longer just sulfuric acid and oil that the railroads are transporting. The increase in volume and variety of hazardous materials being transported has made a major change in the way our people respond. Perhaps at one time you might just pack your ditty-bag and go out and look around, rather with the feeling—'I've got a pair of old rubber boots and an adjustable, open-ended wrench here in my trunk that I am going to take care of it with.' Nowadays, response requires good personal protection and equipment that will not react with the chemical you are working around. Our people have always been cautious, but now they are more and more cautious. For example, we are currently initiating a program that requires continuous medical histories on our inspectors. If they are involved with certain chemicals, we want them checked out, and we want an annual check on their physical condition as well. Emergency response has to be done by a person who is physically able to do it; a person able to be on-scene for 24-hours; a person in such physical condition that he remain alert enough after 24-hours on his feet to know enough to say—'I need rest; I need help.' ''

''The duration of an incident has changed as well,'' adds Miller. ''The companies no longer work 24-hours, and this change has come about in just the past few years. They will say—'I'm not going in there until it is light.' Previously, we used to go in when it wasn't light. After several years, we realized how wrong that was. The current policy of many of the emergency response organizations that are for-hire is that it is a mistake to work at night. You may get somebody hurt.''

''The response itself is becoming more specialized with the advent of E.P.A., O.S.H.A. and the states responding with their own people. No longer does our man have to be *just* a tank car/railroad/chemical/mechanical expert; now he must also be a public relations man, a communicator, and he must have expertise in environmental clean-up and decontamination.''

''For 75 years now we have been protecting the environment and mitigating damage caused by spills; that's what our people do,'' relates Miller. ''If there is a leak, they attempt to plug it and then explain to railroad management how to neutralize it or clean it up. The reason the Bureau has had a Manager of Environmental Services since 1978 is to keep pace with advances in environmental technology. If there is a spill, we clean it up; and clean-up does not mean surface clean-up or strictly emergency-type mitigation operations. It can mean longterm wells, monitoring, and follow-through. It is a natural outgrowth of something we have been doing for 75 years, something that railroads have realized they have to do. If you have a spill, you clean it up; you don't fight with anybody. Ann Mason, Manager of Environmental Services, is a qualified expert able to devote fulltime to advising railroads on environmental concerns. 'If you want to clean-up that spill, here's what you should be doing right now, within five days, within ten days, within six months—here's what is going to happen.' None of our 19 inspectors in the U.S. and Canada can spend six months at a spill site. They will spend all the time that is necessary to get rid of the emergency; but once the spill is no longer an emergency situation, once people's lives and property are no longer threatened, they are going to leave and go about their regular duties. In the majority of cases, Ann Mason's expertise will be called upon for clean-up guidance rather than for emergency response although she also answers questions resulting from an emergency situation. 'I need to know what is going to happen to this particular chemical when it reacts with another chemical; what happens when it gets into this clay, this dirt, or this water? What do we need to do right now in an emergency situation?' ''

''At any major incident, we definitely use a command post,'' continues Mike Miller. ''I can't think of any time when we did not. It may not be the most elaborate thing; I have been at incidents where the command post was a semi-trailer or a nearby school. Sometimes, it is as simple as saying—'Go talk to the man with the orange cap; he is the person we are coordinating all activities through.' Overall, there has been increased use of a command post, or a meeting to make a decision. Unfortunately, this has allowed a lot of people who get involved in emergency response only rarely to make a decision by committee. A derailment is not a democracy, however. There have been people on-scene making decisions, agonizing over making decisions, that experienced response people would have made hours before. In one incident that comes to mind there was a tank car over on its side with the dome down and the safety valve in the dirt. There was no physical damage; what needed to be done was get the safety valve back up in a vapor space. It took seven hours to convince civilian officials that the safest thing to do was right the car. Here was a situation where civilian officials were not being safe; they were jeopardizing public safety. A lot of times the people who have the experience are not allowed to make the decisions or even have a voice in the decision making process. In the past couple of years, decision making has been taken away from the people who should be making the decisions because of the idea of the 'command post decision' and the 'major disaster.' We are getting to a point where I personally feel too many people are involved, yet on the other hand I feel people should be involved. We ought to have somebody on-hand from the state E.P.A., for example, because he is worried about the air and water—we too are worried about the air and water. Sometimes, they do not recognize that we have to get this emergency situation over with, and they do not allow the situation to be handled with the expediency it would have been handled in the past. Our world is becoming a smaller place because of advances in communications; occurances that would not have been viewed as an incident 20 years ago are now viewed as major incidents.''

Field inspectors with the Bureau of Explosives of the Association of American Railroads were the first and are currently one of the most experienced mechanisms for organized response to and control of hazardous materials transportation incidents. Their record of accomplishment in a highly specialized field has spanned a period of 75 years. ''I come from totally an operational background and recognize that certain personal prejudices may result from that experience,'' sums-up Michael Miller. ''Last year, my wife and I watched a television program that was presented as an expose of hazardous materials transportation by a well-known commentator for a national network. 'My God, does this upset me,' I said to my wife. 'What are you upset about?' she asked. 'That is why you have a job. People like this just make your job more secure. You might even get a raise.' That wasn't exactly what I wanted to hear,'' laughs Miller. ''I guess I wanted people to recognize that the safe transportation and handling of hazardous materials is the reason we have 19 field inspectors in the U.S. and Canada. That is the reason Roy Holden and I and all the others are here at Bureau national headquarters in Washington. That is why we have a laboratory. The safe transportation of hazardous materials is what we do for a living, what the Bureau of Explosives has been doing for 75 years. It is what people on railroads are trained to do, why we put money into training programs, why we sit on advisory committees, why we publish *Emergency Handling of Hazardous Materials in Surface Transportation,* and the reason we attempt to talk forthrightly to writers and other interested persons. We would like people to realize that emergency response is not a new thing with us, even though it may all of a sudden have become a new thing, a new 'hype' to all kinds of people. The Bureau of Explosives has been continuously involved in developing and insuring compliance with regulations for the safe transportation of hazardous materials since 1907. Nowadays, there are all kinds of people presenting programs on regulations—'Here are the regulations, and here is what you do'—and I have been *amazed* at their lack of expertise. They have not matured to the point we have just through experience. We are not the ultimate; I recognize that we too are currently getting into new areas as are others. However, there is a problem working with people who are 'instant experts' even though they have no background in the subject matter—and it scares the Hell out of me. You talk about *fear*. I am more fearful now than I have ever been before because I don't know that person who is going to be next to me. I am *very* careful about who I go into a derailment with. Other people with experience in emergency response to hazardous materials incidents are often the same way; they are careful of who is around. If a new man appears at a derailment scene, they are careful of that man and make a point of knowing what he is doing at all times.''

"I would emphasize that a rail incident involving hazardous materials is primarily a problem of that railroad," adds Roy Holde, "and the biggest problem we have today is that too many people think it is their private problem. The question of 'Who's In Charge Here?' continues to be a major concern. We have stated who we feel is in-charge, the senior railroad official on-scene. We look to him, but frequently he has to look to all sorts of people."

The Coast Guard Atlantic Strike Team

The National Strike Force is an outgrowth of the Federal Water Pollution Control Act of 1972 that required publication of a National Oil and Hazardous Substance Pollution Contingency Plan to provide for efficient, coordinated and effective action to minimize damage from oil and hazardous substances. The Act required that the National Contingency Plan include establishment or designation of a strike force of personnel trained, prepared, and available to provide necessary services to carry out the plan. Coast Guard personnel attached to three strike teams form the nucleus of the National Strike Force by providing communications support and advice and assistance for oil and hazardous substance removal. The National Strike Force may also include the Environmental Protection Agency's Environmental Response Team when needed. Coast Guard strike team personnel have expertise in ship's damage control techniques, diving, and pollution removal techniques and methodologies. EPA Environmental Response Team members have the expertise in biology, chemistry, and engineering necessary to provide advice on the environmental effects of oil and hazardous substance discharges, the removal of such materials, and mitigation of the effects of discharges.

The National Strike Force is composed of three Coast Guard regional strike teams. The Atlantic Strike Team, which includes the National Strike Force Dive Team, is based in Elizabeth City, North Carolina and responds throughout the northeastern United States. (including the entire northeast coast from North Carolina to the Canadian border and westward through West Virginia, Ohio, Michigan, Indiana, Illinois, Wisconsin, and Minnesota—including the Great Lakes.)

The Gulf Strike Team stationed at Bay St. Louis, Mississippi covers the southeastern United States including the entire Gulf coast as well as South Carolina, Kentucky, Missouri, Iowa, New Mexico, Oklahoma, Kansas, Nebraska, Puerto Rico, and the U.S. Virgin Islands.

The Pacific Strike Team based at Hamilton Air Force Base in California is responsible for pollution response along the west coast as well as in Arizona, Colorado, Wyoming, South Dakota, North Dakota, Hawaii, and Alaska.

Each strike team maintains a constant state of readiness to be able to provide technical knowledge, supervisory assistance, and deployment of special pollution response equipment in support of a Federal On-Scene Coordinator. Strike teams are on continuous standby alert 24 hours a day, 365 days a year; and utilize Coast Guard C-130 aircraft, helicopters, commercial aircraft, or motor vehicles to transport personnel and equipment to the scene of a pollution incident.

The Atlantic Strike Team is staffed by 38 persons of whom 13 are attached to the National Strike Force Dive Team. The Pacific Strike Team has 16 members while the Gulf Strike Team has 26. Strike team members are specially trained in marine environmental systems and protection, heavy equipment operation, pollution investigation, diving, advanced rescue techniques, and hazardous materials handling.

The Atlantic Strike Team component of the National Strike Force operates out of the U.S. Coast Guard Air Facility at Elizabeth City, North Carolina with Commander Donald S. Jensen as commanding officer and Lieutenant Robert A. Strong as executive officer. As is the case with the Gulf and Pacific teams, the Atlantic Strike Team has four basic missions in order of priority: Response; Training; Planning & Liaison; and Research & Development.

The team's response mission is to provide the federally-designated On-Scene Coordinator with communications support; monitoring and documentation services; advice on oil and hazardous substances control, containment, and recovery techniques; and advice on damage control assessment and diving techniques. The team may provide advice on and monitor the performance of a commercial contractor, but also possesses the specialized equipment and experience necessary to perform required tasks when a qualified commercial contractor is not readily available.

The team's training mission requires that members provide both unit training of strike team personnel to ensure a high state of readiness and the ability to perform Atlantic Strike Team missions, and the training of USCG Marine Safety Offices, and Captain of the Port Offices, in response methods/tools/equipment/materials.

A Planning & Liaison mission requires that team members attend Regional Response Team meetings; assist in preparation of Local and Regional Contingency Plans; and maintain a working relationship with key personnel from local/state/federal agencies, cleanup contractors, and industrial groups involved in environmental protection and pollution response.

With regard to Research & Development responsibilities, the Atlantic Strike Team participates in testing and evaluation of government-developed pollution response hardware and techniques, and provides assessment of the capabilities of such equipment through operational experience.

"National Strike Force response standards require that we be able at all times to immediately dispatch at least one person to the scene of a pollution incident," explains the Atlantic Strike Team's executive officer, Robert Strong. "In addition, we must have four people on their way within two hours. Twelve strike team members must be enroute with all their equipment in no more than six hours. These response standards do not present a problem for us; we can meet them easily."

"Each team member is individually charged with the responsibility for being ready to go," continues Strong, "and we are advanced a fairly sizable amount of money on annual orders as a form of readiness preparation. Each of the enlisted personnel is advanced $500 while officers receive a $1,000 advance—strictly to have money that may be required when dispatched without advance warning. Because we occasionally respond internationally, designated individuals at various levels are required to obtain and maintain valid passports. As new people rotate into the team, we have to obtain additional passports. All of our equipment is either prepackaged or on pallets. If aircraft are available, inside of two hours we can usually have all necessary personnel and required equipment out of the shop."

The minimum of one person who must leave for the scene immediately upon notification of an incident is often used as a "lead person" to evaluate conditions on-site and report back as often as necessary. He or she assesses the situation, notes what other agencies or individuals are already involved, learns what equipment may have already been ordered, determines mitigatory actions that have been taken, and identifies exactly what strike team equipment may be needed. The lead person must have complete knowledge of strike team capabilities and equipment and be able to interact well with the variety of interests that may be represented at an incident location.

The Atlantic Strike Team is activated only as the result of a request from a predesignated Federal On-Scene Coordinator. Although the majority of strike team responses have involved working for either a Coast Guard or Environmental Protection Agency On-Scene Coordinator, the strike team could be activated by an OSC attached to another federal agency. Although strike team personnel may wear Coast Guard blue, on-site they often operate under the direction of a civilian employee of the Environmental Protection Agency.

The Atlantic Strike Team has the capability necessary to respond to a wide variety of assignments. For very labor-intensive assignments such as beach cleanup after a major oil spill in coastal waters; team members may keep track of laborers, insure the cleanup proceeds at an acceptable rate, or maintain cost documentation records so the government can be reimbursed for services provided. For an oil spill at sea, a team and necessary equipment might be put aboard the heaving deck of a tanker, or the helipad of an offshore drilling rig, by a Coast Guard helicopter. Because the strike team must also respond to incidents where oil or chemicals endanger navigable waters of the United States, members may find themselves working at inland hazardous waste sites.

The Atlantic Strike Team has a great deal of experience with oil spill response and is rapidly gearing up for chemical response. "We have a very viable role when it comes to rapid response with all equipment rolling to an oil spill pump job," notes Lieutenant Strong. "Some of our oil spill response gear is prohibitively expensive for commercial contractors, and industry does not have the rapid response capability we are provided by Coast Guard aircraft and vessels. Industry would have to go out and contract for such support; we have it in-house. For a chemical incident on the other hand, industry and commercial contractors are already on-line with much of the necessary equipment, and transportation to the scene does not often present an insurmountable problem for them. Our position is that we are not going to pre-empt industrial and commercial responders if they are capable of doing the job in a safe manner. The National Contingency Plan clearly defines a role for industry and commercial contractors. We look critically at every job assigned to insure that we are not, merely for the thrill or whatever, taking business away from an industrial concern. We will go only if there is a threat to the environment that needs to be abated immediately."

The Atlantic Strike Team maintains an extensive array of tools, equipment, and materials—almost all of which can be transported aboard Coast Guard C-130 aircraft. A 33-foot motorhome is often used as a mobile command post for operations ashore and frequently becomes a very central location on-scene where communications are consolidated, briefings are held, and plans made. The motorhome is equipped with VHF-FM radio gear and a multichannel scanner that allows command post staff to monitor communications of other agencies and organizations involved in a particular assignment, or transmit on basic Coast Board operating frequencies. Normally, even in desolate areas, a telephone can be installed within a few hours. For a two or three day assignment, the team will rely on a portable generator to supply all power needs of the command post. For a longer stay the team would attempt to have commercial power provided, although if necessary the motorhome can operate on generated power for weeks at a time.

Whether or not team members live out of the command post depends on the type of operation. If the job is a 24 hour a day operation, the vehicle is manned around the clock. At a waste site cleanup where the workday may run from 8 A.M. to 6 P.M., the command post vehicle will be shut down for the night.

The strike team is responsible for maintaining and deploying, and testing and evaluating, specialized pollution response equipment.

The Air Deliverable Anti-Pollution Transfer System (ADAPTS) consists of submersible hydraulic pumps powered by a 40 horsepower diesel engines. ADAPTS is prepackaged on pallets at strike team headquarters ready to go on a moment's notice. Two pallets are required. One contains the prime mover, pumps, and hoses. A second pallet carries support equipment such as an emergency lighting system, tools, and first aid gear. ADAPTS is designed for removing oil from stricken tanker vessels and barges. The system is capable of transferring crude oil at a rate of 1,000 gpm against a 60 foot head and through 300 feet of six inch diameter transfer hose, but is not designed for viscous oils.

The Dracone Barge, rather like an immense rubber bladder, is used when it is impossible to get a rigid barge next to a ship, or when wind and water conditions are such that a rigid barge might hole the vessel being offloaded. Three different sizes of Dracone Barge are used by the Atlantic Strike Team. The largest is 300 feet long, weighs 17,000 pounds, and will hold 240,000 gallons of product. Other Dracone Barge units will hold 40,000 gallons and 10,000 gallons of product. The 10,000 gallon model was used recently to offload diesel fuel from a commercial fishing vessel that grounded on an oyster bed off the Maryland coast. The water was too shallow to allow use of a rigid barge, but team members were able to bring in a Dracone Barge at high tide, allow it to settle at low tide, transfer diesel oil from the grounded fishing vessel, then tow the partially loaded barge off the oyster bed as the tide rose again. Four or five trips were required to remove the fishing vessel's fuel from the environmentally sensitive area. When loaded, the Dracone Barge is kept afloat by the difference in specific gravity between its cargo and the surrounding water rather than by any airspace within the bladder. Loaded with oil, the barge will find a neutral bouyancy point. Loaded with light weight diesel oil, for example, the Dracone Barge would be decks awash. A high specific gravity product would sink it. The bladder is inflated for towing at sea, then deflated for use. That is, air is pumped out as product is taken aboard rather than having the product displace air. Deflated, the barge can be towed at a maximum speed of two knots; inflated, it can be towed at speeds up to nine knots. The larger Dracone Barge is stored on a lowboy trailer so it can be transported over-the-road or put aboard a C-130 aircraft.

The Open Water Oil Containment System (OWOCS) is designed to contain, control, or direct oil on the surface in open water. It is designed to contain oil in five foot seas and 20 mph winds and to survive structurally in ten foot seas with 40 mph winds. When deployed, the barrier has a 21 inch freeboard and a 27 inch draft. Sections 612 feet long prepackaged in an 18 foot by nine foot by 62 inch container can be deployed in a matter of minutes from a Fast Surface Delivery System sled. As the barrier is pulled out of its container, ripcoards activate CO_2 cartridges to inflate each float individually. A combination of rigid struts and flexible curtain sections provides very good wave conforming ability, and an external tension line allows the barrier to follow and maintain proper attitude toward the seas. Of the 15 such barriers assigned to all three strike teams, six have been modified to permit oil recovery as well as containment.

The Lockheed Cold Weather Oil Recovery System consists of a welded-steel catamaran-type hull that houses a rotating seven foot by four foot disc drum. An onboard 40 horsepower diesel engine provides power to operate the disc drum, an ice tumbler, and the recovered oil transfer pump simultaneously. Oil adheres to the discs and is wiped off mechanically by plastic wipers, then flows into a central trough and is collected in a 200 gallon tank and is pumped to a support vessel or storage container. The unpowered unit must be towed and is designed for use in semi-protected harbors rather than on the open sea. Its maximum oil recovery rate is roughly 18,000 gallons per hour but much depends on factors such as oil slick thickness, viscosity, temperature, sea conditions, and oil encounter rate. It can be disassembled into seven separate pieces for transport by truck, ship, or aircraft. One disassembled unit will fit into a C-130 aircraft with room to spare.

The Fast Surface Delivery System (FSD) is a 45 foot aluminum planing hull with a 15 foot beam that was developed by the Coast Guard to provide an alternative delivery mechanism for ADAPTS and OWOCS. The sled can deliver or pickup a 20,000 pound floating payload and be unloaded manually. In tests, the 10,000 pound sled plus payload has been safely towed by an HH3F helicopter at speeds of up to 50 knots and by a Cost Guard 82 foot vessel at speeds up to 17 knots in up to six foot seas. The FSD sled is a vessel designed both to sink and float as required. It can be towed to the scene of operations in a planing mode then sunk by opening valves to permit loading or unloading of a floating payload. Taken in tow and brought to a planing mode again, the sled will empty itself of water; the valves are closed, and the sled reassumes the floating characteristics of a boat.

Although the specialized equipment described above is used mainly for waterborne oil containment and recovery, the Atlantic Strike Team also maintains a sizable inventory of chemical response and general support equipment. Team members have to be able to use, maintain, and sometimes repair totally encapsulated chemical suits, respirators, selfcontained breathing apparatus, and vapor/gas monitoring and detection equipment; as well as firm-hulled and soft-hulled (Zodiac) boats, outboard motors, generators, and diving equipment.

The National Strike Force Dive Team, a component of the Atlantic Strike Team, assists in assessing the condition of stranded or damaged vessels and provides other underwater activities needed during a pollution response.

To achieve and maintain a high level of readiness, strike team personnel participate in near-continuous training. "Unit Training" is a highly defined and highly refined in-house activity that began evolving in the summer of 1979 when factors that affect mission performance were studied. Four critical factors were identified. One such factor was a very high personnel turnover rate. Now a four year assignment, the Atlantic Strike Team was until quite recently a three year tour of duty. Another area identified was the very highly specialized and diverse mission of the strike team. One day the team may respond to a major oil spill; while the next day may bring an abandoned waste site, chemical warehouse fire, train derailment, or helicopter salvage assignment. A third high impact area affecting mission performance was infrequent response versus a high state of readiness. Fortunately, Argo Merchants do not go aground or Burma Agates burn everyday, yet when they do there has to be a group of trained personnel ready to respond *immediately* with appropriate response actions. A final factor identified as having major impact on mission performance was peculiar, one-of-a-kind equipment. The Atlantic Strike Team maintains an extensive inventory of open water response equipment that has been generated with federal monies because there is not sufficient need for it in the private sector to provide an economic incentive for its development. The operation, maintenance, repair, and use of such equipment is not taught at military or commercial schools yet such expertise must be developed by team personnel. Achieving acceptable proficiency levels requires a great deal of training time.

Guided by the previously mentioned high impact factors, a unit training program was developed that consists of four basic elements: a unit qualification program; attendance at various government and commercial schools; an inhouse program of classroom lectures and hands-on drills and exercises; and a physical fitness program.

The unit qualification program requires that team members complete certain qualification factors and then demonstrate their proficiency to predesignated team members who sign and date a checksheet upon successful completion of each factor. Once all fctors have been completed for a particular qualification level, the individual appears before a qualification board made up of experienced team members and is quizzed on all aspects of the qualification level to insure that adequate and sufficient training has taken place.

There are separate qualification factors required of response petty officers, response supervisors, and response officers. A response petty officer must understand strike team equipment, philosophy, and policy well enough to be able to support any National Strike Force mission. A response supervisor is an individual who is able to direct a group of response petty officers and turn them into a working unit able to accomplish some end goal whether it be cleanup of five miles of beach area or operation of an ADAPTS pumping system to off-load a damaged tanker. A response officer must be capable of standing alongside a Federal On-Scene Coordinator and advising on behalf of, and speaking for, the National Strike Force; a person able to coordinate an overall response operation.

The Atlantic Strike Team utilizes a variety of training facilities, both government and commercial, to supplement basic competencies required of team members. Selected personnel attend training provided by the Environmental Protection Agency's Environmental Response Team at Edison, New Jersey. In order to be able to provide training of other units and strike team personnel, selected team members are sent to a Coast Guard instructor training facility at Governor's Island in New York City or to a similar Navy program at Norfolk, Virginia. All strike team truck drivers complete a commercial truck driver training course at Charlotte, North Carolina from which they emerge I.C.C.-certified. For hydraulics training, outboard engine repair, hazardous materials response, and other courses; the Coast Guard Reserve Training Center at Yorktown, Virginia is utilized. The U.S. Army trains strike team aircraft loadmasters while Texas A&M is used for marine firefighting instruction. Because strike team assignments often involve working in hazardous environments, Emergency Medical Technicians are always assigned to the team; currently, seven team members are Emergency Medical Technicians. E.M.T. training is provided by the Coast Guard's training center at Petaluma, California. Dive team personnel attend a number of U.S. Navy diving programs, and chemical response team members the U.S. Army Technical Escort School at the Redstone Arsenal in Huntsville, Alabama. Selected officers and senior petty officers are required to spend time aboard a tanker, a barge, and an offshore platform to become familiar with their systems and operations. These training visits are arranged through various commercial companies. Strike team members assigned to maintain and/or repair breathing apparatus and monitoring equipment attend technical schools provided by equipment manufacturers and vendors.

Unit training also includes inhouse lectures, exercises, and drills. Much of the training

U.S. Coast Guard
National Strike Force Equipment Inventory*

Skimming Barriers

Pump Floats

DRACONES (10K, 40K, 240K)

ADAPTS PRIME MOVERS

VOPS (Viscous Oil Pumping System)

Submersible Pumps
    Stripper
    Single stage
    Double stage
    TK-4 (chem)
    TK-5 (chem)
    Thune Eureka
    Sloan

Salvage Equipment
    Salvage pumps (385—610 gpm)
    Single diaphragm pumps
    Double diaphragm pumps
    Non-submersible hydraulic pumps
    Damage assessment kit
    Patching and plugging kit
    Lift bags

Generators
    Onan 2.5, 6 kw
    Honda 2.5, 3.5, 4.5 kw
    Homelite

Air Compressors (service)

Air Compressors (breathing)

Boats
    21' TANB
    22' Boston Whaler Outrage
    Zodiacs
    Assault boats

Communications Equipment

Response Vans
    Dive response van
    Chemical response van

Vehicles
    Mobile Command Post
Tractor Trailers
Others as required

Breathing Apparatus
    Survivair 30 min. SCBA
    Survivair 60 min. SCBA
    Biopak 60 SCBA
    In-line air systems

Fully Encapsulating Suits
    ILC Dover
    Eastwind

Assorted Personal Protective Equipment for HAZMAT Response

Safety Monitoring Equipment
    HNU photoionizer
    Organic vapor analyzer
    Biomarine 900R & RS
    Gastech Mod 1562
    Draeger Colorimetric Detectors

CHEMICAL INFORMATION SYSTEMS
    OHMTADS
Technical library

Diving Equipment
    SCUBA
    USN MK 12 diving system
    USN MK 1 / Superlite 17 diving system
    U/W wireless communications system
    U/W acoustical locator
    Side scan sonar
    Aircraft salvage kit
    U/W cutting and welding equipment
    U/W hydraulic tool kit
    Diver propulsion vehicle

*This list does not represent the complete inventory of the NSF nor does each team have all of the above listed equipment in their inventory.

Figure 1.

relates to operation and maintenance of equipment. A 3rd Class Boatswains Mate may demonstrate the proper use of an emergency generator to supply power to a mobile command post. A Chief Petty Officer may set-up and demonstrate a viscous oil pumping system.

Hands-on drills tend to be frequent and demanding. An on-going training scenario features a training barge moored in the nearby Pasquotank River. The barge is constructed of nine Army pontoon bridge sections and has integrated tanks that can be flooded to force the barge into configurations ranging from decks awash to high in the water; a pronounced list can be added if desired. In one recent training exercise the barge was assumed to have broken away from its tug and grounded causing a chemical-containing drum to roll overboard. A response team is put aboard the barge either by helicopter or Zodiac rubber boat. Maintaining continuous radio contact with a command post ashore, they find two crewmen down. An evacuation of the injured crewmen is completed before damage control assessment can be undertaken. Meanwhile, a dive team performs a survey of the barge hull utilizing an underwater video system. A damage statement has previously been chalked on the hull. The dive team reports the extent and circumstances of damage, and the boarding party must determine how to circumvent the damage. The divers are also tasked with finding the drum that fell off the barge, photographing it so command post personnel can identify the hazard involved, and attaching a sling so the drum can be towed to shore. A shore party will then have to overpack the drum and arrange for disposal. As is the case with actual emergency responses, every training exercise begins with a detailed briefing and ends with a thorough critique.

Individual enlisted team members as well as officers are required to personally conduct training exercises on a regular basis. They must plan the exercise, provide an initial briefing, oversee the response effort, and carry out a critique. At a recent exercise directed by a 1st Class Yeoman, the objective was to identify chemicals stored in a warehouse.

"I am going to break you down into three entry teams, a command post crew, and a suiting crew," explained the Yeoman during a briefing. "We will use two different radio channels; one for the staging area and one for the operating teams. The first entry team, rather than going into the warehouse to begin immediately identifying chemicals, will be used merely to draw a floorplan of the interior layout of the warehouse. When drawing the floorplan, try to get detail and keep it readable. As the first entry team comes out, they are to put their floorplan diagram into a plastic bag, and we will take it to the mobile command post and have copies made. All personnel will be given a copy of the floorplan and we will designate areas on it so that later when we start talking about 'Area A' everyone will know what we are talking about. Using the floorplan drawn by the first entry team, the second entry team will then enter the warehouse and begin identifying chemicals in each area taking a sector at a time. As the primary team moves through the warehouse and reports by radio, they are to pick out landmarks on the floorplan so we will be better able to understand where they are at any given moment. If they go down or have equipment problems, the backup team will be able to get them out by the most expedient means."

"Going into a warehouse, the assumption is we have a static situation," continues the Yeoman responsible for running the exercise. "The warehouse has been there awhile and will be there awhile longer. All we want to do is ascertain what chemicals we are dealing with and determine how best to handle them. It is not as though life or limb were in peril; there is no urgency involved. It is not necessary that the job be done on Thursday rather than Friday. You can't call it undue caution or wasted effort to draw a floorplan; safety has to be foremost in our minds. We have emphasized all along that we must know our exit routes as well as the area we will be working in. Chances are your survey team drawing the floorplan is not going to get into trouble unless they have a problem with their equipment. However, if the primary team is going to get into trouble, there is a strong chance such trouble will occur while they are working with hazardous materials—trying to sample it, contain it, or overpack it. They are going to have to be fully cognizant of everything around them."

"We will use two radio channels because I don't want the support people and the back-up team on the entry party's channel," adds the Yeoman. "If the back-up team can listen to the men inside, they might get psyched. Of course, they will be fully briefed if and when they have to go in. The communications person will have to switch back and forth between radios, but the people inside won't have to worry about someone breaking in trying to get their attention. With this system, if you hear someone on your radio but don't understand the transmission, at least you will know they are talking to *you.*"

"We will use three-man entry teams with each man in Level A protection," (fully encapsulated suits with selfcontained breathing apparatus) concludes the exercise coordinator. "Two men will enter while the third remains at the entrance as a linetender. One man inside will have a radio as will the linetender. Team One will be going in first to make the floorplan; Team Two will be suited up as the stand-by team; and Team Three will be split between the command post and the staging area to get everybody suited up. When Team One rolls out, Team Two will become the entry team and Team Three will become the backup. Because we are short of people today, we are not going to be able to keep the logs we normally maintain, but I still want everyone to report times as we normally do."

Although Atlantic Strike Team personnel tend to be relatively young men and women in good physical condition, continuous physical training is required to safely prepare team members for the extreme physical stress experienced during prolonged operations. "Nothing else I have ever done in my life is as physically demanding as getting into a fully encapsulated suit, breathing compressed air, and attempting to work," notes one experienced team member. More than two years ago, during a regularly scheduled chemical team drill, an Emergency Medical Technician recorded blood pressure, pulse rate and respiration of each participant prior to the drill. The exercise was conducted on a moderately hot day and involved work in fully encapsulated chemical protective suits. It was found that after a short period of exposure a majority of personnel were experiencing physical stress evidenced by greatly elevated blood pressure, pulse rate, and respiration. The Atlantic Strike Team began immediately to design and implement a standardized physical fitness program for all hands. "The program begins with a complete medical monitoring," explains Lieutenant Robert Strong. "Nobody on the team participates in the physical fitness program until he or she has gone through a complete physical examination at the base dispensary and has been seen by a senior physician. We do our outmost to protect our people from undue stress and physical damage that might result from putting them into a program that is beyond their capabilities. Physical training consists of three parts: stretching exercises, body strengthening exercises, and cardiovascular improvement. For the amount of time we have available, running appears to provide the biggest trade-off between body-good and body-damage. There is controversy centered around the benefits to be gained from running; we feel running is better than doing nothing. If you are going to do any damage to your body, the damage is probably minimal compared to the potential gain. Our EMTs do routine spotchecks of all hands to gauge overall physical fitness."

Physical fitness training is scheduled three times a week on Monday/Wednesday/Friday during the first hour of each working day. Personnel are worked into the program slowly over a period of weeks; first running as a team for 1.5 miles each period, and then running their own distance at their own pace. Every other month, all hands participate in an endurance test similar to the U.S. Navy Dive School physical fitness test. Prior to each bimonthly endurance test, unit EMTs record cardiovascular information on each participant. Running times, exercise repititions, and cardiovascular information are recorded upon completion of the test. Documented improvement in team members' physical and cardiovascular endurance has been extensive.

We dedicate and devote a tremendous amount of time to training," emphasizes Strong. "Unlike many units, we can take time for training because training is one of our four primary mission areas. Many units say they don't have time for training; we feel they don't have time not to train. We have been given time to train."

The motto of the Coast Guard is "Semper Paratus." (always ready) For Atlantic Strike Team members charged with maintaining a high state of readiness to respond to actual spills and potential discharges of oil and hazardous substances, Semper Paratus is a way of life.

The Environmental Response Team of the U.S. Environmental Protection Agency

The March 28, 1979 accident at Three Mile Island nuclear power facility in Harrisburg, Pennsylvania focused national attention on the potential destructive capacity of one class of hazardous materials. By coincidence, within a few days E.P.A.'s Environmental Response Team (ERT) was alerted and sent to the same general area; yet for team members Three Mile Island was the least of their concerns . . . they were far too busy preventing a major catastrophe down the Road in Hagerstown, Maryland. In so doing, they demonstrated why hazardous materials response experts needing technical advice and assistance often call for the ERT.

Early in the morning of March 22, 1979; six days before the incident at Three Mile Island, a tractor-trailer enroute from the Electro Phosphorus Company in Mulberry, Florida to Metallurgical Products Company in West Chester, Pennsylvania stopped at a redlight in downtown Gettysburg, Pennsylvania and burst into flames. The truck carried nearly 20 tons of white phosphorus—a poisonous, pyroforic, flammable solid—packed in 30-gallon drums. Toxic fumes released by explosion and fire sent more than 100 persons, including an estimated 60 firefighters, to area hospitals with eye irritation, abdominal pain, vomiting, dizziness, coughing, and abnormal thirst. The only way firefighters were eventually able to extinguish the pyroforic cargo was to completely cover the entire tractor-trailer with truckloads of wet sand.

An industrial response team sent by the manufacturer overpacked each of the fire damaged 30-gallon drums in 55-gallon drums that were then filled with water and sealed. During this operation, a drum exploded and at least six additional persons were injured. On March 28, as disaster struck Three Mile Island 35 miles to the northeast, the overpacked drums were loaded aboard two trailer trucks for what was supposed to be a trip back to the manufacturing facility in Florida. Instead, the trucks were driven 30 miles south to the carrier's terminal in Hagerstown, Maryland while the manufacturer and carrier haggled over who should take possession of the now extremely hazardous cargo.

Ultimately, the manufacturer informed the carrier that it was abandoning the cargo. Explosive and toxic gases being generated within the overpacked drums had become a ticking timebomb, so ultrahazardous that even the company that manufactured the phosphorus would not now have anything to do with it—and with good reason. It was recognized that the steel drums were corroding from within due to complex chemical reactions that resulted in generation of explosive hydrogen gas and poisonous phosphine gas.

When the State of Maryland became aware that the highly hazardous cargo was sitting at a truck terminal in Hagerstown, and neither the manufacturer nor the carrier was taking adequate action to eliminate a recognized threat to public safety, assistance was requested from the U.S. Environmental Protection Agency's regional office. E.P.A.'s regional On-Scene Coordinator activated a regional response team and requested assistance and advice from E.P.A.'s national Environmental Response Team. ERT members were asked by the On-Scene Coordinator to assess the degree of hazard present, determine how large a geographic area was threatened, and make recommendations as to proper corrective actions.

ERT personnel identified the most serious hazards as a potential release of extremely toxic phosphine gas, and a potential for explosion of hydrogen/phosphine within the drums that would adsorb water and form phosphoric acid. After extensive consideration of a number of inter-related factors, the ERT advised the On-Scene Coordinator that residents within a five to ten mile radius of the truck terminal were exposed to extreme risk, and that either the two trailer trucks had to be immediately moved to a remote disposal site or the population within that area would have to be evacuated. The danger to area residents would increase every hour the chemical reaction occurring within the drums was allowed to continue. Team members further recommended detonation of the barrels at a remote site.

The On-Scene Coordinator then attempted to gain access to a remote site within a reasonable distance of Hagerstown. After intercession by the Governor of Maryland and the

White House, the U.S. Department of Defense agreed to allow use of Fort A.P. Hill located 170 miles south in Virginia.

After midnight on April 6, 1979 one of the strangest convoys in history left Hagerstown, Maryland bound for the nation's capital enroute to Fort Hill. The 22-vehicle procession included Maryland State Police cruisers, U.S. Army explosive ordinance disposal teams, the two trailer loads of overpacked phosphorus drums, fire trucks, water trucks, dump trucks carrying 200 tons of sand, lowboy trailers loaded with bulldozers, and miscellaneous vehicles carrying response personnel from E.P.A., the Maryland Department of Natural Resources, and O.H. Materials.

Without sirens the convoy followed Interstate-70 south to I-495, (the Capital Beltway) I-495 around the western edge of Washington, D.C., then I-95 south to connect with U.S. 17 near Fredericksburg, Virginia. As dawn was breaking, the convoy entered Fort A.P. Hill via U.S. 301. There, the Environmental Response Team along with other government, military, and commercial personnel developed and implemented safe operating procedures to be used during destruction of the distended, extremely sensitive drums. Between April 8 and April 18 all overpacked, fire damaged drums of white phosphorus were successfully detonated, one at a time.

E.P.A.'s Environmental Response Team was initiated in 1978 as one component of a National Contingency Plan required by Section 311 of the 1972 Water Pollution Control Act designed to coordinate Federal hazardous materials response and cleanup efforts, particularly those related to spills of oil and hazardous substances. E.P.A. is responsible for on-scene coordination at inland incidents while the U.S. Coast Guard is responsible for spills of oil and hazardous materials in coastal waters and the Great Lakes. Passage of the Comprehensive Environmental Response, Compensation, And Liability Act ("Superfund") in December of 1980 greatly broadened Federal initiatives in responding to hazardous waste emergencies and brought about a major rewrite of the National Contingency Plan.

In the early days of the National Contingency Plan, the Environmental Response Team was mainly involved in emergency response to spills of oil and hazardous substances; nowadays, the ERT is more often called in for planned response to uncontrolled and abandoned hazardous waste sites. That is, although the ERT's basic responsibility remains the same . . . to provide technical assistance to Environmental Protection Agency regional and program offices during environmental emergencies . . . current assignments tend to deal with hazardous wastes rather than with spills of oil/hazardous materials/hazardous substances.

As of mid-1981 E.P.A., the states, and private interests had identified approximately 8,800 dumpsites, evaluated the dangers of 5,400 of them, and completed investigations of 2,300 of these. E.P.A. estimates there are currently 2,000 or so uncontrolled or abandoned hazardous waste sites that represent situations serious enough to require Federal intervention. It is recognized that many additional sites may be creating problems.

"Superfund" created a trustfund of up to $1.6 billion over a five year period beginning in 1981 to provide emergency and longterm cleanup by the U.S. government of chemical spills and abandoned hazardous waste sites that threaten people or the environment. Approximately 87 percent of the money in this trustfund will be acquired from taxes imposed on oil and 42 specific chemical compounds, while 13 percent will be paid directly by the taxpayers. "Superfund" will generate roughly $22 million a month to cover hazardous materials spill/hazardous waste dumpsite cleanup.

A new office, the Office of Solid Waste and Emergency Response, (OSWER) was created within E.P.A. to administer "Superfund," Section 311 of the Water Pollution Control Act of 1972, and the Resource Conservation & Recovery Act (RCRA) in an effort to bring about a comprehensive, well coordinated attack on uncontrolled hazardous waste sites and other environmental emergencies.

Eleven personnel assigned to the Environmental Response Team within OSWER and stationed at Edison, New Jersey and Cincinnatti, Ohio have training and experience in biology, ecology, chemistry, and chemical engineering, environmental health, sanitary engineering, and industrial hygiene. Team members both plan and conduct emergency operations placing special emphasis on application of new technology and equipment, especially safety equipment and decontamination procedures. The team provides advice and assistance to E.P.A. regional and program offices in the areas of chemical, biological, and physical treatment and monitoring techniques; control, restoration, disposal, and contingency planning during emergencies; installation, operation, and evaluation of instrumentation and field response systems; sampling and analysis of air, water, and soil; water pollution biology and toxicology; environmental radiation; occupational health and safety; and computerized gas chromatography/mass spectrometry. Currently, the ERT is averaging 50 on-site responses a year.

The Environmental Response Team has ten major functions.

1. Maintain an around-the-clock activation system.

2. When requested, dispatch team members to emergency sites to assist E.P.A. regional and program offices.

3. Provide consultation on water and air quality criteria, toxicology, interpretation and evaluation of analytical data, and engineering and scientific studies.

4. Develop and conduct site-specific safety programs.

5. Provide technical experts for a Public Affairs Assist Team.

6. Supervise the work of contractors.

7. Provide specialized equipment to meet specific site requirements such as monitoring, analytical support, waste treatment, containment and control.

8. Provide assistance in the development of technical manuals, policies, and standard operating procedures.

9. Assist in developing new technology for use at environmental emergencies and uncontrolled hazardous waste sites.

10. Train federal, state and local government officials and industry representatives in the latest technology for environmental emergencies at hazardous waste responses.

In performing such functions, ERT personnel get involved in a wide variety of specific tasks that require a high level of technical expertise. They must be completely familiar with varied data and information sources that can be tapped to indicate physical, chemical, and toxicological properties of substances in a given situation in order to provide technical assistance and advice to an On-Scene Coordinator relative to specific hazards present and effective mitigation efforts. They must completely understand the capabilities and applicability of a variety of data sources such as OHMTADS, (the Oil and Hazardous Materials Technical Assistance Data System created by the U.S. Environmental Protection Agency/National Institutes of Health) CHRIS, (the U.S. Coast Guard's four-volume Chemical Response Information System) industrial resource systems and other information sources.

ERT members often deal with industrial experts in order to obtain recommendations for the handling of a specific chemical, and must be able to evaluate alternative mitigatory techniques in order to provide specific recommendations to an On-Scene Coordinator. They may obtain necessary information, assemble it into a form suitable for presentation, and in other ways assist an On-Scene Coordinator in preparing and presenting briefings to various groups relative to the status of operations on-site.

A major concern of the ERT, and a common assignment, is to devise and conduct on-site safety programs to insure safe and effective response to environmental incidents involving potential or actual releases of hazardous materials or wastes. Because the ERT is a group of highly trained specialists with extensive experience gathered at a wide variety of incidents, they are seldom called in to provide advice on run-of-the-mill cleanup situations. Rather, a "routine" or "normal" assignment for the Environmental Response Team will often involve a high degree of hazard, complex technical problems, and/or the possibility of severe environmental damage. Often the team must devise and implement one-of-a-kind solutions for situations that cannot be adequately controlled with conventional procedures. Over the years ERT personnel have developed a number of different methods for treating water contaminated by soluble toxic substances, removal of insoluble substances from streambeds, removal of toxic materials from groundwater, treatment of contaminated soils, and containment and disposal of numerous toxic materials. Oftentimes more than one procedure could be applied to a given situation, and team members assist the On-Scene Coordinator in determining the most acceptable methods, analyzing potential risks associated with various abatement techniques and balancing them against the hoped-for environmental benefits, and determining the most cost effective use of manpower and equipment. ERT personnel may also be called upon for recommendations as to the degree of mitigation required, to identify sensitive and critical environmental areas that require special protective measures, to identify practical methods for dealing with contaminated wildlife and waterfowl, to advise on revegetation or erosion control necessary for restoration of a damaged area, or to recommend methods and coordinate disposal of chemicals and contaminated debris.

The Environmental Response Team forms a focal point for three groups that supply personnel and expertise: the Operational Support Section; the Environmental Impact Section; and the Analytical Support Section. The Operational Support Section develops and implements site-specific strategies for both short-term and longterm responses at uncontrolled hazardous waste sites and other environmental emergencies. This section assists E.P.A. regional personnel to conduct engineering feasibility studies for containment, cleanup and disposal activities; and develops alternative strategies for both on-site and off-site disposal of hazardous materials and wastes. Operational Support Section personnel may evaluate existing technical approaches, develop and implement new technical approaches required by a particular situation found on-site, perform cost analyses, and apply the latest safety and health techniques to various operations such as site access control, personnel monitoring, decontamination, and respiratory protection. The section also provides and supervises on-site contract support, and ensures compliance with all applicable regulations.

A very central, and quite visible, component of the Operational Support Selection is an arsenal of state-of-the-art response equipment available through the Environmental Emergency Response Unit. (EERU) The EERU is basically the E.P.A.'s model nationwide hazardous material spill response and waste site control organization for situations where the use of complex cleanup equipment and techniques are required. The objective of field demonstrations of such equipment by the Environmental Emergency Response Unit is to encourage commercialization and broader use of equipment similar to that developed with public funds.

The EERU is a cooperative effort between and among research and operations personnel within E.P.A., the Environmental Response Team, and commercial contractors. The unit is not used for routine cleanup operations nor is it intended to compete with commercial cleanup contractors; rather, emphasis is placed on use of prototype equipment to spur further development of spill prevention and control equipment and technologies by the private sector. For that reason, response by the EERU is generally limited to instances where a particular type of technology or capability possessed by the unit is not available commercially.

The newest piece of one-of-a-kind equipment unveiled by E.P.A.'s Environmental Emergency Response Unit came on-line as recently as April, 1982 . . . a Mobile Incineration System that can perform on-site detoxification of particularly difficult to handle substances such as PCBs, kepone, malathion, and TCDD. The $2.2 million incinerator, mounted on three flatbed trailers, adds a priceless component of mobility to E.P.A.'s response efforts by permitting incineration of extremely dangerous chemicals at abandoned waste dumps without having to transport them. Prior to the availability of E.P.A.'s mobile incineration system, there were only three incinerators in the entire United States that could thermally detoxify certain substances. The stationary incinerators that are able to handle such substances have a six to nine month backlog, and transportation costs are often prohibitive. The mobile incinerator system is capable of producing the 2,200 degree Fahrenheit temperature needed to destroy PCBs and certain pesticides, and can process approximately 100 tons of dry material or six tons of liquid wastes in a day. Ash residue consisting of inert materials and heavy metal residues must be collected for disposal at a secured landfill facility, and the unit cannot detoxify heavy metal contaminants.

The EERU also assisted in developing and uses two Mobile Physical/Chemical Treatment Trailers that perform on-site flocculation, sedimentation, filtration, and carbon adsorption. Contaminated water can be processed at flow rates of between 100-600 GPM by the larger of the two treatment trailers, and 30 GPM by the smaller unit.

A Mobile Laboratory has been developed to provide analytical services during cleanup operations at spill sites and uncontrolled waste dumpsites so as to avoid delays involved in shipping samples to a central laboratory. The mobile laboratory can analyze virtually all organic and inorganic hazardous substances, including PCBs, pesticides, and heavy metals. Instrumentation onboard the 35-foot semitrailer includes a gas chromatograph/mass spectrometer, two gas chromatographs equipped with flame ionization and electron capture detectors, an atomic absorption spectrometer, infrared and fluorescence spectrometers, an emission spectrograph, and a total organic carbon analyzer.

The EERU uses a Spill Assessment Van on-site to identify needed treatment measures when the full analytical capabilities of the mobile laboratory are not warrented. A 25-foot commercial step-van was equipped with a fume hood, refrigerator, sink, water tank, storage cabinets, bench space, and a five kilowatt gasoline-powered generator. The spill assessment van also carries appropriate glassware, reagents, test kits, and instruments so that treatability testing and sample analysis can begin as soon as the van arrives at an uncontrolled hazardous waste site.

Additional major pieces of equipment utilized by Environmental Emergency Response Unit personnel include a Mobile Stream Diversion System, a Mobile Decontamination System, a Mobile Beach Sand Cleaner, and Mobile Activated Carbon Treatment Trailers. A Hazardous Materials Spills Warning System housed in a 27-foot automotive-type trailer provides an in-stream system capable of detecting a variety of spilled hazardous materials. A submersible pump in a water course supplies continuous water samples to three instrument consoles in the trailer, and the system can operate unattended without maintenance for 14 days. Instrumentation can detect acids and bases, ionic compounds, oxidizing and reducing substances, organic compounds, and aromatic compounds.

Not all prototype equipment used by the EERU requires a trailer for transport. An Acoustic Emission Monitoring Device is available to evaluate the stability of earthen dams around waste ponds that can be found at many industrial sites and chemical waste disposal areas, locate the area of any instability, and detect/locate seepage. This simple-to-operate device that can be used either periodically or continuously operates on the principle that noises are generated by earth movement. Intensity and frequency of acoustic emissions has been correlated with stress levels for many soils and can be used to estimate stability of dam structures.

Another automatic spill detection alarm system used by EERU is the Cyclic Colorimeter capable of detecting a wide variety of heavy metal pollutants. By injecting sodium sulfide into a sample stream the apparatus can produce either a quantitative indication of the pollutant or provide an alarm when a threshold level is reached.

A Hazardous Materials Spills Detection Kit about the size of an average suitcase was designed to provide necessary instrumentation, equipment, and reagents needed by field personnel to detect and trace contaminants in waterways. Included within the kit are a pH meter, conductivity meter, spectrophotometer, filter assembly, and additional components.

A Hazardous Materials Identification Kit was designed to identify 36 representative hazardous materials (toxic metals, anions, organic compounds) and related substances. The kit is intended to allow identification of groups of contaminants rather than quantification of specific substances. The identification kit was designed to be used in conjunction with the detection kit since use of both kits improves identification capabilities, particularly for inorganic materials. Cyanide and fluoride, for example, cannot be distinguished by the identification kit alone but identification is possible when both kits are used concurrently. Equipment supplied with the identification kit includes an inverter/shortwave ultraviolet lamp unit for photochemical and thermal reactions; plus reagents and auxiliary equipment such as test papers, detector tubes, spot test supplies, and thin layer chromatography materials.

The EERU has also been active in testing a multipurpose gelling agent used to transform liquids into a semi-solid material more easily handled by mechanical apparatus; pesticide detection apparatus used to monitor water for the presence of organophosphate and carbamate insecticides; and a foam dike system that allows a single operator to quickly spray a rigid foam that is resistant to chemical attack, nontoxic, disposable, and nonflammable and thus create an artificial dike to contain chemical spills.

Certain major pieces of equipment are currently under development or being tested by EERU. A mobile In-Situ Containment/Treatment System transported aboard a 43-foot lowboy trailer is being tested as a means of performing on-site detoxification of contaminated soil by chemical reaction in situations where a large volume of contaminated soil is involved. Grouting material is injected into the soil to form a curtain around the contaminated area, then the soil is treated in place by oxidization/reduction, neutralization, or precipitation.

A Mobile Independent Physical/Chemical Wastewater Treatment System developed and currently being modified by the EERU provides an on-site means of treating large volumes of contaminated wastewater during extended periods of cleanup activity.

In addition, a Mobile System For Detoxification/Regeneration Of Spent Activated Carbon housed in a 45-foot semitrailer is used to cleanse contaminated activated carbon at the worksite.

The EERU is currently experimenting with a Mobile Reverse Osmosis Treatment System, originally designed for E.P.A. as a pilot plant to test the feasibility of treating acid mine wastewater, that is being modified for field use at incidents involving concentrated solutions of hazardous materials such as leachate from uncontrolled hazardous waste sites.

E.P.A.'s Environmental Response Team is responsible for activating the Environmental Emergency Response Unit when such highly specialized, state-of-the-art response equipment is needed.

A second organizational group within the Environmental Response Team, the Analytical Support Section, specializes in developing data by designing and implementing sampling and analytical programs. The Analytical Support Section operates and staffs mobile laboratory equipment, provides nonroutine analytical support both on and off-site, collects/samples/interprets data, and implements sensing programs as needed in order to prepare site-specific analytical and safety plans. For example, this section has designed procedures for the safe bulking of barreled wastes in order to cut down on handling and transportation costs. The procedure involves the use of spot tests for pH, oxidation-reduction potential, radiation, flammability, volatile gases, and water reactivity. When it is necessary to test for specific hazardous materials, the procedure is modified to add rapid screening measures for each barrel. Often, the Analytical Support Section is called in to develop safety plans that involve determining zones of contamination, personal protective equipment required for operations within selected areas of a specific site, workplace air monitoring requirements, and proper means of personnel and equipment decontamination.

A third organizational group within the Environmental Response Team, the Environmental Impact Section, provides hazard assessment and toxicity testing assistance. In order to make correct environmental decisions in formulating a response strategy for a specific situation, this section first determines the degree of risk to people and the environment from air, surface water, ground water, soil, and biomass contamination. In making a hazard assessment, personnel first consider the hazard itself then determine the direction and speed of the contaminant's migration in order to identify or develop alternative methods for achieving control, removing contamination, and bringing about the restoration of a site.

As of early 1982, the Environmental Response Team had responded to more than 170 environmental emergencies and provided direct technical assistance for an additional 250 incidents since its formation in 1978. Lessons learned through this extensive experience are transmitted to response personnel from federal, state, and local agencies through formal training programs developed and provided by ERT members. Classroom training, laboratory exercises, and hands-on field simulations are used to assist participants in recognizing hazards associated with specific materials, determining risk to persons and the environment, reducing or preventing hazardous consequences, and protecting response personnel. In this manner methods, tools, equipment, and strategies developed by the Environmental Response Team are made available to a wide range of environmental emergency response personnel.

6 COMMERCIAL RESPONSE ORGANIZATIONS

IT Corporation

Although IT Corporation of Wilmington, California is known to emergency services people throughout the Golden State primarily for its Environmental Emergency Response Teams, the company is actually a full-service environmental management outfit heavily involved with a wide variety of environmental pollution control systems. IT personnel develop, build, and operate on-site treatment facilities for industry to detoxify and/or neutralize wastes that are generated within a particular facility. Company engineers work with industry to reduce hazardous waste streams by increasing the efficiency of the manufacturing process, and design and build high-temperature incinerators used by a number of major industrial corporations. IT Corporation also transports hazardous waste materials, and just recently obtained the necessary permits to construct an $84 million hazardous wastes detoxification and treatment facility in the State of Louisiana just south of Baton Rouge. These broader aspects of company operations are not well known to the general public who tend to think of IT Corporation as a *response* organization . . . and with good reason. When hazardous materials hit the fan any place within California, chances are that IT will be called in to handle clean-up. There are a lot of incidents in California, and the company tends to get a lot of ink. IT's Environmental Emergency Response Teams have been responding since 1973, have handled approximately 500 incidents over the years, and are currently receiving from three to five calls a week. They have a solid reputation of doing the job right.

"We obtain our response team members from existing employees who have been doing routine work with IT Corporation for a couple of years," reports John Theiss, a corporate vice-president who often serves as media communicator during incidents. "We select those we feel are the best, the most safety conscious, those who have a particular respect for hazardous working conditions. We have approximately 650 total employees of whom 150 are response personnel. We select potential team members from among the elite of our employees, and quite a few wash-out during training. In addition to their normal on-the-job training, they receive three weeks of concentrated instruction on handling of hazardous materials in an uncontrolled environment. All are required to complete training in first aid, C.P.R., confined space entry, and emergency response training. By selecting team members from among our regular, day-to-day operating personnel; our manpower capability is somewhat unique in that we can call upon quite large numbers of trained personnel to handle nearly any size incident."

"We presently have 150 team members scattered throughout the State of California," adds Ken York, who is no doubt familiar to a member of readers from his rousing presentations at training seminars around the country. "Each team consists of a bare minimum of four people: an industrial hygienist/toxicologist, a chemically-trained foreman, and two trained team members. Any time a team is called out, we automatically activate what we call 'CERAG' . . . Corporate Emergency Response Advisory Group, consisting of staff members who are chemical engineers, industrial hygienists, environmentalists, analytical chemists, training specialists, and operations specialists. This behind-the-scenes organization gathers and absorbs all the information obtainable about the spill scene prior to our team arriving on the scene. The CERAG group is tied-in to the response team at the site by telephone or radio."

"We respond to all types of hazardous materials incidents *except* radioactives and explosives," continues Ken York. "Dispatch is handled through a 24-hour emergency telephone number, 800-262-1900, for all of California. All of our people at key locations throughout the state are equipped with response vehicles. The newest unit that just went on-line in March consists of a 32-foot gooseneck, enclosed trailer pulled by a fifth-wheel diesel, bobtailed truck. It carries everything under the sun: twelve 300 cubic foot bottles of air for recharging, eight SCBA spare bottles, hoselines, total environment suits made of both butyl rubber and PVC . . . a total of eight different grades of protective clothing . . . and all sorts of support equipment. We can do entire highway closings with barricades, signs, and lights. We also carry chlorine kits, plugging and patching materials, lifting and patching airbags, . . . the whole ball of wax. We have devised in-house a dozen or more patching and plugging methods that we use for different types of containers and products. We run a total of 127 vacuum trucks and have Hydro-Blast units for removing materials from pavements as well as high-temperature pressure washers. There are roughly 100 assorted gear trucks and pick-ups, a 'Super Sucker' for dry-vacuuming materials, pumps, hoses, and all sorts of lighting equipment."

"We had a spill the day before Thanksgiving on Interstate Highway 5 outside of Fresno. Now just picture the logistics of this situation" adds Ken York with a grin. "We have just shut down the major artery between northern and southern California at the beginning of one of the biggest weekends of the year. My God, we have isolated Disneyland. In order to get the situation resolved quickly we flew team members from Wilmington; we drove vacuum trucks from Taft; we drove an emergency response unit from Martinez; and drove supervisory personnel from Bakersfield. At this particular incident, as we have done a number of times in the past, the company plane carrying our crew from Wilmington landed right on the freeway. They found a truck that had been totally involved in fire. All identification data was gone, the placards destroyed, the driver dead. The only option was to record the license plate number and trace it back through the registered owner to learn what the truck was hauling. Fortunately, in this instance it was done very, very quickly. The commodity was carbolic acid/phenol."

"An incident near Bakersfield last September was interesting in that it involved a commodity that was NOT a regulated hazardous material," adds York. "The product was 'Bolls-Eye,' a cotton defoliant that did not meet the LD-50 requirements set by CFR-49. Therefore, the truck was not placarded, and the bills-of-lading were not in special order or specially marked. A fireman started reading container labels, however, and noted the material released 15 percent arsenic. The fire chief, having been trained, decided that everyone at the scene should immediately take a shower with the fire hoses. Twelve of the 14 people there took a shower. Two others were a bit shy and decided to drive some place else to take their shower. The two vomited until their stomachs were empty, then had dry heaves and severe headaches for approximately 12 hours thereafter. This was the result of exposure to a material that is NOT a regulated hazardous material."

"Our people do a lot of training on access to tank cars and highway trucks in an overturned position so that when we arrive at an incident like the Air Cal crash yesterday (An Air California 737 jet crashed at John Wayne Airport in Orange County on February 17, 1981.) we have a pretty fair idea how to gain access," explains Ken York. "It would have been embarassing yesterday if we had stood around saying . . . 'Oh my God. This is an airplane; it's got wings. Where's the gas cap?' We were called in, basically, to handle the fuel. Run-off from a combination of leaking fuel and fire-fighting efforts was in danger of running into Newport Bay. We ran down a half mile and built a dike to contain that threat, then positioned a vacuum truck and started sucking. We then moved back relatively close to the scene and positioned another vacuum truck between the wreck and a storm drain and began sucking it up in that area. Once the fire was out and the foam began to decompose, we began to skim off the fuel that rose to the surface. We continued skimming until they hauled away the aircraft and we could get underneath and finish the job.''

Ken York was asked recently to identify the most worrisome general type of situation that IT crews are called in to deal with. "I would have to say our most difficult situations involve unidentified mixed chemicals and post-fire residues," he replied without hesitation. "Post-fire residues are probably the most difficult problem we handle because the material is not ALL burned obviously. There could be water reactives, flammables, poisons . . . all with the labels destroyed. It sounds silly to say but many products can be hazardous yet control and containment procedures can become quite academic if you know what you are dealing with. Once you have the necessary training and experience, and once you can trust your team members, many situations become academic even though in reality the product may be ultra hazardous. Conversely, the situation can get very, very hairy if you don't know what you are dealing with. If we can identify the product, we feel fully prepared and equipped to handle it.''

Hazardous Materials Technology & Services, Inc.

A Florida-based response organization known as "HAZARDS" (Hazardous Materials Technology & Services, Inc.) may have the most highly-trained, most experienced hazardous materials incident response staff of any commercial organization in the country. All ten incorporators of the three year old company are now, or were previously, members of the original Jacksonville Fire Department Hazardous Materials Team. By doing double-duty over a period of years as members of both the fire department team and the "HAZARDS" team, personnel tend to have spent a tremendous amount of time in acid/gas/entry suits at actual incidents. Although the average age of team members is quite young, probably no more than 27 years, their combined hands-on experience is truly amazing; members live and breath incident preplanning, response, and control.

In addition to 11 initial response specialists on 24-hour beeper call, "HAZARDS" total staff of 25 includes a scientist and chemist, a radioactive materials technician, an explosives authority, hazardous materials instructors, and transportation vehicle specialists. Also available for incident assistance is Dr. Jay Edelberg, "HAZARDS" Medical Director, who is a well known specialist in the field of emergency medicine with a special interest in the study of respiratory problems associated with hazardous materials incidents.

"HAZARDS" offers a quick-response team on 24-hour call to respond to spills, leaks, and fires involving hazardous materials at transportation and industrial accidents within 300 miles of Jacksonville. The team guarantees that within one hour of a call they will have a three to five man team enroute by over-the-road vehicle, plane, or jet helicopter . . . whichever is most expedient. Although a "HAZARDS" team carries much specialized equipment specifically related to hazardous materials containment and control, they depend on local fire service organizations at the scene for fire apparatus and conventional equipment. "HAZARDS" specialized equipment includes regular protective and uniform clothing, special protective equipment such as Eastwind acid/gas suits, proximity flame resistant suits, and SCBA. Additional equipment includes explosimeters, regular and heavy-duty tools, (wrenches, sockets, come-alongs, etc. for railcars, trucks, cylinders, and tanks) chlorine kits, special leak-stopping equipment, an extensive reference library, pH kits, sorbent materials, high expansion and AFFF foam, eductors and nozzles, and several vehicles. In addition to incident response, "HAZARDS" offers various seminars for industrial and transportation organizations as well as for public safety agency personnel.

"Our geographical area of coverage is within 300 miles of Jacksonville for incident response, and 500 miles for training," noted Don Tullis of "HAZARDS." As 300 miles is approached, time is working against us as far as effective incident response is concerned. We do use a jet helicopter or plane as needed, but the first five or six hours on-scene is usually critical. Although we have responded as far south as Vero Beach and as far north as the area around Thomasville, Georgia; there are definitely problems if you get too far from home.''

"In addition to incident response," added Tullis, "we do preplanning, inspections, employee training, management training, and determination of resources available. We do a great deal of pre-incident planning with paper companies, chemical companies, and general industrial organizations . . . everything from set-up and training of their fire brigades in the handling of routine Class A, B, C fires to training in hazardous materials response. During the past year we have also worked with a number of fire departments. We also work in the area of disposal of hazardous wastes and materials," concluded Don Tullis. "This is becoming such a critical area that we have had to become proficient in it."

There have been some transportation incidents the team has worked that received extensive coverage in the media, but "HAZARDS" personnel attempt to keep their involvement quite low-key. Industrial organizations call on "HAZARDS" when suddenly faced with a serious problem, and they look to the team to handle the situation in a minimum amount of time with minimum publicity.

Team members stress discipline, coordination, and strict adherence to predetermined procedures. They have worked together at many incidents and know what to do without being told. As one observer noted: "There are definitely certain steps and procedures used. Atop a hydrogen chloride car when the fittings are iced over, commodity is cascading down over you, your vision is totally obscured, and the noise precludes any communication . . . you need a procedure to follow. You can't have a team like this if the men have to be assigned to 'pick-up the shovel, dig the hole, and put the dirt over there.' They have to know instinctively what to do and have the desire to do it better than anybody else.''

Environmental Emergency Response by O.H. Materials

For a major commercial environmental response organization such as O.H. Materials Co. there are seldom any easy assignments. During the past ten years, the company has successfully responded to spill emergencies throughout the United States and Canada containing and controlling several thousand different hazardous and non-hazardous materials involving spills of a few gallons up to several million gallons. It is a demanding and skilled trade that requires total commitment.

At an uncontrolled hazardous waste site in an industrial area of Elizabeth, New Jersey; O.H. Materials Co. safely removed over 60,000 drums of waste chemicals after an explosion and fire left the facility a smoldering environmental time bomb. Uncontrolled wastes sites are a ticklish problem in any case; cleanup of extremely unstable fire residues at such a site can be one of the most hazardous undertakings known to man. To alleviate the hazards, mitigate the damages, and restore the environment in such a situation requires highly specialized equipment, experienced personnel, detailed safety procedures, and rather unique know-how.

First, the drums were removed from haphazard discard piles by a mechanical grappler. Drums were staged according to contents—liquids, sludges, lab packs, and empties—for final analysis to determine toxicity, reactivity, and compatibility. Compatible liquids were then bulked into a specially designed compatibility chamber, treated, loaded into tank trucks and transported to an approved incinerator for proper disposal. Bulking of compatible liquids greatly reduced the volume of materials and the time required to complete the job by eliminating multiple handling of individual drums. Without O.H. Materials' compatibility chamber concept, standard practice would have required overpacking of each drum. Barrels of waste were emptied onto a concrete pad where another bulking operation took place. The sludges were mixed with sand and fly ash then loaded into visqueen-lined dump trucks for transportation to a federally-approved landfill. Empty barrels were crushed to further reduce bulk. By reducing bulk, time, and transportation costs, O.H. Materials was able to assist in the mitigation of an intolerable environmental and health hazard cost effectively.

O.H. Materials has its roots in the construction industry. The original corporation, Kirk Bros., Inc., (a construction company) founded in Findlay, Ohio in 1969 engaged mainly in the construction of water and wastewater treatment plants. A new subsidiary, Ohio Hygienics, evolved in 1971 to provide spill clean-up services in the Midwestern United States. The name Ohio Hygienics was later changed to O.H. Materials Co. as the company became a major clean-up contractor providing services throughout the North American Continent.

"Today, about 90 percent of our business is with the private sector as differentiated from government," notes James S. Walker, Vice President-Government & Industry Relations for O.H. Materials Co. Five or six years ago, the majority of our business involved working with the transportation industry—primarily rail and trucking. Since 1977, we have been involved with more than the classical emergency response problems. We have helped clients clean up uncontrolled hazardous waste sites, responded to incidents at fixed facilities involving hazardous materials, provided lagoon and continuous groundwater treatment, and decontaminated structures and facilities. Many problems have been around for years—such as, abandoned hazardous waste sites and contaminated structures involving asbestos, PCB's, and other toxic contaminants."

"We have approximately 200 full-time people, 40 of which are scientists specializing in microbiology, chemistry, hydrology, engineering, or analytical sciences," adds Walker. "The majority of our scientific and technical staff, and most of our response equipment, is located in Findlay, Ohio. We have offices in Atlanta, Georgia; Ottawa, Illinois; Trenton, New Jersey; Lexington, Kentucky; and Orlando, Florida." Over the past ten years we have developed a reputation for being able to stabilize all kinds of emergency situations," continues James Walker.

Although O.H. Materials has been involved in some highly visible projects such as Crewtview and Molino, Florida; the Valley of the Drums in Kentucky; Love Canal; Seymour, Indiana; and the disposal of a fire-damaged and abandoned truck load of white phosphorus at Hagerstown, Maryland; the company does not seek publicity for its work. "Generally, the generator or spiller hires us," explains Walker. "He is dependent on O.H. to provide a service on his behalf. We provide the service, do it in a safe and cost-effective manner, while trying to remain very low-key, giving the client his just credit for resolving the problem. Although we work closely with many interests on-site, we report to and take direction from only our client. At a transportation accident, there may be 15 or 20 different government agencies involved. Our primary concern is to get in there and resolve the crisis on our client's behalf. The clients are far better served when we confine ourselves to the incident and let them deal with the media and/or other agencies."

The specifics of an actual assignment differ depending on whether the problem is an emergency or non-emergency situation according to Walker. "With an emergency situation, the available information early-on is often fragmented, incomplete and sometimes inaccurate. In the case of an emergency, we generally are called in to stabilize the situation. Such an assignment may require patching tank cars or containing a spill. It may include minimizing the horizontal or vertical spread of material or the safe transfer of product from one container to another. The tasks may well depend on who else is at the scene as well as existing local capabilities.

For a commercial response contractor, rapid deployment of resources and personnel is a critical factor. O.H. Materials Co. must be ready to respond, often over great distances, 24-hours a day, 365 days a year. "A fairly high percentage of the time when we receive an emergency call, we have never met the person who is on the other end of the line," says Walker. "Our immediate actions depend upon his knowledge of the situation. Obviously, we first want to know who is calling. Many times the person calling us is making the call on behalf of someone else. Basically, we want to know what is spilled, how much is spilled, where it is spilled, how well it is contained, when the incident occurred, what they want done, who we can meet on-site, and the projected weather conditions for the area. The caller usually wants to know from us what equipment we plan to bring, and how long it will take us to get there.

Our initial information may be very limited. Quite often people at the scene cannot give us complete information; they may have other potentially leaking cars and won't know the full extent of the problem until morning. However, the more accurate information we can obtain from that initial call the better we can evaluate the numbers and types of people and equipment that should be dispatched immediately. If the caller is not at the scene, we will attempt to establish communications with someone on-scene to obtain first-hand information while we start calling in technical and operations people for a review of the situation. At the same time, we develop a general idea of what we are going to need, so we start identifying the necessary equipment. Because of the fragmented nature of initial information from the scene of an emergency, we believe it is important for us to get on-scene immediately to accurately assess the overall situation. We need chemical/physical data relative to the substances involved to contain the spill in a safe manner. We also need container-related information; the number of tanks, their size and shape, whether or not it is possible for us to get at them, and how badly they are leaking. We need very specific data, and this is often difficult to obtain over the phone.''

"Accordingly, we get some of our people to the scene as quickly as possible. They are able to use their training to evaluate the situation objectively and take initial containment measures," continues Walker. "If time is critical, we have two corporate aircraft, a prop plane and Citation II jet with five full-time pilots. We often immediately dispatch people and some response equipment on an aircraft to establish needs and take initial actions. Over-the-road equipment can be dispatched after this review. All our equipment is mobile and kept in ready-to-roll condition. By the time our trucks get to the scene, the situation will have been reasonably contained or stabilized by the fly-in crew. While our people are enroute, our future actions are determined by what additional information we can obtain from the scene. We establish and maintain contact with someone on-scene to obtain progress reports of an incident, while our crews are in transit. We also attempt to determine for each incident, the manufacturer of the product, so we can talk with their product specialists or technical people regarding potential hazards, personal protective equipment, information on the container, and their safety recommendations."

O.H. Materials Co. is organized into a number of functions including: Operational Services, Engineering, Equipment Fabrication & Repair, Technical Services, Accounting and Purchasing. The supervisors, foremen, and recovery technicians of the Operational Services Division form the on-scene teams that work in various locations, climates, and terrains to contain and control oil and hazardous materials spills. They take action to minimize the impact of the spilled materials on the surrounding people, property, and the environment. They handle the nitty-gritty of response such as venting and burning of tank cars; flaring of tank car contents; collecting, transporting, and disposing of explosives; and tracing and isolating pollutant flows for on-site treatment or disposal.

The Technical Services Division operates a main laboratory plus a number of field laboratories capable of providing qualitative and quantitative analyses. This knowledge is crucial to provide safe and effective decontamination activities. In situations where spills or waste sites result in contamination of soil, water or sediment; the biological section within the division may use biodegradation techniques to reduce the contaminants. Through application of microbial additives, faster and more thorough cleanup is often possible. Biochemists monitor the microbial population through adjustment of nutrient levels, dissolved oxygen concentrations, pH, and the checking of temperatures.

The Engineering Division—staffed by chemical, civil, and environmental engineers—designs systems, assists in determining equipment and systems needs in environmental cleanups, and applies containment and clean-up processes to field situations. The design of underground recovery and treatment systems; use of biodegradation to aid in the clean-up of oil and chemical spills; the design of compatibility chambers for bulking of hazardous chemicals to reduce transportation and disposal costs; and the construction and design of field laboratories are typical assignments for this division.

Personnel of the Equipment Fabrication & Repair Division develop, construct and maintain a wide variety of spill control equipment including mobile laboratories, decontamination trailers, fume scrubbers, pneumatic transfer receivers, transfer stations, clarifiers, compatibility chambers, vacuum receivers, galley trailers, underground recovery systems, and carbon filters. They also modify standard production trucks and vehicles for emergency response and support operations. Maintenance of hundreds of licensed vehicles and all other equipment operated by the Company is accomplished by this Division.

To maintain a competitive edge within its field, O.H. Materials Co. has a sizable inventory of specialized equipment including vacuum units, transfer and decontamination trailers, watercraft, mobile centrifuges and clarifiers, high pressure washdown pumps, water sampling apparatus, safety gear, and portable lighting, as well as treatment equipment mentioned above.

Skid-mounted vacuum units can be transported to remote sites by truck, barge, or flat car and moved on site by bulldozer if necessary. The diesel-powered vacuum units are equipped with vapor scrubbers to capture hazardous materials vapors. They can load and unload products simultaneously.

O.H. Materials' transfer units are designed for use in remote areas and are equipped with portable hydraulic power packs, viscous materials pumps, acid pumps, nitrogen purging systems, pneumatic transfer systems with air dryers and fume scrubbers, steamheating equipment, temperature monitoring devices, metering equipment, vapor control systems, power entry tools, and tank patching equipment. The transfer units, which are totally self-contained semi-trailers, also carry stainless steel and teflon-lined hoses and acid hoses as well as stainless steel, teflon, and PVC pipe components.

Each of the numerous decontamination trailers used in the field supports up to 14 men with their protective equipment. Each trailer has running water, wash basins, eye wash units, and a shower. Diesel generators operate lights, pumps, and electrical systems. Safety equipment on board the decontamination trailers includes self-contained breathing apparatus, respirators, replacement cartridges and cannisters for special atmospheres, protective clothing, chemical and acid suits, sample gloves and sampling equipment, basic handtools, non-sparking tools, and miscellaneous project safety and material handling equipment. The decontamination units are designed to provide maximum safety for employees but also to prevent hazardous chemicals from being spread. At the beginning of each shift, employees enter the decontamination trailer and change from street clothes to work clothes, then move to a tent or building to don protective clothing before proceeding to work stations. At the end of each shift workers decontaminate protective clothing and equipment at a transition area outside of the unit, then enter the unit and change into street clothes.

Mobile clarifiers with a capacity of 15,000 gallons, and capable of handling a flow rate of 200 gallons a minute, are used as settling tanks. The company also uses a 10,000 gallon mobile compatibility chamber for bulking compatible liquid hazardous materials and wastes from drums found at hazardous waste sites in order to eliminate the overpacking of damaged drums. This substantially reduces transportation and disposal costs. The compatibility chamber is built to withstand heat from reaction when hazardous materials are bulked.

"Almost every unit of equipment we have from small toolboxes to major pieces has supplementary back-up equipment," stresses James Walker. "We try to be ready for any contingency, ready to replace any part that may wear out, or any equipment that may wear out, or any equipment that is critical to the operation. All our toolkits are packaged taking 'worst case' hazards into consideration. We use containerized packaging of tools and equipment that can be put aboard a vehicle in a matter of seconds. Our crews are supplied with decontamination trailers completely geared up so that we have the necessary personnel ready to go when they hit the job. The inventory is continuously maintained and equipped with everything that initial responders could reasonably need (as identified by us in over a decade's experience)."

"Our patching kits are extremely complete," adds Walker. "If you run into chlorosulfonic acid, for example, what you need to use for patching material can be affected by the type of

Equipment List

This list reflects equipment owned and operated by O. H. Materials Co. It does not reflect the quantity of equipment available.

Vacuum Equipment
3,500-gallon Vacuum Trucks
1,800-gallon Vacuum Trucks
1,500-gallon Vacuum Skid Units

Trucks and Trailers
Over-the-Road Diesel Tractors
Lowboy Trailers
Pollution Control Equipment Trucks
Four-wheel Drive Vehicles
One-ton Trucks
Two-ton Trucks
Decontamination Trailer with Generators
Pickup Truck/Van/Cars
Crew Trailers
Office Trailers
Box Trailers
Galley Trailers

Protective Clothing and Equipment
Chemical Suits
Fire Entry Suits
Proximity Suits
Protective Clothing
Regulated Manifold Air Supply System
Breathing Apparatus - Self-contained
Emergency Oxygen Inhalation Systems
Portable Eye Showers
Full-face Respirators
Half-face Respirators
Explosion Meters
Cascade Systems
Bomb Suits
Bomb Blankets
Geiger Counters

Recovery and Treatment Equipment
Pneumatic Recovery Units
Hydrostatic Injection Units
Fume Scrubbers
100,000-gallon Portable Storage Vessel
8,000-gallon Transfer Stations
Floating Aerators
Mobile Clarifier with Sludge Collection
Storitainer (Portable Storage Tank)
Portable Buildings
Portable Pools

Flow Meters
Recovery/Injection Systems
1-cell Mobile Activated Carbon
    Filtration Units (Hydraulic Tip-up)
1-cell Mobile Activated Carbon
    Filtration Units
2-cell Mobile Activated Carbon
    Filtration Units
3-cell Mobile Activated Carbon
    Filtration Units
High Capacity Mixed Media Prefilters
Medium Capacity Mixed Media Prefilters
Low Capacity Mixed Media Prefilters

Analytical Equipment
Portable Laboratories
Gas Chromatograph/Mass Spectrometer
    (GC/MS)
Liquid Chromatograph/Mass
    Spectrometer (CC/MS)
High Performance Liquid Chromatograph
    (HPLC)
Gas Chromatographs
Fixed Wavelength Ultraviolet Detectors
Variable Wavelength Ultraviolet Detectors
Mobile Infrared Air Analyzers (Miran)
Fluorescence Detectors
Electron Capture Detectors (ECD)
Flame Ionization Detectors (FID)
Nitrogen/Phosphorus Detector (NPD)
Micro Hall Dectectors
Total Organic Carbon Analyzers
Spectrophotometers (Hach)
Specific Ion Meters
Flash Point Analyzers
Portable Air Samplers
Head Space Analyzers
Dissolved Oxygen Meters
pH Meters
Auto Samplers
Gas Chromatograph Specific Detectors
    (NPD, ECD, etc.)
Colony Counters
Data System for Mass Spectrometry
    (Including 31,000 Compound
    EPA/NIH Data Base

continued . . .

Figure 1.

Chemical Transfer Equipment
Transfer Equipment Trailer
Hot Tap Machines
Pneumatic Transfer Receivers
Non-sparking Tool Sets
Patch Kits
2" and 3" Chemical Transfer Hoses
Chemical Transfer Pumps
Chemical Transfer Compressors
Hydraulic Power Packs

Construction Equipment
Tandem Dump Trucks
32' Cat 225 Backhoes
955 Cat Loaders with 4-1 Buckets
17' Ford 5500 Backhoe
D-6 Cat Dozers with Winch
941 Track Loaders
D-3 Dozer with Swamp Pads
Fork Lift Tractors
Field Tractors
Barrel Grapplers

Lights and Generators
Light Plants - 5,000 Watt
Satellite Lights
150 KW Generators
100 KW Generators
50 KW Generators
15 KW Generators
10 KW Generators
6.5 KW Generators
5 KW Generators

Compression Equipment
Hot Water High Pressure Lasers
500,000 BTU Hot Water Heaters
500,000 BTU Steamers
250 CFM Air Compressors
185 CFM Air Compressors
Pressure Washers
250 CFM Positive Displacement Air
Pumps

Water Equipment
Work Boats
30' Pontoon Boats
16' Jon Boat with Motors
16' Jon Boats
Oil Skimmer Heads (Manta Ray)
6" Containment Boom
4" Containment Boom

Pumping Equipment
Portable Controlled Flow Samplers
2" Hydraulic Driven Pumps

4" Electric Submersible Pumps
3" Electric Submersible Pumps
3" and 4" Trash Pumps
1"/1½"/2" Electric Pumps
1½" High Pressure Pumps
2" Trash Pumps
2" Suction Hoses
3" and 4" Suction Hoses
4" Discharge Hoses
3" Discharge Hoses
2" Discharge Hoses
1½" Discharge Hoses
1½" Air Hoses
4" Vacuum/Pressure/Pumps
6" Vacuum/Pressure/Pumps
8" Vacuum/Pressure/Pumps
6" Aluminum Transfer Pipes
Peristaltic Test Well Pumps
Vacuum Pumps (1/8 Horse Power)
¾" Electric Pumps

Miscellaneous Tools and Equipment
Split Spoon Soil Samplers
Hurst Power Tools
Cooling System-Heat Exchangers
Brush and Weed Cutters
Sandblasters
Survey Instrument Sets
Cutting Torches
Welders
Metal Saws
Snowmobiles
Electric Hammers
½" Impact Wrenches
Skill Saws
Chain Saws
Barrel Carts
Angle Grinders
Pipe Threaders
175,000 BTU Space Heaters
100,000 BTU Space Heaters
100,000 BTU Furnaces
Scaffolding
1,100-gallon P.T.O. Drive Vacuum
Units
Hi-rail Equipped Pickup
Hi-rail Cart
Bio-activation Vessels
Autoclaves

Aviation Equipment
Citation II Jet
King Air 200 Turbo Prop

Figure 1. Concluded.

container, the location of the leak, the size and shape of the hole, amount of product in the tank, whether the car is insulated or noninsulated, and the time of year. If the temperature is 84 degrees, the patching material you use is quite different than that you would use if the temperature were five degrees—even though you are patching the same commodity in the same container. We have prepackaged kits of over 50 different types of patching materials so no matter what the situation we can apply the appropriate technology. Although materials are important, more critical is experience with different types of patching, knowing when or when not to use specific techniques or materials, predicting their degree of effectiveness, and recognizing their limitations. Successful leak stopping is often more a function of procedure and technique than it is of materials.''

"We have several thousand reference manuals plus computerized search services that assist us in identifying the hazards and protecting our people," notes Walker. "What it comes down to is not so much having the manuals but the knowledge of which manual applies and the location of necessary information. Even more important than manuals, are people who specialize in a particular container or a particular product. Over the past decade we have built up many contacts that we can call upon for specific information and help. In addition, we have internal manuals, and procedures that we use constantly. Outside manuals and reference materials are valuable if you have the right piece of information, but the specific information you are seeking is seldom in just one book. In some cases, the guidance we have found in books and manuals may not be the best way to handle a given problem; it may not be the safest way. We are constantly evaluating procedures based on actual experience."

"Obtaining basic information about the chemicals involved in an incident is necessary before you respond," emphasizes Walker. "You must have the know-how to protect your people. There must be a determination made of body and respiratory protection needs. If you have to transfer a tank car, you have to determine the type and size of off-loading equipment you will need. You must determine the distance you are going to pump, how much and what type of hose will be needed, and the type of fittings needed. Quite a bit of information is required before action is initiated. That is why true professionals are the people who manufacture the chemical. They have been in the business of safely manufacturing, transporting, storing, and disposing of the chemical. We ask for their assistance. We often end up talking to an individual who is a product specialist or has been working with a particular type of tank car or tank truck for a good many years, someone who has years of experience with the particular commodity or container."

The variety of environmental emergency response assignments handled by a company such as O.H. Materials is both extensive and demanding. A train derailment in a rural area spilled 140,000 gallons of Acrylonitrile, a soluble Class B poison, that seeped into the soil and contaminated the groundwater. Standard technology would have required excavation to a depth of seven feet, a width of 100 to 200 feet, and a length of from a quarter mile to a half mile. Approximately four 100-car trains would have been required to haul the contaminated soil to a disposal site, and disposal of the contaminated soil in a landfill would not have precluded future environmental problems. Costs for excavation, transportation, and disposal of the saturated soil would have been enormous. In order to recover toxic material from the soil in a manner that would eliminate or minimize the amount of material requiring disposal, O.H. Materials Co. developed a process for underground recovery and treatment. A flushing medium was pumped through injection points into the soil, creating a cone of depression. The contaminant was recovered by a vacuum pump. Contaminated material collected through the recovery system was deposited in retention ponds built on-site using portable retention pools, aerated to remove the recovered material from the water, and then treated through a sand and carbon filtration unit. The cleansed water was constantly sampled by personnel operating out of mobile laboratories. Once concentration of toxic materials in the soil was significantly reduced, samples of the contaminant were examined and a special strain of bacteria developed and injected into the soil using the underground recovery system. To maximize the effectiveness of the bacteria, nutrients and oxygen were added. The entire leaching/biodegradation process took 55 days to bring toxic concentrations within regulatory acceptance levels.

When a transportation accident in the midwest resulted in spillage of 100,000 gallons of mixed chemicals, alcohols, and solvents; O.H. Materials' environmental response personnel initially patched tank cars and transferred materials to undamaged containers. Soil borings were taken to determine soil characteristics, and extent of contaminated area. The characteristics of the local hydrology indicated the spilled materials had migrated through sandy clay soil to a shallow artesian system that fed private wells in the area. Test wells were installed to monitor underground flow and concentrations of chemicals in the groundwater. An on-site mobile laboratory tested 120 soil and water samples per day for the five contaminants, and air quality was monitored continuously. A pneumatic recovery system was installed to stem further migration of contaminants. As chemical concentrations in the recovery system influent were reduced to acceptable levels, flushing and injection rates were increased to cleanse the soil layers above the water table. From 50,000 to 80,000 gallons-per-day of contaminated water were processed through a multi-stage system of separators, clarifiers, sand filters, and carbon adsorption systems, air spargers, and fume scrubbers. Finally, bacteria were injected into the soil to complete treatment and the spill site restored.

A damaged storage tank in Missouri spilled 4,000 gallons of pentachlorophenol and oil that migrated to a farm pond and posed the threat of a major fishkill in the Big Piney River. An O.H. Materials crew repaired a pylon support to the storage tank and began immediate containment measures. The edge of the farm pond was sandbagged to prevent over-flow, and oil film was vacuumed from the pond's surface. Analysis of the pond water by an on-site laboratory indicated concentrations of pentachlorophenol far above acceptable discharge levels. Approximately one million gallons of pond water was processed through mobile carbon filters. After sampling and analysis, this stream was discharged to an adjacent stream. The pond was drained and the pond bottom and bordering contaminated soils were excavated and hauled to an approved, secured landfill. As a final step in the clean-up and restoration process, microbiologists established a mutant strain of non-pathogenic bacteria to degrade pentachlorophenol that might leach into the pond in the future.

At an unauthorized landfill in the Missouri River Valley; oils, PCBs, and various unidentified chemicals were found to be leaching into a stream which flowed into the Missouri River. A carbon filtration system was installed on-site to remove organic contaminants from the stream, the stream's flow was diverted away from the leaching area, and the landfill was excavated. Crews wearing fully enclosed chemical gas suits and self-contained breating apparatus unearthed waste oils containing PCBs, trichlorobenzene, tetrachlorobenzene, chlorobenzene, xylene, phenol, paint thinners, solvents, and various unidentified chemicals. Nearly 6,000 sealed drums were required to ship the contaminated soil to a federally approved landfill for disposal. To prevent surface water drainage from forcing remaining contamination into the groundwater table, the area was covered with an impervious layer of bentonite.

When a pesticide storage warehouse in Mississippi that contained more than 100 different types of pesticides required decontamination, an O.H. Materials crew wearing protective chemical suits drummed or overpacked the pesticide material and shipped it to a federally approved facility for disposal. The final decontamination procedure involved stripping the interior walls, decontaminating the walls by sandblasting, and sealing the walls using a special epoxy paint.

"The primary lesson we have learned over the past ten years is something that many people have learned together," concludes James Walker. "At the scene of an incident everybody has a piece of the problem, and everybody has a piece of the solution. It takes a united effort by all interests involved—local, state, federal, and the private sector—to contain and cleanup a major incident in a safe and cost-effective manner. You need the assistance of all people involved to do the best work possible for those who may be affected by the incident."

"Another lesson we have learned is that it takes one person to coordinate activities. You can't have three different people in charge at an incident. You need a focal point, an experienced person, to coordinate overall activity and delegate authority.''

"A third lesson we have learned applies to anyone who might respond; you cannot take anything for granted. At an incident there are so many related factors that if you make a mistake on one, or make an assumption that is incorrect or inaccurate, the consequences can be extremely serious. It pays to double-check your information even when you are dealing with people who have been handling that product or container for a long period of time. There is a need to consider alternative methods and procedures. On-scene, we attempt to first define the problem in its entirety, identify alternatives for action to obtain control or containment, and implement the best solution. Then, we continually reevaluate each of those actions, constantly reassessing the developing situation. Actions which worked in the early stages of an accident may not be the best method several days into the incident. Only by repeated evaluation of the overall situation can we identify the best way to properly allocate our resources. Our success has largely been related to this flexibility and being inventive enough to apply a creative solution to a difficult problem.''

Cook's Marine & Chemical Service

Until quite recently when he expanded operations, Jerry Cook was a one-man traveling chemistry laboratory providing analysis and representation services at hazardous materials incidents, particularly in the southeast. Although Cook has represented government agencies and others, he most often works for carrier organizations, particularly railroads. In today's world of hazardous materials/wastes/substances, increased regulation, and stringent liability considerations; it is no longer good enough to say "something is creating somewhat of a problem." You must be able to say what it is, how much there is, and what it can do . . . EXACTLY. Jerry Cook is often called in to answer these questions as a representative of the carrier. "The thing about railroads is they railroad," says Cook. "They often know not about environmental considerations, parts-per-million, or on-site laboratory analysis." Cook provides his employers with expert advice and guidance as to the seriousness of a given situation as well as proper corrective measures to be taken, and acts as an intermediary with the regulatory people. He cringes at use of the word "expert," however. "I don't use the word 'expert' in this business at all. All the experts are in the graveyards; they learned the hard way.''

Cook has a portable laboratory and reference library that can be trailered to an incident scene. He uses gas chromatography, detector tubes, Ph kits, flame ionization, flammable conductivity, electron capture, and a wide variety of analytical methods and techniques to determine identity, concentration, and neutralization factors. "My specialties are cyanide derivatives and/or chlorine, and a number of different gases. With my laboratory facilities I can operate down as low as the three parts-per-billion range. In many situations, however, I have to use the 'Q.A.D. Method,' '' (QUICK and Dirty) laughs Jerry Cook. "One of my major objectives is to save my employer money, but if the situation is bad I am going to say so. I am not going to tell people a concentration is less than it actually is, although I will try to put people's minds at ease if the situation warrants. I attempt to achieve a balanced view of the situation. There is often so much confusion at an incident scene with sometimes more than 20 viewpoints represented. So many times people overreact. By the same token, you have the group that under-reacts. They are just as bad as the ones who over-react because they may get somebody hurt.''

Cook makes his analysis and recommendations and deals with both the carrier and all other interests. He has to be as diplomatic as a man can be, and maintain his cool under the most nervewracking conditions. "There are confrontations," admits Jerry Cook with a grin. "There was a tense situation recently. I told the man, 'Mister, I don't want to be rude, crude, or anything; but I've got a roundtrip ticket. If you don't want to do what I say you should do, I'll just get back on that big bird and go.' In handling people, it's different strokes for different folks. The average guy has his rice bowl; he's got his own ideas of what he is supposed to do in an emergency situation. Yet you can't just tell a person what he wants to hear. If he's wrong, you have to have a smile on your face when you tell him he's wrong in a nice way. You *cannot* say, 'Hey Man, Shut up. You're not doing any good.' You might be able to tell him how they handle a similar incident back in your home state—he might listen to that."

Jerry Cook recently initiated a second company. "E.C. of Laurel," (Enviro-Chem) Mississippi; a clean-up and disposal company equipped with a response van and trailer, acid suits, chemical suits, booming apparatus, chemical pumps, and a wide variety of spill control equipment. E.C. of Laurel handles both containment/control and/or clean-up/disposal for three main forms of transportation: truck, railroad, and barge. However, a surprising amount of work for the new company has involved clean-up and disposal at chemical laboratories attached to universities, colleges, and high schools. "A typical recent case involved a university laboratory," remembers Cook. "They had tried to dispose of 25 pounds of picric acid on their own, got scared, and decided to call for professional assistance. Like many labs, they had become way oversupplied during the years. They had accumulated 24 different explosives, 1200 various poisons, 50–100 pounds of cyanide, and some PCP." "We also handle disposal of PCB transformers and similar materials," concludes Jerry Cook.

Roberts Environmental Services

The story of Roberts Environmental Services, headquartered in Eugene, Oregon; is a tale of transition—how a small company changed, adapted, expanded, and applied modern technology to meet the needs of its clients. "The company was formed as Roberts Septic Tank & Sewer Service in 1942 and supplied sewer services and septic tank pumping within Wayne County until taken over by present management in 1974," remembers company president, Rich Reiling. "Within our county area there are 42 different types of mills in the wood products industry ranging from pulp and paper mills to sawmills. They use different chemicals and refining processes that have lots of residues, and we were asked to take care of these. We started expanding our knowledge in these specific areas and began purchasing equipment to deal with it on a technical level. Then, of course, our former service area was not able to support vast capitalization necessary for the equipment, so we had to move outside of the geograpical area we had traditionally served."

The company moved heavily into industrial tank cleaning to the point where it now handles approximately 200 such jobs a month. Because of expertise developed in the field of industrial cleaning, the company started being asked from time-to-time to handle clean-up of spills. This caught the interest of company management who began to place emphasis on spill recovery, in the process developing an additional level of expertise. Although industrial cleaning (vacuum loading, high-pressure water blasting, high-velocity line cleaning, hazardous environment cleaning, waste removal and disposal, sewage disposal, groundwater decontamination, etc.) is still Roberts Environmental's bread-and-butter, and gives them an edge in equipment/experience/technology over some pure response organizations, the company handles an average of four hazardous materials emergency response calls a month. Their area of response coverage for equipment is anywhere west of the Rocky Mountains; for personnel alone, or if equipment can be leased locally, they will respond nationwide. In addition to the corporate office at Eugene, Oregon there are branch offices in Salt Lake City, Boise, Seattle, and Portland.

A recent incident involved a gasoline tank truck overturned on the freeway 30 miles south of Eugene. The weather was cold for western Oregon; when Roberts response personnel arrived on-scene fumes were being held down in a roadside ditch, and motorists with tire chains on their vehicles were driving by creating a source of ignition. After getting the freeway closed down and convincing massed spectators to evacuate the area, Roberts personnel proceeded with clean-up that included cleaning out a storm drain and the handling of soil removal.

Another incident involved groundwater recovery. Over a period of two years, a service station had lost approximately 6,000 gallons of unleaded gasoline through a pump with a cracked casing. The fuel entered the water table and spread out over eight to ten acres of ground. Roberts personnel surveyed the area to locate the edges of the plume, then drilled 17 wells and installed pumping equipment that over a six month period recovered the bulk of the fuel.

"One of the larger mills in the area had from 5-6,000 gallons of bunker fuel oil explode from tanks and cover all their stainless steel equipment," adds Rich Reiling. "It was quite a clean-up process to go in there and chemically clean their tanks. Another major mill out on the coast required a clean-up process that was quite similar. Industry provides the largest segment of our business. With that knowledge and background, responding to spills like overturned tankers is relatively easy for us because we have the equipment to go after it. We handle similar types of materials every day in our industrial cleaning, so handling them in the field was not a major step for us to take."

"For emergency response, we receive calls through our 24-hour service," notes Reiling. "Depending on where the incident is and the type of spill response required, we can dispatch a crew of from one individual to as many as 35. Generally, there are four persons who initially handle emergency response calls. Whoever receives a particular call stays in touch with the calling party to learn the nature of the spill, how long the situation has been in effect, and to estimate the type of equipment and personnel he needs to mobilize. He then goes in advance of the response party to size-up the spill site and request additional resources. We have several different types of vacuum equipment such as standard, 140-barrel tankers with large vacuum pumps that are D.O.T.-approved for caustics; and a dry/wet mode industrial vacuum loader, brand name 'Guzzler,' with the capability of moving either dry or liquid materials from as far away as 1,000 feet and as far down as 200 feet. We have used it for spills down over the side of a banking where traditionally you had to go down and pump it into barrels and airlift it out."

"Moving from industrial cleaning to emergency response required a couple of major changes," reflected Rich Reiling recently. "When we got into spill response, regulations became a very, very important aspect of the business. When you are doing work in a plant, you can meet the OSHA requirements, the plant requirements; but when you leave the plant site and get out in the open highway, the regulations change greatly so we had to make changes and buy equipment that satisfied the new regulations. There were major changes in our equipment, in our attitude toward the work we did, and in addition of qualified personnel who had backgrounds in spill response. Spill response was a natural progression for us, but movement into spill response required a lot of changes. There may be two sides to every issue, but when you go out on a spill response you are the guy in the middle. I think any commercial spill responder feels this pinch; it is his responsibility to meet all the regulations of all the agencies from the highway patrol to E.P.A. to the highway department and any other agency that may have jurisdiction. He has to coordinate all these interests at the same time he is dealing with the 'owner' of the problem. To do this effectively was a major learning experience for us."

"Safety has been the second area where we have really changed our attitude," continues Reiling. "In order to do the work and meet the government regulations, a thorough safety program is essential. We have safety meetings before we go on-scene and during an incident, then we critique the incident when it is over. I think this has given us a lot of credibility as far as the agencies are concerned. Also, the 'owner' of the problem feels a lot more comfortable when he observes that methods utilized to do the clean-up are safe. One outcome of our concern for safe operation, and a reinforcement of it, was in bringing Bill Henle on-board. He has a strong background in training, including work with the Wisconsin state haz. mat. response team. He has built within the company a strong internal training program so that our people are being trained constantly."

Training of first responders to hazardous materials incidents has become a further extension of services provided by Roberts Environmental. Such training is being directed toward public safety agency responders. (fire, police, civil defense) and personnel such as managers connected with the Port of Portland. Currently under development are training programs for industry in terms of complying with RCRA.

Rich Reiling was asked recently if he could provide any tips that might benefit other response organizations. "One of the really basic, but important, tips is to have a comprehensive outline available at the telephone so that when a person calls you for response, you ask *all* the right questions," he responds. "Such persons are normally excited; a lot of times they are calling you from a phone booth or by means of a patch-through call. They feel an urgency to get something done immediately, and you have to pin them down and really ask the questions. I would recommend that right from the very beginning you have a detailed list of questions next to the telephone that you use as a guide to fill-in the answers."

"Another necessity is a good communications system; without a good communications system you might as well forget about being a fast responder. We have extensive, portable communications gear that we carry with us when making an initial response."

"Safety has to speak for itself. You've got the safety of the people who are on-site, and the safety of the community to be concerned with. We all get carried away sometimes in trying to accomplish what we want to do, so we have procedures for keeping foremost in our minds general safety requirements."

"When we first arrive on-scene," adds Reiling, "we don't go charging in. What we try to do is stay back for awhile, make some assessment, and try to determine who is in command of the scene before you actually start operations. This is one of the biggest problems I have experienced. You may be talking to one guy, and he thinks he is in charge; you get further on into the job and you find that someone else is really in charge—it can be a really messed-up situation. For this and many other reasons, your analysis or size-up of the job is critical. When you finally go in to do your work, you want to be sure you have done your assessment in all areas and have established a good communications procedure with everyone involved."

"One thing we like to do," adds Reiling, "is set up a command center *away* from the area that is directly involved. This way reporters coming on-scene have an area to go to, and you can have meetings with all involved agencies—at a distance from the involved area. I'm not saying you have to place your command center miles away, but try to set it up a bit out of the way."

"In addition, a lot of times when we are called to a scene, firefighters are already there. We are not going to go barging in and take over work that they have already done. It is *very* important that you walk in there and learn what they have done, at least spend the time to ask them about the direction they feel the incident is headed. Most of the time when you are asking them, they turn around and ask you for directions. If you go in and start giving orders—people are human, and human considerations have to be taken into account."

During a discussion of the various types or response organizations an individual or organization might call on when suddenly faced with a serious problem involving hazardous materials, Rich Reiling noted: "One whole philosophy is that by using specially engineered equipment we can get the job done quicker, more economically, with less personnel. By using high technology equipment, we expose less personnel to the problem area and can move greater volumes of material. By using a higher level of technology, which can be expensive, ultimately the cost to the 'owner' of the problem is often lower. Much of the work we do involves cleaning nasty little holes for industry, basically a handling kind of problem. Say they have to get a residue out of their boilers; they put it into a bucket and hand it through a small hole to another guy who puts it into a wheelbarrow and wheels it outside; then they take a loader and load it into a truck and haul it away. We, on the other hand, take one piece of equipment, run an eight inch hose into the tank, and load it from there. The 'carry' can be 400 feet, up over a 100 foot barrier and down the other side—we reduce the handling. Now the only personnel involved is the one guy running the truck. Any place there is a tricky handling problem, that is where we are likely to be seen doing a lot of work. We use heavy equipment, high technology equipment; we have personnel operating and setting up the job who understand the technology. In the long run, our customers look at the cost effectiveness of the equipment we use. Nowadays, every company is looking at costs. They are going to get the most for the least, and that is what we can provide."

Peabody Clean Industry

Disposal of hazardous waste is one of the more critical problems currently facing industry, yet the Hazardous Materials Division of Peabody Clean Industry had absolutely no problem in disposing of hazardous waste from one recent cleanup operation. After fire and explosion devastated a precious metals recovery/plating plant, Peabody personnel were called in to handle cyanides and acids within the rubble. Fifty-two drums of recovered gold and platinum cyanides, some containing as much as 60 ounces of precious metal, were estimated to be worth $36,000–$42,000 *per drum*. This was not exactly a bucket and mop cleanup operation, yet neither was it a particularly unusual assignment for the company known throughout the northeast as "Peabody Clean." Although the end product of the operation, contaminated industrial waste, was unusual due to its inherent value; the organization, techniques, and methods used to respond to an assignment of this nature were the result of experience and extensive planning within the Peabody organization.

That experience, and the internal organization and planning developed to respond in a professional manner, has been both extensive and varied. Control and containment of hazardous waste sites has been a common assignment for Peabody personnel. An illegal hazardous waste site in northern New England required site organization, analysis, categorization, repackaging and treatment of an estimated 5,000 drums of hazardous wastes. In southern New England, the Peabody pros had to excavate, prepare, and dispose of an "estimated" 16,000 to 20,000 drums of hazardous waste illegally dumped on a farm. The total volume disposed of at this site was strictly an estimate because the dumper's method of operation had been extremely cost efficient; he had eyeballed every drum taken in; drums that appeared to be in reasonably good condition were dumped and sold to a reconditioner, while less serviceable drums were emptied and then crushed to make room for more drums. The end result was an environmental nightmare requiring highly developed technical expertise to achieve control and abatement.

Not all uncontrolled hazardous waste sites are located in rural areas. In one New England city Peabody handled cleanup of drums and decontaminated soil and buildings. The one-time illegal operator had been storing drums of PCBs, heavy metals, waste solvents and other hazardous materials in his garage, and renting garages from neighbors—RIGHT DOWNTOWN.

One "small" job proved to be anything but. Called in to handle "a few drums," Peabody equipment operators thought they detected suspiciously soft ground. A few picks with excavating equipment led to uncovery of an eventual 2,000 plus drums and 3,000 plus yards of contaminated earth.

Another abandoned, illegal chemical dump contained 10,000 plus drums of various hazardous wastes requiring extensive cleanup and disposal. Yet another site, the former KinBuc Landfill in New Jersey, required containment, collection, and disposal of 2,000 to 4,000 gallons-per-day of leachate outfall from oils, PCBs, acids, pesticides, solvents and biologically active agents. At the former Chemical Control site in New Jersey, Peabody participated in the cleanup of 55,000 exposed and buried drums plus contaminated soil and several large bulk storage tanks.

Experience gained by the Hazardous Materials Division of Peabody Clean Industry has not been limited to hazardous waste site cleanup however. Additional assignments requiring a high degree of technical expertise have included

- PCB decontamination at a vandalized building.
- Cleanup of a phosphorus pentasulfide spill.
- Cleanup of a 5,500 gallon spill of toluene diisocyanate along an Interstate highway.
- Collection and disposal of PCB-contaminated gas condensate.
- Decontamination of soil and fixtures at a pesticide warehouse, including disposal of hazardous waste.
- Containment, neutralization, and disposal of a 2,700 gallon spill of pyridine from along a major tollroad.

• Beryllium and lead dust decontamination of an industrial building, and disposal of the residues.

• Emergency excavation and removal of 4.3 miles of dirt road illegally coated with PCBs, heavy metals, and waste solvents.

• Decontamination of an office building and laboratory after a fire involving PCBs.

• Removal of PCB-contaminated ballast from along 15 miles of railroad track.

• Containment and cleanup of a major rail incident involving a ruptured tank car of phosphorus trichloride.

• Cleanup of a pesticide manufacturing facility after an explosion and fire.

• Cleanup of drums and bulk tanks from a burned out chemical plant.

Assignments such as these, plus an average of four to five emergency responses to hazardous materials incidents per month, added to a day-to-day routine of industrial cleaning/asbestos decontamination/oil spill response/tank maintenance requires a company with great depth in the areas of trained personnel, specialized equipment, and organizational capabilities. Founded in 1958, Peabody Clean Industry has grown from an oil spill containment firm to one of the largest hazardous materials containment firms in the northeast, along the way becoming a leader in the management of "Superfund" hazardous waste sites. Its parent organization, Peabody International Corporation, employs 5,600 people in 61 locations engaged in providing equipment and services for the energy conversion, power generation, pollution control, industrial and agricultural markets. For example, Peabody Galion in Galion, Ohio manufactures solid waste mobile and stationary compactors, transfer stations and trailers, dump truck bodies, hydraulic hoists and pumps, and power lift gates. Peabody Myers in Streater, Illinois produces mobile vacuum loaders and sewer cleaning equipment. In 1981 Peabody International had annual revenues of $424 million. Peabody Clean Industry provides hazardous materials, oil spill, and industrial cleanup services plus hydroblasting throughout New England and the middle Atlantic states from offices in East Boston, Massachusetts; Linden, New Jersey; Davisville, Rhode Island; and Long Island City, New York. Peabody Clean also has operations associated with Peabody VIP with offices in Baton Rouge, Louisiana; Houston, Texas; Corpus Christi, Texas; Chicago, Illinois; and Paulsboro, New Jersey.

Peabody Clean has handled an estimated 10,000 incidents involving hazardous waste and toxic chemical spill cleanup/recovery/removal; and has long been active in the areas of oil spill management, tank cleaning and industrial maintenance. Clients include a broad range of Federal, state and local regulatory agencies; varied industrial concerns, and fire and other public safety agencies. Its Hazardous Materials Response Team is backed by a tremendous assortment of containment/mitigation/personal protective· equipment including high strength/high volume industrial vacuum systems, hydrolasers for removing polymerized materials, stainless steel and explosion proof pumps and hose systems, and assorted sorbents and booms. (See Figure 2 for a detailed listing of response equipment.)

Peabody Project Team And Project Management

The Peabody Project Team of Peabody Clean Industry, Peabody VIP, Wegman Engineering, Peabody New England, (construction engineers) Hayden Harding & Buchanan Consulting Engineers, and Peabody Process is composed of subsidiaries of Peabody International Corporation to offer a unique blend of disciplines within a single parent company. Each individual company has extensive experience in all phases of air, water and solid waste handling technology be it of a hazardous nature or industrial waste management.

Previous joint venture experience has encompassed hydrogeological studies; industrial waste landfill evaluations; hazardous waste generation studies; secure landfill planning, design, construction and operation; on-site analytical and air monitoring studies; flyash and flue gas desulfurization system evaluation, design and construction; on-site treatment system

**Partial Listing of Response Equipment Available To
The Hazardous Materials Division of Peabody Clean Industry**

Vehicles:

Heavy Duty Vactor
Vacuum Pump Trucks (Tractor/trailer &
  straight)
Mobile Field Office & Communications Center
Tractor Trailer Units
Utility Trucks (Pickup, Econovans, Autos)
Rack Trucks, Crewcabs, and Heavyduty Utility
  Vehicles
All-Terrain Vehicles
Bulk Tanker w/o Tractor
Box-type SPill Trailer
Skid Loader ("Bobcat")
Response Van
Trailer-mounted, Portable Diesel Lighting
  Tower

Equipment:

Mobile Catalyst Screening/Cleaning Unit
  (Cracking tower catalysts, etc.)
Mobile Life Support Trailer (Supports/
  monitors breathing air and work
  environment via TV and analytical
  instruments)
Decontamination Units (Fully outfitted)
Weather Station Modules
Mobile Analytical Laboratory (Outfitted on a
  case-by-case basis)
Mobile De-watering systems
Swiss Skimmers
Slurp Skimmer (Esso Research)
Power Work Boats & Assorted Small Craft
Boom Float
Mudcat Dredge
Hydro-Lasers (Up to 10,000 psi; and up to
  30,000 psi.)
Steam Unit
Steam Boiler
"HOTSY" High Pressure Hot Water Washer

Crisafulli Pump (1400 gpm)
High Pressure Water Herding Pumps (Hale)
Utility Pump
Centrifugal Pump (2 inch)
Pneumatic Pumps (2,3, and 4 inch)
Air Compressor and Hose.
Air-operated Drum Pump
Acid and General Purpose Hose
Air Operated Vacuum
Portable Radio
Portable Photoionizer/Vapor Detector
Radiation Survey Meter
Metal Detector
Explosimeter
Oxygen Meter
pH Meter
Air Sampler and Tubes
Radiation Audibile Alarm
Portable Generator

**Protective Equipment
and Clothing:**

Acid (Chemical Protection) Suits
Fire Entry Suits
Selfcontained Breathing Apparatus (MSA-401)
Cascade Air System (Single Outlet & Four
  Outlet)
Assorted Half-face and Full-face Respirators
Cartridges, Cannisters, and Filters for
  Respirators
Disposable Entry Suits
Disposable Coveralls (Sarynex coated and
  Polycoated)
Disposable Raingear
Assorted Aprons
Heavyduty Boots
Assorted Gloves (Viton, Nitrile, Butyl, Vinyl
  coated, Surgeons)
Goggles, Ear Protectors, and Splash Guards

Figure 2.

planning, design, construction and operation; pollution liability and hazard assessment engineering studies; and hazardous material/waste handling, removal, treatment, transportation and ultimate disposal planning.

Greg Heath, Environmental Services Manager for northern New England, is responsible for coordination of all hazardous materials programs for the Boston office. Prior to joining Peabody, Heath worked on development, shakedown, and emergency field applications of prototype spill cleanup equipment for the U.S. Environmental Protection Agency's Emergency Environmental Response Unit, and the Oil and Hazardous Materials Simulated Environmental Test Tank.

"Peabody International acquired Coastal Services five years ago," explains Greg Heath. "Altogether, Coastal/Peabody has been in the hazardous material/waste business for four years. Until 18 months ago, our hazardous materials management program was coordinated primarily through our Linden, New Jersey office. For the past year-and-a-half we have offered response capabilities from our Linden, New Jersey; Davisville, Rhode Island; and East Boston Massachusetts offices.''

"There are many ways in which hazardous material cleanup does not involve emergency response," emphasizes Heath. "You have to keep your employees working; you have to have business coming in all the time. Our normal industrial maintenance activities involving hazardous materials give us a continuous background in working with such materials; also, regular industrial services including cleanup is one way we keep up with our training."

"In order to adequately serve industry," continues Heath, "we must stay on top of regulations. There is one key person within the company who coordinates safety and environmental affairs by maintaining anticipatory compliance with Federal, state, and interstate regulations governing the handling, treatment, packaging and transport of hazardous materials. This person insures that pertinent information pertaining to the Resource Conservation & Recovery Act, Toxic Substances Control Act, Spill Prevention & Control Countermeasures Plans, 'Superfund,' and D.O.T. and O.S.H.A. regulations filters down to us in the field.''

Due to the hazardous nature of many assignments handled by Peabody Clean Industry, personnel develop and use a written protocol for each job. A written protocol is used to protect employees, the public, industry and the environment from undue exposure to hazardous situations; and to effectively and economically recover and/or dispose of hazardous wastes in accordance with all pertinent regulations. Because the average reader might not be aware of the great amount of planning that is required for a commercial response contractor to safely and successfully handle even a "routine" job involving hazardous materials or wastes, Greg Heath agreed to deliniate the various factors Peabody personnel routinely consider during any such assignment. Company personnel feel strongly that to handle the variety and volume of hazardous material they commonly deal with there must be a detailed plan to insure that all critical factors are dealt with. To "wing it" would be to invite disaster. Commercial response contractors are not materials handlers; they are called in to contain and control truly hazardous materials, some of which can injure and kill by ingestion, absorption, inhalation, or radiation.

Client Contact—"A client may have a clearcut emergency response need or may just need some industrial cleaning services," notes Heath. "Initially, we perform a *Site Survey* which may involve spill personnel having to wear proper protective equipment. First, we find out exactly what the client needs—what is the emergency situation; what are the hazards involved? What situation needs to be addressed first? We then attempt to prioritize the various components of the job. We try to break down each job by exactly what has to be done for operations. We develop a *Hazard Assessment;* what exactly is the hazard level? Should the product sit for awhile; will it set-up? Do we snuff it up with a vacuum truck or use an alternative method? If time is available, we pull together a 'think tank' group of people from within the company. As a result of this meeting, we draw up a written *Technical Approach* that considers the relative feasibility of various options. We also devise a *Manpower & Equipment Utilization Plan,* identify *Disposal Options,* and perform a quick *Cost Analysis.''*

Operations—"As we go on-scene to begin operations, we follow a *Work Plan* developed previously that spells out exactly what is to be done. There is also a written *Safety Plan* that is an important consideration in company operations. To just walk into a situation, put your hand on a valve, and close it—that is not the way we do it," adds Greg Heath. "First, there may have to be a sampling program to determine proper protective clothing to be used, possible reactivity of the material, required respiratory protection, explosion hazard if any, immediate and longterm dangers represented by the product, and a myriad of additional factors. For any major job, there is *always* a written Safety Plan."

"Another important consideration for us is a *Site Plan,*" adds Heath. "We use what is called a 'zonal concept.' We deliniate various areas as a Hot Zone, Buffer Zone, and Safety Area. We may have to consider where to have communications equipment installed and where to stage our tools and equipment. In addition, we outline contacts with coordinating agencies.

We do hundreds of jobs a year and feel we know the factors that need to be considered in working with hazardous materials and wastes. We may well have a better approach to a given situation than other agencies or organizations represented on-site that do not have such varied experience."

"The time required to create such written plans for specific jobs can vary widely," says Greg Heath. "If we have full knowledge of a company's general and chemical operations, it may take just a few hours. In other situations, we may have to do some detailed research. There may be hazards present that we have never previously encountered. We will never put anyone into a situation where their lives are going to be endangered."

"Normally, if we are called in by E.P.A. or Coast Guard, they will have determined a lot of information needed for our planning before they call us," notes Heath. "There is quite a bit of lag time in some situations. Generally, the more agencies involved the more lag time there will be. A final component of the written Safety Plan that we develop for each job is an identification of *Regulations That Override Operations.* That is, we have to identify and evaluate all D.O.T., E.P.A., NIOSH, OSHA, state, local, or industrial regulations that apply. Then we provide a *Safety Plan Orientation,* or briefing, so that everyone will know why we are on-site and what we are expected to do. This can be done very quickly for an emergency response, or consume more time if time is available."

"The next step is *Mobilization.* We consider use of personnel, supplies, and incidentals; and direction of subcontractors. We may not have all equipment available that will be required for a job and may have to subcontract some of the work to our competition due to our equipment being dedicated to specific uses or the fact that we are often working spills concurrently. We often deal with two spills at one time; it is a very realistic situation for us to have to tell potential clients that at the moment our equipment is tied up."

"Next we move into *Operations On-Scene* where eight different factors have to be considered. First is *Client Representation/Contact,* working with the client to make certain he knows exactly what is going on, the methods we are using, and the problems/results we are encountering. There must be continuous communication. Next we consider *Site Assessment & Air Monitoring,* and then follow-through on our previously determined *Zonal Deliniation* to establish and maintain clean/buffer/contaminated zones during the actual operation. We also deal with *Deliniation of Responsibilities to Personnel,* and *Personnel Supervision/Command Decision Coordination.* Further, *Site Specific Decontamination* requires that we determine and enforce exactly how the decontamination procedure is going to be run. If someone has to go to the privy, decontaminate him; as nominal an operation as this may be, decontaminate him. Then, as a job is being concluded, there is *Completion of Work Sheets,*" continues Heath. "You review with the client exactly what was done and the client provides *Approval of Services Performed.*"

The next major step is *Demobilization of Manpower & Equipment.* This includes *Ultimate Decontamination* and *Reload.* "You don't sit back and say 'That was a fine job' or go celebrate," grins Greg Heath. "RIGHT NOW, you reload your equipment so you are ready for the next job."

Disposal Options—Whether an operation represents an emergency or nonemergency situation, the product that has now been contained and controlled needs to be safely and properly disposed of. Any major commercial response organization has to have complete and total knowledge of all available disposal options and be able to implement proper disposal of a particular hazardous material or waste. No job is complete until material has been properly manifested and disposed of in accordance with state and Federal regulations. Peabody arranges disposal through any of the four major techniques currently available: reuse/recovery, treatment detoxification, incineration, or secured landfill. Any method or combination of methods may have to be arranged depending upon the nature and composition of the waste to be handled. "Disposal" may often include identification and classification of materials as well as containment, physical removal, and actual disposal. Disposal of hazardous materials definately requires special care and handling. Some materials with a BTU content can be burned, but PCBs, for example, can only be disposed of by high temperature incineration. Chlorinated solvents, on the other hand, can be incinerated at highly regulated facilities with flue gas emission control equipment, or by redistillation.

Debriefing—The final component of the model used by Peabody hazardous materials responders, *Debriefing,* includes evaluation of the operation by involved personnel, the completion of a written *Summary of Operations,* written *Renovation Recommendations,* and consideration of future *Marketing Potential* with the client. The model can be modified depending upon a particular situation; it can be added to or parts can be deleted. The model is actually a means to insure absolute thoroughness, but it also represents an attitude within the company with which difficult tasks are approached. With Peabody, there is no such thing as "good enough." Hazardous materials response demands thoroughness, extreme care, and a high level of technical expertise.

Greg Heath differentiates between those responses that represent "uncontrolled operations" and those that are more stable. "With an uncontrolled operation, we may have to pull back and regroup," he says. "We are willing to regroup if the situation warrants; you have to be realistic. You must continually evaluate the level of control you have in any operation. If you feel you are losing control, you have to identify routes to rectify the situation. Some examples of an uncontrolled situation would be if you are unable to stop materials from leaking, if the spilled material has generated a toxic cloud, or if you are currently unable to contain the material that has spilled. You may not yet be in full control of the situation for whatever reason—things are still moving on. You may have succeeded in minimizing harmful effects but may not yet be in full control. Another facet of an uncontrolled operation might be represented by an abandoned or uncontrolled waste site. Quite often, you initially really don't know what is there. You can't rely on drum labelling in a waste dump situation. Reused drums have led to a number of accidents. You need analytical equipment backup to safely control the situation."

In-depth support within the company is an important plus for Peabody Clean Industry. "We build and modify a lot of our own vacuum trucks," says Greg Heath. "Peabody Galion makes garbage trucks and dumptrucks. Peabody Myers manufactures what is called a Vactor, a major piece of equipment used in sewer cleaning. It has been modified recently to handle industrial hazardous waste cleanup. It can pickup cinder blocks—there is that much power. A 400 h.p. engine provides the power and a specially designed blower and filtration system handle contaminated air and particulate matter."

Clean Caribbean Cooperative—Because of a longstanding, and somewhat unique, contractual arrangement; Peabody response personnel are on continuous stand-by ready to load specialized equipment aboard commercial cargo aircraft and fly to a spill of "Mexican Crude" or other petroleum product or hazardous material anyplace in the Caribbean Sea. Back in 1973, a number of oil companies (Amoco Marine Transportation Company; Bahamas Oil Refining Company; Bonaire Petroleum Corporation, N.V.; Burmah Oil Bahamas, Ltd.; Empresa Columbiana de Petroles; Esso; Mobil Oil; Petro-Canada; Phillips Petroleum; Shell Curacao, N.V.; Sun Transport, Inc.; Texaco International Trader, Inc.; Trinidad and Tobago Oil Company, Ltd.) engaged in tanker transportation of hydrocarbons or in operating petroleum hydrocarbon handling or producing facilities in the broad Caribbean (roughly from the Central American peninsula in the west, the Bahamas in the north, the Leeward Islands and Barbados to the east, and Tobago/Trinidad/Aruba to the south) formed the Clean Caribbean Cooperative (CCC) to enhance existing capabilities to promptly and efficiently respond to oil spills in the area. Clean Caribbean Cooperative member companies agreed to share the cost of providing a source of materials, equipment, and services that could be used in responding to an incident. Because the immense surface area of the Caribbean precludes total reliance on surface transportation of equipment and materials to a spill which may occur at great distance from those Caribbean locations where oil spill cleanup capability is in place, the Cooperative leased a stockpile of specialized response equipment which is stored at Peabody's Davisville, Rhode Island facility for prompt dispatch by air. Since 1973, Peabody Clean In-

**Air-Deliverable Oil Pollution Response Equipment
Dedicated to Use by Clean Caribbean Cooperative**

2   Slickbar Transvacs with accessories. (Vacuum/Positive Displacement Pump Oil Recovery System.)

12 Slickbar 250 Foot Section Mark 9 Boom.

1   Fully-equipped On-Scene Command Center Module. (Self-contained 42 foot by 8.2 foot, 16-wheel trailer designed to support a 27-man response team in remote areas. Includes an Executivetone Systems telephone system with ten trunklines; a radio system and repeater capable of longrange communications that is selfcontained and able to operate on 110VAC or 12VDC; and two air-inflatable buildings that can be equipped with lights, telephones, and furniture from the command center module to provide additional enclosed work space.)
     Various Aircraft/Marine/Portable Radios.

1   Slickbar Trimaran Boat with Trailer and Motor. (Tri-hull, 27.5 foot by 14 foot/8 inch folding work boat towable in travel mode by light truck or four wheel drive vehicle.)

12 Kemper Sea Containers. (Towable, inflatable transfer tank with a capacity of 1200 gallons of recovered/salvaged oil.)
     Various handtools.
     Various Navigation Tools.

2   Avon Inflatable Boats.

1   Griphoist.

1   Grove Ten Ton Crane.

2   Bird Alarm Systems.

1   Portable Electronic Weather Station.

1   Onan Portable Generator.

1   Barrel Hook with Chain.

1   Trimaran Dispersant Spray System and Boom.

7   Load Levelers and Chains.

1   Air Compressor.

1   Hydraulic Pump.

1   Bobcat Loader.

2   Harding Jacks.

2   Boston Whaler Boats.

1   Extension Antenna.

1   Side-mount Dispersant Spray System.

1   Hale Pump.
     Discharge and Suction Hoses.

1   Barrel Pump.

2   Portable Holding Tanks.

1   Blackhawk Porta Power.

6   Worklights.
     Corexit Dispersant. (100 drums are stored at Miami Airport with additional drums on-hand in Rhode Island. Also, CCC has a dedicated four-engine dispersant spray plane located in Mesa, Arizona that when fully loaded with dispersant can effectively treat 420 acres of oil spill in 20 minutes.)

Figure 3.

dustry has contracted with Clean Caribbean Cooperative to lease, maintain, and periodically test necessary equipment and materials; train Peabody employees in the use and operation of the leased equipment; package and store all such equipment so that it can be immediately transported to an airfield and loaded aboard a cargo aircraft; and provide supervisory personnel during the shipment and operation of leased equipment and material. Peabody's Clean Caribbean Response Team is a minimum 35-man strike force drawn from various disciplines within the company and assigned to facilities at East Boston, Massachusetts and Davisville, Rhode Island ready to mobilize, respond, control and cleanup spillage of oil as directed by a member company of the Clean Caribbean Cooperative on a 24-hour a day, seven day a week basis.

Upon activation of a detailed Clean Caribbean Cooperative (CCC) Emergency Response Procedure, equipment is loaded onto flatbed trucks and taken to a nearby airport for transfer to cargo aircraft. All CCC equipment was selected because it is air transportable and can be used effectively in sea and weather conditions common to the Caribbean area. (See Figure 3 for List of Equipment Dedicated to Clean Caribbean Cooperative Response.) Because fast response to an oil spill is crucial if environmental damage is to be minimized, and the transport by air and then by land or water of men and equipment to the actual spill area requires the cooperation of various governments whose national territory and waters may be threatened by a spill; Peabody's CCC Emergency Response Procedure includes a prearranged system designed to insure prompt issuance of special permissions and permits as follows.

• Landing permits for chartered aircraft. (Chartered aircraft do not take off until a telexed confirmation of the landing permit is received.)
• Clearance by Customs of equipment and materials.
• Clearance by Immigration for non-national oil spill cleanup personnel.
• Granting of work permits for non-national personnel.
• Authorization to make contracts with local organizations and individuals for transportation, food, labor, lodging, security, communications, equipment rentals and other services necessitated by the oil spill cleanup operation.
• Use of special radio equipment.
• Use and operation of specialized marine equipment on national waters.

Summary

Oil spill cleanup, hazardous materials response and control, hazardous waste site containment and control, and industrial cleaning and handling; through extensive planning, use of specialized equipment and trained/experienced personnel, plus attention to detail, Peabody Clean Industry is able to handle it all. As "Superfund" cleanup of hazardous waste sites gains momentum around the country, Peabody Clean is one of a limited number of companies that by dint of broad experience over a number of years is capable of handling the unique demands of such work. Greg Heath firmly believes such expertise is required as part of an interdisciplinary effort to properly manage hazardous materials handling at inactive waste sites.

"Regardless of which regulatory or enforcement agency—be it Federal, state, or local—is assigned the responsibility for observing cleanup in an inactive hazardous waste site; it must be emphasized that the actual management and operational undertaking ultimately resides with a professional hazardous material handling contractor," he stresses. "The regulatory and enforcement agencies set their parameters for a general approach to the cleanup based on established general industry guidelines, but it is the response team members of the professional contractor that implement not only those recommendations but the contractor's own hazardous material/waste management program. The contractor's management program of operationally sound principles for mitigating the hazards often associated with hazardous waste sites is developed not only through the contractor's experience in the field, but also from the technical approach of his interdisciplinary effort common to the new breed of full service hazardous material handling contractors."

"Inactive waste site assessment is just a portion of the through-put model the contractor must mold and manipulate into a safe and cost efficient design," adds Heath. "The professional contractor is characterized by his approach and methodology for remedial action utilizing information from site history, climatological and geographic studies, data from hydrogeological studies, random and selective material sampling, field analysis techniques, and air quality monitoring techniques. Also, the contractor must provide an immediate response capability for containment and storage system failure. He must maintain ongoing operations and contend with arising emergencies. His decisions include the best approach to accomplish the most cost efficient disposal options. Thus, he is realistic and in-tune with the financial restrictions that may be mandated by the current general state of the economy. And now, more than ever, the professional contractor is fully cognizant of safety management and realizes that each operational performance must adhere to an established safety plan.''

"The professional hazardous material handling contractor is not a person or group of adventuresome entrepreneurs," concludes Greg Heath, "but rather a collective effort of engineers of professional disciplines working with available technology. From the training of all hands-on personnel to conformance with all pertinent regulatory agency standards, the professional hazardous material handling contractor has developed not just a marketable service but the overall provisions for an effective hazardous material/waste management program.''

7 TRAINING AND INSTRUCTION IN HAZARDOUS MATERIALS RESPONSE AND CONTROL

The numbers of different hazardous materials, as well as the total volume, being manufactured, processed, transported, and disposed of in this country have increased by geometric proportions in recent years. According to the chemical manufacturers Association, there are 350,000 minor formulations or chemical mixtures introduced into commerce every year. The Chemical Abstract Service has catalogued four million unique chemicals. In 1975 the United States produced 155 billion pounds of 8,000 different synthetic organic chemicals with an estimated value of $25 billion. The National Institute of Occupational Safety and Health has listed 28,000 *toxic* chemicals and identified 2,200 of that number as suspected carcinogens. It is estimated there are currently 100 billion shipments per year by all modes (highway, rail, pipeline, air, water) of hazardous materials within the United States. Each day, more than 200,000 *bulk* shipments of acids, corrosives, gases, pesticides, and other dangerous materials are in transit across the nation.

In the United States, we are faced daily with hazardous materials incidents, the great bulk of which occur during handling related to transportation. Maintaining our immense industrial capacity requires the production, processing, handling, transportation, and disposal of vast quantities of fuels, chemicals, petrochemicals, and hydrocarbons. The more often a hazardous material is handled, the greater the opportunity for disaster. The Department of Transportation currently requires that approximately 1600 different materials be placarded as hazardous when transported in commerce.

The proliferation of hazardous materials on our roads and rails, and innumerable incidents that followed, brought a new dimension to local government responsibilities, as well as a recognition of the need for specialized training in hazardous materials emergency response. In years past, a local fireman could be reasonably well prepared if he possessed a basic understanding of the chemistry related to fire, air, and simple comubutible materials such as wood. That is no longer good enough.

Today's first responder is repeatedly called upon to deal with toxic fumes, flammable gases, corrosives, oxidizers, poisons, organic peroxides, cryogens, and reactive chemicals. Even a "small" hazardous materials incident can require a vast number of operational decisions from the on-scene commander. The difficulties involved in providing *effective* hazardous materials emergency response training can be enormous. There is no "average" hazardous materials incident due to an infinite number of variables that can interact at any incident. Any attempt to identify training needed to respond to an "average" incident requires consideration of an unending geometric progression of variables similar to the following. Multiply 18,000 hazardous material data cards in the CHEMTREC tubfile *by* five modes of transportation *by* a variety of shipping containers and vehicles *by* the location of the incident *by* weather conditions on-site *by* population density in the area *by* reactive characteristics of other commodities in the load *by* knowledge and training retention of fire service personnel who turn over every five years *by* . . . and on and on until some small understanding is obtained of the potential variables an incident commander may find on-site at any incident.

The decision-making process required of an incident commander can be extremely complex yet for some years now the Federal government has been conspicuous by its absence in providing hazardous materials information and training to local responders. First men in have pretty much had to learn what they know about hazardous materials from agencies and organizations not a part of the Federal government. Ask any ten firemen to identify "N.F.P.A." or "CHEMTREC" and eight out of ten will at least come close. Ask the same ten firemen to identify "M.T.B." and you will get eight or nine blank stares. Firemen are attuned to the two private organizations, The National Fire Protection Association and the Chemical Transportation Emergency Center operated by the Chemical Manufacturers Association, but many have never heard of the tax-supported Materials Transportation Bureau of the U.S. Department of Transportation.

In its September 29, 1978 report, *Emergency Response To Hazardous Materials Transportation Accidents,* the Committee On Government Operations of the U.S. House of Representatives noted:

• "Emergency response personnel from local communities, often firefighters, are usually the first to be called to the scene of an accident and have a heavy responsibility for the results of the accident.

• Local firefighters and other emergency response personnel are often entirely unassisted by any outside agency or organization in the handling of a hazardous materials incident.

• Accidents involving the transportation of hazardous materials frequently present special problems unique to the experience of local firefighters or other emergency personnel.

• Many local firefighters lack adequate training in the handling of hazardous materials emergencies. This lack of training is a danger to their lives and to the lives of other personnel and civilians near the scene of an accident. The highest risk is commonly to firefighters and other emergency response personnel. The need to reduce this risk is immediate.

• No documented evidence exists to demonstrate the adequacy of firefighter training programs on hazardous materials transportation incidents now available from any source."

Congress recognized that many local response personnel were placed in the position of battling hazardous materials incidents with a shoeshine and a smile, and that, perhaps, it might be a Federal responsibility to provide some assistance in the form of training. In the past, local emergency response organizations have generally had to develop their own hazardous materials training programs, utilize state programs, or send personnel off to distant locations to attend training provided by industrial organizations or private trainers. For the average department, effective *advanced* training in hazardous materials emergency response was an impossible dream. Until recently, there had been little assistance available from any agency of Federal government.

In Denver, Colorado, during 1980 however, the Materials Transportation Bureau of the U.S. Department of Transportation partially funded local efforts to develop a "model" in-depth hazardous materials emergency response course. Pilot two-week, 80-hour presentations made possible through cooperative efforts of a number of local and state government agencies and industrial organizations were conducted. Admission was limited to personnel who had a basic knowledge of hazardous materials and were assigned management, supervisory, or training responsibilities in the emergency response field or in the handling of hazardous materials. The contractor, Colorado Training Institute, was required to develop an 80-hour course of instruction in the handling and management of hazardous materials transportation emergencies; conduct three sessions of the course; and provide the Materials Transportation Bureau a Course Coordinator/Instructor Kit along with written guidance on organizing and administering the course that could serve as a model for use by other organizations in other areas.

The Denver area has long had extensive cooperation between and among agencies with hazardous materials responsibilities. The Colorado Committee on Hazardous Materials Safety, formed in 1973, includes representatives from area fire departments, police agencies, chemical companies, state and local agencies, carriers, and other industrial organizations. Since 1973, the Committee has sponsored many hazardous materials seminars throughout Colorado ranging in length from one to three days.

Three members of the Committee ramrodded the early operations of the Colorado Training Institute, and were instrumental in pulling together the 80-hour program in Advanced Hazardous Materials Emergency Response. Darrel J. Behrendsen, a towering, soft-spoken detective with the Denver Police Department; has been involved with hazardous materials in one way or another since 1968. Charles Heister is a Trooper with the Colorado Highway Patrol. Bruce Swaisgood is a Fire Prevention Specialist with Rockwell International Cooperation at their Rocky Flats facility in Golden, Colorado. Some idea of the level of commitment to hazardous materials information and training held by some organizations represented on the Colorado Committee on Hazardous Materials Safety can be gained from the fact that the Denver Police Department and the Colorado Highway Patrol donated the services of Behrendsen and Heister to Committee efforts for one year. In similar fashion, Rockwell International donated the services of Bruce Swaisgood for 26 weeks during the year.

The 80 hours of training was broken into 20 basic segments. Gar Haag of Shell Chemical Company reviewed Section 49 of the Code of Federal Regulations. Robert Yoder of Rockwell International Corporation dealt with radioactive materials, demonstrated a wide variety of detection devices, and brought along a Radiation Response Team from Rockwell's Rocky Flats facility. LiquidAir of Chicago sent a team to demonstrate the handling of cryogenics, complete with giant thermos bottles of the more common cryogens. James Cibella of Amerigas gave an excellent presentation on chlorine compounds and anhydrous ammonia, and made sure trainees had an opportunity to apply ''A'' and ''B'' chlorine kits to cylinders during breaks. Poisons were handled by a member of the Army's Technical Escort Detachment from Rocky Mountain Arsenal, while William Pappas of Martin Marietta Corporation dealt with explosives. Roy Peck and a team from Phillips Petroleum Corporation in Bartlesville, Oklahoma handled compressed gases and flammable and combustible liquids. Peck repeatedly made himself available for one-on-one discussions and demonstrations with trainees regarding ways to control the more common leaks in rail tankcars. Representatives from Shell Chemical and Balcom Chemical made a combined presentation on pesticide control, while Gene Kaufman of Shell covered the handling of organic peroxides. Robert DeGreve and a team of specialists from Arapahoe Chemical Company gave an extremely effective presentation on industrial chemicals, including a demonstration of potential chemical reactions. Walt DeFreece of Ward Transport and Ralph Graham of the Denver office of the Bureau of Motor Carrier Safety joined forces to deal with tank truck incidents and corrective measures. Rail tanks were treated by Fred Bryant of the Association of American Railroads and Jack Fortier of the Federal Railroad Administration. Pipeline transportation and gas installations were well handled by James Shomaker of the Public Service Company of Colorado, while Sidney Shore, a fulltime pilot for Frontier Airlines and a parttime representative of the Federal Emergency Management Agency, delineated various agencies available for assistance in a hazardous materials emergency; and Major Kenneth Powell of the Colorado State Patrol presented basic guidelines for effective command post operation, citing examples from the Patrol's lengthy experience with command post operation during the Big Thompson flood. Chuck Heister changed hats from course administrator to instructor to provide a top-notch treatment of evacuation considerations and techniques, and David Henderson from the C.B.S. affiliate in Denver used videotape to help trainees better understand how to deal with the media back home. John Marshall of the Denver Fire Department's Hazardous Materials Emergency Response Team had everyone's attention as he demonstrated protective gear and response apparatus and equipment used by the three-man Denver team. Finally, Ken York from the IT Corporation in California, a showman and entertainer if ever there was one, covered containment of hazardous materials spills.

The Colorado Training Institute program definitely lived up to its billing as indepth, advanced training in the management and control of hazardous materials transportation incidents. It is easy to say the course is a good one, but a bit more difficult to explain exactly why it is good. First of all, the course administrators had a commitment that could not be questioned. As experienced, Gung-ho public safety people who would take it as a personal insult if any trainee felt he had wasted his time in Denver, they did their homework and preparation extremely well. From providing a bottomless coffeepot for trainees to requiring course outlines and printed handout materials from every instructor, they organized sessions for a minimum amount of delay.

Secondly, a number of industrial organizations participating as trainers went all out, sending teams of personnel in a number of instances. Rockwell International, Phillips Petroleum, Shell Chemical, Liquid Air, Amerigas, and Arapahoe Chemicals all really made an effort. They brought training aids, gear, apparatus, detection devices, mock-ups, and control and containment implements that trainees could play with to their hearts' content. There appeared to be a real effort by industrial people to stay clear of "corporate image polishing" and tell it like it is.

In addition, a number of the handout materials prepared especially for the course were firstrate, almost collector's items. At the end of the day, you could seldom find an extra handout lying around; they had been carefully squirreled away by trainees who instantly recognized their value.

Lastly, the course was very productive due to the interest and interchange among trainees, many of whom had considerable experience with and exposure to hazardous materials. A certain level of experience and responsibility was required of trainees so that when they returned to their home organizations they would have enough clout to put into use what they learned in Denver about the management and control of hazardous materials incidents.

In addition to the indepth Hazardous Materials Emergency Response course described above, Colorado Training Institute also offers three-day courses in Hazardous Materials Awareness, a four-day course in Commercial Vehicle Inspection, a five-day course for shippers and carriers, and a five-day program on Cargo Tank Compliance/Inspection.

For information on current programs, *Contact:* Colorado Training Institute, 1001 East 62nd Avenue, Denver, Colorado, 80216. Tel. 303-289-4891.

The pages that follow provide a brief look at training institutions, programs, and facilities that currently offer instruction related to hazardous materials response and control and/or regulatory compliance. The majority of these programs admit response and compliance personnel from various levels of government, industry, and commercial organizations; although a few limit enrollment to specified groups. For information on enrollment, scheduling, and current offerings; contact a specific training facility at the address provided.

Safety Systems, Inc.

On May 2, 1981 350 persons, most of them on their own time and their own money, spent their Saturday attending a nine-hour seminar on hazardous materials response and control in the small town of Clifton Park, New York 15 miles north of Albany. Even a national demonstration in support of motherhood would be hardpressed to draw that many people, but a Florida-based organization known as Safety Systems, Inc. does it regularly.

Safety Systems was formed three years ago by Captain Ron Gore who initiated and currently directs the Hazardous Materials Response Team of the Jacksonville, Florida Fire Department that has handled in excess of 1000 incidents since it was formed in January of 1977. During 1982, Safety Systems scheduled 30 nine-hour seminars around the country plus one 40-hour seminar in St. Augustine, Florida.

"Safety Systems is a company dedicated to sharing vital information concerning hazardous materials leak/fire/spill control techniques that we have learned through our own experience or from the experience of others," notes Ron Gore. "There are approximately 12 members of Safety Systems, who normally work parttime with us and fulltime with various fire service, law enforcement, and military organizations. We draw heavily on these people and their experiences in developing and presenting our training programs.''

''Approximately 60 percent of the attendees at our seminars are members of the fire service,'' adds Gore, ''with about 30 percent from industry and ten percent representing law enforcement/transportation/environmental and similar groups. We attempt to identify and share with these people vital information from our experience that will help them do their job better out in the field. We are practical, very nuts-and-bolts type people who put on nuts-and-bolts type seminars. We believe in, and we teach, that there is a need for technical advice . . . we definitely need molecule-Vs-molecule type people, but our goal is to train our people who are going to be down in the mud, actually working to stop the leak or control the spill. We don't have a technical program, nor will we ever have. People are worried about the XYZs of hazardous materials when we haven't even learned the ABCs as yet. Our program teaches you where to put the plug. We advise people where they can go to get a technical type of program.''

''Firefighters are invariably the first responders,'' continues Captain Gore. ''If they are armed with a certain amount of information and have the proper tools, they can prevent or decrease a lot of damage to life, property, and the environment. We find people who attend our seminars are thirsty for practical knowledge, specific information that will help them better handle the emergency. Often times they have been taught, and rightly so, to 'stay away' . . . but have not been taught many positive techniques. I definitely believe in the 'no fight' fire or hazardous materials situation yet believe there are many incidents that can be contained, leaks that can be stopped or lessened, spills that can be picked-up. We attempt to show trainees the different techniques that are available.

The Safety Systems crew is highly organized and ultra professional. Their program never seems to lag, and they would be personally embarassed if a single trainee felt he had wasted his time at their seminar. They travel in a large motorhome with an attached trailer to carry their extensive demonstration tools/equipment/materials, and when trainees arrive at 7:30 in the morning the equipment has already been laid out on the temporary training ground in the order it will be required for various hands-on scenarios during the day. Team members are professionally dressed and totally safety-conscious. They tightly control the day; there is absolutely no fooling around, yet trainees are entertained as well as informed.

''Training that is dull will obscure very important information.'' explains Ron Gore who has the ability to put across vital training information by sometimes coupling it with often hilarious stories taken from 17 years experience as a firefighter. ''I try to tell stories and use examples my audience can identify with. Definitely, we try to share our experiences so they can see exactly what we are talking about. Feedback indicates trainees like the demonstrations. They enjoy the indoor/outdoor combination of classroom/hands-on training. They like to be told information, but they also want to see that information demonstrated under close-to-real conditions. The outdoor sessions reinforce the classroom sessions, and vice-versa. A lot of people come up to me and say . . . 'Boy, that was a fine show! They know it is a good program, but they like the show part of it as well. The overall purpose is communication.''

''Traveling around the country as trainers, we ourselves learn so very much,'' concludes Captain Gore. ''Perhaps that is why our respective employers are so tolerant of our operation. We always try to return home with new ideas or samples of new equipment and materials that are available elsewhere but perhaps little known in the Jacksonville area. The experience keeps us abreast of what's happening, new tools, different strategies, and varied techniques. There is a strong motivation factor to be derived from this work. The experience excites and motivates me to want to do more, to contribute more to my own community. Americans are competitive by nature. We get out there and try to outperform each other, and that drive is necessary in this type of activity. You try to do better.''

When you attend a Safety Systems training session, you bring your bunker gear. During at least 50 percent of each day, trainees respond to leak/fire/spill scenarios where they are forced to turn theory learned in the classroom into action. The wealth of actual experience on which

Safety Systems personnel are able to draw has been translated into a number of training devices that replicate actual incident situations, situations designed so trainees can gain a true appreciation for the proper tools/equipment/knowledge. One example from among the number is known as the "space project;" a collection of valves, pipes, hoses, couplings, tubing, gaskets and connectors that can be activated by a fire hose to replicate approximately 15 different types of vapor and liquid leaks one at a time. With one "space project" at either end of the temporary fire ground, teams of three trainees and one instructor are repeatedly sent in to attempt to bring under control leaks that seem only too real to an observer. Trainees are thus given a "live" challenge, and when a team gets in there and does it right, invariably a cheer goes up from the bystanders. Motivation runs high, and trainees don't hang back.

At the beginning of the day, Ron Gore had told the trainees . . . "This program is designed for the grunt, to provide him as much assistance as we can give in nine hours." At the end of the day, trainees seemed to feel the promise had been kept.

Contact: Ronald G. Gore, Safety Systems, Inc., P.O. Box 8463, Jacksonville, Florida 32239, Tel. 904-725-3044.

Fire Rescue Consultants

Warren E. Isman, Director of Fire & Rescue Services for Montgomery County, Maryland; and Gene P. Carlson, Editor-Fire Protection Publications, Oklahoma State University; (co-authors of the extremely well done 1981 book, *Hazardous Materials,* published by Glencoe) conduct two-day seminars around the country on "Handling Hazardous Materials Incidents." They use a combination of over 2,000 slides, three films, 20 overhead transparencies, tactics problems, recognition drills, utilization of reference materials, and two manuals. Part I (Two days) covers Introduction, Identification, Reference Sources, Preplanning, Train Transportation, Threat Assessment, Storage Facilities, Pipelines, Decision Making, and Problem Solving. Part II (Two days) covers Explosives & Oxidizers, Cryogenics, Chlorine, Advanced Train Transportation, Decision Making, Case Histories, Recognition Drill, Reference Source Drill, CHEMTREC Evolution, and Case Problems.

Isman and Carlson also offer a separate training program dealing with "Management of Hazardous Materials Incidents for EMS Personnel."

For additional information, *Contact:* Fire Rescue Consultants, P.O. Box 5703, Rockville, Maryland, 20855.

Texas A & M

The Oil and Hazardous Materials Control Training Division of Texas A&M University System is known throughout the world for providing realistic response training. The Division currently offers three major programs: an Oil Spill Control Course, a Hazardous Material Control Course, and a Tank Truck Rollover Training Course. Courses are offered regularly througout the year, devote at least 50 percent of instructional times to hands-on exercises, utilize extensive training scenarios, and are periodically evaluated by industrial oversight or advisory committees to insure pertinent instruction for both industry and emergency response personnel. Courses are open to any interested registrant.

Oil Spill Control Course:

The five-day Oil Spill Control Course, offered more than 20 times a year at Galveston, Texas; is designed to provide participants training and information necessary to handle an oil spill with normally available equipment and manpower. By means of both classroom training and field exercises, students are enabled to recognize potential spill situations; modify and update existing oil spill contingency plans; establish a supervisory team to implement a contingency plan; organize, train, and direct a response team; recommend and direct the use of proper oil spill control equipment such as skimmers, booms, and sorbents; and preplan arrangements for additional support equipment and supplies not readily available. Students are also taught to establish an effective communications network during a drill, meet legal requirements for properly reporting oil spills, and effectively handle public relations considerations at a spill scene.

To develop necessary skills, students are taught contingency planning, organization, and training of an oil spill response team, and the necessary tools/equipment/materials to get the job done. They study movement, containment, and clean-up of oil spills; disposal of oily debris; use and availability of oil spill clean-up contractors; and prevention of oil spills. They learn to properly use communications equipment, booms, skimmers, and sorbents and chemical agents.

To emphasize lessons learned in the classroom, enrollees are required to participate in a number of field exercises. They evaluate effectiveness of different types of sorbent materials, demonstrate basic competency in small boat handling, deploy and tend booms, operate skimmers, use sorbents and chemical agents, operate on-scene communications equipment, and engage in contingency planning.

Hazardous Material Control Course:

Texas A&M's five-day Hazardous Material Control Course is held repeatedly throughout the year at the Brayton Firemen Training Field in College Station, Texas. The course provides participants with the basic knowledge and information necessary to respond to or assist with hazardous materials incidents. Extensive training aids and scenarios developed for the course permit students to work with the latest equipment and technology available for hazardous materials control. Classroom coverage includes properties of hazardous materials, initial response actions, protective clothing and equipment, monitoring and detection devices, and sources of aid and assistance. Enrollees study containment, recovery, and treatment of hazardous materials; patching and stabilizing of leaking containers; proper use of foams, sorbents, and chemical agents in the control of hazardous materials spills; and basic technology. They engage in contingency planning problem sessions, and learn methods for the handling of liquefied petroleum gas emergencies and tank truck incidents.

Using training scenarios including trailer trucks, drums, tank cars, pipelines, and industrial facility mock-ups; trainees participate in field exercises that require working in a hazardous environment and use of self-contained breathing apparatus and encapsulated suits. They are required to use actual response equipment; such as, patches and plugs, monitoring equipment, booms and dikes, overpack drums, and recovery gear to respond to simulated emergency situations. On Thursday of each course week, trainees organize response groups to handle a simulated incident; and efforts are videotaped for use during an end-of-week critique.

Tank Truck Rollover Training Course:

The two-and-a-half day Tank Truck Rollover Training Course is also held at the Brayton Firemen Training Field in College Station, Texas using training aids and scenarios developed specifically for the course to allow students to gain hands-on experience with up-to-date equipment and technology for responding to tank truck rollovers. The first day is devoted to tank truck design and appurtenances while the final day-and-a-half emphasizes field exercises. The course is aimed at industrial representatives, fire and public safety personnel, contractors, and others involved in the operations required to respond effectively to tank truck rollovers involving volatile liquids.

Students study valves, venting systems, and equipment common to tank truck operations; as well as personal safety, monitoring and detection equipment, and control and off-loading strategies and methods. Through field exercises using a specially outfitted tank trailer, participants learn techniques for off-loading, stabilizing, and uprighting disabled tank trucks. Finally, students are required to actually respond to problems encountered in monitoring, off-loading, containment, and recovery of products.

In addition to courses described above, the Texas Engineering Extension Service's Oil and Hazardous Materials Control and Fire Protection Training Divisions jointly offer 20-hour extension courses in hazardous materials control for industrial personnel, state agency representatives, and emergency responders throughout the State of Texas.

For additional information, *Contact:* Oil & Hazardous Material Control Training Division, Texas Engineering Extension Service, The Texas A&M University System, F.E., Drawer K, College Station, Texas 77843, Tel. 713-845-3418.

California State Fire Marshal

The California State Fire marshal, in cooperation with the California Highway Patrol and local fire and police agencies, has provided an eight-hour training course in hazardous materials control to more than 30,000 California police and fire officers. The course is designed to acquaint first responders from various disciplines (police, fire, public works) with the information they need to quickly and accurately assess a hazardous materials incident, and is presented in joint training sessions to provide interaction and common understanding of each agency's responsibilities and concerns. There is no cost for qualified participants, but enrollment is limited to California Public Safety Agency personnel.

Contact: Office of the State Fire Marshal, Edward W. Bent, Supervisor, Fire Service Training & Education Program, 7171 Bowling Drive, Suite 600, Sacramento, California 95023, Tel. 916-427-4161.

National Fire Academy

The Federal Emergency Management Agency's National Fire Academy, a component of the National Emergency Training Center in Emmitsburg, Maryland; was established by Congress in 1974 to advance the professional development of the fire service and other persons engaged in fire prevention and control activities. The National Fire Academy provides hazardous materials related training by three different means: regular residence courses, weekend residence educational opportunities, and outreach or field training programs.

The 110-acre National Fire Academy campus in Emmitsburg, Maryland located 50 miles northwest of Baltimore and 75 miles north of Washington, D.C.; has 39 fully equipped classrooms that can accommodate up to 1,000 students at a time. Four dormitories can house 300 students in single occupancy rooms. Additional facilities include two auditoriums, a lecture hall, a science building, and a library.

Two-week residency courses pertaining to hazardous materials are offered throughout the year.

"Hazardous Materials I-Basic" is designed for fire officers who may become involved in a hazardous materials incident or in the inspection of occupancies, vehicles, watercraft or aircraft containing hazardous materials. It provides the basic knowledge required to evaluate the potential hazards and behaviors of materials considered as hazardous for one or a combination of reasons. Instruction centers on the underlying reasons for certain recommended actions so as to lead to improved decision making and safer operations and handling.

"Hazardous Materials II—Operational Considerations" is for fire officers and other emergency services personnel who may be called upon to command an incident involving hazardous materials. It is intended to provide the student with an ability to properly manage such incidents.

"Hazardous Materials III—Planning" is for senior level fire and rescue officers who have or will have planning and operational responsibilities for hazardous materials incidents and fire department personnel assigned the responsibility for development of a hazardous materials incident control plan. Major subject areas include the community's potential for disaster, government and private sector capabilities for assistance, interagency relations, command post organization, emergency response teams, and the preparation of a hazardous materials incident management manual.

The weekend educational opportunities program is designed to increase the availability of Academy educational opportunities to the volunteer fire service and to maximize use of the Academy's facilities by scheduling on-campus weekend training.

Academy field or outreach programs support state and local fire training efforts by delivering at various locations around the country under sponsorship of local and state training organizations courses not available at the local level. Each program consists of one or more courses, and each course covers two days of training. Instruction is provided by National Fire Academy personnel.

Courses related to hazardous materials that are offered at the Academy or around the country on weekends are as follows.

"Hazardous Materials Incident Analysis" provides the fire suppression unit member with techniques for safely managing hazardous materials incidents. Eight broad classes of hazardous materials are analyzed including flammable liquids and gases, oxidizers, poisons and toxic materials, acids and corrosive cryogenics, and water reactive materials and metals.

"Pesticide Fire and Spill Control" provides training in safe handling of a pesticide fire or spill from arrival on the scene through clean-up operations. Local exposure to pesticide incidents and related fire command decisions are taught through simulations and case history exercises. Topics include hazard recognition, labeling, exposure, symptoms, first aid, information resources, environmental impacts, spill control, fireground operations, and pre-incident planning. For addtional information, *Contact:* National Fire Academy, National Emergency Training Center, 16825 South Seton Avenue, Emmitsburg, Maryland 21727, Tel. 301-447-6771.

David Frank Associates

M. David Frank specializes in seminars pertaining to "Railroad Emergencies Involving Hazardous Materials" and has provided more than 200 such sessions for railroads and fire and police agencies throughout the eastern U.S. during the past four years. He is a Fire Control Consultant to the Chessie System, and Associate Professor & Coordinator of the Fire Protection Technology Career Program at Catonsville Community College in Maryland.

Frank stresses initial evaluation and size-up of the derailment site within the first five minutes on-scene. He covers:
• Types of Hazardous Materials Shipped By Rail.
• Guidelines for the Evaluation of Liquefied Petroleum Gas Emergencies Involving Tank Cars.
• Scenario of Events at a Derailment Scene, the Function of Railroad Crews, and Anticipated On-Scene Communications Problems.
• Methods Used to Identify Contents of Rail Cars.
• Incident Approach Safeguards.
• Train Make-Up, Crew Duties, Waybill Emergency Instruction Information, and Fire/Rescue Problems Involving Diesel Engines.
• Identification of Rail Cars. (Markings, words, symbols, color, strains, stenciling, or dome configuration.)

For additional information, *Contact:* David Frank Associates, 416 South Rolling Road, Catonsville, Maryland 21228, Tel. 301-455-4510.

Transportation Safety Institute

The Department of Transportation's Transportation Safety Institute in Oklahoma City provides a variety of training related to hazardous materials regulation and enforcement. *Such courses are available only to state and federal employees.* Although courses change periodically as needs are identified, some of the courses offered recently are as follows:
- Hazardous Materials Safety & Enforcement. (Two weeks.)
- Rail Transportation of Hazardous Materials. (Two weeks.)
- Motor Carrier Transportation of Hazardous Materials-Advanced. (Four days.)
- Air Transportation of Hazardous Materials. (Two weeks.)
- FAA Multimodal Transportation of Hazardous Materials-Shipper's Course. (One week.)
- Attorney's Short Course On Hazardous Materials Regulations-Air. (One week.)

For information on specific course offerings, *Contact:* Program Manager, Hazardous Materials, D.O.T.—Transportation Safety Institute, 6500 S. MacArthur Blvd., Oklahoma City, Oklahoma 73125, Tel. 405-749-4824.

National Fire Protection Association

The National Fire Protection Association periodically schedules hazardous materials training seminars around the country and/or sells complete training materials for use by local instructors employed by various types of emergency services organizations. N.F.P.A.'s eight-unit, 20-hour packaged programmed instruction unit, "Handling Hazardous Materials Transportation Emergencies," has received wide usage around the country. Composed of slides, student performance manuals, and training aids; the unit includes formal instruction followed by individual and group problem solving exercises in eight areas.
- Hazardous Materials In Transit.
- Definitions & Classes of Hazardous Materials.
- Identifying Hazardous Materials.
- Obtaining Technical Assistance In Hazardous Materials Incidents.
- Command & Control of Hazardous Materials Incidents—I.
- Command & Control of Hazardous Materials Incidents—II.
- Terminating & Reporting Hazardous Materials Incidents.
- Planning for Hazardous Materials Emergencies.

For additional information, *Contact:* National Fire Protection Association Educational Technology Unit, Batterymarch Park, Quincy, Massachusetts 02269.

Environmental Safety & Design (ENSAFE)

ENSAFE provides industrial safety training of which hazardous materials training is a major component. Currently active in 26 states and Puerto Rico, the company custom designs in-house training to meet compliance and safety requirements, mainly for industry and government. Training modules are developed and applied to such areas as the handling, use, transportation, and disposal of hazardous materials/wastes/substances. ENSAFE identifies, develops, and provides training for various levels of people within a user company. Depending upon specific needs identified through an in-plant survey, individual training modules may address . . .
- Executive Orientation & Familiarization.
- Identification of Hazardous Materials & Wastes Within the Operation.
- Training of Operating Management.
- Development of Contingency Plans.
- Refinement of Emergency Procedures.
- Packaging Considerations.
- Safety & Personal Protective Equipment.
- Packaging, Marking, & Labeling for Supervisors and Workers.

• Emergency Equipment Related to the Specific Facility. (Valves, shut-offs, mechanical controls, etc.)

ENSAFE also provides periodic seminars directed toward persons having responsibility for preparing hazardous materials for shipment and accepting hazardous materials for transportation. For additional information, *Contact:* ENSAFE, P.O. Box 34207, 5705 Stage Road, Suite 224, Memphis, Tennessee 38134, Tel. 901-372-7962.

National Draeger's Training Gallery For Respiratory Protection Devices

The National Draeger training program is designed to permit self-contained breathing apparatus (SCBA) users to train under conditions of extreme physical and mental stress. Consequently, each trainee must be in good health to participate and must sign a release.

National Draeger's SCBA training gallery located near Pittsburgh, Pennsylvania is designed to provide respirator training under real-life conditions by using a series of chambers and other devices to allow evaluation under conditions of extreme physical and mental stress. The training gallery includes a work room, training room, airlock, and control desk.

The *work room* contains apparatus such as a work-measuring device, endless ladder, bicycle-type ergometer, and moving-belt ergometer for determining the behavior of respirator users under defined conditions that permit measurement of physical stress. The *training room* is used for improving circulation, practicing maneuvers, and sharpening reactions under conditions similar to those encountered in real life. The respirator user must master an obstacle course while carrying on specific tasks. Conditions can be made more difficult by generating heat or smoke, creating darkness or releasing irritants. The *lock* allows the training room to be entered or exited without escape of smoke or irritants. The *control desk* allows training to be controlled and monitored from a single central point.

Installation and rescue work in cramped conditions can be practiced in a work tunnel comprised of horizontal, vertical and sloping sections. The vertical sections contain a rigid steel ladder. A pipeline with elbows, flanges and valves runs along the inside of the tunnel and can be pressurized with compressed air for a number of training applications. A training tank consisting of one section with an entry hatch and tank cover, and a second section with a lattice window for supervision and/or rescue; is used to practice working in and rescuing people from tanks, containers, and narrow shafts. An orientation route is used for training in orientation, maneuvering and behavior under conditions of daylight, darkness and smoke. Obstacles include a tube for crawling through, steps, a passage restrictor, a joist simulator, and a confined-space hurdle. For search and rescue training sessions trainees may be required to rescue a weighted dummy from the orientation route. Training in the orientation route can be complicated by introduction of hazards common to the fire fighting environment. A smoke generator can be used to produce non-corrosive smoke; should the need arise the smoke-filled room can be entered with no respiratory protection without breathing being impaired. Noise is transmitted into the training room from the control desk by means of a cassette recorder and loudspeaker. The training room can be heated to simulate the hot environment encountered in a burning structure. The glow of a fire can be simulated via a lighting console.

All components required for control and monitoring of training sessions are operated from a central control desk; the entire session is viewed through a TV monitoring system and can be recorded for play back to the trainees. The training gallery is able to accommodate eight persons at one time and each group requires two hours to complete a training session. The fee for such training as of 1982 was $25.00 per person and training is available to any interested individual or group although the majority of trainees are firefighters or industrial fire brigade personnel.

National Draeger personnel are currently developing a mobile training gallery utilizing a trailer truck that will include a maintenance and service facility. For complete information, *Contact:* National Draeger, Inc., 101 Technology Drive, P.O. Box 120, Pittsburgh, Pennsylvania 15230. Tel. 412-787-8383 Telex. 86-6704.

Government Services Institute

Government Services Institute, a commercial training operation based in Florida, offers separate five-day Basic and Advanced courses in "Transportation of Hazardous Materials" designed and presented by Gary J. Groman to insure compliance with both Department of Transportation and Environmental Protection Agency regulations pertaining to hazardous materials, substances, and wastes. The focus here is on understanding applicable regulations and performing the functions necessary to insure regulatory compliance when preparing, offering, and transporting such items. Groman, an attorney and former employee of the Department of Transportation, is extremely well-versed in applicable regulations and has designed a number of instructional materials and work projects used in the two courses. Both the Basic and Advanced courses are being scheduled through 1984.

For additional information, *Contact:* Government Services Institute, P.O. Box 5212, Spring Hill, Florida 33526, Tel. 904-683-8553.

American Trucking Associations

The Operations Council of the American Trucking Associations, Inc. offers one-day seminars across the country on Handling Hazardous Materials and Wastes for managers and supervisors who direct the handling and transportation of such items. For additional information, *Contact:* The Operations Council, American Trucking Associations, 1616 P Street, N.W., Washington, D.C. 20036, Tel. 202-797-5438.

National Spill Control School

The National Spill Control School on the campus of Corpus Christi State University located on a 240-acre island ten miles from downtown Corpus Christi, Texas; offers separate five-day resident courses on "Oil Spill Prevention & Control" and "Hazardous Materials Spill Prevention & Control."

"Oil Spill Prevention & Control" covers . . .
- History of Oil Spills & Resultant Legislation.
- Physical & Chemical Reactions, Chemical Collecting Agents, Dispersants.
- Sources & Causes of Oil Spills.
- Prevention Programs.
- Uses of Spill Containment Devices.
- Visit to Corpus Christi Area Oil Spill Control Association to observe deployment of booms, use of skimmers and sorbents, methods of corraling oil spills, and use of chemical herding agents.
- Rescue & Rehabilitation of Waterfowl.
- Media Relations.
- Changes & Effects of Introducing Oil into the Biological Environment.
- Alternatives Available to Provide Oil Spill Response.
- Contingency Planning.

"Hazardous Materials Spill Prevention & Control" covers . . .
- Review of Hazardous Materials Spills & Resultant Legislation.
- Classification/Identification of Hazardous Materials.
- Chemical & Physical Properties.
- Environmental Health.
- Personnel Protection.
- Media Relations
- Prevention Programs.
- Containment & Removal Tools & Methods.
- Information Systems.
- Laboratory Demonstrations.

- Disposal.
- Chlorine.
- PCBs.
- Agricultural Chemicals.
- Contingency Planning.

For additional information, *Contact:* National Spill Control School, Corpus Christi State University, 6300 Ocean Drive, Corpus Christi, Texas 78412, Tel. 512-991-8692.

LNG-LPG Training Facility—Massachusetts Firefighting Academy

The Massachusetts Firefighting Academy dedicated a liquefied natural gas (LNG) and liquefied petroleum gas (LPG) fire training facility at Hopkington, Massachusetts in April of 1979 designed to provide realistic training on liquefied gas leaks and fires. Initially, training was limited to firefighters and gas industry personnel from the New England area; but eventually the site will accommodate first responders from a wider geographical area. Training aids include a 2000-gallon propane tank enclosed within an outer tank; the space between the two tanks can be filled with water so that by adjusting water flow and thus the heat from an impinging fire, either insulated LNG tank or noninsulated LPG tank conditions can be simulated.

The Massachusetts Fire Academy, the National Fire Protection Association, and gas industry personnel worked cooperatively for seven years to raise the necessary funds, lease a site, purchase equipment, develop a curriculum, and design the facility to train firefighters and other personnel in safe, effective response to incidents involving LNG and LPG. For additional information, *Contact:* The Massachusetts Firefighting Academy, 59 Horse Pond Road, Sudbury, Massachusetts 01776, Tel. Thomas Rinaldo, Coordinator, LNG/LPG Firefighting School, 617-443-8926.

Hazardous Risk Advisory Committee of Nashville

Early each year, the Hazardous Risk Advisory Committee of Nashville, Tennessee; a nonprofit organization which has as its objective the training and education of first responders and others involved in hazardous materials incidents, sponsors a three or four-day Seminar On Hazardous Materials Safety. For additional information, *Contact:* Hazardous Risk Advisory Committee, Metro Civil Defense, Floor 7-M, Metro Courthouse, Nashville, Tennessee 37201, Tel. Eric Foster at 615-259-6145.

California Specialized Training Institute

The State of California established the California Specialized Training Institute (CSTI) in 1971 at Camp San Luis Obispo to provide hands-on training in such areas as Civil Emergency Management, Terrorism, Hazardous Materials, Corrections, School Violence, and Prison Gangs. The stated objectives of the Institution are . . .

1. To enhance, through education, the ability of the government and the private sector to cope with and manage emergency situations to insure survival of persons and institutions during periods of social, economic and environmental stress.

2. To promote among participants a mutual understanding of the responsibilities, policies, capabilities and legal limitations of various agencies involved in emergency operations.

3. To improve the level of professionalism of the operational law enforcement sector in order to minimize the risk to themselves and reduce overall the incidence of violence in society.

4. To provide a library of information relating to current planning, concepts and emergency operations methodology.

During its first ten years of operation, CSTI enrolled in excess of 25,000 students from 49 states and a number of foreign countries. Classes are structured to insure a cross section of all disciplines of federal, state, local governments and industry that have a role in mitigating the effects of unusual occurrences.

A 44-hour Hazardous Materials course covers special management aspects of hazardous materials incidents and utilizes a role-playing exercise so that trainees deal with the consequences of an unexpected or uncontrolled incident, accident, spill or leak.

A 44-hour Nuclear Incident/Accident program addresses reactor types and sites; Nuclear Regulatory Commission responsibilities and concerns; fuels, fuel loading, and storage; fuel and waste transportation and storage; evacuation and relocation; and seismic and geological considerations.

For additional information, *Contact:* State of California Military Department, California Specialized Training Institute, Camp San Luis Obispo, California 93406, Tel. 805-544-7101.

Flying Tigers

Two-day seminars on "Transporting Hazardous Materials By Air" designed to instruct shippers, forwarders, and consolidators how to comply with the various government and air carrier regulations and requirements are offered throughout the year by Flying Tigers, the world's largest air freight airline. Subject matter includes identifying, packaging, marking, labeling, and documenting hazardous materials shipments by air.

For additional information, *Contact:* Flying Tiger Line, 7401 World Way West, P.O. Box 92935, Los Angeles, California 90009, Tel. 213-646-4086.

Ansul Fire Protection's Fire Training School

Ansul Fire Protection, a pioneer in the development of "dry chemical" fire extinguishing agents, has operated a Fire Training School since the early 1940's to instruct students in the proper operation, application, inspection and maintenance of fire suppression equipment manufactured by the company, especially dry chemical. The School in Marinette, Wisconsin runs three-days each week, throughout the summer months between April and November. The Ansul School is designed to instruct in the use of "first aid" type fire suppression equipment such as hand portable extinguishers, wheeled, foam and 1½ inch water hoselines. Each summer the school trains approximately 1,000 students in regularly scheduled classes as well as special classes for individual groups.

The Ansul Fire School operates four separate types of training programs: a petroleum/chemical fire school; an industrial fire school; a mining fire school; and a utilities fire school. During 1982 the fee for each three-day school was $650. per participant.

Although each type of class is somewhat different from others, generally a limited amount of time is spent in a classroom while most of the training is conducted on the Fire Test Field with each trainee extinguishing a wide range of actual fires with a variety of first aid equipment. Four basic principles are stressed: An understanding of extinguisher capabilities and limitations; proper operation of extinguishers; proper application techniques; and the importance of maintenance.

Trainees are taught to control liquid spill fires under and around vehicles, compressors, turbines, metalworking machines, storage drums, transformers, filling stations, bulk storage areas, batch processing areas, warehouses, and paint mixing areas. They attack and control flammable-liquid-in-depth fires related to dip tanks, drainboards, solvent cleaning tanks, drip pans, quench tanks and flammable liquid tank dikes. They learn how to combat multi-level gravity flow liquid fires where a spill fire must be extinguished before the gravity flow portion of the fire can be brought under control such as would be the problem with ruptured tanks, overflows in filling operations, or broken valves with fuel running down any object. Typical flammable-liquid-under-pressure situations students are faced which include seal or packing failures in pumps, compressors or turbines; flange failures in piping systems, and valve stem packing failures. Students must also learn to handle fires featuring flammable gas under pressure, and class A fires of common combustible materials.

During the month of March Ansul Fire Protection conducts a series of "winter" schools each year at the Lamar University/Beaumont Fire Department Training Grounds in Beaumont, Texas. In addition, Ansul instructors are available on a contract basis to conduct schools for customers on their sites anywhere in the world. Because of their highly developed expertise in a specialized field, Ansul instructors have conducted training schools in Singapore, Borneo, Algeria, Scotland, Saudi Arabia, South Africa and a number of other countries. For complete information, *Contact:* Ansul Fire Protection, One Stanton Street, Marinette, Wisconsin 54143, Telephone: 715-735-7411, Telex: 26-3433.

Hazardous Materials Advisory Council

The Hazardous Materials Advisory Council, (HMAC) a Washington-based, privately funded nonprofit organization representing those involved in transporting hazardous materials/hazardous wastes/hazardous substances; offers seminars, training programs, and conferences throughout the year for both members and nonmembers. HMAC members tend to be manufacturers and shippers of chemicals, plastics, petro-chemical products, and other hazardous materials such as railroads, carriers by highway/water/air, and those who supply containers and packaging supplies for shipping and transporting hazardous commodities of every type and hazard class.

Specific training offered each year varies based upon needs identified within the industry. For up-to-date information, *Contact:* Hazardous Materials Advisory Council, 1100 17th Street, N.W. Suite 908, Washington, D.C. 20036 Tel. 202-223-1271.

Transportation Skills Program

Each year Transportation Skills Programs, Inc. offers an extensive, coast-to-coast series of one-day seminars on hazardous materials, substances, and wastes training and compliance for shippers, carriers, managers, agents, and industrial personnel. The course is a compact but comprehensive combination of slides/lecture/handouts; a good one-day orientation to regulations and compliance requirements for busy people. Training booklets and compliance tools developed for the course are well done and easy to follow. Each one-day session generally covers . . .
- Introduction to EPA/DOT Hazardous Materials, Substances, & Waste Regs.
- Hazardous Materials, Substances, & Wastes Shipping Papers & Manifests.
- Labels, Marking, and Placarding.
- Empties, Leakers, Shipping Containers & Emergency Responses.
- Review: Comprehensive Environmental Response, Compensation And Liability Act of 1980-Superfund. (PL-96-510)
- ''Right To Know.''
- Preparing for a RCRA Inspection.
- Training of Company Personnel.
- ''Waste Exchanges.''
- Pertinent State Regulations.

The training materials and aids developed by Transportation Skills Program represent a significant effort to digest voluminous material in order to highlight important aspects of regulations that impact on shippers, carriers, and industrial personnel.

For additional information, *Contact:* Transportation Skills Programs, 320 West Main Street, Kutztown, Pennsylvania 19530, Tel. 215-683-5098.

E.P.A.'s Environmental Response Team

The Environmental Protection Agency's Environmental Response Team provides training directed mainly toward E.P.A. employees and state and local government personnel whose responsibilities mesh with those of E.P.A.. There is no fee for such persons. Employees of other federal agencies, and industrial personnel, can be admitted on a space-available basis for a fee that during 1982 varied from $110. to $170. depending upon the course attended. During 1982, the Environmental Response Team offered the following types of courses.

Hazardous Materials Incident Response Operations: (14 five-day classes at Edison, New Jersey.)

Field Monitoring & Analysis of Hazardous Materials: (Three five-day classes held at various locations.)

Incident Mitigation & Treatment Methods: (Two five-day courses offered during 1982.)

Respiratory Protection: (Three three-day courses offered during 1982.)

Hazard Evaluation & Environmental Assessment: (Two five-day courses offered during 1982.)

Personal Protection & Safety: (Two five-day courses offered during 1982.)

Advanced Respiratory Protection: (Two three-day courses offered during 1982.)

Courses are changed periodically to respond to identified needs.

For up-to-date information, *Contact:* Tom Sell, U.S.—E.P.A., Environmental Research Center, 26 West St. Clair Street, Cincinnati, Ohio 45268, Tel. 513-684-7537.

Emergency Action, Inc.

A training operation based in South Carolina, Emergency Action, Inc.; specializes in customized training for fire, rescue, EMS, and industrial fire brigade personnel. The two principals of the corporation are young fire service officers with extensive, hands-on experience in hazardous materials response and control and emergency medical services operations. The company is presently active in North Carolina, South Carolina, and Georgia; but will travel further afield as needs warrant. Emergency Action, Inc. has worked with EMS organizations in North Carolina on a program dealing with "Emergency Medical Response To Hazardous Materials Incidents." Rather than providing standardized training of a general nature, Emergency Action normally works with a specific fire department, industrial organization, EMS group, or rescue squad to identify the response equipment and materials they have on hand, determine the specific hazards or problems the organization is most likely to deal with, and consider levels of prior training; then they develop a program to meet the needs of the specific organization. For additional information, *Contact:* Emergency Action, Inc., P.O. Box 10661, Charleston, South Carolina 29411, Tel. 803-767-0585 or 803-553-2672.

ALM Enterprises

Three-hour seminar/workshop sessions on hazardous materials, directed mainly toward fire/police/EMS/industrial fire brigade personnel, are provided in the far west by Al Mozingo, a fire service training officer from southern California. The program covers Recognition/Hazards/Placarding/Use of Emergency Action Guide/Evacuation/Use of Resources. Uses slides, handouts, Information Accumulation Form, and a simulated incident. For additional information, *Contact:* ALM Enterprises, P.O. Box 20912, El Cajon, California 92021, Tel. 714-447-2828.

J.T. Baker Chemical Company

J.T. Baker Chemical Company, a producer of laboratory and specialty industrial chemicals, annually conducts two separate series of two-day seminars across the country.

"Management & Disposal of Hazardous And Chemical Wastes" provides information on what generators, transporters, storage and disposal facilities must do to conform to the Resource Conservation And Recovery Act. (RCRA) Covered are D.O.T./E.P.A. requirements for hazardous wastes including state and federal waste tracking manifests, drum labeling and marking, shipping documents, mandatory training programs, containerization, placarding and fines. Additional areas covered include:

• Standards for generators and haulers of wastes, standards and permits for offsite storage and design/operation standards and permits for disposal facilities.

• Insurance and legal requirements imposed by RCRA.

• Establishment of in-house waste management control program including layout and safety guides on chemical storage and chemical interactivity.

• Special disposal problems such as asbestos, PCBs, and highly reactive materials that are regulated by other environmental legislation.

• Waste exchanges, information clearinghouse, on-site recovery and new technology.

• Existing disposal sites and technologies as they apply to specific waste streams i.e., specific secure chemical landfills, incinerators, acid/alkaline neutralization facilities, aqueous and chemical treatment sites, solvent recovery, solvent/fuel burning sites, deep well injection, etc.

A "Hazardous Chemical Safety Seminar" stresses recognition of potential hazards, correct product usage and accident prevention techniques, and conduct of in-house training programs. Day one covers:

• Hazard Analysis

• Flammable Materials Workshop (vapor phase explosions, static electricity, flash point determination, and grounding equipment)

• Corrosive Chemicals.

• Eye and Face Protection and shielding

• Corrosive Chemicals Workshop (chemical burn treatment, acid spill control, and corrosion testing.)

Day Two Covers:

• Insidious Hazards

• Toxic Chemicals

• Chemical Storage, Building Design, Ventilation, Safety Cabinets, and Effective Separation of Incompatible Materials.

• Sources of Information/Technical Data

• Chemical Waste Disposal/RCRA

• Cryogenic Liquids and Compressed Gases

• The Safety Audit

• First Aid Responses To Emergency Situations, Decontaminating & Emergency Equipment

For additional information, *Contact:* J.T. Baker Chemical Company, Office of Safety Training, 222 Red School Lane, Phillipsburg, New Jersey 08865, Tel. 201-454-2500.

UNZ & Co.

UNZ & Company periodically conducts two-day compliance seminars throughout the northeast. Day 1 covers Hazardous Materials Transportation:

Use of Code of Federal Regulations—49, individual responsibility, spill and incident reporting, compliance and safety, hazardous wastes, etc.

Day 2 can be taken as one of two options. Option A covers "Hazardous Wastes Handling & Disposal."

RCRA, definitions, standards for generators and storers, standards for treatment/storage/disposal facilities, etc.

Option B for Day 2 covers "Shipping & Handling Hazardous Materials Internationally"

Determinations, Classifications, regulations, responsibilities, prevention, packaging/marking/labeling, etc.

Contact: UNZ & Company, P.O. Box 308, 190 Baldwin Avenue, Jersey City, New Jersey 07703, Tel. 800-631-3098.

U.S. Coast Guard Hazardous Chemicals Training Course, Yorktown, Virginia

A two-week Hazardous Chemicals Training Course (HCTF) conducted by the Marine Safety School at the U.S. Coast Guard Reserve Training Center located in Yorktown, Virginia is designed to train Coast Guard and other selected emergency response personnel to plan for, and respond to, hazardous chemical emergencies. The course stresses "hands-on" training and emphasizes the practical aspects of planning, response, and use of appropriate equipment for containment and control of hazardous substances.

"Students here are 99 percent Coast Guard personnel," explains Lieutenant Commander Gerald Ranes, Senior Instructor for the HCTC program. "They are normally assigned to a Coast Guard Marine Safety Office under a Captain of the Port, or possibly a Coast Guard Marine Inspection Office. All students, who range in rank from Second Class Petty Officer to Captain, are in positions where they may have to be utilized for response to a hazardous chemical. Over the past 12 years or so the Coast Guard has become quite expert at responding to oil spills; more recently, legislation and regulations call for the Coast Guard to assume major responsibilities in the area of chemical spills. This course was initiated to train Coast Guard personnel for a specific job—response to hazardous chemical spills."

"When the course was first started it was more theoretical," remembers Ranes. "It was for commissioned officers only, and we went into great detail on toxicology and chemistry. We had second thoughts about the direction we had initially taken and decided we were not in the business of teaching theory. Students are interested in RESPONSE—what they can do, what they have to do, how they can protect people and environment endangered by a spill. We moved from the theoretical aspect to emphasis on the hands-on, practical aspects of incident response and control."

"In the HCTC we deal strictly with chemical spills and stay away from oil spills," explains Commander Ranes. "There are *other* courses here at the Marine Safety School that go quite deeply into oil spills. During Fiscal Year 1982 we are teaching 12 regular classes plus two classes for active duty reserve personnel, or a total of 14 classes in hazardous chemicals. Each class enrolls 20 students, so we are talking about 280 trainees during the year. Currently, students are about evenly divided between officers and enlisted."

To become familiar with the chemical, physical, and toxic properties of hazardous materials; students are required to use a wide variety of standard reference sources including the Coast Guard's own *Chemical Hazards Response Information System,* (CHRIS) the NIOSH *Guide For Respiratory Equipment,* the Department of Transportation's *Emergency Response Guide,* resource materials from the American Conference of Industrial Hygienists, and the National Fire Protection Association's *Fire Protection Guide On Hazardous Materials.* The CHRIS manual is covered in depth so students learn to effectively assess a chemical hazard.

Trainees are taught to use, calibrate, and maintain numerous detection and monitoring devices such as combustible gas indicators and organic vapor analyzers. In a laboratory fitted with ventilation booths, different meters are demonstrated then students are provided with meters that have been purposely thrown out of calibration. They must recalibrate the meters and use them to take samples. Students are assigned to various response teams, and each team must then explain a particular measuring device to the rest of the class.

After an introduction to local contingency plans, students are given a contingency plan written at the Marine Safety School for a mythical Marine Safety Office, "Hiatusport." The Hiatusport plan is based on requirements and components of both the National Contingency Plan and subregional plans to familiarize students with the basic composition of such plans. "Hiatusport" located on Chesapeake Bay is complete with oil terminals, chemical facilities, and an LNG terminal. Response teams are presented with hypothetical problems and must utilize the Hiatusport contingency plan to develop recommendations that can be used to correct the problem. Each team must present their recommendations in class to fellow trainees who are not shy about pointing out problem areas.

Students are taught to select proper respiratory equipment for a given situation. In the laboratory they don and use respirators ranging from a five-minute emergency escape breathing device to selfcontained breathing apparatus. "We have found that by taking it step-by-step we are able to give them individualized attention, check them out carefully to insure proper usage, so they gain confidence in the equipment," notes Ranes. "We go through a standard predonning test, then provide a short lecture on regular maintenance of the equipment. We tell them what maintenance they can perform back at their unit as well as what they cannot handle. After a fit test for each student, we first use a sample of bananna oil. We remove that and inject a little irritant smoke—they let us know immediately if they do not have a good fit."

Extensive training is provided on selection, use, and compatibility of various types of protective clothing ranging from simple splash gear to fully encapsulated chemical suits.

"We start with less complicated respirators, move to breathing apparatus, and then to encapsulated suits," says Commander Ranes. "We do not throw them directly into encapsulated suits. We recognize there will be some people who just will not be able to get into a fully encapsulated suit. A person may have a real fear of being enclosed in such a suit. When that zipper hits the top, you are inside. We feel that this is the time and the place to learn if a student just cannot get into a suit. Once they leave here and return to their regular units, their commanding officer accepts the fact that they have gone into the suit, they are comfortable in it, and should the need arise they will be able to work in it. We tell them—'If you can't get into the suit, let us know *now*.' We make it clear they do not have to wear an encapsulated suit at a spill. A person can wear protective clothing with an SCBA worn outside, say at a decontamination area, and be just as valuable to the On-Scene Coordinator. We really don't push them into encapsulated suits. I've been teaching here a year-and-a-half, and during that time I've had just two students who were unable to wear fully encapsulated suits."

"Next, we move into rail cars. A *lot* of people ask why we teach Coast Guard personnel about rail cars," adds Ranes with a grin. "If you look at the routes used by railroads, there are many miles of trackage next to navigable waters. There is a very good chance we may have to respond to an overturned tank car spilling its contents into navigable waters. We use fullscale mock-ups of the standard pressurized tank cars that show arrangement of valving and guages. We also cover tank trucks. For tank cars and tank trucks, we provide strictly an overview; we do not go into great detail."

Trainees learn proven methods of stopping leaks in 55 gallon drums, and the use of chlorine "C" kits on tank cars. They practice site entry, decontamination, and sampling techniques; and run through a series of emergency response exercises where they are given an opportunity to practice what they have learned.

The culmination of the course is a response drill featuring three different scenarios. "They are given three separate tasks to perform while dressed in totally encapsulating suits and SCBA," explains Commander Ranes. "They go in as two-man teams with a two-man back-up, and enter and exit through a decontamination area. The purpose of the drill is not to make the students experts by any means, but rather to familiarize them with working in encapsulated suits and let the protective On-Scene Coordinators know what they can expect from people wearing such suits. The difference in reaction times between summer and winter is amazing. We can put a man into a suit with a 30-minute air supply during hot weather, and he will be able to work perhaps ten minutes before his alarm bell goes off. The combination of exertion and extreme heat causes him to breathe harder and use air more rapidly. In the winter time, we've had people go as long as 45-minutes *doing the same tasks.*"

"The first scenario is a mock-up of a manifold on the dock. The valve is supposedly stuck open, and the team has to cap it off with a blank flange. We apply base fire pressure to the piping, pretending the water gushing out is actually a hazardous chemical. The students are completely dressed out and supplied with the necessary tools, nuts-and-bolts, and a blank flange. They have to stop the leak. They could probably do it in two or three minutes dressed in street clothes, but wearing an encapsulated suit they lose a lot of dexterity. They find themselves dropping the nuts and bolts and getting a bit frustrated.''

''In another scenario we have a simulated leak in a 55-gallon drum. The trainees must utilize a couple of hoses and valves, an empty 55-gallon drum, and a drum carrier to transfer product from the leaking drum to a sound one by gravity feed. The third scenario calls for repair of a 55-gallon drum. They have to patch the drum, stop the leak, and put the damaged drum into an overpack drum. While it might have been very easy to mix patching material in the lab, doing it while completely dressed out is another matter.''

Since many of the units trainees will be returning to after training will be acquiring emergency response vans, the Hazardous Chemicals Training Course secured and equipped a response van that is used as a training aid. Radio equipment within the van permits communications from an incident scene to other Coast Guard units and small boats, and to vehicles or base stations equipped with citizen's band transceivers. A ''Command and Control Area'' within the van provides space for display of charts and maps of the area in which the van may be operating. Spare 30 minute SCBA air bottles are stored in a charger so they can be used over and over again. A three-bottle air cascade system is used both to supply air directly to personnel through an air line up to 300 feet in length, and to refill air bottles. Larger air bottles contain 308 cubic feet of air compared to approximately 45 cubic feet in the smaller tanks. A refrigerator for storage of samples, an air conditioning system, and lights can all be run off a portable AC generator that is set-up on the ground some distance from the van. Perhaps the most useful ''tool'' in the van is a library of reference materials. Metal lockers are used to store protective clothing, tape and talcum powder used when donning encapsulated suits, lifelines and harnesses, respiratory devices, and airbags used for lifting/leveling/moving. Also carried on-board the van is access control gear such as barriers, barrier tape, and traffic cones; a variety of sorbent materials, and a large selection of handtools. A metal workboat is carried atop the van with an outboard motor in a bracket at the rear. There is also a wash station, or decontamination shower/eyewash, stored in sections for assembly on a wooden platform that when rigged with a plastic liner catches washdown water coming off a man being decontaminated and is then held for safe disposal at a later time. Plexiglass-covered display boards can be erected both inside and on the outside of the van as a ''status board'' to provide people on-scene with a constant updating of the situation. The plexiglass-covered ''status board'' inside the van folds down into a bunk. A multi-colored awning on the side plus the high visibility of the van itself provides an instantly recognizable command post location on-scene.

The Hazardous Chemicals Training Course of the Coast Guard Marine Safety School has been designed to provide realistic, hands-on training so students develop an understanding of the chemicals, reference materials, protective clothing, respiratory equipment, general tools/ equipment/materials—and problems—they will have to deal with when they actually respond to an incident as a Federal On-Scene Coordinator or member of a Coast Guard support group providing assistance to an On-Scene Coordinator.

For additional information, *Contact:* Chief, Marine Safety School, U.S. Cost Guard Reserve Training Center, Yorktown, Virginia 23690.

Chemical Manufacturers Association

The Chemical Manufacturers Association periodically sponsors two-day Emergency Response Team Workshops at various locations that are organized and presented under the guidance of the CHEMTREC Advisors Group made up of extremely knowledgeable industrial emergency response managers. The workshops provide guidance and information for company emergency response teams for ''on-scene'' handling of emergencies.

After introductory comments pertaining to industry responsibility, CHEMTREC safety procedures, and methods of dealing with the news media; trainees are split into groups of 30 and each group is rotated through seven separate training modules.

1. Breathing Equipment & Protective Equipment.
2. Field Analytical Techniques—Air & Water.
3. Keeping Hazardous Materials Contained in Tank Cars.
4. Tank Cars—Capping, Patching, & Transferring.
5. Tank Trucks—Capping, Patching, & Transferring.
6. Drums & Cylinders—Patching & Handling.
7. Cleanup Techniques—Land & Water.

This training is currently scheduled through August of 1983. For up-to-date information and scheduling, *Contact:* Chemical Manufacturers Association, 2501 M Street, N.W. Washington, D.C. 20037, Tel. 202-887-1115.

Roberts Environmental Services, Inc.

Roberts Environmental provides both classroom and hands-on training related to hazardous materials for government (fire, police, civil defense, military) and industrial personnel nationwide. As a commercial oil and hazardous materials response contractor, Roberts' training emphasizes techniques and lessons learned by commercial responders that are directly transferable to industry and government response and control efforts.

For additional information, *Contact:* Roberts Environmental Services, Inc., P.O. Box 10093 Eugene, Oregon 97440, Tel. 503-688-4531.

Environmental Hazards Management Institute

Environmental Hazards Management Institute develops and provides hazardous materials/hazardous wastes training for both industrial personnel and first responders. During 1981/1982 principals of the Institute participated in the development of three different examples of such training.

1. *Hazardous Waste Detection Specialist Training:*

In cooperation with the Small Business Development Program of the University of New Hampshire and the New Hampshire Private Industry Council, and representatives of both the private and public sectors; recruited and trained CETA-eligible individuals (Comprehensive Employment and Training Act) as Hazardous Waste Detection Specialists using a 4½ month curriculum specially developed for the program.

2. *Planning For Compliance:*

In cooperation with the University System of New Hampshire developed and administered a ten-day program to train line supervisors and middle and upper-level managers in manufacturing organizations using and/or producing hazardous materials/toxic substances/hazardous wastes in issues and legislation relating to the handling, storage, transportation, and disposition of such materials. (Resource Conservation & Recovery Act, Toxic Substances Control Act, Hazardous Materials Transportation Act, Emergency Response, Hazard Mitigation, Toxicology, Compliance Planning, Haz Mat Audits, Safety/Detection/Monitoring Equipment, Medical Monitoring, Reduction/Destruction/Disposal, etc.)

3. *Railroad Incident Response Seminar:*

Developed and presented a program designed for first responders that covered railroad involvement with hazardous materials, role of the Bureau of Explosives, emergency response within the railroad industry, industry notification, railroad procedures during derailments, sources of information, etc.

For additional information, *Contact:* Environmental Hazards Management Institute, P.O. Box 283, Portsmouth, New Hampshire 03801, Tel. 603-436-3950.

The Celanese Fire Training Center

Funded and engineered by Celanese Corporation, the Celanese Fire Training Center located on six acres near the corporation's Celriver manufacturing plant was originally intended to serve only the emergency and fire training needs of Celanese personnel. In actual practice, however, the Center offers eight two-day advanced fire technology seminars throughout the spring and fall for industrial fire brigade members from a number of companies, and municipal firefighters. In addition, up to 500 Celanese personnel from east of the Mississippi River are trained there during an average year.

The Center uses pit fires, mock-ups, and scenarios to provide hands-on training in dry chemical extinguishment, propane fires, pipe rack fires, loading terminal situations, incinerator fires, and the use of high expansion foam. More than 4,000 students, mainly industrial fire brigade personnel, have been trained at the facility. Administrative tasks related to operation of the training center are handled by York Technical College. For additional information, *Contact:* Celanese Fire Training Center, Dean of Extension Service, York Technical College, Rock Hill, South Carolina 29730, Tel. 803-555-1212.

International Society of Fire Service Instructors

The International Society of Fire Service Instructors (ISFSI) conducts general fire service training programs throughout the United States, including hazardous materials training for fire service responders. By means of contracts with sponsoring fire service organizations or open-enrollment seminars, ISFSI currently offers both a basic hazardous materials familiarization course and a more advanced 12-hour seminar dealing with the tactical considerations of handling a hazardous materials incident. (Size-up, attack, pesticide spills, chlorine, etc..) For additional information, *Contact:* International Society of Fire Service Instructors, 20 Main Street, Ashland, Massachusetts 01721, Tel. 617-881-5800.

The Materials Transportation Bureau of the U.S. Department of Transportation has compiled and periodically updates four separate lists of facilities offering hazardous materials transportation courses and seminars.

List 1: Colleges and universities offering degree and non-degree courses and seminars in packaging and transportation of hazardous materials.

List 2: Colleges and other institutions offering hazardous materials courses as part of a fire science or other emergency service training curriculum.

List 3: Corporations and other business organizations which offer courses, seminars, and training material dealing with packaging, marking, labeling, shipping and transport of hazardous materials.

List 4: Federal, state and local government agencies offering courses on the packaging, handling and transportation of hazardous materials.

Although each of the facilities listed was contacted by Materials Transportation Bureau personnel to verify the offering of hazardous materials related training, no attempt was made to list fees, course content, specific locations, schedules or frequency of offering; and no attempt was made to monitor or evaluate the courses and seminars offered. For details concerning the programs offered, individual facilities should be contacted directly.

To add new programs, courses and offerings to the various lists, *Contact:* Department of Transportation, Research & Special Programs Administration, Materials Transportation Bureau, Information Services Division (DMT-11), Washington, D.C. 20590, Tel. 202-426-2301.

US Department of Transportation
**Research and Special Programs
Administration**

COLLEGES AND UNIVERSITIES OFFERING DEGREE AND NON-DEGREE COURSES AND SEMINARS IN PACKAGING AND
TRANSPORTATION OF HAZARDOUS MATERIALS

ALABAMA

University of Alabama
Dept. of Commerce & Business
Dr. James Kefler
Box J
University, Alabama 35486
(205) 348-6010

University of South Alabama
Dept. of Marketing and Trans-
portation
Dr. Joseph Horsely
Mobile, Alabama 36688
(205) 460-6411

CALIFORNIA

University of California-
Berkeley
Institute of Transportation
Studies
Mr. Henry Bruck
109 McLaughlin Hall
Berkeley, California 94720
(415) 642-5672

Golden Gate University
School of Transportation
Mr. Monroe Sullivan
536 Mission Street
San Francisco, California 94105
(415) 442-7000 ext. 7422

COLORADO

Metropolitan State College
School of Professional Studies
Mr. William B. Rourke, Jr.
1006 11th Street, Box 30
Denver, Colorado 80204
(303) 629-2982

CONNECTICUT

Norwalk Community College
Dept of General Business
Mr. Milton Goldstein
333 Wilson Avenue
Norwalk, Connecticut 06854
(203) 853-2040

FLORIDA

Florida International Univ.
School of Business
Mr. J. A. F. Nichols
SBOS-DM 346
Miami, Florida 33199
(305) 554-3320

Florida Junior College
Kent Campus
Transportation Department
Mr. Paul A. Halloran
Jacksonville, Florida 32205
(904) 387-8167

Miami-Dade Community College
South Campus
Aviation Department
Mr. Gerard Pucci
11011 S.W. 104th St.,Rm.3324
Miami, Florida 33176
(305) 596-1154

University of Miami
Ryder Prog. in Trans.
Dr. Anthony Cantanese
1541 Brescia
Miami, Florida 33144
(305) 284-2211

IOWA

Iowa State University
Safety Education Dept.
Dr. Jack Beno
Safety Education Lab.
Building 208-C
Ames, Iowa 50011
(515) 294-5945

Northern Iowa Area Comm.
College
Dept. of Trade & Industry
Mr. Bill Frelund
500 College Avenue
Mason City, Iowa 50401
(515) 421-4355

KANSAS

Kansas State University
Dept. of Civil Eng.
Mr. Eugene R. Russell
Seaton Hall
Manhatten, Kansas 66506
(913) 532-5662

University of Kansas
Radiation Safety Office
Dr. Benjamin Friesen
Lawrence, Kansas 66045
(913) 864-4089

LOUISIANA

Louisiana State Univ.
Nuclear Science Center
Dr. William F. Curry
Baton Rouge, Louisiana 70803
(504) 388-2163

MICHIGAN

Michigan State Univ.
School of Packaging
Dr. Hugh Lockhart
East Lansing, Michigan 48824
(517) 355-3604

NEW YORK

Franklin D. Roosevelt Inst.
of Maritime Studies
Ms. Elizabeth Norberg
15 State Street
New York, New York 10004
(212) 269-2710

Suffolk County Comm. College
Mr. Joseph E. Galvin
533 College Road
Selden, New York 11784
(516) 451-4278

Syracuse University
School of Marketing and
Transportation
The Franklin Program
Dr. Theodore O. Wallin
129 College Place
Syracuse, New York 13210
(315) 423-2962

University of Niagara
Institute of Transportation
Travel and Tourism
Mr. Ben Perry
Niagara University, NY 14109
(716) 285-1212 ext. 373

OHIO

Ohio State University
Continuing Education Dept.
Mr. Bernard J. LaLonde
Columbus, Ohio 43210
(614) 422-0331

Univ. of Cincinnati
Evening College
Mr. Kenneth Dickens
102 McMicken Hall
Cincinnati, Ohio 45221
(513) 475-4431

UTAH

LDS Business College
Special Courses and
Conferences
Mr. Ross Derbridge
411 East South Temple
Salt Lake City, Utah 84111
(801) 363-2765

WASHINGTON

Seattle Community College
Business & Commerce Div.
Mr. Edward Spalding
9600 College Way North
Seattle, Washington 98103
(206) 634-4436

WISCONSIN

Northeast Wisconsin Technical
Institute
Dept. of Marketing & Business
Mr. E. R. DeRoche
2740 West Mason Street
Green Bay, Wisconsin 54303
(414) 497-3125

Univ. of Wisconsin-Stout
Dept. of Packaging
Mr. Marvin Kufahl
Menomonie, Wisconsin 54751
(715) 232-2426

INFORMATION SERVICES DIVISION, DMT-11
MATERIALS TRANSPORTATION BUREAU
DEPARTMENT OF TRANSPORTATION
WASHINGTON, D.C. 20590

REVISED APRIL 1982

LIST #2

COLLEGES AND OTHER INSTITUTIONS OFFERING
HAZARDOUS MATERIAL COURSES AS PART OF A
FIRE SCIENCE OR OTHER EMERGENCY SERVICE TRAINING CURRICULUM

ALABAMA

Alexander City State Jr. College
Fire Science Dept.
Paul Blackwell
Cherokee Road P. O. Box 699
Alexander City, AL 35010
(205) 234 6346

George C. Wallace State
 Community College
Fire Science Dept.
Michael Houghland
P. O. Drawer 1049
Selma, AL 36701
(205) 875 2634 Ext. 31

ALASKA

Anchorage Community College
Fire Science Program
James Evans
2455 Providence Dr.
Anchorage, AK 99504
(907) 279 6602

ARIZONA

Arizona College of Technology
Fire Science Dept.
William Buttery
Route 97
Winicelman, AZ 85292
(602 356 7864

Cochise College
 Douglas Campus
Fire Science Dept.
Richard Seals
Douglas, AZ 85607
(602) 364 7943

Cochise College
 Sierra Vists Compus
Fire Science Dept.
Richard Seals
901 Columbo
Sierra Vista, AZ 85635
(602) 934 2211

Eastern Arizona College
Fire Science Dept.
Ralph Orr
Thatcher, AZ 85552
(602) 428 1133

Glendale Community College
Fire Science Dept.
Renault Catalano
600 West Oliver Ave
Glendale, AZ 85301
(602) 934 2211

Mohave Community College
Vincent Salmon
1971 Iagerson Ave
Kingman, AZ 86401
(602) 757 4331

Phoenix College
Fire Science Dept.
Robert F. Noll
1202 West Thomas Road
Phoenix, AZ 85013
(602) 264 2492

Pima Community College
Fire Science Dept.
Ignacio Garcia
2202 West Anklam Road
Tuscon, AZ 85709
(602) 884 6693

Scottsdale Community College
Fire Science Dept.
Ed Gates
9000 E. Chaparral Road
Scottsdale, AZ 85253

CALIFORNIA

Allan Hancock College
Fire Science Dept.
Robert Pile
800 South College Dr.
Santa Maria, CA 93454
(805) 922 6966

American River College
Fire Science Dept.
Louis Quint
4700 College Oak Dr.
Sacramento, CA 95841
(916) 484 8316

American River College
 Placerville Campus
Fire Science Dept.
Art Scott
106 Placerville Dr.
Placerville, CA 95667
(916) 622 7575

Antelope Valley College
Fire Science Dept.
Frank C. Roberts
3041 West Ave K
Lancaster, CA 93534
(805) 943 3241

Bakersfield College
Fire Science Dept.
Joseph Angelo
2101 K St Mall
Bakersfield, CA 93305
(805) 395 4481

Barstow College
Fire Science Dept.
Jack Sherman
2700 Barstow Road
Barstow, CA 92311
(714) 252 2411

Butte College
Fire Science Dept.
Fred Allen
Route 1, Box 183A
Oroville, CA 95965
(916) 895 2401

Chabot College
Fire Science Dept.
Glenn Bass
25555 Hesperian Blvd.
Hayward, CA 94545
(415) 782 3000

Cosumnes River College
Fire Science Dept.
Cecie Fontanoza
8401 Center Parkway
Sacramento, CA 95823
(916) 421 1000

East Los Angeles College
Fire Science Dept.
M. S. Pena
1301 Brooklyn Ave
Monterey Park, CA 91754
(213) 265 8650

Glendale College
Fire Science Dept.
Dave Leek
1500 North Verdugo Road
Glendale, CA 91208
(213) 240 1000

Imperial Valley College
Fire Science Dept.
W. D. Rudolph
P. O. Box 158
Imperial, CA 92251
(714) 352 8320

Long Beach City College
Fire Science Dept.
Denny Pace
4901 East Carson Street
Long Beach, CA 90808
(213) 599 2421

Los Angeles Valley College
Fire Science Dept.
George Yochum
5800 Fulton Ave
Van Nuys, CA 91401
(213) 781 1200

Merced College
Fire Science Dept.
Lee McCabe
3600 M Street
Merced, CA 95340
(209) 723 4321 Ext. 282

Modesto Junior College
Fire Science Dept.
Eldon L. Helm
College Ave
Modesto, CA 95350
(209) 524 1451 Ext. 226

Mount San Jacinto College
Fire Science Dept.
Benton Caldwell
21-400 Highway 79
San Jacinto, CA 92383
(714) 654 7321

Cabrillo College
Fire Science Dept.
David Barbin
6500 Soquel Dr
Aptos, CA 95003
(408) 425 6447

Chaffey College
Fire Science Dept.
Eddie Smith
5885 Haven Avenue
Alta Loma, CA 91701
(714) 987 1737

Cuesta College
Fire Science Dept.
Edwin M. Pearce
P. O. Box J
San Luis Obispo, CA 93406
(805) 544 2943

El Camino College
Fire Science Dept.
Ed Muraski
16007 Crenshaw Blvd.
Via Torrance, CA 90506
(213) 532 3670

Grossmont College
Fire Science Dept.
Dave Lien
8800 Grossmont College Dr
El Cajon, CA 92020
(714) 465 1700

Indian Valley Colleges
Fire Science Dept.
Howard Stillwell
1800 Ignacio Blvd.
Novato, CA 94947
(415) 883 2211

Los Angeles City College
Fire Science Dept.
Stanley Schall
855 North Vermont Ave
Los Angeles, CA 90029
(213) 663 9141 Ext. 207

Los Medanos College
Fire Science Dept.
Carlton Williams
2700 East Leland Road
Pittsburg, CA 94565
(415) 439 2181 Ext. 273

Merritt College
Fire Science Dept.
K. L. Giles
12500 Campus Dr
Oakland, CA 94619
(415) 531 4911 Ext. 296

Monterey Peninsula College
Fire Science Dept.
Jim Cardwell
980 Fremont Blvd.
Monterey, CA 93940
(408) 649 1150 Ext. 405

Nappa College
Fire Science Dept.
Calvin Russell
2277 Napa-Vallejo Highway
Napa, CA 94558
(707) 252 8211

Cerro Coso Community College
Fire Science Dept.
James Sirman
Ridgecrest, CA 93555
(714) 375 5001

Columbia Junior College
Fire Science Dept.
J. Amundsen
P. O. Box 1849
Columbia, CA 95310
(209) 532 3141

College of the Desert
Fire Science Dept.
Bill Kroonen
43-5000 Monterey Ave
Palm Desert, CA 92260
(714) 346 8041

Fresno City College
Fire Science Dept.
Roy Edwards
1101 East University Ave
Fresno, CA 93741
(209) 442 4600 Ext. 8517

Hartnell College
Fire Science Dept.
Thomas C. Campbell
156 Homestead Ave
Salinas, CA 93901
(408) 758 7261

Lake Tahoe Community College
Fire Science Dept.
Jim Leavitt
P. O. Box 14445
South Lake Tahoe, CA 95702
(916) 541 4660

Los Angeles Harbor College
Fire Science Dept.
Van G. Waring
1111 Figueroa Place
Wilmington, CA 90744
(213) 835 0161

College of Marin
Fire Science Dept.
Joseph E. Berruezo
Kentfield, CA 94909
(415) 454 3962 Ext. 298

Miramar College
Fire Science Dept.
R. H. Rossmaessler
10440 Black Mountain Road
San Diego, CA 92126

Mount San Antonio College
Fire Science Dept.
John G. O'Sullivan
1100 North Grand Ave
Walnut, CA 91789
(714) 598 2811 Ext. 252

Oxnard College
Fire Science Dept.
John Dell
534 South A St
Oxnard, CA 93030
(805) 486 7315

Palomar College
Fire Science Dept.
R. A. Jackson
1140 West Mission
San Marcos, CA 92069
(714) 744 1150 Ext. 456

College of Redwoods
Fire Science Dept.
Don Peterson
Eureka, CA 95501
(707) 443 8411

San Bernardino Valley College
Fire Science Dept.
Ed Temby
701 South Mt. Vernon Ave
San Bernardino, CA 92403
(714 885 0231

San Jose City College
Fire Science Dept.
Robert G. Egan
2100 Moorpark Avenue
San Jose, CA 95128
(408) 298 2181 Ext. 435

Santa Barbara City College
Fire Science Dept.
Melvin Elkins
721 Cliff Dr
Santa Barbard, CA 93109
(805) 965 0581

College of the Sequoias
Fire Science Dept.
Leroy A. Berg
915 South Mooney Blvd.
Visalia, CA 93277
(209) 733 2050 Ext. 280

College of the Siskiyous
Fire Science Dept.
Bill Rowe
800 College Avenue
Weed, CA 96094
(916) 938 4463

West Hills College
Fire Science Dept.
James Butterworth
300 Cherry Lane
Coalinga, CA 93210
(209) 935 0801

Pasadena City College
Fire Science Dept.
Victor E. Stephens
1570 East Colorado Blvd.
Pasadena, CA 91106
(213) 578 7123

Rio Hondo College
Fire Science Dept.
Eugene Mahoney
3600 Workman Mill Road
Whittier, CA 90608
(213) 692 0921

City College of San Francisco
Fire Science Dept.
Martin Kilgariff
50 Phelan Ave
San Francisco, CA 94112
(415) 239 3359

College of San Mateo
Fire Science Dept.
Bob Dawson
1700 West Hillsdale Blvd.
San Mateo, CA 94402
(415) 574 6162

Santa Monica College
Fire Science Dept.
Paul Stein
1815 Pearl St
Santa Monico, CA 90405
(213) 396 5277

Shasta College
Fire Science Dept.
John White
1065 North Old Oreyon Trail
Redding, CA 96001
(916) 241 3523

Solano Community College
Fire Science Dept.
Chuck Kimball
P. O. Box 246
Suisun City, CA 94585
(707) 864 7000

West Los Angeles College
Fire Science Dept.
Marilyn Brock
4800 Freshman Dr
Culver City, CA 90230
(213) 836 7110

Porterville College
Fire Science Dept.
Edward M. Buckles
900 South Main St
Porterville, CA 93257
(209) 781 3130

Riverside City College
Fire Science Dept.
Bob Holstrom
4800 Magnolia Ave
Riverside, CA 92506
(714) 684 3240

San Joaquin Delta Community
 College
Fire Science Dept.
Joe Daly
5151 Pacific Ave
Stockton, CA 95207
(209) 478 2011 Ext. 201

Santa Ana College
Fire Science Dept.
Bill Ogden
Seventeenth At Bristol
Santa Ana, CA 92706
(714) 835 3000 Ext. 255

Santa Rosa Junior College
Fire Science Dept.
John Healy
1501 Mendocino Ave
Santa Rosa, CA 94501
(707) 527 4441

Sierra College
Fire Science Dept.
Frank Strong
5000 Rocklin Road
Rocklin, CA 95677
(916) 624 3333 Ext. 312

Victor Valley College
Fire Science Dept.
Charles A. Peterson
P. O. Drawer OO
Victorville, CA 92392
(714) 245 4271 Ext. 255

Yuba College
Fire Science Dept.
Don Vedo
2088 North Beale Road
Marysville, CA 95901
(916) 742 7351 Ext. 215

COLORADO

Aims Community College
Fire Science Dept.
Bill Adamson
Box 69
Greely, CO 80631
(303) 356 9600

Community College of Denver
 Redrocks
Fire Science Dept.
Bruce Birza
12600 W 6th Ave
Golden, CO 80401
(303) 988 6160

El Paso Community College
Fire Science Dept.
2200 Bott Ave
Colorado Springs, CO 80904
(303) 471 7546

CONNECTICUT

State Tech. Colleges of
 Connecticut
Fire Science Dept.
Lawrence M. Ford
401 Flatbush Ave
Hartford, CT 06106
(203) 238 6587

University of New Haven
Fire Science Dept.
Peter Desio
300 Orange Avenue
West Haven, CT 06516
(203) 934 6301

DELAWARE

Delaware Tech & Community College
Fire Science Department
Mr. Buchanan
Box 1260, Stanton Campus
Newark, DE 19711
(302) 368-6900

Delaware Tech & Community College
  Kent
Fire Science Department
Lowell Barrett
1823 N. DuPont Highway
Dover, DE 19901
(302) 678-4665

DISTRICT OF COLUMBIA

University of District of Columbia
Fire Science Department
Mr. Ortiz
4200 Connecticut Avenue, NW
Washington, DC 20008
(202) 282-7387

FLORIDA

Broward Community College
Fire Science Department
Mr. Charles Redmond
3501 S.W. Davies Road
Ft. Lauderdale, FL 33314
(305) 581-8700 Ext. 230

Miami-Dade Community College
Fire Science Department
Mr. James J. Guthrie
11380 N.W. 27th Avenue
Miami, FL 33167
(305) 577-6870

Edison Community College
Fire Science Department
Mr. R. V. Concilio
College Parkway
Ft. Myers, FL 33901
(813) 481-2121 Ext. 252

Florida Jr. College
Fire Science Department
Mr. Nat Cole
11991 Beach Blvd.
Jacksonville, FL 32216
(904) 646-2060

Gulf Coast Community College
Fire Science Department
Mr. Lester Morley
5230 West Highway 98
Panama City, FL 32401
(904) 769-1551

Indian River Community College
Fire Science Department
Mr. Henry T. Christen
3209 Virginia Avenue
Ft. Pierce, FL 33450
(305) 464-2000 Ext. 449

Jacksonville Fire Academy
Fire Science Department
Mr. Max Hendrix
2345 Roselle Street
Jacksonville, FL 32204
(904) 633-5588

Palm Beach Jr. College
Fire Science Department
Mr. Donald C. Witmer
4200 Lyngress Avenue
Lake Worth, FL 33460
(305) 965-8000

Seminole Community College
Fire Science Department
Mr. James King
Sanford, FL 32771
(305) 323-1450 Ext. 266

St. Petersburg Jr. College
Fire Science Department
Mr. E. Z. Jackson
2465 Drew Street
Clearwater, FL 33515
(813) 546-0011 Ext. 442

Valencia Community College
Fire Science Department
Rudolph During
P. O. Box 3028
Orlando, FL 32802
(305) 299-5000

GEORGIA

Clayton Junior College
Fire Science Department
Mr. Roy Dobyns
Morrow, GA 30260
(404) 363-7590

Georgia Fire Academy
Fire Science Department
Mr. William Lynch
534 Clay Street
Marietta, GA 30060
(404) 424-7315

Southern Technical Institute
Fire Science Department
Mr. J. R. Lee
534 Clay Street
Marietta, GA 30060
(404) 424-7371

HAWAII

Hawaii Community College
General Education & Public
  Service Division
Fire Science Department
Mr. Rex Yamasaki
1175 Manono Street
Hilo, Hawaii 96720
(808) 961-9311

IDAHO

Boise State University
Fire Service Training
Mr. Tom Tyree
1910 College Blvd.
Boise, ID 83725
(208) 385-1011

ILLINOIS

Black Hawk College
Fire Science Department
Mr. Simon Roberts
6600 34th Avenue
Moline, IL 61265
(309) 796-1311

State Community College of
    East St. Louis
Fire Science Department
Mr. Willard Maytubley
417 Missouri Avenue
East St. Louis, IL 62201
(618) 875-9100 Ext. 356

Illinois Institute of
    Technology
Fire Science Department
Mr. Boyd Hartley
3300 South Feptral
Chicago, IL 60616
(312) 567-3150

College of Lake County
Fire Science Department
Mr. John Shelton
19351 W. Washington St.
Grayslake, IL 60030
(312) 223-6601

Moraine Valley Com. Col.
Fire Science Department
Mr. Art Stoike
10900 South 88th Avenue
Palos Hills, IL 60465
(312) 974-4300

Prairie State College
Fire Science Department
Mr. Eddie O'Connor
200 E. 197th Street
Chicago Heights, IL 20411
(312) 782-5965

Triton College
Fire Science Department
Mr. Leo E. Flynn
2000 Fifth Avenue
River Grove, IL 60171
(312) 456-0300

City of Naperville Fire
    Prevention Bureau
Captain William Kuhrt
133 W. Jefferson
Naperville, IL 60540
(312) 420-6143

Carl Sandburg Community Col.
Mr. William D. Masters
P.O. Box 1407
Galesburg, IL 61401
(309) 344-2518

Elgin Community College
Fire Science Department
1700 Spartan Drive
Elgin, IL 60120
(312) 697-1000

Joliet Junior College
Fire Science Department
Mr. Lawrence Walsh
1216 Houbolt Avenue
Joliet, IL 60436
(815) 729-9020

Lewis & Clark Com. Col.
Fire Science Department
Mr. R. H. Patterson
Godfrey, IL 62035
(618) 466-3411

Oakton Community College
Fire Science Department
Mr. F. Salzberg
7900 N. Nagle Avenue
Morton Grove, IL 60053
(312) 967-5120

Rock Valley College
Fire Science Department
Mr. Carl Cascio
3301 North Mulford Road
Rockford, IL 61101
(815) 226-3704

William R. Harper College
Fire Science Department
Mr. Charles Henrici
Algonquin & Roselle Roads
Palatine, IL 60067
(312) 397-3000

Fire Service Institute
University of Illinois
Mr. Gerald Monigold
301 South Wright Street
Champaign, ILL 61820
(217) 333-3800

College of Du Page
Mr. John Seffner
22nd Street at
    Lambert Road
Glen Ellyn, IL 60137
(312) 858-2800

Illinois Central College
Fire Science Department
Mr. Daryl Hartwig
P.O. Box 2400
Peoria, IL 61635
(309) 694-5580

Kishwaukee College
Fire Science Department
Mr. Ernest Harfst
Malta Road
Malta, IL 60150
(815) 825-2086

Lincoln Land Com. Col.
Fire Science Department
Mr. O. R. Vanderwater
Shepherd Road
Springfield, IL 62700
(217) 786-2269

Parkland College
Fire Science Department
Mr. Fred Johnson
2400 W. Bradley Avenue
Champaign, IL 61820
(217) 351-2200

Sauk Valley College
Fire Science Department
Mr. A. R. Crowe
Rural Route #1
Dixon, IL 61021
(815) 288-5511 Ext. 358

Police Training Institute
University of Illinois
    Continuing Education
Mr. Clifford Van Meter
Armory Bldg., Room 359
Champaign, IL 61820
(217) 333-2337

INDIANA

Indiana Vocational Technical Col.
Fire Science Department
Mr. Robert Ruff
1534 West Sample Street
South Bend, IN 46619
(219) 289-7001

IOWA

Des Moines Area Com. College
Fire Science Department
Mr. Burges Shriver
2006 Ankeny Blvd.
Ankeny, Iowa 50021
(515) 964-6200

Iowa State University
Fire Science Department
Mr. Keith Royer
Ames, IA 50010

Kirkwood Community College
Fire Science Department
Mr. Donald Fuller
6301 Kirkwood Blvd., SW
Cedar Rapids, IA 52406
(319) 398-5496

KANSAS

Wichita State University
Fire Science Department
1845 Fairmount
Wichita, KS 67208
(319) 689-3425

KENTUCKY

Eastern Kentucky University
Fire Science Department
Mr. F. Dale Cozad
Richmond, KY
(606) 622-1454

Jefferson Community Col.
Fire Science Department
Mr. Martin Brown
P. O. Box 1036
Louisville, KY 40201
(502) 584-0181

Western Kentucky Univ.
Fire Science Department
Mr. Charles J. Wright
Bowling Green, KY 42101
(502) 745-4797

LOUISIANA

Louisiana State University
Division of Continuing Education
Mr. Dan Delanger
Pleasant Hall
Baton Rouge, LA 70803
(504) 766-0600

MAINE

Southern Maine VO/TECH Institute
Mr. Josiah Staples
Fort Road
South Portland, ME 04106
(207) 799-7303

Kennebee Valley Vocational
   Technical Institute
Fire Science Department
Ms. Barbara Lanman
Gilman Street
Waterville, ME 04901
(207) 873-6133

MARYLAND

Catonsville Community College
Fire Science Department
Mr. David Frank
800 South Rolling Road
Catonsville, MD 21228
(301) 455-4444

Charles County Com. Col.
Fire Science Department
Mr. Steve Maxwell
P.O. Box 910, Mitchell Rd.
La Plata, MD 20646
(301) 934-2251

Montgomery College
Fire Science Department
Mr. Richard Ulrich
Rockville, MD 20850
(301) 762-7400

Prince Georges Com. Col.
Fire Science Department
Mr. Edwin Beller
301 Largo Road
Largo, MD 20870
(301) 336-6000

University of Maryland
Fire & Rescue Institute
Mr. John Herr
College Park, MD 20742
(301) 454-2416

University of Maryland
Fire Protection Engineering
College of Engineering
Dr. John L. Bryant
College Park, MD 20742
(301) 454-2424

University of Maryland
Maryland Fire & Rescue Institute
Western Maryland Regional Office
Mr. Joseph W. McDaniel, Jr.
P.O. Box 3303
LaVale, MD 21502

MASSACHUSETTS

Berkshire Community College
Fire Science Department
Mr. Gene Kamp
West Street
Pittsfield, MA 01201
(413) 449-4660

Cape Cod Community College
Fire Science Department
Mr. Robert Tucker
West Barnstable, MA 02669
(617) 362-2131

Mount Wachusett Com. Col.
Fire Science Department
Mr. Herman Gelbwasser
Green Street
Gardner, MA 01440
(617) 632-6600

Springfield Technical Com. Col.
Fire Science Department
Mr. Albert W. Valentine
1 Armory Square
Springfield, MA 01105
(413) 781-6470

Bristol Community College
Fire Science Department
Mr. Robert Sherman
77 Ellsbree Street
Fall River, MA 02720

Massasoit Community College
Fire Science Tech Dept.
Prof. Phillip E. Blye
Brockton, MA 02402
(617) 588-9100

North Shore Community Col.
Fire Science Department
Mr. Francis M. Ryan
3 Essex Street
Beverly, MA 01915
(617) 927-4850

Bunker Hill Community Col.
Fire Science Department
Mr. Joseph G. Von Handorff
Austin Street
Middleton, MA 01949
(617) 241-8600

Middlesex Community Col.
Fire Science Department
Mr. Lawrence Rice
21 Springs Road
Bedford, MA 01730
(617) 275-8910

Quinsigamond Com. Col.
Fire Science Department
670 W. Boylston Street
Worcester, MA 01606
(617) 853-2300

MICHIGAN

Henry Ford Community College
Fire Science Department
Mr. Thadeus C. Matley
5101 Evergreen Road
Dearborn, MI 48128
(313) 271-2750

Lansing Community College
Fire Science Department
Mr. Robert Ogilvy
419 North Capitol Avenue
Lansing, MI 48914
(517) 373-7013

C. S. Mott Community College
Fire Protection
Mr. Fred Lamb
Flint, MI 48503
(313) 762-0281

Kellogg Community College
Fire Science Department
Mr. Allen Vosburg
450 North Avenue
Battle Creek, MI 49016
(616) 965-3931 Ext. 212

McComb County Com. Col.
Fire Science Department
Mr. Art Kingsbury
Center Campus, P.O. Box 309
Warren, MI 48093
(313) 286-2058

St. Clair County Com. Col.
Fire Science Department
Mr. Krug
Port Huron, MI 48060

Lake Michigan College
Fire Protection Tech.
Lieutenant Devine
Benton Harbor, MI 49022
(616) 927-3571

Madonna College
Fire Science Department
Mr. James Lynchey
36600 Schoolcraft
Livonia, MI 48150
(313) 591-1200

Washtenaw Community Col.
Fire Protection Technology
Mr. Dean Hackney
Ann Arbor, MI 48107

MINNESOTA

Metropolitan Community College
Mr. William Quirk
1419 Harmon Place
Minneapolis, MN 55403
(612) 341-7061

University of Minnesota
Fire Science Department
Mr. Roger Young
3000 University Ave., SE
Minneapolis, MN 55415
(612) 373-3844

St. Clair County Com. Col.
Fire Science Department
Mr. C. F. Knight
323 Erie Street
Port Huron, MI 48060
(313) 984-3881

MISSISSIPPI

Hinds Junior College
Fire Science Department
Mr. Norman Preston
Raymond, MS 39154
(601) 857-5261

Jackson State University
Dept. of Industrial Tech
Mr. J. T. Smith
1235 Lynch
Jackson, MS 39203
(601) 968-2466

MISSOURI

Central Missouri State Univ.
Fire Science Department
Mr. Robert Semonisck
Warrensburg, MO 64093
(816) 429-4111

Columbia College
Extended Studies Division
Director of Fire Science
Columbia, MO 65201

Drury College
Fire Science Program
Ms. Rosa Lee White
900 N. Benton
Springfield, MO 65802
(417) 865-8731

East Central Junior College
Fire Science Program
Mr. Ed Conway
Union, MO 63084
(314) 583-5193

Jefferson College
Fire Science Department
Mr. Raymond Walsh
Hillsboro, MO 63050
(314) 789-3951 Ext. 140

Penn Valley Com. Col.
Fire Science Department
Mr. Richard Lehmann
3201 Southwest Trafficway
Kansas City, MO 64111
(816) 756-2800 Ext. 257

St. Louis Community College
   at Florrisant Valley
Fire Science Department
Mr. Herbert V. McMahon
3400 Pershall Road
St. Louis, MO 63135
(314) 595-4200

St. Louis Community College
   at Forest Park
Fire Science Department
Mr. D. B. Miller
5600 Oakland Avenue
St. Louis, MO 63110
(314) 644-9285

University of Missouri
Fire Training
Mr. Bill Westhoff, Jr.
1001 Ashland Gravel Rd.
Columbia, MO 65201
(314) 882-6498

MONTANA

Montana Board of Public Education
Fireman Training Program
33 S. Last Chance Gulch
Helena, MT 59601
(406) 449-2785

NEBRASKA

Southeast Community College
Fire Science Department
Mr. Don H. Venter
1309 N. 17th Street
Lincoln, NE 68508

University of Nebraska-Omaha
Fire Science Department
Mr. H. A. Dahlquist
P.O. Box 688
Omaha, NE 68132
(402) 554-2543

NEVADA

Clark County Community College
Fire Science Department
Mr. Dave Hoggard
3200 E. Cheyenne Avenue
North Las Vegas, NV 89030
(702) 643-6060

Northern Nevada Com. Col.
Fire Science Department
Mr. William J. Berg
921 Elm Street
Elko, NV 89801
(702) 738-8493

NEW HAMPSHIRE

New Hampshire VOC-TEC College
Fire Protection Program
Mr. Thomas Dawson
Prescott Hill - Route 106
Laconia, NH 03246
(603) 524-3207

Lily Pond Fire School
Lakes Region Mutual Fire
   Aid Association
Mr. Edward Warfield
64 Court Street
Laconia, NH 03246
(603) 624-2386

NEW JERSEY

Atlantic Community College
Fire Science Department
Mr. E. J. Fottrell
Mays Landing, NJ 08330
(609) 625-1111 Ext. 243

Essex County College
Fire Science Department
Mr. Charles Lowollo
31 Clinton Street
Newark, NJ 07102
(201) 877-3000

Passaic County Com. Col.
Fire Science Department
Mr. Mark F. Schaffer
170 Paterson Street
Paterson, NJ 07505
(201) 279-5000

Bergen Community College
Fire Science Department
Mr. Horace Chandler
400 Paramus Road
Paramus, NJ 07652
(201) 447-1500

Jersey City State College
Fire Science Department
2039 Kennedy Blvd.
Jersey City, NJ 07305
(201) 547-3311

Somerset County College
Fire Science Department
Ms. Carol Murtaugh
P.O. Box 3300
Somerville, NJ 08876
(201) 526-1200

Camden County College
Fire Science Department
Mr. John Tenbrook
P.O. Box 200
Blackwood, NJ 08012
(609) 227-7200

Mercer County College
Fire Science Department
Mr. Al Porter
1200 Old Trenton Road
Trenton, NJ 08690
(609) 586-4800

NEW MEXICO

University of Albuquerque
Fire Science Department
Dr. Stevenson
Saint Joseph Place
Albuquerque, NM 87106
(505) 831-1111 Ext. 330

NEW YORK

Broome Community College
Continuing Education
Fire Protection Program
Mr. Ogden Clark
P.O. Box 1017
Binghamton, NY 13902
(607) 772-5005

Monroe Community College
Fire Science Department
Mr. John T. Maher
1000 E. Henrietta Rd.
Rochester, NY 14600
(716) 442-9950

Schenectady County Com. Col.
Fire Science Department
Dr. Irma R. Chestnut
Washington Avenue
Schenectady, NY 12305
(518) 346-6211

Central Texas College
  Overseas Europe
Fire Protection Technology
Hanau
APO New York, NY 09165

Onondaga Community Col.
Fire Science Department
Mr. Larry Linch
700 East Water Street
Syracuse, NY 13215
(315) 469-7741 Ext. 5225

Suffolk County Vocational,
  Education & Extension Bd.
Fire Training Center
P.O. Box 128
Yaphank, NY 11980
(516) 265-7269

John Jay College of
  Criminal Justice
Fire Science Department
Captain Richard Abbott
445 W. 49th Street
New York, NY 10019
(212) 489-5183

Rockland Community Col.
Fire Science Department
Mr. Thomas Goldrick
145 College Road
Suffern, NY 10901
(914) 356-4650

Westchester Com. Col.
Fire Science Department
Mr. Charles Crowley
75 Grasslands Road
Valhalla, NY 10595
(914) 347-6800

NORTH CAROLINA

Central Piedmont Com. Col.
Fire Science Department
Mr. George W. Wright
Elizabeth Ave. at N. King's Dr.
Charlotte, NC 28204
(704) 373-6705

Guilford Technical Institute
Fire Science Department
Mr. Harold J. Fegan
P.O. Box 309
Jamestown, NC 27282
(919) 292-1101

Durham Technical Institute
Fire Science Department
Mr. Joseph Wade
1637 Lawson Street
Durham, NC 27703
(912) 596-9311

Richmond Technical Institute
Fire Science Department
Mr. Richard McIntyre
P.O. Box 1189
Hamlet, NC 28345
(919) 582-1980

Forsyth Technical Inst.
Fire Science Department
Mr. Larry Weaver
2100 Silas Creek Parkway
Winston-Salem, NC 27103
(919) 723-0371

Rowan Technical Institute
Fire Science Department
Mr. Larry Gibson
P.O. Box 1595
Salisbury, NC 28144
(704) 637-0760 Ext. 46

James Sprunt Technical Inst.
Fire Science Department
Mr. Emmel Coggins
P.O. Box 398
Kenansville, NC 28349
(919) 296-1341

Tech. Inst. of Alamance
Fire Science Department
Mr. Jerry Harris
411 Camp Road
Burlington, NC 27215
(919) 578-2002

Western Piedmont Com. Col.
Fire Science Dept.
Mr. Jerry Rowland
1001 Burkemont Avenue
Morgantown, NC 28655
(704) 437-8688

NORTH DAKOTA

North Dakota Fireman's Assoc.
Mr. D. E. (Monk) Gilman
Beach, ND 58621
(701) 872-4392

OHIO

University of Akron
Fire Science Department
Mr. Harrington
302 E. Buchtel Avenue
Akron, OH 44325
(216) 375-7906

University of Cincinnati
Fire Protection
Mr. D. F. Pinger
100 E. Central Parkway
Cincinnati, OH 45210
(513) 475-6567

Columbus Technical Inst.
Fire Science Department
Mr. Glenn Clark
550 E. Spring Street
Columbus, OH 43215
(614) 221-6743

Cayahoga Community College
Fire Science Department
Mr. Fred C. Sutton
2900 Community College Ave.
Cleveland, OH 44115
(216) 241-5966

Hocking Technical College
Fire Science Department
Mr. William Fennestofle
Route 1
Nelsonville, OH 45764
(614) 753-3591

Lorain County Com. Col.
Fire Science Department
Mr. Walter McGreedy
1005 N. Abbe Road
Elgrin, OH 44035
(216) 365-4191

Ohio State Fireman's
 Training Academ-
Ms. Tina Hazlett
8895 E. Main Street
Reynoldsburg, OH 43068
(614) 864-5510

Owens Technical College
Fire Science Department
Mr. William Russell
30335 Oregon Road
Perryburg, OH 43551
(419) 666-0580

Stark Technical College
Fire Science Department
Mr. Joseph L. Hafer
6200 Frank Avenue, NW
Canton, OH 44720
(216) 994-6170

OKLAHOMA

Oklahoma State Tech. Inst.
Fire Science Department
Mr. Bob Mowles
900 N. Portland
Oklahoma City, OK 73107
(405) 947-0771

Oklahoma State University
Fire Science Department
Mr. Dale F. Janes
Stillwater, OK 74074
(405) 624-5000

Tulsa Junior College
Fire Science Department
Mr. McElyea
909 S. Boston
Tulsa, OK 74119
(918) 587-6561 Ext. 175

Western Oklahoma State College
Fire Science Department
Mr. Cecil Chesser
2801 N. Main
Altus, OK 73521
(405) 477-2000

OREGON

Chemeketa Community College
Fire Science Department
Mr. Cecil Dill
P.O. Box 14007
Salem, OR 97309
(503) 399-5163

Clackamas Community College
Fire Science Department
Mr. Durwood Thomas
19600 South Molalla Avenue
Oregon City, OR 97045
(503) 656-2631

Portland Community College
Fire Science Department
Mr. John Koroloff
120005 West 49th Avenue
Portland, OR 97219
(503) 244-6111

Rogue Community College
Fire Science Department
Mr. Mark Burns
3345 Redwood Highway
Grants Pass, OR 97526
(503) 579-5541

PENNSYLVANIA

Butler County Fire Chiefs
  Association
Butler County Fire School
Mr. John Stokes
124 W. North Street
Butler, PA 16001
(412) 382-4200

Pennsylvania State Univ.
Fire Science Department
Mr. Charles R. Meck
209 Keller Bldg.
University Park, PA 16802

Delaware County Com. College
  Fire Academy
Mr. Walter Omlor
Media, PA 19063
(215) 353-5400 Ext. 427

Com. Col. of Philadelphia
Fire Science Department
Mr. Paul Ruhne
1600 Spring Garden St.
Philadelphia, PA 19130
(215) 972-7436

Northampton County Area
  Community College
Fire Science Department
Ms. Regina Tauke
3835 Green Pond Road
Bethlehem, PA 18017
(215) 865-5351

RHODE ISLAND

Providence College
Fire Science Department
Mr. Roger L. Pearson
River & Eaton Streets
Providence, RI 02918
(401) 865-1000

Rhode Island Jr. College
Fire Science Department
Mr. John Marmaras
400 East Avenue
Warwick, RI 02886
(401) 825-2145

SOUTH CAROLINA

Midlands Technical College
Fire Science Department
P.O. Drawer Q
Columbia, SC 29250

South Carolina Fire Academy
Mr. Paul W. Risher, Jr.
Illinois Avenue
W. Columbia, SC 29169
(803) 758-8420

SOUTH DAKOKA

Fire Service Training
Mr. Jim Slippin
222 West Pleasant Dr.
Pierre, SD 57501
(605) 773-3876

TENNESSEE

Chattanooga State Techical
  Community College
Fire Science Department
Mr. Leslie Owen
4501 Amnicola Highway
Chattanooga, TN 37406
(615) 622-6262

University of Tennessee
Fire Science Department
Mr. Mike Solecki
Charlotte Avenue
Nashville, TN 37219
(615) 251-1341

Roane State Community College
Fire Science Department
Mr. William C. Marshals
Harriman, TN 37748
(615) 354-3000

Walter State Community Col.
Fire Science Department
Mr. Ronald Lemke
Morristown, TN 37814
(615) 581-2121

Shelby State Com. Col.
Fire Science Department
Mr. Clem Weinrich
P.O. Box 22026
Memphis, TN 28122
(901) 382-0504

TEXAS

Del Mar College
Fire Science Department
Mr. E. E. Walters
Baldwin at Ayers
Corpus Christi, TX 48404
(512) 881-6425

El Paso Community College
Fire Science Department
Mr. Gerald B. Money
6601 Dyer Street
El Paso, TX 79904
(915) 778-7117

Galveston College
Fire Science Department
Mr. James Frazier
4015 Avenue "Q"
Galveston, TX 77550
(713) 763-2661

Midland College
Fire Science Department
Mr. Mil Goodwin
3600 N. Garfield
Midland, TX 79701
(915) 684-7851

South Plains College
Fire Science Department
Mr. B. P. Robinson
2404 Avenue Q
Lubbock, TX 79405
(806) 747-0576

Tyler Junior College
Fire Science Department
Mr. R. T. Minter
Henderson Highway
Tyler, TX 75701
(214) 593-4401

Odessa College
Fire Science Department
Mr. O. Nordmarken
P.O. Box 3752
Odessa, TX 79760
(915) 337-5381 Ext. 238

Temple Junior College
Fire Science Department
Mr. S. W. Churchill
2600 South First Street
Temple, TX 76501
(817) -73-9961, Ext. 51

Texarkana Community College
Social Science Division
Mr. Bob Bell
2500 N. Robinson Rd.
Texarkana, TX 75501
(214) 838-4541

San Antonio College
Fire Science Department
Mr. Mike Pickett
1300 San Pedro
San Antonio, TX 78284
(512) 734-7311 Ext. 209

Texas A & M Univ. System
Engineering Extension Svc.
Fire Protection Trng. Div.
Mr. David White
F. E. Drawer K
College Station, TX 77843

UTAH

Utah Technical College - Provo
Fire Science Department
Mr. G. D. Evans
1395 North 150 East
Provo, UT 84601

VERMONT

Southeastern Vermont
  Emergency School
Fire Science Department
Mr. Mark B. Rivers, Director
P.O. Box 44
Brattleboro, VT 05301

Vermont Fire Fighters Assoc.
Mr. Walter Read
East Dorset, VT 05253
(802) 362-1369

VIRGINIA

George Mason University
Dr. John M. Smith, Gen Studies
4400 University Drive
Fairfax, VA 22030
(703) 323-2405

Tidewater Community College
Fire Science Department
Mr. A. B. Corley
1700 College Crescent
Virginia Beach, VA 23456
(804) 427-3070

J. Sargent Reynolds Com. Col.
Fire Science Department
Mr. George Kitchen
P.O. Box 12084
Richmond, VA 23241
(804) 264-3301

Northern Virginia Com. Col.
Fire Science Department
Mr. Robert L. Smith
8333 Little River Turnpike
Annandale, VA 22003
(703) 323-3253

WASHINGTON

L. H. Bates Voc. Tech. Inst.
Fire Science Department
Mr. J. F. Wilbert
1101 So. Yakima
Tacoma, WA 98405
(206) 597-7257

Edmonds Community College
Fire Science Department
Mr. Gary Isham
Lynnwood, WA 98036

Columbia Basin College
Fire Science Department
2600 N. 20th Avenue
Pasco, WA 99301
(509) 547-0511

Whatcom Community Col.
Fire Science Department
Ms. Barbara Merriman
5217 Northwest Road
Bellingham, WA 98225
(206) 676-2170

Commission for Vocational
  Education
Fire Service Training
Mr. Edward Prendergast
Airdustrial Park, Bldg 17,LS-10
Olympia, WA 98504

Yakima Valley College
Fire Science Department
P.O. Box 1647
Yakima, WA 98907

WEST VIRGINIA

Community College of Marshall U.
Fire Science Technology Program
Mr. Larry Artrip
Huntington, WV 25701
(304) 696-3646

Shepherd College
Fire Science & Safety Tech.
Dr. Howard Carper
Shepherdstown, WV 25443
(304) 876-2511 Ext. 275

Fairmont State College -
  Community College
Fire Science Program
Mr. Jack Clayton
Fairmont, WV 26554
(304) 367-4000

West Virginia Northern Com.Col.
Fire Science Program
Mr. Richard Sambuco
Wheeling, WV 26003
(304) 233-4900

Parkersburg Community Col.
Fire Science Program
Mr. Pat Alford
Parkersburg, WV 26101
(304) 424-8290

West Virginia State College
Fire Protection Technology
Mr. Gwinn
Institute, WV 25112
(304) 766-3192

WISCONSIN

Fox Valley Technical Inst.
Fire Science Department
Mr. Charles Bavry
P.O. Box 2277
Appleton, WI 54911
(414) 739-0831

Moraine Park Tech. Inst.
Fire Service Training
Mr. Bob Bruce
235 N. National
Fond du Lac, WI 54935
(414) 922-8611 Ext. 413

Gateway Tech. Institute
Fire Science Department
Dr. Nevala
3520 30th Avenue
Kenosha, WI 53141
(414) 656-6900

Northeast Wisc. Tech. Inst.
Fire Service Department
Mr. William T. Schmidt
2740 West Mason Street
Green Bay, WI 54303
(414) 497-3003

Milwaukee Area Tech. Col.
Fire Science Department
Mr. Robert L. Wolf
1015 North 6th Street
Milwaukee, WI 53203
(414) 278-6428

Southwest Wisc. VO-Tech
  Institute
Fire Service Department
Mr. Don Covert
Bronson Blvd.
Fennimore, WI 53809

WYOMING

University of Wyoming
Fire Science Department
Dr. E. G. Meyer
Laramie, WY 82071

DEPARTMENT OF TRANSPORTATION
RESEARCH AND SPECIAL PROGRAMS ADMINISTRATION
MATERIALS TRANSPORTATION BUREAU
INFORMATION SERVICES DIVISION [DMT-11]
WASHINGTON, D.C. 20590

SEPTEMBER 1978

U.S. Department
of Transportation

**Research and
Special Programs
Administration**

List #3

CORPORATIONS AND OTHER BUSINESS ORGANIZATIONS WHICH OFFER COURSES, SEMINARS AND OTHER TRAINING MATERIAL DEALING WITH PACKAGING, MARKING, LABELING, SHIPPING AND TRANSPORT OF HAZARDOUS MATERIALS

Academy of Advanced Traffic
Mr. Anthony Matero
One World Trade Center, Suite 5457
New York, New York 10046
(212) 466-1980

Academy of Advanced Traffic
Mr. Lee Thomas
211 South Broad Street
Philadelphia, Pennsylvania 19107
(215) 545-0801

J. T. Baker Chemical Company
Ms. Carol Morris
222 Red School Lane
Phillipsburg, New Jersey 08865
(201) 859-2151

Center for Professional Advancement
Ms. Talia Caterina
P.O. Box H
East Brunswick, New Jersey 08816
(201) 249-1400 ext. 200

Chemical Manufacturers Association
Mr. John Zercher
2501 M Street, N.W.
Washington, D.C. 20037
(202) 887-1255

Conrail
Mr. Horace Bothum
Six Penn Center Plaza, Room 315
Philadelphia, Pennsylvania 19104
(215) 977-4559

ENSAFE
Environmental and Safety Design, Inc.
Mr. Wendell Knight
P.O. Box 34207
Memphis, Tennessee 38134
(901) 372-7962

Federal Express
Mr. George Truesdale, Dept. 1611
P.O. Box 727
Memphis, Tennessee 38194
(800) 238-5355 ext. 3917

Flying Tiger Airlines
Mr. Alan Hollander
Safety Department HO8
7401 World Way West
Los Angeles, California 90009
(213) 642-4082

J. J. Keller and Associates, Inc.
Mr. Mark Catlin
145 W. Wisconsin Avenue
Neenah, Wisconsin 54956
(414) 722-2848

National Fire Protection Association
Education Technology Unit
Batterymarch Park
Quincy, Massachusetts 02269
(617) 328-9290

Lion Technology, Inc.
Mr. William P. Taggart
466 Mount Hope Avenue
Dover, New Jersey 07801
(201) 366-3200

Americal Trucking Association, Inc.
Operations Council
Mr. Brent Grimes
1616 P Street, N.W.
Washington, D.C. 20036
(202) 797-5439

Radiation Service Organization, Inc.
Mr. Timothy Osborne
P.O. Box 419
Laurel, Maryland 20707
(301) 792-7444
(202) 953-2482 (Washington, D.C.)

Safety Systems, Inc.
Mr. Ronald G. Gore
P.O. Box 8463
Jacksonville, Florida 32239
(904) 725-3044

Seaboard Coast Line Industries, Inc.
Mr. Peter Gill, Manager
Hazardous Materials Training
500 Water Street
Jacksonville, Florida 32202
(904) 359-1587

Southern Pacific Transportation Company
Mr. Robert Andre
One Market Plaza
San Francisco, California 94105
(415) 541-1182

List #4

U.S. Department
of Transportation
**Research and
Special Programs
Administration**

FEDERAL, STATE AND LOCAL GOVERNMENT AGENCIES OFFERING COURSES AND SEMINARS
ON THE PACKAGING, HANDLING AND TRANSPORTATION OF HAZARDOUS MATERIALS

Ammunition School
Mr. Zigler
DARCOM Ammunition Center
ATTN: SARAC-ASA
Savannah, Illinois 61074
(815) 273-8515
Autovon 585-1110

Department of Transportation
Transportation Safety Institute
Mr. David Goodman
6500 S. MacArthur Blvd.
Oklahoma City, Oklahoma 73125
(405) 686-4824

Joint Military Packaging Training Center
Ms. Elsie M. Clark
ATTN: DRXPT-A
Aberdeen Proving Grounds, Maryland 21005
(301) 278-5185 or 2230
Autovon 283-5185 or 2230

Naval School of Physical Distribution
Oakland Army Base
Oakland, California 94625
(415) 466-3325
Autovon 836-5969

Sheppard Air Force Base
ATTN: STTC/TTGXT
Mr. C. W. Formby, Training Manager
Sheppard Air Force Base, Texas 76311
(817) 851-2075 or 6759
Autovon 736-2075 or 6759

NOTE: Generally, the schools listed above accept Federal and State employees and persons
working for organizations that have contracts to furnish goods or services to the Government.
Some of the schools offer both in resident and on site training at specific locations.

The California Specialized Training Institute *
Brig. General Neil G. Allgood, U.S.A.
Duilding 904, Camp San Luis Obispo
San Luis Obispo, California 93406
(805) 544-7100

Colorado Training Institute *
Mr. Walter DeFreece
1001 East 62nd Avenue
Denver, Colorado 80216
(303) 389-4891

Maryland Department of Transportation
State Aviation Administration
Ms. Elizabeth Matarese
3rd Floor, Terminal Building
Baltimore-Washington International Airport
Baltimore, Maryland 21240
(301) 859-7062 or 7111

Port Authority of New York and New Jersey
Ms. Eunice C. Coleman, Program Manager
The World Trade Institute
One World Trade Center, 55th Floor
New York, New York 10048
(212) 466-3170

State of North Carolina *
Department of Insurance
Mr. Phil Riley
Fire and Rescue Services Division
P.O. Box 26387
Raleigh, North Carolina 27611
(919) 733-2142

State of Washington
State Criminal Justice Training Commission
Mr. Phillip Shave
Olympia, Washington 98504
(206) 459-6342

* Also offer courses in Hazardous Materials Emergency Response

NOTE: Check with the agencies above for details concerning scheduling of courses and other
details.

REVISED APRIL 1982

Training Services, Inc.
Mr. Leonard J. Smith
130 Orient Way
Rutherford, New Jersey 07070
(201) 933-5880

Transportation Skills Programs
Mr. Robert J. Keegan
320 W. Main Street - P.O. Box 286
Kutztown, Pennsylvania 91530
(215) 683-5098

UNZ and Company
Mr. Brian Scerry
190 Baldwin Avenue
Jersey City, New Jersey 07306
(800) 631-3098
(201) 795-5400 (New Jersey)

THE ORGANIZATIONS LISTED ABOVE OFFER BOTH COURSES AND SEMINARS. CONTACT THOSE ORGANIZATIONS FOR SCHEDULING AND OTHER DETAILS.

E. I. Dupont de Nemours and Co., Inc.
Dr. Arthur C. Santora
Applied Technology Division
Barley Mill Plaza - Marshall Mill Building
Wilmington, Delaware 19898
(302) 772-5998

Video Systems Network, Inc.
Mr. Robin Mitchell
12530 Beatrice Street
Los Angeles, California 90066
(800) 421-6521
(213) 870-1231

National Agricultural Chemicals Assoc.
Director of Regulatory Affairs
1155 15th Street, N.W.
Washington, D.C. 20005
(202) 296-1585

NOTE: THE ORGANIZATIONS LISTED ABOVE OFFER TRAINING MATERIALS ONLY.

REVISED APRIL 1982

8 RESEARCH SOURCES AND RESOURCES

The volume and variety of hazardous materials being manufactured, processed, transported, stored, used and disposed of has mushroomed so rapidly in recent years that it has been difficult for persons involved in this changing field to keep fully informed. This chapter identifies sources of information, resource materials, training aids, technical advice and assistance, and data related to hazardous materials response and control, regulation, and planning. Such informational sources are divided among the following categories:

- Agencies, Organizations, Associations, & Institutes.
- Computerized Data Systems.
- Conferences.
- Manuals. (Handbooks)
- Periodical Publications.

AGENCIES, ORGANIZATIONS, ASSOCIATIONS, & INSTITUTIONS

American Petroleum Institute

The American Petroleum Institute was established in 1919 as the first national trade association in the United States to encompass all branches of the petroleum industry. The Institute provides technical assistance and publications to the petroleum industry and is a source of resource materials and training aids for emergency first responders concerned with safe operation of inland bulk plants, firefighting in and around petroleum storage tanks, underground leaks, emergency planning and mutual aid for products terminals and bulk plants, storage tank construction and operation, and similar concerns. Prices of Institute technical publications are kept low to make them available to a maximum range of interested individuals and organizations. A *Publications And Materials* catalogue that is published annually provides descriptions and prices for an extensive array of monographs, technical reports, industry standards, books, films, and brochures.
Contact: American Petroleum Institute, Publications & Distribution Section, 2101 L Street, N.W., Washington, D.C. 20037, Tel. 202-457-7160.

American Trucking Associations

The American Trucking Associations, one of the nation's largest industrial associations, is a confederation of 51 affiliated state trucking organizations and 13 national conferences for special types of motor truck operations (such as the National Tank Truck Carriers, Inc.). A.T.A. currently employs a staff of 250 to provide services to member associations. Hazardous materials related services include training, training aids, publications, and related materials.

Contact: American Trucking Associations, Inc., 1616 P Street, N.W., Washington, D.C. 20036 Tel. 202-797-5336 (Safety Department).

Chemical Manufacturers Association

The Chemical Manufacturers Association, (CMA) founded in 1872 as the Manufacturing Chemists Association, is a trade association of chemical manufacturers representing more than 90 percent of the production capacity for basic industrial chemicals in the U.S. and Canada. Member companies must be manufacturers of chemicals who sell to others a substantial portion of the chemicals which they produce. "Chemicals" is interpreted to exclude products of mixing, formulating, or compounding operations which do not involve a change in chemical structure.

CMA has long been active in programs to improve the safety of shipping containers, both package and bulk units, in order to minimize failures and leakage of contents. A broad range of activities are handled by an Association staff of 85, and by committees formed by industry specialists drawn from member companies.

The Chemical Manufacturers Association established and operates the Chemical Transportation Emergency Center (CHEMTREC) in Washington, D.C.

Recognizing And Identifying Hazardous Materials, a tape/slide program developed by the Interindustry Task Force on the Rail Transportation of Hazardous Materials, is available from CMA for free showing throughout the country. The well-done program is aimed at volunteer fire departments to assist in planning for emergencies involving hazardous materials in communities.

Contact: Chemical Manufacturers Association, 2501 M Street, N.W., Washington, D.C. 20037, Tel. 202-887-1255.

The Chlorine Institute

The Chlorine Institute, formed in 1924, is an association of more than 130 firms engaged in chlorine production or the manufacture of chlorine-related equipment in the United States, Canada, and overseas. Institute member firms account for 99 percent of the chlorine manufactured on the North American continent. The Institute provides safety recommendations for chlorine manufacturing, shipment, handling, and storage; and works with government agencies responsible for establishing transportation regulations and specifications for chlorine shipping containers. It publishes and distributes a wide variety of manuals, pamphlets, drawings and other technical materials to a worldwide audience. More than a half-million copies of its *Chlorine Manual,* a guide to safe chlorine handling, are currently in circulation.

The Chlorine Institute developed the Chlorine Emergency Plan (CHLOREP) under which 62 trained teams are available 24 hours a day to respond to any chlorine emergency in the U.S. or Canada, and the Institute's CHLOREP Committee holds regular training seminars.

Chlorine Institute emergency kits and component parts are manufactured to Institute specifications and distributed by Indian Springs Manufacturing Company. (P.O. Box 112, Baldwinsville, New York 13207. Tel. 315-635-6101) The kits (Kit A for chlorine cylinders; Kit B for chlorine ton containers; Kit C for chlorine tank cars/tank trucks.) operate on the principle of capping off leaking valves and/or fusible plugs and, in the case of cylinders and ton containers, of sealing off a leak in the container wall itself. Chlorine emergency kits are widely used by response personnel; currently, there are nearly 7,500 such kits throughout the U.S. and Canada.

Extensive technical resource materials and training aids available from the Institute are described in its *General Publications List.*
Contact: The Chlorine Institute, 342 Madison Avenue, New York, New York 10173, Tel. 212-682-4324.

Disaster Research Center of Ohio State University

The Disaster Research Center engages in a variety of sociological research studies on the relations of groups and organizations in community-wide emergencies, particularly natural and technological disasters. The major research focus is on emergency organizations and their planning and response to large-scale community crises. Current emphasis is on delivery of mental health services and emergency medical services in large scale casualty-producing situations as well as socio-behavioral responses to acute chemical hazards, and the problems involved in mass evacuation. Center personnel have studied legal aspects of governmental response in disasters, the diffusion of knowledge about emergency preparations, the operation of rumor control centers, mass media reporting of community crises, methodological problems in observational studies of stress situations, and the handling of the dead in catastrophes. Besides collecting its own data, the Center serves as a repository for documents and materials collected in research by other agencies and researchers. The Center's research library is open to all interested persons involved in emergency planning. The Center has its own book, monograph, and report series, and publishes a quarterly newsletter, *Unscheduled Events.*

The Disaster Research Center has recently completed a four-year project, funded by a National Science Foundation grant, on *Sociobehavioral Responses To Chemical Hazards: Preparations For and Response To Acute Chemical Emergencies at the Local Community Level.* (A copy of the 133-page Final Report is available for $7.50 within the United States/$10.00 outside the U.S.) The focus of the research was on organizational and community preparations for, responses to, and recovery from relatively sudden chemical disasters. Field studies were made of police and fire departments, civil defense offices, hospitals, relief agencies, mass media units, organizations involved in the producing/transporting/storage/use of dangerous chemical substances, and other groups that would be involved in dealing with sudden mass emergencies. Three separate phases of the study dealt with planning and preparation for disasters involving chemical agents, actual incidents involving chemical hazards, and longer run consequences of any sudden chemical disaster. Publications list available.
Contact: Disaster Research Center, The Ohio State University, 128 Derby Hall, Columbus, Ohio 43210, Tel. 614-422-5916.

The Fertilizer Institute

The Fertilizer Institute is the national association for the fertilizer industry and is active in governmental as well as industry affairs. The Institute is affiliated with 47 state, regional, and international associations for the purpose of mutual information exchange and legislative/regulatory liaison. A prime concern is the safe production, handling, transportation, and use of chemicals and materials common to fertilizer plant operations; such as, anhydrous ammonia, aqua ammonia, ammoniating solutions, liquid and solid sulfur, sulfuric acid, phosphoric acid, and fertilizer grade ammonium nitrate.

The Fertilizer Institute supplies at low cost a broad offering of fertilizer-related materials: books, brochures, and folders; motion pictures and color slides; decals, posters, and signs. Such materials range in content from basic information on fertilizers and their use to technical and safety data. Selected publications of interest to emergency response personnel include . . .
- Fertilizer Safety Guide.
- Safety Requirements for the Storage and Handling of Anhydrous Ammonia.
- Agricultural Ammonia Safety.

- Fertilizer Safety Talks.
- Shipping Papers, Markings and Placarding for Hazardous Materials Transportation.
- Operational Safety Manual for Anhydrous Ammonia.
- Agricultural Ammonia Handbook.

Publications list available.
Contact: The Fertilizer Institute, 1015 18th Street, N.W., Washington, D.C., 20036, Tel. 202-861-4900.

Hazardous Materials Advisory Council

The Hazardous Materials Advisory Council (HMAC) is a Washington-based, privately-funded, nonprofit organization that represents and serves those interested and involved in transporting hazardous materials, hazardous wastes, and hazardous substances. Its objectives are safe transportation, reasonable regulations, effective compliance, and consistent enforcement. HMAC members are manufacturers and shippers of chemicals, plastics, petro-chemical products and other hazardous materials; member organizations include railroads, motor and water carriers and airlines as well as those who supply containers and packaging supplies for shipping and transporting hazardous commodities of every type and hazard class. Services to members include HMAC Advisory Bulletins; meetings, workshops, and seminars to develop informed viewpoints and take appropriate action to respond to regulations that impact on the business of members; provision of a forum for the exchange of ideas among shippers, carriers, freight forwarders, insurers, and container manufacturers; provision of information, assistance, and program materials to develop safety and compliance training programs within member companies; and presentation of members' viewpoints to Congressional committees and Federal agencies.
Contact: Hazardous Materials Advisory Council, 1100 17th Street, N.W.; Suite 908, Washington, D.C. 20036, Tel. 202-223-1271.

Institute of Makers of Explosives

The Institute of Makers of Explosives is the safety association of the commercial explosives industry. Publications list available.
Contact: Institute of Makers of Explosives, 1575 Eye Street, N.W. Suite 550, Washington, D.C. 20005, Tel. 202-789-0310.

International Association of Fire Chiefs

The International Association of Fire Chiefs through efforts of its Committee On Hazardous Materials, training sessions, publications, recommended practices and procedures, and similar efforts keeps members informed relative to hazardous materials concerns, and represents the interests of members with government and industry.
Contact: International Association of Fire Chiefs, 1329 18th Street, N.W., Washington, D.C. 20036, Tel. 202-833-3420.

National Agricultural Chemicals Association

The National Agricultural Chemicals Association through its member companies operates a national pesticide information and response network, The Pesticide Safety Team Network, (PSTN) that is activated through CHEMTREC.
Contact: National Agricultural Chemicals Association, The Madison Building, 1155 Fifteenth Street, N.W., Washington, D.C. 20005, Tel. 202-296-1585.

National Fire Protection Association

The National Fire Protection Association is a nonprofit organization which acts as a clearinghouse for fire protection and prevention information. Its informational services include publications and films, technical assistance programs, a public education program, and the development and dissemination of codes and standards.

The NFPA library collection consists of 3600 bound volumes, 15,000 pamphlets and documents, and approximately 600 periodicals titles in subject areas such as arson investigation, fire prevention and protection, fire service management, flammability of materials, model building codes, and voluntary industry standards related to fire safety. The library is open to association staff and members, and to the general public by appointment.

NFPA is the major publisher of print and audiovisual materials on fire protection and prevention, including a wide variety of materials specifically related to hazardous materials concerns. The Association regularly publishes four separate catalogs:
* Catalog of Fire Safety Films & Audiovisuals.
* Fire Service Catalog—Training & Educational Materials.
* Business & Industry Catalog—Fire Safety Educational And Training Materials.
* NFPA Codes And Standards Catalog.

Contact: Publications Sales Division, National Fire Protection Association, Batterymarch Park, Quincy, Massachusetts 02269, Tel. 617-328-9230, Telex 94 0720.

National Transportation Safety Board

The National Transportation Safety Board (NTSB) is an independent Federal agency that serves as the overseer of U.S. transportation safety. The Safety Board maintains a public docket at its Washington, D.C. headquarters which contains records of all accident investigations, all safety recommendations, and all safety enforcement proceedings. These records are all available to the public and may be copied, reviewed, or duplicated for public use. The Safety Board also undertakes special studies of safety problems in air and surface transportation. Original factual records of all accidents investigated by NTSB are on file in the Washington, D.C. office. Upon request and payment of small fees, any portion of these records of investigations will be reproduced and mailed to the requester.

The purpose of the NTSB is to identify the cause and recommend corrective action relative to transportation accidents investigated. The Board is headed by five presidential appointees, two of whom must have technical backgrounds, and employs 360 people at headquarters and in 12 small field offices. The National Transportation Safety Board determines the cause of all civil aircraft accidents that occur in the United States; and investigates all air carrier accidents, most fatal light plane accidents, and selected additional aviation accidents. In surface transportation, the Board is required to investigate and determine the cause of all fatal railroad and pipeline accidents, and is authorized to investigate and determine the cause of highway accidents selected in cooperation with the states. In addition, the Board is required to determine the cause of all major marine accidents and investigate those major accidents in which U.S. Coast Guard operations and functions are so involved as to require independent investigation.

A major accident involving hazardous materials in any mode of transportation (Aviation, Intermodal, Highway, Marine, Pipeline, Railroad) is likely to be investigated by an NTSB team and result in a detailed report identifying the accident cause and recommending corrective actions. All actions and decisions of the Safety Board in the form of accident reports, special studies, statistical reviews, safety recommendations, and press releases are made public. Single copies of most such publications can be obtained at no charge as long as copies are available. Once the supply of a particular accident report on hand at NTSB has been exhausted, the report can often be purchased from either the National Technical Information Service (NTIS) or the U.S. Government Printing Office.

Contact: National Transportation Safety Board, 800 Independence Avenue, S.W., Washington, D.C. 20594.

Natural Disaster Recovery And Mitigation Resource Referral Service

The Academy for Contemporary Problems funded by the Federal Emergency Management Agency and the National Science Foundation operates the Natural Disaster Recovery And Mitigation Resource Referral Service to distribute information to state and local officials; improve their ability to recover quickly from natural disasters; and mitigate the effect of natural hazards on people and structures. The service is part of a project to improve the dissemination of research results on natural disaster recovery and mitigation to state and local officials. (The focus here is on natural disasters—dams, drought, earthquakes, hurricanes, floods, tornadoes, volcanoes, etc.—rather than on hazardous materials emergencies, but much of the information available has applicability to hazardous materials planning.)

The Service operates a resource center containing information on many issues related to natural disasters.

- Building Codes
- Structural Standards
- Disaster Assistance
- Evacuation
- Land Use
- Mitigation
- Fed. Emergency Mgt. Agency
- Utilities & Sewer Systems
- Litigation & Liability
- Planning
- Presidential Disaster Declaration
- Public Information & Awareness
- Reconstruction
- Recovery Activities
- Relocation
- Training Courses & Exercises
- Water Resource Management

Also available are materials on the following specific hazards:

- Agriculture (Risks to)
- Coastal Zones
- Cyclones
- Droughts
- Earthquakes
- Fires & Explosions
- Floods
- Hurricanes
- Landslides
- Snow
- Subsidence
- Thunderstorms
- Tornadoes
- Tidal Waves
- Volcanoes
- Weather

The Center attempts to link disaster specialists, researchers, public interest groups and other interested parties; and project staff will attempt to answer questions regarding natural disaster recovery and mitigation. The Center maintains an active exchange library of documents, articles, and materials.

Contact: Resource Referral Service, Academy for Contemporary Problems, 400 North Capital Street, N.W.; Suite 390, Washington, D.C. 20001, Tel. 202-638-1445.

Natural Hazards Research & Applications Information Center

The Center deals only with natural events, not man-made hazards such as hazardous materials. The Center is a national clearinghouse for research data dealing with economic loss, human suffering, and social disruption caused by earthquakes, floods, hurricanes, and other natural disasters. The Center publishes a free, quarterly newsletter; *Natural Hazards Observer,* that provides well done reports on new research and findings from completed projects; pertinent legislation; applications of research at Federal, state, and local levels and by private agencies; and announcements of recent publications and future conferences.

The *Natural Hazards Observer* does periodically have some coverage of interest to hazardous materials professionals.

Contact: Natural Hazards Research & Applications Information Center, Institute of Behavioral Science, #6, University of Colorado, Boulder, Colorado 80309, Tel. 303-492-6818.

Spill Control Association of America

The Spill Control Association of America is an international, nonprofit trade association representing the interests of the oil and hazardous materials spill control industry. Membership includes spill cleanup contractors, manufacturers/distributors of specialized spill control and cleanup equipment, individuals in government/education/industry, and companies that share a concern for proper management of spill problems. The aims and objectives of the Spill Control Association have been to (1) provide information as to the spill control industry's practices, trends, and achievements; (2) establish liaison with local, state, and Federal government agencies responsible for laws and regulations regarding spills of oil and hazardous materials; and (3) cooperate in the development of industry programs and efforts so that spills are efficiently handled. Also, the Association periodically sponsors seminars on subjects of interest to members.

Contact: Spill Control Association of America, 1515 North Park Plaza, 17117 W. Nine Mile Road, Southfield, Michigan 48075, Tel. 313-552-0500.

The Transportation Systems Center

The Transportation Systems Center is the multimodal research, analysis, and development center of the U.S. Department of Transportation. The Center has a number of projects ongoing at any given time pertaining to the evaluation and development of solutions to specific urban, rural, intercity, and international transportation problems. The Center is also a major technology sharing resource of D.O.T. for the dissemination of transportation information and statistical data to state and local governments, private industry, and academia. TSC's Center for Hazardous Materials Transportation Research provides coordination and a focal point for all D.O.T. efforts in this area. The Transportation Systems Center complex consists of six buildings, 950 people, over 30 laboratories, and a complete reference library.

Contact: Transportation Systems Center, U.S.-D.O.T., Research & Special Programs Administration, Kendall Square, Cambridge, Massachusetts 02141, Tel. 617-494-2000.

U.S. Government Sources

The two primary reference sources for keeping track of U.S. hazardous materials regulations are the *Federal Register* and the *Code of Federal Regulations,* both of which are normally available in larger libraries around the country such as university and state libraries or federal depository libraries. The *Federal Register,* published daily, makes available to the public Federal agency regulations, *including proposed changes in regulated areas,* and other legal documents covering a wide range of government activity. The *Code of Federal Regulations* is the annual compilation and cumulation of executive agency regulations published in the daily *Federal Register* combined with previously issued regualations that are still in effect. An alphabetical listing by agency of subtitle and chapter assignments in the *Code* is provided at the rear of each volume.

Many reports, monographs, research studies, etc. produced with public funds may be purchased through either the National Technical Information Service, or the U.S. Government Printing Office.

The U.S. Department of Commerce's National Technical Information Service also provides moderate cost searches of government published research in hundreds of different fields. Each published search (for example, Natural Gas Marine Transportation; Railroad Freight Transportation; Waste Processing & Pollution In The Chemical & Petrochemical Industries; Emergency Medical Services, etc.) for a specific topic consists of as many as 200 different research summaries (abstracts). NTIS is the only central source of research reports and other technical information from the vast Federal network of departments, bureaus, and agencies. Write to NTIS for the latest available *Current Published Searches From The NTIS Bibliographic Data Base* listing more than 1500 published searches. Such searches are bibliographies containing full bibliographic citations developed by information specialists in the subject areas most often requested.

Contact: National Technical Information Service, 5285 Port Royal Road, P.O. Box 1553, Springfield, Virginia 22161, Tel. 703-487-4700; or

Contact: Superintendent of Documents, U.S. Government Printing Office, Washington, D.C. 20402.

U.S. Department of Transportation

The Department of Transportation is the Federal agency mainly responsible for development and enforcement of regulations pertaining to the transportation of hazardous materials and for the investigation of accidents in which hazardous materials were involved. Within the overall department . . .

• The Materials Transportation Bureau formulates and issues hazardous materials regulations and exemptions.

• The modal agencies (Federal Aviation Administration, Federal Highway Administration, and the Federal Railroad Administration) conduct research and development, assist in the formulation of regulations, and handle enforcement of hazardous materials transportation regulations.

• The Coast Guard has established the National Oil & Hazardous Substances Pollution Contingency Plan; the Coast Guard operated National Response Center in Washington, D.C.; the Chemical Hazards Response Information System; (CHRIS) the Hazarrd Assessment Computer System; (HACS) and a hazardous chemicals training course. Coast Guard personnel man the National Response Center 24-hours a day (Tel. 800-424-8802. Emergency Use Only) to receive notification of pollution discharges and ensure that current information is passed along to proper operational and administrative commands for response action, and staff a National Strike Force consisting of persons trained to provide communications support/ advice/assistance for oil and hazardous substance removal.

Telephone numbers for various information sources within D.O.T. are as follows.

• Office of Public Affairs, Office of the Secretary of Transportation. Tel. 202-426-2144.

• Office of Public Affairs, Federal Highway Administration, Tel. 202-426-0660

• Bureau of Motor Carrier Safety. Tel. 202-426-1808

• Office of Public Affairs, Federal Railroad Administration. Tel. 202-426-0881

• National Highway Traffic Safety Administration. Tel. 202-426-9550

• "Railroad Safety." Office of Safety Programs, Federal Railroad Administration. Tel. 202-426-0898

• "Pipeline Safety." Office of Pipeline Safety Operations, Research and Special Programs Division. Tel. 202-426-2392

• "Transportation of Hazardous Materials." Office of Hazardous Materials Operations, Research & Special Programs Division. Tel. 202-426-0656

• U.S. Coast Guard. Tel. 202-426-9568

Also, information about D.O.T. regulations on truck and truck driver safety and the highway transport of hazardous materials can be obtained by calling 800-424-9158. (426-1724 in the Washington, D.C. metropolitan area) From the service, staffed by Bureau of Motor

Carrier Safety personnel and accessible from the contiguous 48 states, operators of trucking companies and the general public are able to request motor carrier safety publications and receive answers to questions regarding Bureau of Motor Carrier Safety functions and responsibilities.

Federal Emergency Management Agency

The Federal Emergency Management Agency is primarily responsible for coordinating the Federal response to all types of disasters. Within FEMA, the U.S. Fire Administration operates the National Fire Academy in Emmitsburg, Maryland where a number of hazardous materials training programs are originated and offered.
Contact: Federal Emergency Management Agency, 2400 M Street, N.W., Washington, D.C. 20472, Tel. 202-634-7654.

U.S. Environmental Protection Agency

The Environmental Protection Agency has both regulatory and response functions and responsibilities.

For information about specific E.P.A. regulations, use the telephone numbers provided below.

Air: Tel. 919-541-5343	Effluent: Tel. 202-426-2522
Drinking Water: Tel. 202-755-5643	Pesticides: Tel. 202-755-4854
Toxic Substances: Tel. 202-755-0535	Noise: Tel. 202-557-7743
Solid Waste: Tel. 202-755-9157	Radiation: Tel. 703-557-9710

The Office of Emergency and Remedial Response within the Environmental Protection Agency was initiated to provide a focal point for E.P.A.'s Emergency Response Program required by "Superfund." (Comprehensive Environmental Response, Compensation, and Liability Act of 1980) Under Superfund, emergency and longterm cleanups of hazardous substances and inactive waste sites are financed by a trust fund collected through taxes paid by manufacturers, producers, and exporters and importers of oil and 42 chemical substances. Within E.P.A.'s Office of Emergency and Remedial Response, there are separate divisions for Emergency Response, Hazardous Response Support, and Hazardous Site control.
Contact: Office of Emergency and Remedial Response, U.S. Environmental Protection Agency, Washington, D.C. 20460.

COMPUTERIZED DATA SYSTEMS

Hazard Assessment Computer System

The Hazard Assessment Computer System (HACS) is the computerized counterpart of two of the four manuals comprising the U.S. Coast Guard's Chemical Hazards Response Information System: (CHRIS) Manual 2, *Hazardous Chemical Data;* and Manual 3, *Hazard Assessment Handbook.* Computer terminal displays can illustrate the relationships among spill concentration, thermal radiation, location, and time. Also, HACS can be utilized for emergency discharge advance planning, and the development and testing of approved hazard assessment methods.

HACS was designed and implemented to rapidly and quantitatively answer the following types of questions:

• When will the air/water concentration of a discharged material reach a specific level of toxicity at a given location?

• When will the air/water concentration return to a specified safe or nontoxic level?

• What is the concentration of discharged material at a specified location and time?

HACS currently contains the necessary chemical and physical data to allow hazard assessments for 900 commonly shipped chemicals.

Department of Defense "Hazardous Materials Information System"

A hazardous materials information system (HMIS) developed by the Department of Defense is supplying transportation and safety data to the department's employees and other users who handle, ship and store hazardous materials. Established 40 months ago, the program is operated from the Defense General Supply Center in Richmond, Virginia. Information received on Material Safety Data Sheets (MSDS) required of suppliers and contractors is stored in computers and distributed quarterly to as many as 6,000 DOD worksites. Presently there are more than 14,000 items listed in the system, and additional items are added daily.

The typical entry presents the common name of the item, the name of the manufacturer, an emergency telephone number, the chemical name and formula, the NIOSH number, radioactivity data if applicable, health and physical property data, safety/storage/handling and firefighting procedures, and spill-and-leak procedures. Although the system was developed and is used by the Department of Defense to minimize the health and environmental risks associated with worker-contact with these materials, anyone may purchase the database on approximately 100 microfiche sheets (reduction size 48x only) from the Superintendent of Documents—U.S. Government Printing Office, Washington, D.C. 20402. (Each fiche equals 270 pages of printed matter; there are approximately 100 fiche in the set. A set, consisting of basic system data and three updates a year, is available for $70 to domestic addresses and $87.50 to foreign addresses. When ordering, use subscription I.D. Number "DHMIS" and Stock Number 008-007-90201-6.)

The major goals of the HMIS are:

• To centrally store, for hazardous materials used by DOD, the material contained on the Material Safety Data Sheets, (MSDS) and to distribute this information to personnel responsible for insuring proper and safe handling, storage, and use of hazardous materials.

• To centrally store and distribute sufficient data to insure shipment in compliance with the Department of Transportation, the Air Force, the Inter-Governmental Maritime Consultive Organization, and the International Air Transport Association.

• To centrally provide data that will assist in the safe disposal of these same hazardous materials. (Information on hazardous materials disposal has *not* yet been included in the system).

The database maintained at the Defense General Supply Center in Richmond, Virginia is being built-up overtime by input received from all DOD focal points. The procuring activity obtains a MSDS from the supplier/contractor as part of the procurement process. The focal point reviews the data for technical accuracy and conformance with automated data processing program constraints. Also the focal point will prepare a Transportation Data Sheet for each hazardous item regulated for shipping. During the data entry process at DGSC in Richmond, data is screened for legibility and edited for conformance to field size and character configuration requirements. In addition to routine entry of new records into the system; additions, corrections, and deletions can also be accomplished. The primary output is a composite microfiche publication of all the data in the system that is produced annually with quarterly cumulative updates. Weekly hardcopy updates are provided to DOD focal points to assist with inquiries between publication cycles.

The HMIS can be interrogated by hazardous ingredients using the NIOSH code, by the storage compatibility code, (a one or two-position alpha or alpha-numeric code assigned to items in the HMIS system to insure that incompatible chemicals are not stored next to each other) by a specification number, by the hazard class for each mode of transportation, by the FSCM number, (the "Federal Supply Code for Manufacturers" is a five-digit code assigned to any contractor who does business with the Government. The code represents the specific company name and plant address for each code that is assigned) or by a specific stock number. All interrogations provide either a list of stock numbers meeting the interrogation condition or a listing of full file data.

Tom Reese who supervises the HMIS operation at Richmond, noted in a recent interview that the center is careful to honor requests for protection of proprietary data. "Basically, there are two versions of the microfiche," he says. "An 'LR' version contains all the data for a given entry whereas the 'L' version deletes the ingredients and formula for those items designated as proprietary by the manufacturers. The balance of the data for the proprietary items is displayed. The 'L' version was developed primarily to enable us to comply with Freedom of Information Act requests in which proprietary data are not released. However, I would estimate that only five percent of all entries are treated as proprietary data."

"It is important to emphasize that we collect this data for dissemination to the various users in the field so individual Safety Officers can develop their own local safety operating procedures," adds Russ Van Allen who works with Reese. "Previously, local Safety Officers, Industrial Hygienists, and Environmtal Officers may have had only a stock number on a drum—they didn't necessarily know what the drum contained. Now we associate the stock number with a body of information so the local Safety Officer can establish procedures for the safety of his employees."

"Specific transportation data we have in the system consists of the proper shipping name, hazard class, and label requirements for the five major modes of transportation used by DOD: surface, domestic surface and air, international water and air, and military air," continues Tom Reese. "With this information, a transportation specialist finds the entry in the commodity table of the appropriate shipping regulation. This table (Table 172.101 of the D.O.T. regs. printed in Code of Federal Regulations—49) is used to certify the shipment; we are not interested in having people certify off the data system—we want them to certify from the regulations. We take them to the correct starting point within the regulations because from a legal standpoint the regulations are the governing criteria."

Contact: Hazardous Materials Information Systems Defense General Supply Center, U.S. Department of Defense, Richmond, Virginia 23297. Tel. 804-275-3104.

Oil and Hazardous Materials Technical Assistance Data System (OHMTADS)

The OHMTADS computerized information retrieval file is available to spill response personnel by telephone hookup to a computer terminal. OHMTADS stores detailed information on hundreds of chemical compounds. The information, numerical data as well as interpretive comments, has been assembled into the computer from literature. It emphasizes the effects the materials can have when spilled, but much more information can be provided including trade names, synonyms, chemical formulas, major producers, common modes of transportation, flammability, explosiveness, potential for air pollution, methods of analysis, and chemical/physical/biological/toxicological properties. In less than 15 minutes, OHMTADS can relay procedures for safe handling and cleanup of spilled materials. OHMTADS also allows identification of unknown materials. After key characteristics of the unknown material are furnished to the system, OHMTADS screens for candidate substances with similar physical and chemical properties. For example, if the computer is given the color, odor, density, etc. of an unknown material, it will generate a list of candidates. Continued elimination of substances on this list will lead to identification of the material.

OHMTADS, a joint project of the U.S. Environmental Protection Agency, the National Institute of Health, and Information Sciences Corporation; is a component of the broad NIH/EPA Chemical Information System (CIS). The OHMTADS component of CIS consists of a program that reads a wide variety of physical, chemical, biological, toxicological, and commercial data on 1029 substances. There are potentially 126 fields of data for each of the 1029 substances. (See Illustration 1 for a listing of the 126 fields of data potentially available.)

Since hardware devices needed to use OHMTADS are limited to a telephone, a modem, and a computer terminal; any response vehicle equipped with a mobile telephone could achieve direct access to the system whenever needed. A 13-pound portable computer terminal that operates off a $200. battery pack is smaller than an electric typewriter. Placing the telephone receiver in a receptacle on the portable terminal allows direct, two-way communication with the OHMTADS computer. For those on a more limited budget, the OHMTADS database can be purchased on microfiche for under $25. Portable, battery-operated microfiche viewers can be purchased for between $200. and $300. Access to the OHMTADS system is by payment of a modest annual subscription fee (under $350.) plus a payment of $45. per connect hour. *Contact:* OHMTADS, Information Sciences Corporation, 2135 Wisconsin Avenue, N.W. Suite 401, Washington, D.C. 20007, Tel. 202-298-6200, Tel. 800-424-2722 (Toll-free).

"HAZMAT" And "Pipes"

A commercial vendor has been given the responsibility of maintaining, updating, and improving the Hazardous Materials Incident Reporting System (HAZMAT) and the Pipelines Safety Reporting System (PIPES).

The HAZMAT database currently contains over 100,000 reported incidents of hazardous materials spills occurring in transportation modes other than pipeline. Data put into the HAZMAT system is taken from "Hazardous Materials Incident Report" forms (Form D.O.T. F 5800.1) filed with the U.S. Department of Transportation, (See Figure 2) and from telephone reports of incidents received by the National Response Center in Washington, D.C. The database includes information on such things as cause-for-leak, number of deaths or injuries, amount of property damage, locations of incidents, origin of shipments, and other related information.

PIPES is a database resulting from processing of four different data sources:
• Transmission and distribution system leak reports.
• Transmission and distribution system annual reports.
• Leaks involving the piped transmission of fluids.
• Reports telephoned into the National Response Center.

For additional information about either HAZMAT or PIPES, *Contact:* National Data Corporation, 2135 Wisconsin Avenue, N.W. Suite 300, Washington, D.C. 20007, Tel. 202-298-6200.

Conferences

One of the quickest methods of being brought up-to-date on new techniques, programs, methods, tools, equipment, and strategies in the field of hazardous materials response and control is to attend any one of a number of well established conferences held throughout the country each year. Most such conferences publish all papers presented at the conference in *Proceedings,* the cost of which is normally included in the registration fee for the conference. The six conferences listed below are probably the best known.

1984 Hazardous Materials Spills Conference

April 9–12, 1984; Nashville, Tennessee.
A biennial national conference, sponsored every other year by the Bureau of Explosives, Chemical Manufacturers Association, U.S. Coast Guard, U.S.—E.P.A., and a dozen other major organizations.
Contact: Hazardous Materials Spills Conference, Suite 700, 1629 K Street, N.W., Washington, D.C. 20006, Tel. 202-296-8246.

#	CODE	DEFINITION
1	ACC	OHM-TADS Accession Number: A unique, computer assigned identifier for the data file.
2	CAS	Chemical Abstracts Service Registry Number: .) unique, international identifier for material of interest.
3	SIC	Standard Industrial Code: Industry-employed codes which can be used to identify manufacturers of material.
4	MAT	Material Name: Generally, the common name for the materials.
5	SYN	Synonyms: Alternative identifiers of the material for which the data is valid.
6	TRN	Company Trade Names: Lists commercial trade names and the associated manufacturer whenever possible.
7	FML	Chemical Formula: Gives most common formula or describes nature of materials included in the general heading such as components of an industrial blend or mixture.
8	SPC	Species in Mixture: Identifies typical product purity in cases of single constituent materials, or specific major components of heterogeneous mixtures.
9	USS	Common Uses: Enumerates common uses of materials.
10	RAL	Rail(%): Percentage shipped by rail (estimate).
11	BRC	Barge (%): Percentage shipped by barge (estimate).
12	TRK	Truck (%): Percentage shipped by truck (estimate).
13	PIP	Pipeline (%): Percentage shipped by pipeline (estimate).
14	CON	Containers: Lists type of shipping containers normally used or required by law. Typical shipment size when available.
15	STO	General Storage Procedures: Relates to precautions to be taken when storing the material. Rationale for these measures varies from safety considerations to precautions designed to prevent degradation of the material.
16	HND	General Handling Procedures: States the precautions to be taken when handling the material. Information relates to both safety considerations and practices designed to prevent degradation of the material.
17	PRD	Production Sites: Lists major producers and their plant locations.
18	HYD	Hydrolysis Product of: Lists hazardous materials which decompose to the material of reference when contacted with water.
19	ADD	Additive (%): Lists typical stabilizers and inhibitors added to the base material.
20	BIN	Binary Reactants: Lists materials known to react when put in contact with the material of reference.
21	COR	Corrosiveness: General statement of observations on corrosive action to materials commonly used for packaging, or equipment that might be required at a spill site.
22	SGM	Synergistic Materials: Lists other materials and water quality parameters whose presence can increase the toxicity of the material of interest.
23	ANT	Antagonistic Materials: Lists other materials and water quality parameters whose presence can reduce the toxicity of the material of interest.
24	FDL	Field Detection Techniques, Limit(ppm), Ref: A three part segment listing potential field detection techniques, the lower sensitivity limit, and the literature reference where more data can be obtained. Field test generally refers to any gross identification method that can be used at the spill site without elaborate or non-portable equipment. It normally assumes that the material or the chemical class has been identified so that general tests for aldehydes or phenols, etc. are applicable. The two major types of tests listed are inorganic colorimetric reactions and organic spot tests.

Figure 1. Each of these 126 fields of data is potentially available for any of the 1029 substances covered

25 LDL   Laboratory Detection Techniques, Limit (ppm), Ref: Follows format of previous segment for specific tests that can be used for positive identification of
material. These tests are generally reliant on sophisticated laboratory
analysis equipment, such as atomic absorp- tion units and gas chromatographs.

26 STD   Standard Codes: Enumerates the National Fire Protection Association
codes for materials as well as pertinent transportation codes.

27 FLM   Flammability: Summarizes potential for fire at a spill site. Uses the NFPA
ranking system described by one of the following modifiers:

> very
> quite
> moderate
> slight
> non-flammable

28 LFL   Lower Flammability Limit (%): Listed value is % of material in air which is
the lower limit of flammability.

29 UFL   Upper Flammability Limit (%): Listed value is % of material in air which is
the upper limit of flammability.

30 TCP   Toxic Combustion Products: Occasionally lists specific materials or classes
of materials released when compound of concern is burned or heated to
decomposition.

31 EXT   Extinguishing Methods: Notes fire fighting techniques and outlines unique
precautions to be taken, if any.

32 FLP   Flash Point (degrees C): Listed open cup value when available, otherwise
closed cup.

33 AIP   Auto Ignition Point (degrees C): Listed value at which auto ignition occurs
in the presence of adequate air.

34 EXP   Explosiveness: Summarizes potential for violent rupture or vigorous reaction
at a spill site.

35 LEL   Lower Explosive Limit (%): Listed value is % of material in air which is the
lower explosive limit.

36 UEL   Upper Explosive Limit (%): Listed value is % of material in air which is the
upper explosive limit.

37 MLT   Melting Point (degrees C): Accepted value under standard conditions
unless otherwise noted below in segment 38.

38 MTC   Melting Characteristics: Decomposes, ignites, etc.

39 BLP   Boiling Point (degrees C): Accepted value under standard conditions unless
noted below in segment 40.

40 BOC   Boiling Characteristics: Reduced pressure, etc.

41 SOL   Solubility (ppm 25 degrees C): Typically the listed value for standard reference conditions.

42 SLC   Solubility Characteristics: Slightly and moderately are used when a specific
value is not given.

43 SPG   Specific Gravity: Listed value for material in the state in which it is most
often shipped. For materials whose boiling point is near ambient temperatures, the liquid state was usually referenced.

44 VPN   Vapor Pressure (mm Hg): The pressure characteristic (at any given temperature) of a vapor in equilibrium with its liquid or solid form.

45 VPT   Vapor Pressure Text: Indicates conditions under which measurement is
made.

46 VDN   Vapor Density: A value derived by dividing the mass of the vapor by its
volume and measuring at a specific temperature. A value < 1 indicates that
the vapor is lighter than air, > 1 is heavier than air and will give the appearance of a fog, hugging the ground.

Figure 1. Continued.

47 VDT   Vapor Density Text: Indicates temperature and any other conditions under which measurement is made.

48 BOX   Biochemical Oxygen Demand (BOD lb/lb): Describes relative oxygen requirements of wastewaters, effluents, and polluted waters. Lists biochemical oxygen demand of pure substance on a lb/lb, or % of theoretical demand basis.

49 BOD   Biochemical Oxygen Demand Text: Displays same information listed in segment 48 and includes duration of the test and source of information.

50 PER   Persistency: Interprets BOD and chemical data to estimate material life span in a free aquatic system. When possible, degradation products are specified.

51 PFA   Potential for Accumulation: Recounts data on ability of various organisms to accumulate a material and the specific organs in which concentration is most pronounced.

52 FOO   Food Chain Concentration Potential: Indicates potential for material to be concentrated to toxic levels while it is passed up the food chain. Where possible, data is given on findings in predator species.

53 EDF   Etiological Potential: Enumerates diseases and ailments initiated or accelerated by exposure to the material of interest.

54 CAG   Carcinogenicity: Relates results of work directed to isolating carcinoma in test animals. Human data is used when available.

55 MUT   Mutagenicity: Cites finding of tests for mutagenicity.

56 TER   Teratogenicity: Cites finding of tests for teratogenicity.

57 FTX   Freshwater Toxicity Number (ppm): This segment indicates the concentration in parts per million at which test results were reported.

58 FTB   Freshwater Toxicity Text:

Column 1   Concentration in ppm at which test results were reported.
Column 2   Time of exposure expressed in hours.
Column 3   Species tested, usually a common name.
Column 4   Effect on organism tested, often given as TLm or LD 50.
Column 5   Test environment, includes data on water quality and other controlled conditions.
Column 6   Source of information.

59 CAT   Chronic Aquatic Toxicity Limits (ppm): Maximum level in ppm found to be safe for extended exposure of fish to the material of interest.

60 CAR   Reference for Chronic Aquatic Toxicity: Source of information.

61 STX   Salt Water Toxicity: Indicates toxicity to estuarine or marine animals in parts per million.

62 STB   Salt Water Toxicity Text: Follows same general format as segment 58.

63 ATX   Animal Toxicity: Displays doses reported in milligrams of material per kilogram of body weight of

64 ATB   Animal Toxicity Text:

Column 1 _Doses in mg of Material pr kg body weight of test animal.
Column 2 -Time of exposure.
Column 3 -Special lists animal of reference - typically lab animals - rats, guinea pigs, mice, pigs, dogs, and monkeys.
Column 4 -Parameter, description of exposure. Terms indicate whether dose caused death or other toxic effects, and whether it was administered as a lethal concentration, or toxic concentration in the inhaled air. Refer to Appendix for abbreviations.

Figure 1. Continued.

Column 5 -Route, lists mode of application. Refer to Appendix for abbreviations.

Column 6 -Reference, source of data.

65 ATL   Chronic Animal Toxicity Limits(ppm): Maximum level reported in ppm thought to be the threshold for extended use on livestock.

66 ATR   Reference for Chronic Animal Toxicity Limits: Source of in- formation.

67 LVN   Livestock Toxicity(ppm): Lists recommended or safe levels of concentration in ppm for use on livestock.

68 LVR   Reference for Livestock: Source of information.

69 WAN   Acute Waterfowl Toxicity (ppm): Concentration in ppm considered to be hazardous to waterfowl upon acute exposure.

70 WAR   Reference for Acute Waterfowl Toxicity: Source of information.

71 CWF   Chronic Waterfowl Toxicity Limits (ppm): Concentration in ppm considered to be maximum permissible in water inhabited by waterfowl.

72 CWRT   Reference for Chronic Waterfowl Toxicity: Source of information.

73 AQN   Aquatic Plants (ppm): Concentration in ppm found to be injurious to aquatic flora listed.

74 AQR   Reference for Aquatic Plants: Source of information.

75 IRN   Irrigable Plants (ppm): Concentration expressed in ppm found to be injurious to crop listed.

76 IRR   Reference for Irrigable Plants: Source of information.

77 CPT   Chronic Plant Toxicity Limits (ppm): Threshold level expressed in ppm for extended use as irrigation water.

78 CPN   Reference for Chronic Plant Toxicity Limits: Source of information.

79 TRT   Major Species Threatened: This segment was originally designed to spotlight individual species especially susceptible to the material of interest. Data such as this are very rare. Consequently, the segment includes specific data on tests run with different species.

80 TIC   Taste Imparting Characteristics (ppm): Level in ppm at which material will impart a taste to the flesh of fish living in the affected waters.

81 TIR   Reference for Taste Imparting Characteristics: Source of information.

82 INH   Inhalation Limit (Value): Generally the accepted threshold limit value (TLV) which is that level acceptable for industrial exposure over an eight hour period. May sometimes be the LC50 for inhalation.

83 INT   Inhalation Limit(Text): Units and source of information for the above segment.

84 IRL   Irritation Levels(Value): Level at which skin and mucous membrane irritation occurs.

85 IRT   Irritation Levels (Text): Reference and explanatory comments for above segment.

86 DRC   Direct Contact: Summary statement indicating corrosiveness or irritation value of material in direct contact with skin, mucous membranes, or eyes.

87 JNS   General Sensation: Designed to identify some of the reactions people might have (symptoms and effect on body) when exposed to the designated material, sensation upon breathing the vapors, vapor concentration levels at which noticeable reactions occur, warning properties, and miscellaneous toxicological observations.

88 LOT   Lower Odor Threshold (ppm): Listed value in ppm.

89 LOR   Lower Odor Threshold Reference: Source of information.

90 MOT   Medium Odor Threshold (ppm): Listed value in ppm.

91 MOR   Medium Odor Threshold Reference: Source of information.

92 UOT   Upper Odor Threshold (ppm): Listed value in ppm.

93 UOR   Upper Odor Threshold Reference: Source of information.

94 LTT   Lower Taste Threshold (ppm): Listed value in ppm.

Figure 1. Continued.

95 LTR Lower Taste Threshold Reference: Source of information.

96 MTT Medium Taste Threshold (ppm): Listed value in ppm.

97 MTR Medium Taste Threshold Reference: Source of information.

98 UTT Upper Taste Threshold (ppm): Listed value in ppm.

99 UTR Upper Taste Threshold Reference: Source of information.

100 DHI Direct Human Ingestion (mg/kgwt): Notes toxic dose levels via human consumption in milligrams toxicant per kilogram body weight.

101 DHR Reference for Direct Human Ingestion: Source of information. IP "102 DRK" 12 Recommended Drinking Water Limits (ppm): Cites Public Health Service Drinking Water Standards whenever available.

103 DRR Reference for Recommended Drinking Water Limits: Source of information.

104 BCE Body Contact Exposure (ppm): States acute contact threshold limits in water where available.

105 BCR Reference for Body Contact Exposure: Source of information.

106 PHC Prolonged Human Contact (ppm): States safe level for bathing and swimming (prolonged) in parts per million.

107 PHR Reference for Prolonged Human Contact: Source of information.

108 SAF Personal Safety Precautions: Lists equipment to be employed when working in a spill area. Refers to disaster conditions and as such often presupposes fire or intense heat. Response teams should use their own judgment in deciding when stated precautions are no longer necessary. For most circumstances, eye protection, hard hats, and gloves are recommended.

109 AHL Acute Hazard Level: Attempts to indicate level of hazard resulting from a spill. Relates to inhalation, ingestion and contact with material. Also lists specific water use hazard level such as fish toxicity and irrigation water toxicity.

110 CHL Chronic Hazard Level: Interprets chronic toxicological-biological hazard to life forms subjected to material of interest for extended periods of time.

111 HEL Degree of Hazard to Public Health: Interpretive summary of data from previous segments. This segment focuses on those toxicological chemical hazards directly affecting public health.

112 AIR Air Pollution: Summarizes degree of hazard to people in the vicinity of a spill. May refer to fumes, vapors, mists, or dusts of the material spilled or its combustion and/or decomposition products.

113 ACT Action Levels: An interpretive segment designed to aid in initiating response activities. Suggests notification of fire and air authority if material poses flammability or air hazard. Recommends alerting Civil Defense if explosion hazard exists. When explosion or severe air pollution exists, evacuation is indicated. If the material in question is highly corrosive or can be absorbed through the skin at toxic levels, affected waterways should be restricted from public access. When flammable materials are involved, ignition sources should be removed. Air contaminants require entry from upwind. If the spill involves solids, attempts should be made to prevent suspension of dusts in the air. If the material is one that will form a slick on water before dissolving, early attempts at containment will be quite beneficial. It is assumed that these actions will be complemented by general defensive responses. These include, notifying downstream water users of the spill, stopping all leaks or diverting their flow from reaching surface waters, and removing all bags, barrels and/or other containers that may still be leaking to the water body.

114 AML In Situ Amelioration: Lists potentially effective treatment methods which could be applied to the body of water for removal of the spilled material. Methods deemed to include hazards equal to or greater than that of the contaminant were systematically excluded. The term carbon refers to activited carbon in granular or powdered form.

115 SHR Beach and Shore Restoration: This segment is used mainly to indicate if material can be safely burned off beaches. Occasionally, a recommendation is made to wash affected area with a neutralizing solution.

Figure 1. Continued.

115 AVL   Availability of Countermeasures Material: Lists major materials required for countermeasures recommended in segment 114 (in situ amelioration) and possible local sources for those materials.

117 DIS   Disposal Methods: Describes recommended techniques for disposing of spilled materials.

118 DSN   Disposal Notification: Lists local authorities who should be notified before disposal methods in segment 117 are initiated.

119 IFP   Industrial Fouling Potential: Relates potential problems from use of water contaminated by the material of interest. Generally refers to use in boiler feed and cooling water. Materials with flash points below 50 degrees C are listed as potential rupture hazards when included in boiler feed or cooling water.

120 WTP   Effect on Water Treatment Process: Describes potential interaction with typical water and wastewater treatment facilities. Most frequent entries concern effect of chlorination on the aesthetic properties of contaminated water, and the effect of high concentration on sewage organisms.

121 WAT   Major Water Uses Threatened: Lists water uses imperiled by a spill and consequently indicates what type of downstream water users should be notified of the spill.

122 LOC   Probable Location and State of the Material: This is an interpretive segment of physical data designed to assist personnel in identifying the material spilled and its whereabouts. The data attempts to describe the physical appearance of the material as shipped (e.g., a dark red powder, etc.) and its probable location if the spill occurs in or near surface water.

123 DRT   Soil Chemistry: A general description of the behavior and exchange capacity of various cations and anions in soil.

124 HOH   Water Chemisty: A general description of the behavior of the material of interest in aqueous solution.

125 COL   Color in Water: Identifies the color or appearance of concentrated solutions of the material of interest. In many cases, dilution and material coloring will minimize the visibility of the color listed here.

126 DAT   Adequacy of Data: A simple classification was used to indicate the availability of data.

Poor -   indicates toxicological data are sparse if they exist at all.

Fair -   indicates toxicological data were found but no aquatic toxicities are listed.

Moderate -   indicates toxicological data were found along with some information on toxicity towards fish.

Good -   indicates both toxicological and aquatic toxicity data were found.

Limited Reference -   identifies those materials for which a complete literature survey was not run.

Figure 1. Concluded.

1983 National Oil Spill Conference

February 28–March 3, 1983; San Antonio, Texas

A biennial national conference held alternate years from the Hazardous Materials Spills Conference.

Contact: National Oil Spill Conference, Suite 700, 1629 K Street, N.W., Washington, D.C. 20006, Tel. 202-296-8246.

DEPARTMENT OF TRANSPORTATION

Form Approved OMB No. 04-5613

HAZARDOUS MATERIALS INCIDENT REPORT

INSTRUCTIONS: Submit this report in duplicate to the Director, Office of Hazardous Materials Operations, Materials Transportation Bureau, Department of Transportation, Washington, D.C. 20590, (ATTN: Op. Div.). If space provided for any item is inadequate, complete that item under Section H, "Remarks", keying to the entry number being completed. Copies of this form, in limited quantities, may be obtained from the Director, Office of Hazardous Materials Operations. Additional copies in this prescribed format may be reproduced and used, if on the same size and kind of paper.

A | INCIDENT

1. TYPE OF OPERATION
   1 ☐ AIR 2 ☐ HIGHWAY 3 ☐ RAIL 4 ☐ WATER 5 ☐ FREIGHT FORWARDER 6 ☐ OTHER *(Identify)*_____

2. DATE AND TIME OF INCIDENT *(Month - Day - Year)*
   _____ a.m.
   _____ p.m.

3. LOCATION OF INCIDENT

B | REPORTING CARRIER, COMPANY OR INDIVIDUAL

4. FULL NAME

5. ADDRESS *(Number, Street, City, State and Zip Code)*

6. TYPE OF VEHICLE OR FACILITY

C | SHIPMENT INFORMATION

7. NAME AND ADDRESS OF SHIPPER *(Origin address)*

8. NAME AND ADDRESS OF CONSIGNEE *(Destination address)*

9. SHIPPING PAPER IDENTIFICATION NO.

10. SHIPPING PAPERS ISSUED BY
    ☐ CARRIER   ☐ SHIPPER
    ☐ OTHER *(Identify)*

D | DEATHS, INJURIES, LOSS AND DAMAGE

DUE TO HAZARDOUS MATERIALS INVOLVED

11. NUMBER PERSONS INJURED

12. NUMBER PERSONS KILLED

13. ESTIMATED AMOUNT OF LOSS AND/OR PROPERTY DAMAGE INCLUDING COST OF DECONTAMINATION *(Round off in dollars)*

14. ESTIMATED TOTAL QUANTITY OF HAZARDOUS MATERIALS RELEASED

$

E | HAZARDOUS MATERIALS INVOLVED

15. HAZARD CLASS (*Sec. 172.101, Col. 3)	16. SHIPPING NAME (*Sec. 172.101, Col. 2)	17. TRADE NAME

F | NATURE OF PACKAGING FAILURE

18. *(Check all applicable boxes)*

(1) DROPPED IN HANDLING	(2) EXTERNAL PUNCTURE	(3) DAMAGE BY OTHER FREIGHT
(4) WATER DAMAGE	(5) DAMAGE FROM OTHER LIQUID	(6) FREEZING
(7) EXTERNAL HEAT	(8) INTERNAL PRESSURE	(9) CORROSION OR RUST
(10) DEFECTIVE FITTINGS, VALVES, OR CLOSURES	(11) LOOSE FITTINGS, VALVES OR CLOSURES	(12) FAILURE OF INNER RECEPTACLES
(13) BOTTOM FAILURE	(14) BODY OR SIDE FAILURE	(15) WELD FAILURE
(16) CHIME FAILURE	(17) OTHER CONDITIONS *(Identify)*	**19. SPACE FOR DOT USE ONLY**

Form DOT F 5800.1 (10-70) (9/1/76)
*-Editorial change to incorporate redesignation per HM-112.

Figure 2. Continued.

G	PACKAGING INFORMATION - *If more than one size or type packaging is involved in loss of material show packaging information separately for each. If more space is needed, use Section H "Remarks" below keying to the item number.*			
	ITEM	#1	#2	#3
20	TYPE OF PACKAGING INCLUDING INNER RECEPTACLES (*Steel drums, wooden box, cylinder, etc.*)			
21	CAPACITY OR WEIGHT PER UNIT (*55 gallons, 65 lbs., etc.*)			
22	NUMBER OF PACKAGES FROM WHICH MATERIAL ESCAPED			
23	NUMBER OF PACKAGES OF SAME TYPE IN SHIPMENT			
24	DOT SPECIFICATION NUMBER(S) ON PACKAGES (*21P, 17E, 3AA, etc., or none*)			
25	SHOW ALL OTHER DOT PACKAGING MARKINGS (*Part 178*)			
26	NAME, SYMBOL, OR REGISTRATION NUMBER OF PACKAGING MANUFACTURER			
27	SHOW SERIAL NUMBER OF CYLINDERS, CARGO TANKS, TANK CARS, PORTABLE TANKS			
28	TYPE DOT LABEL(S) APPLIED			
29	IF RECONDITIONED — A REGISTRATION NO. OR SYMBOL			
	OR REQUALIFIED, SHOW — B DATE OF LAST TEST OF INSPECTION			
30	IF SHIPMENT IS UNDER DOT OR USCG SPECIAL PERMIT OR EXEMPTION, ENTER PERMIT OR EXEMPTION NO.			

H	REMARKS - Describe essential facts of incident including but not limited to defects, damage, probable cause, stowage, action taken at the time discovered, and action taken to prevent future incidents. Include any recommendations to improve packaging, handling, or transportation of hazardous materials. Photographs and diagrams should be submitted when necessary for clarification.

31. NAME OF PERSON PREPARING REPORT (*Type or print*)	32. SIGNATURE
33. TELEPHONE NO. (*Include Area Code*)	34. DATE REPORT PREPARED

Reverse of Form DOT F 5800.1 (10-70)

Figure 2.

8th Annual Inland Spills Conference

Late September or early October, 1983 at a city in Ohio.
An annual conference sponsored by the Ohio E.P.A.
Contact: Office of Training & Safety, Ohio E.P.A., 361 East Broad Street, Columbus, Ohio 43216, Tel. 614-466-8820.

5th Annual Southwestern Spill Conference On Prevention And Control of Spills of Oil and Hazardous Materials

Fall of 1983 at a location in South Texas.
Contact: George Oberholtzer, Director, National Spill Control School, Corpus Christi State University, 6300 Ocean Drive, Corpus Christi, Texas 78412, Tel. 512-991-8692.

1983 National Conference And Exhibition on Management of Uncontrolled Hazardous Waste Sites

To be held the week following Thanksgiving, 1983; Sheraton Washington Hotel, Washington, D.C.
Fourth annual conference administered by Hazardous Materials Control Research Institute; HMCRI also sponsors other conferences related to environmental concerns and wastes throughout the year.
Contact: Hazardous Materials Control Research Institute, 9300 Columbia Blvd., Silver Spring, Maryland 20910, Tel. 301-587-9390.

1983 Annual Conference & Hazardous Materials Transportation Exposition

May 11–13, 1983; Houston, Texas.
Annual conference sponsored by the Hazardous Materials Advisory Council; a national, nonprofit membership organization of shippers/container manufacturers/carriers designed to promote safety in the domestic and international transportation and handling of hazardous materials, substances, and wastes. HMAC also sponsors a packaging conference each year in March or April.
Contact: Hazardous Materials Advisory Council, Suite 908, 1100 17th Street, N.W., Washington, D.C. 20036, Tel. 202-223-1271.

MANUALS

Hazardous materials response personnel rely heavily on manuals to guide their actions in attempting to bring hazardous materials incidents under control. The five manuals described below are probably the best known and most widely used, but each is a bit different from the others and response personnel will often consult two or three different manuals when dealing with a specific chemical.

Chemical Hazards Response Information System (CHRIS)

The *Chemical Hazards Response Information System* (CHRIS) is an official publication of the U.S. Coast Guard developed for use by Federal On-Scene Coordinators. CHRIS is designed to provide timely information essential for proper decision making by responsible Coast Guard personnel. A very brief description of each of the four manuals included within CHRIS is provided as follows.

Manual 1; *A Condensed Guide to Chemical Hazards*

Intended for use by response personnel who may be the first to arrive at the site of an accidental discharge or fire to assess the dangers and consider the appropriate large-scale response necessary to safeguard life and property. Manual 1 lists 1,000 chemicals.

Manual 2; *Hazardous Chemical Data*

This manual is a central component of CHRIS. It lists the specific chemical, physical, and biological data for 1,000 chemicals needed for the preparation and use of other components of the system. It is intended for use primarily by the On-Scene Coordinator and by regional and National Response Centers for devising, evaluating and carrying out response plans.

Manual 3; *Hazard Assessment Handbook*

Describes procedures to be used for estimating the quantities of a hazardous chemical discharge onto a navigable waterway during shipment. It also describes how to estimate its concentration in air and water as a function of time and distance from the discharge. Methods for predicting the resultant toxicity, fire, and explosion effects are also described.

Manual 4; *Response Methods Handbook*

Describes cautionary and corrective response methods for reducing and eliminating hazards that result from chemical discharge. Manual 4 is written specifically for Coast Guard personnel with training or experience in pollution response.

Available from: Superintendent of Documents, U.S. Government Printing Office, Washington, D.C. 20402, Tel. 202-783-3238

Emergency Handling of Hazardous Materials In Surface Transportation

Published by the Bureau of Explosives/Association of American Railroads; edited by Patrick J. Student. (600 pages; Revised 1981; Size 8½ by 11.) *Emergency Handling of Hazardous Materials In Surface Transportation* contains nearly 2,400 commodity-specific guides for firefighters, railroad employees, industry personnel, transportation workers and other emergency response personnel who must deal with the hazards of transporting dangerous commodities. Information provided includes both general information about hazardous materials incidents (general recommendations for response, recommendations for response keyed to 15 D.O.T. hazard classes, an explanation of terms used for emergency responders, etc.) and commodity-specific emergency response information for each hazardous materials regulated by the Department of Transportation and listed in Title 49 of the *Code of Federal Regulations.* For Environmental Protection Agency designated hazardous substances, commodity-specific emergency environmental damage mitigation information is also provided. One cross reference index shows the D.O.T.-required four-digit identification number and the product(s) and page(s) to which that number applies. A second cross reference index shows the seven-digit Standard Transportation Commodity Code number, (the STCC number) the particular commodity to which that number applies, and the page number on which information about that commodity appears.

Available from: Elizabeth P. Rabben, Supervisor-Publication Services, Bureau of Explosives/Association of American Railroads, 1920 L Street, N.W., Washington, D.C. 20036, Tel. 202-293-4048.

Fire Protection Guide On Hazardous Materials

Published by the National Fire Protection Association; a nonprofit, educational, voluntary membership organization recognized internationally as a clearing house for information on fire prevention, firefighting procedures, and means of fire protection. NFPA's *Guide* is widely used by fire department personnel to determine the proper procedures to prevent fires and other emergencies during the use, storage, and transportation of chemicals; and to make informed decisions on the procedures to be followed during an emergency involving chemicals. The NFPA *Guide* is divided into five sections.

1. *Flash Point Index of Trade Name Liquids.* The flash points of more than 8,800 trade name products are listed. (The flash point of a liquid is a good indication of its relative flammability) The listing is alphabetical by trade name and includes a section for numerical trade names.

2. *Fire Hazard Properties of Flammable Liquids, Gases, and Volatile Solids.* The fire hazard properties of more than 1,300 flammable substances are listed alphabetically by chemical name. Hazard Index markings (See #5 below re: NFPA-704 System for the Identification of the Fire Hazards of Materials) are included for most entries.

3. *Hazardous Chemicals Data.* Provides data for approximately 416 chemicals arranged alphabetically as to their fire, explosion, and toxicity hazards.

4. *Manual of Hazardous Chemical Reactions.* Includes 3,550 mixtures of two or more chemicals reported to be potentially dangerous in that they may cause fires, explosions, or detonations at ordinary or moderately elevated temperatures. Arranged alphabetically by chemical name.

5. *Recommended System For the Identification of the Fire Hazards of Materials.* The "NFPA-704 System" simplifies determining the degree of health, flammability, and reactivity hazard of materials. The system also permits identification of reactivity with water, radioactivity hazards, and fire control problems.

Available from: National Fire Protection Association, Batterymarch Park, Quincy, Massachusetts 02269.

Hazardous Materials Emergency Response Guidebook

The Department of Transportation's *Emergency Response Guidebook* developed by the Materials Transportation Bureau for all personnel involved in the handling and transportation of hazardous materials details initial response actions to be taken by first responders. The *Guidebook* is composed of five different sections:

• Numerical Index of Hazardous Materials. (Four-digit identification number, number of proper emergency response guide, and name of the hazardous material.)

• Alphabetical Index of Hazardous Materials. (Name of material, number of proper emergency response guide, and four-digit identification number.)

• 55 Separate Emergency Response Guides Numbered from 11 Through 66. (Each separate "Guide" provides emergency information relative to fire or explosion, health hazards, general emergency actions, fire/spill/leak, and first air procedures for a group of chemicals with similar characteristics.)

• Table of Isolation and Evacuation Distances for Selected Materials.

• Placards and Applicable Guide Numbers. (Specific placards are related to specific "Guides." This section is to be used *only* if no four-digit identification number or shipping name can be determined.)

Available from: The D.O.T. *Hazardous Materials Emergency Response Guidebook* is available for purchase from the following commercial suppliers.

American Trucking Association
Customer Service Section
1616 P Street, N.W.
Washington, D.C. 20036
Tel. 202-797-5384

Labelmaster
7525 No. Wolcott
Chicago, Illinois
60626
Tel. 800-621-5808

J.J. Keller
145 Wisconsin Avenue
Neenah, Wisconsin
54956
Tel. 800-558-5011

UNZ & Co.
190 Baldwin Avenue
Jersey City, New Jersey
07306
Tel. 800-631-3098

N.I.O.S.H./O.S.H.A. Occupational Health Guidelines For Chemical Hazards

Published by the U.S. Department of Health and Human Services, Public Health Service, Centers for Disease Control, and the U.S. Department of Labor; edited by Frank W. Mackison, R. Scott Stricoff, and Lawrence J. Patridge, Jr. (Three volumes; January, 1981. Size 8½ × 11. Looseleaf binder format.)

Occupational Health Guidelines for Chemical Hazards summarizes information on permissible exposure limits, chemical and physical properties, and health hazards. Also provided are recommendations for medical surveillance, respiratory protection, and personal protection and sanitary practices for specific chemicals that have Federal occupational safety and health regulations applied to them. (Approximately 400 chemicals) The "Guidelines" are intended primarily for the industrial hygienist and medical surveillance personnel responsible for initiating and maintaining an occupational health program. It is also used by others, including workers, for obtaining summary information about specific chemical substances found at the worksite.

A specific "Guideline," for each chemical includes data for the following categories.
 1. *Substance Identification:* (Formula, synonyms, appearance, and odor.)
 2. *Permissible Exposure Limit:*
 3. *Health Hazard Information:* (Routes of exposure, effects of overexposure, reporting signs and symptoms, recommended medical surveillance, and summary of toxicology.)
 4. *Chemical & Physical Properties:* (Physical data such as molecular weight, boiling point, specific gravity, vapor density, melting point, solubility in water, vapor pressure, and evaporation rate; reactivity, incompatibilities, hazardous decomposition products, special precautions, flammability, and warning properties.)
 5. *Monitoring & Measuring Procedures:*
 6. *Respirators:*
 7. *Personal Protective Equipment:*
 8. *Common Operations and Controls:*
 9. *Emergency First Aid Procedures:*
 10. *Spill, Leak, & Disposal Procedures:*

Available from: Superintendent of Documents, U.S. Government Printing Office, Washington, D.C. 20402, Tel. 202-783-3238, (D.H.H.S./N.I.O.S.H. Publication No. 81–123).

PERIODICAL PUBLICATIONS

Dangerous Goods Newsletter

Transport of Dangerous Goods
Transport Canada
Tower B, Place de Ville
Ottawa, Ontario
K1A ON5 Canada
Subscription Fee: No Charge. (Specify French or English language)
Frequency of Publication: Monthly.

Areas of Coverage: Covers legislation and regulations in Canada, CANUTEC, (Canadian Transportation Emergency Center) meetings and conferences, international news, packaging, evaluation and analysis of hazardous materials programs in Canada, IMCO publications, and similar news and information.

Fire Service Publications

Because the overwhelming majority of first responders to hazardous materials incidents are firefighters, the fire magazines tend to carry a number of articles dealing with response to and control of hazardous materials incidents. The magazines listed below are among the best known for regular use of such articles.

Emergency (E.M.S. orientation)
P.O. Box 159
Carlsbad, California
92008

Fire Engineering
875 Third Avenue
New York, New York 10022
Tel. 212-489-2200

Fire Service Today
National Fire Protection Assoc.
Batterymarch Park
Quincy, Massachusetts 02269
Tel. 617-328-9290

Western Fire Journal
9072 E. Artesia Blvd., Suite 7
Bellflower, California 90706
Tel. 212-866-1664

Fire Chief
625 N. Michigan Avenue
Chicago, Illinois 60611
Tel. 312-642-9862

Firehouse
515 Madison Avenue
New York, New York 10022
Tel. 212-935-4550

International Fire Chief
International Assoc. of Fire Chiefs
1329 18th Street, N.W.
Washington, D.C. 20036
Tel. 202-833-3420

Hazard Monthly
Research Alternatives
705 New Mark Esplanade
Rockville, Maryland 20850
Tel. 301-424-2389
Subscription Fee: $18.00 year individuals; $37.00 year institutions.
Frequency of Publication: Monthly.
Areas of Coverage: Covers information about the problems created by and solutions to man-made technological and natural hazards—prevention and mitigation of disasters, preparedness for and response to emergencies, and longterm recovery from catastrophe. Includes articles, news reports, and special features. The focus here is on both man made and natural disasters including coverage of hazardous materials and toxic concerns. Each issue carries a special, one-page "Hazardous Materials Report." Established May, 1980. Newspaper format with photos; 16 or more pages per issue.

Hazardous Materials Intelligence Report (Adam Finkel, Editor)
World Information Systems
P.O. Box 535, Harvard Square Station
Cambridge, Massachusetts 02238
Tel. 617-491-5100

Subscription Fee: $295. year U.S. subscribers; $385. foreign.
Frequency of Publication: Weekly.
Areas of Coverage: Provides up-to-date incident reports. Covers regulation and rulemaking, news items pertaining to both hazardous materials and hazardous wastes, reports of recent research and technological concerns, publications, and new products and equipment.

Hazardous Materials Newsletter
U.S. Department of Transportation
Materials Transportation Bureau
Research & Special Programs Administration
Washington, D.C. 20590
Subscription Fee: No Charge.
Frequency of Publication: Bimonthly.
Areas of Coverage: Covers notices and amendments pertaining to hazardous materials published since previous issue; training scheduled by various agencies and commercial training companies for upcoming months; penalty actions imposed on carriers and shippers; notice of government hearings related to hazardous materials; and similar news and information.

Hazardous Materials Newsletter (John R. Cashman, Editor)
P.O. Box 204
Barre, Vermont 05641
Tel. 802-479-2307
Subscription Fee: $26. year U.S., Canada, & Mexico; $28. year foreign surface mail; $31. year foreign air mail.
Frequency of Publication: Bimonthly.
Areas of Coverage: Emphasis is on response to and control of hazardous materials incidents. Regular segments include Calendar, People in Hazardous Materials, Incident Reporting, Legislation & Rulemaking, Tips From The Pros, Spotlight On Response Teams, Industrial Perspective, Watch On The Potomac, Research Sources & Resources, Training Ideas, and Building a Library.

Hazardous Materials Transportation (Richard S. Sexton, Editor)
Cahners Publishing Company
221 Columbus Avenue
Boston, Massachusetts 02116
Tel. 617-536-7780
Subscription Fee: $195. year.
Frequency of Publication: Monthly.
Areas of Coverage: Covers what to expect from Congress, upcoming international packaging standards, regulatory trends at E.P.A./D.O.T./I.C.C./O.S.H.A., rules and regulations, local ordinances, etc. for persons responsible for the shipping and handling of hazardous materials.

Hazardous Waste News
Business Publishers, Inc.
951 Pershing Drive
Silver Spring, Maryland, 20910
Tel. 301-587-6300
Subscription Fee: $257. year plus $10. postage.
Frequency of Publication: Weekly.
Areas of Coverage: News and information directed toward hazardous waste generators, managers, regulators, and equipment vendors.

Hazardous Waste Report
Aspen Systems, Inc.
1600 Research Blvd.
Rockville, Maryland 20850
Tel. 301-251-5000
Subscription Fee: $260. year.
Frequency of Publication: Biweekly.
Areas of Coverage: Covers "Superfund" implementation, state requirements, industrial liability, new technologies, EPA policies, litigation and enforcement, congressional activity, announcements, RCRA, etc. Subscription includes periodic *Trends And Analyses,* (separately published individual reports on current political, economic and legal issues pertinent to hazardous waste concerns) a "Hotline" reference service, an annual index, and a binder for maintenance of newsletter issues.

Journal of Hazardous Materials
Elsevier Scientific Publishing Company
P.O. Box 211
1000 AE Amsterdam
The Netherlands
Subscription Fee: $84. year. (Printed in English)
Frequency of Publication: Quarterly.
Areas of Coverage: Original journal articles covering all environmental problems that can arise from the manufacture, use, and disposal of potentially hazardous materials. Emphasis on procedures that minimize risk. Coverage includes properties of hazardous materials, safety and health hazards, legislation, incidents, and assessment. Includes *Hazbits,* a review of print media coverage of hazardous materials concerns.

Nuclear Waste News
Business Publishers, Inc.
951 Pershing Drive
Silver Spring, Maryland 20910
Tel. 301-587-6300
Subscription Fee: $237. year plus $5.20 postage.
Frequency of Publication: Biweekly.
Areas of Coverage: Covers the collection, packaging, transportation, storage, processing and ultimate disposal of high-level, low-level and transuranic wastes generated by . . .
• Nuclear power generating facilities, including spent fuel processing.
• Medical, laboratory and industrial uses of isotopes and other radiation sources.
• Uranium mining and milling operations and other mineral activities.
• Weapons development, testing, and related military applications.

Spill Technology Newsletter
Environmental Protection Service
Department of the Environment
Ottawa, Ontario
K1A 1C8 Canada
Tel. 819-997-3921
Subscription Fee: No Charge.
Frequency of Publication: Bimonthly.
Areas of Coverage: Each issue contains an average of four journal-type articles on oil spill concerns, identification of recent reports and publications in the field of oil spill countermeasures, and a listing of upcoming conferences of interest to readers. The focus here is on *oil spill countermeasures within Canada* rather than on hazardous materials in general.

State Regulation Report
Business Publishers, Inc.
951 Pershing Drive
Silver Spring, Maryland, 20910
Tel. 301-587-6300
Subscription Fee: $157. year plus $5.20 postage.
Frequency of Publication: Biweekly.
Areas of Coverage: Covers state legislative and regulatory initiatives and their effect on toxic substances control, hazardous waste management, pesticide certification and enforcement, hazardous materials transportation, and consumer/occupational/environmental health.

Toxic Materials News
Business Publishers, Inc.
951 Pershing Drive
Silver Spring, Maryland 20910
Tel. 301-587-6300
Subscription Fee: $257. plus $10. postage.
Frequency of Publication: Weekly.
Areas of Coverage: Covers E.P.A.'s toxic substances control program, pesticide and hazardous waste programs, toxic air and water pollutants, workplace and household product carcinogens, and transportation of hazardous materials.

Toxic Materials Transport
Business Publishers, Inc.
951 Pershing Drive
Silver Spring, Maryland 20910
Tel. 301-587-6300
Subscription Fee: $157. year plus $5.20 postage.
Frequency of Publication: Biweekly.
Areas of Coverage: Covers D.O.T. and E.P.A. regulation and rulemaking, NTSB investigations, Federal/state/local regulations affecting routing and response, and industrial involvement in laws/regulation/technology/costs/compliance. Directed toward shippers, carriers, and regulators.

9 IMPACT OF HAZARDOUS MATERIALS CONCERNS ON SELECTED COMMUNITIES

Jefferson County Kentucky HAZ-MAT: (Hazardous Materials Mutual Aid Team)

Jefferson County HAZ-MAT, a cooperative effort by approximately 19 agencies of government and industrial organizations, is a hazardous materials mutual assistance group designed to minimize the effects of incidents within a major chemical manufacturing/petroleum refining area near Louisville. Southwestern Jefferson County is home to an extensive chemical/plastics/petroleum processing complex including 13 major chemical plants operated by B.F. Goodrich, DuPont, Borden Chemical, Stauffer Chemical, and other large industrial corporations. This complex of industries created a mutual aid group some 38 years ago, called Rubbertown Mutual Aid Association, under which member organizations assist one another in the event of a serious incident. By using this long-standing industrial association as a base, and adding representatives from a number of public safety agencies in recent years; public/private interests in Jefferson County have joined forces—equipment, personnel, resources, expertise, and determination—to insure the safe transportation and handling of hazardous materials within their community.

Officer Bill Wetter of the Jefferson County Police Department is a former Chairman of Jefferson County HAZ-MAT Mutual Assistance Team. Also represented within the HAZ-MAT effort are 23 separate fire departments, the Metropolitan Sewer District, the Agency of Civil Preparedness, the Kentucky State Police, and a number of industrial organizations. "The structure of the organization is strictly mutual aid," notes Wetter, an Emergency Medical Technician and Paramedic, "none of the participants are paid to be a member of HAZ-MAT. I represent the Jefferson County Police Department, the provider of emergency medical services within the county. Our main responsibility at an incident is provision of E.M.S. for firefighters and people at the scene as well as for perimeter security, and for evacuation in conjunction with the responding fire department. We might also establish an emergency medical field hospital if necessary, decide if air evacuation is needed using police and military helicopters, secure additional ambulances, or make a call-up of additional physicians. The various fire departments have their own operation, including response vehicles and monitoring and control equipment. Civil Preparedness, as a further example, has their own responsibility for evacuation and the establishment of emergency shelters. The Health Department is responsible for environmental concerns, while the Metropolitan Sewer District handles matters involving sewers and waterways. Each organization has its own 'bag', but we meet monthly as a collective group to talk over each agency's responsibilities so they can be dovetailed into an overall plan."

"Each HAZ-MAT representative as he/she arrives at the incident scene reports immediately to a designated command post," continues Wetter. "The main On-Scene Commander is the local fire chief for that district. He has the ultimate authority, bar none, for control and containment efforts. As we report to the Fire Chief acting as On-Scene Commander, he begins gathering information from representatives of the individual HAZ-MAT agencies and organizations, and we get our on-scene orders from him. We know each other because we meet together each month, but also because our individual protective gear is clearly marked with the designation of the specific organization we are representing.''

''Within the past couple of months we have been able to develop a common radio frequency so that representatives of the various fire departments, police, E.M.S., Civil Preparedness, Health Department, and industrial corporations are on a common radio network,'' adds Bill Wetter. ''We are using the 155.265 frequency that is normally set aside by the F.C.C. for strictly emergency and disaster coordination. The Rubbertown Mutual Aid Association has their own radio net that is hooked into our county-wide net so we can talk, for example, to the senior chemist at Borden Chemical, the health physicist at DuPont, or a radiological expert at the University of Louisville Medical School. These people are normally contacted by radio page and told to get on the radio. They come up on the radio and provide us with any advice or assistance we may need. Because of the radio network, we can often have a pretty good idea of what is happening on-scene before we arrive, and we are able to move ahead to activate key people. For example, if the product is identified as belonging to DuPont, we will immediately contact the DuPont representative and have him move to the scene, either in his private vehicle or by sending a police vehicle to pick him up.''

Now that they have instituted a common radio network, the HAZ-MAT group is in the process of publishing a resource document that will identify every piece of specialized response equipment and every person with special expertise in this area—from the nearest location of three ton of sand to a nuclear physicist or an expert in vinyl chloride.

''We try to get everyone involved who has a particular expertise,'' notes Officer Wetter, ''so we get to know each other on a first-name basis. We sit down together monthly to talk over common concerns, critique actual operations, hold training exercises, and develop specialized training. We have four basic committees: Communications, Training, Resource Document, and Operations. People on the Training Committee, for instance, tend to be training officers with fire departments or training officrs employed by specific industries in the area. They put on seminars dealing with general concerns; such as S.C.B.A., acid-gas-entry-suits, use of resource materials; and with quite specific concerns such as methods of dealing with acrylonitrile, sodium cyanide, or chlorine. We found by just checking around that we had a vast resevoir of knowledge in the area and a great many super people. We have not had a single organization say they could not contribute time, personnel, or equipment. For example, we in the Jefferson County Police Department have an E.M.S. response vehicle that probably has $15,000 worth of equipment on board. We haven't paid one cent for that equipment; it has all been donated by local people with private industry who can easily see the benefit of having a mutual aid organization such as HAZ-MAT. Industrial people definitely play a major role in HAZ-MAT. When I finished my term as Chairman, I was replaced by the Director of Health and Safety at DuPont.''

HAZ-MAT participants meet the third Wednesday of every month at a different location within the county, normally a fire house, police station, or industrial facility. Attendees critique incidents they have responded to, engage in preplanning, work on a hazardous materials annex to the overall county disaster plan, run training exercises, and so forth. A recent monthly meeting was held at the railroad museum in the eastern part of the county. A tank car was positioned on a siding, a leak was simulated, and a response exercise was held by all participating agencies and industrial organizations. The exercise was videotaped, a critique was held, some problem areas were identified, and participants are now working cooperatively to correct such problems.

''A particular strength of HAZ-MAT is the ability this community has demonstrated in bringing together extremely knowledgeable people who then maintain open lines of communication,'' concludes Bill Wetter. ''We have visited other communities where hazardous materials response was the sole responsibility of one agency. I can't see how a community the

HAZARDOUS MATERIALS ACCIDENT
1st ON-SCENE — CHECK LIST

1. *Report* the incident as a possible H/M accident. Give exact location and request assistance.
2. Stay up-wind and up-grade.
3. Isolate the area of non-essential personnel.
4. Avoid contact with liquid or fumes.
5. Eliminate ignition sources (smoking—flares—combustible engines).
6. Rescue injured only if prudent.
7. Identify materials & determine conditions (spill—fire—leak—solid—liquid—vapor; single or mixed load; waybills—bills of lading. Shipper-owner—manufacturer & carrier). *Report*
8. Initiate evacuation—Downwind first—Where to? *Report*
9. Establish command post location—upwind a safe distance. *Report* exact location & give approach route.

A. Circumstances of Hazardous Materials (HM) incidents vary so widely, it is impossible to establish specific guidlines to cover all incidents.
B. The goal is to remove the threat to public health or welfare, safety, and property which may result from a hazardous materials incident.
C. Do not compound the existing problem by creating a disaster out of an emergency.
D. The senior fire ground commander of the juristiction is the ON-SCENE Commander (OSC). As such he makes all the decisions.
E. Until the fire ground commander is on-scene, you must take charge and set the scene for a coordinated response and recovery.
F. *No one* is an expert in *all* problems associated with hazardous materials. Experts in specific fields provide needed pieces of information to the (OSC) to solve the overall problem.
G. You may have to delay attending to the injured in order to *save* the lives of many others.
H. Do not concern yourself with saving the H/M product or the carrier, it can be replaced.
I. Keep your dispatcher advised at all stages of your actions. He must advise other responding units and agencies.
J. Isolate the area of everyone not directly involved with incident until on-scene commander arrives.
K. *Do not* become part of the problem yourself by attempting irresponsible rescues or heroics.

Figure 1. Jefferson County Hazardous Materials Mutual Aid Team.

size of Jefferson County could have just one agency shoulder the responsibility for hazardous materials preparedness and response—it calls for a massive effort. The magnitude of the problem indicates you can't use traditional methods. The overall approach to hazardous materials has to be a combined effort from as many people as are necessary. If a community has the resources, it is imperative that they be tapped. With hazardous materials, I don't believe you can be overprepared, nor can you have too many resources available.''

Fire Department Controls the Transportation of Hazardous Materials Within and Through the City of Boston

There is a high level of interest in the pros and cons of *local* regulation of the transportation of hazardous materials. The City of Boston, because of its aggressive action on such questions and issues, has become a focal point of attention. The geographical location of the city presents a unique situation, and helped to bring about regulation and enforcement of hazardous materials transport by the Boston Fire Department.

To review the situation briefly, on March 3, 1981; one day after initial implementation, Boston's regulations controlling the transportation of hazardous materials within the city were suspended by a temporary restraining order issued by the federal district court in Boston on a complaint filed by national and state trucking associations. However, on April 6, 1981, the federal district court denied the major elements of a request by the trucking associations for a temporary injunction on the regulations, and thus opened the door for the Boston Fire Department to implement enforcement of the major thrust of the regulations; that is, to *prohibit* the transportation of dangerous chemicals through the city when there is neither a point of origin nor destination (delivery point) in the city, and to *restrict* the use of city streets in the downtown area between 6 A.M. and 8 P.M. daily *not* including Saturdays, Sundays, and holidays.

The regulations apply only to the transportation in bulk of large, specified quantities of LNG, LPG, liquefied hydrogen, certain other flammable liquids, certain flammable solids, generally so large as to require use of a tank truck. Also covered is the transportation in any quantity of certain explosives, poisonous gases, and radioactive materials.

Edward V. Clougherty, Ph.D. has been a chemist with the Boston Fire Department for the past 20 years, possibly the only fulltime chemist employed by a fire department in the country. His family has been involved with the Boston Fire Department since around 1900. His father was a former Deputy Chief, and Ed Clougherty represents the third or fourth generation of his family to serve. He has been involved with many phases of department operations: development of standard operating procedures pertaining to hazardous materials transportation and control, design and specifications for protective clothing for firefighters, fire prevention activities involved in the storage and handling of all types of chemicals, and matters of chemistry and flammability related to control of furnishings within the City of Boston.

"The geographic location of the city presents a unique situation," says Dr. Clougherty. "Located on the coast, downtown Boston is bisected north and south by Interstate Highway 93. The Massachusetts Turnpike (I-90) enters from the west and deadends with I-93 close to the heart of downtown. Thus, traffic passing through Boston from the north, west, or south has to go through the downtown section of the city: the very busy Quincy Market area, the highrise section of the city, etc. Because of a unique road pattern downtown where truck traffic has to leave the Interstate in order to avoid a 250 foot tunnel, and travel on city streets for a distance of 300 feet in one area plus an additional short distance in another area; carriers of hazardous cargo find themselves in one of the most densely populated areas of the city, an area of extremely high traffic."

"A further aspect of hazardous materials transportation within the City of Boston is the variety and amount of such cargos brought through the city," adds Clougherty. "Boston is bordered on the north by the cities of Everett and Chelsea. Everett is one of the major LNG port facilities on the east coast as well as an LPG port facility. Chelsea, Everett, and parts of

East Boston contain large gasoline/fuel handling facilities. In addition, Boston within its own city limits has a major LNG storage facility, an area south of downtown in the Dorchester section known as Commercial Point. A great number of LNG trucks, anywhere from 3,000 to 4,000 a year, move from the Port of Everett through the heart of the city to the storage facility at Commercial Point. The Interstate Highway, of course, is not a city street and therefore we have no authority on it. However, because the trucks have to come off the Interstate to avoid a tunnel, they come onto Boston streets in the center of the city where we do have jurisdiction."

In 1978, the Mayor of Boston requested Fire Commissioner George H. Paul to form and chair a Special Committee On Emergency Preparedness composed of representatives of city departments, community groups, and industrial organizations in order to recommend methods for safeguarding the storage and transportation of hazardous materials within the city. In September of 1979, the committee provided a report in which it described the nature of the problem, made specific recommendations about the storage/use/transportation of hazardous materials within the city, and suggested an ordinance to control the highway transport of hazardous materials. The Report also identified five major hazardous materials storage sites; outlined Boston Fire Department Standard-Operating-Procedures for "Extreme Emergency Incidents," response to the LNG storage plant at Commercial Point, and the handling of hazardous materials incidents involving rail cars or road vehicles; and designated routes for non-local transportation of LNG. From the overall report came a recommended ordinance for control of hazardous materials transportation by truck within the city. The ordinance, after proper legal procedures, was adopted by the Boston City Council.

"The ordinance itself did *not* create any regulations," explains Ed Clougherty. "It gave recommendations; it empowered the Fire Commissioner and the Commissioner of Health and Hospitals to write regulations; and it stated what the committee wanted accomplished through the ordinance. . . but it did not provide specific requirements. The actual regulations were written primarily by the fire department and published December 15, 1980 following a public hearing held September 15, 1980. After due notification to the public, the regulations were implemented on March 2, 1981. The thrust of our regulation is to prevent a possible catastrophe; we are not seeking to over-regulate an already heavily regulated industry. We feel the City of Boston in its unique geographical position has a potential for a catastrophic incident which the Fire Commissioner as the principal public safety officer in the city had to address. we had a mandate from the Mayor, we had the ordinance as a mandate from the citizenry, and the regulations were passed."

On March 3, 1981; one day after the regulation went into effect, the City was summoned to appear in federal district court and served with a temporary restraining order to cease and desist from the implementation of the regulations, an order obtained by representative trucking associations based on affidavits they had presented to the court. At that time, the question of an inconsistency ruling was raised by the truck associations. The federal judge involved indicated that if a ruling was not forthcoming from the U.S. Dept. of Transportation, he would rule on the request for inconsistency. That, evidently, was enough pressure to make D.O.T. act, and approximately March 19, 1981 D.O.T. issued its inconsistency ruling which prohibited a portion of the Boston Fire Department regulations. Basically, D.O.T. said Boston could not have a non-local ban on hazardous materials traffic, but could have a permit system for downtown because the streets in question were city streets. Overall, however, D.O.T. made no ruling on a large part of the regulations.

On April 6, 1981 the city was back in federal court on a motion for a temporary injunction, also brought by the trucking associations. At that hearing, Boston successfully defended its regulations against the court challenge by the trucking industry. U.S. District Judge Massone lifted the temporary restraining order on most of Boston's regulations including the one barring interstate shipments of hazardous materials through the city. Only a few minor regulations were enjoined.

"The federal judge did *not* allow three things," adds Dr. Clougherty. "Although we could have a permit system for the downtown area, we could not require carriers to place a decal on their trucks indicating they had obtained a permit, nor could we require them to carry a permit in the vehicle. Also, we had asked that an 'empty' truck, a truck carrying residual quantities of commodity, be marked so that for response purposes we would be able to distinguish between residual quantities and full quantities; but the judge did not allow this. The original regulations also called for use of the United Nations four-digit identification code which became mandatory nationwide on February 1, 1982. The judge said we could not use this; he did not say 'at this time,' but I am pretty sure that is what he meant. With the exception of these three original components, the regulations went into effect on April 6, 1981 and have been in effect since. We expect to go to trial again but have no indication of when the trial is coming. We are proceeding with the implementation of the regulations.''

The Boston regulations currently in effect apply to the following materials and quantities.

Class A Explosives: Any quantity.
Class B Explosives: Any quantity.
Poisonous Gases (A): Any quantity:
Flammable Solids:
   (which require the 2500 pounds or more "Dangerous When Wet" Label.)
Radioactive Material: Any quantity.
   (with Radioactive Yellow Three Label, excluding radioactive materials that are packaged in USA-DOT-7A Type A containers and are intended for use in, or incident to, research or medical diagnosis or treatment.

Liquefied Petr. Gas: 2500 pounds or more.
Methane (Liquefied): 400 pounds or more.
Liquefied Hydrogen: 400 pounds or more.
Flammable Liquids:
   (transported in bulk quantities of 1,000 gallons or more with flashpoints of 73 degrees F or less, with the exception of 140 proof or less.

Use of the city streets in the downtown area for the transportation of the regulated hazardous materials in the quantities specified is prohibited during the hours of 6 A.M. and 8 P.M. daily except Saturdays, Sundays, and holidays (except that permits for transportation in the downtown area on restricted use days during certain hours can be obtained upon application to the fire department).

Use of the city streets for transportation of regulated hazardous materials excluding the flammable liquids is *prohibited* where there is neither a point of origin or destination (delivery point) within the city (that is, flammable liquids are allowed through the city within certain areas even though they may not originate or terminate within the city. On the other hand, a commodity such as propane gas originating in Everett north of Boston and headed south, must go around the City of Boston seven days a week, 24 hours a day). However, where no alternative route outside the city is considered practicable, a permit to use city streets may be requested.

"We have been enforcing the regulations jointly with the Boston Police Department,'' continues Dr. Clougherty. The fire department is *not* empowered to stop moving vehicles, but of course the police are. By having a joint police/fire group on the street we can stop vehicles, control traffic if necessary, look at bills-of-lading, identify drivers, and generally check for conformance with the regulations. Initially, there was a certain amount of surveillance work. We would go out at six in the morning and stay until eight noting the type of traffic coming through. We sent out notices to companies we noted that had not yet applied for a permit and provided information on how they should apply. If a company had a permit but was in violation; such as, being on the road in the downtown area at 7:30 A.M., we sent them a notice stating they were subject to fines and revocation of their permit. This initial surveillance was merely an observation of moving vehicles, observation of placards, etc. We attempted to distinguish between trucks carrying gasoline (regulated) and trucks carrying fuel oil (not regulated) by bottom-loading or by our knowledge of the companies operating such vehicles. We did see a few propane trucks and are currently attempting to address this problem. Generally, however, we have had good cooperation from the affected parties. The non-local ban in particular, loads other than flammable liquids not originating or terminating within the city, has had very noticeable effect. Prior to installation of the regulations, a large quantity of LNG and propane went through the city. The LNG has stopped; there is no LNG going through the city now except that going to the Dorchester facility which is within the city. There is a very small amount of propane still moving through the city; our enforcement effort is addressing this situation and we are confident our efforts will be effective.''

"Regarding the overall question of response to and control of hazardous materials incidents," concludes Dr. Clougherty, "the Boston Fire Department follows individual written guidelines, or 'Standard Operating Procedures.' There is an SOP for LNG fires, rail car incidents, extreme emergency incidents, and one for radiological incidents, as well as a written contingency plan for the LNG storage facility in Dorchester. We also have prefire plans for all the major industrial occupancies, and a citywide communications system for emergency notification. The department has within its structure a large number of conventional ladder companies that play a key role at chemical emergencies. They have a variety of special tools and equipment including chemical protective suits and special breathing apparatus that allows respiratory protection in very high concentrations of hazardous substances. The officers and men of the Rescue companies are specially selected, train constantly, and bear the brunt of chemical emergency response on their shoulders.''

Tucson-Pima County Joint Hazardous Materials Emergency Response Team

When members of the Tucson-Pima County Joint Hazardous Materials Emergency Response Team respond to an incident scene, the *FIRST* thing they have to do is find parking spaces. It is entirely possible the eight responders representing four separate local/county/state government jurisdictions may arrive in eight different vehicles. The response procedure may sound unusual, but then the team itself is unusual — and innovative.

To respond effectively requires trained and experienced personnel, specialized tools/equipment/materials, and a high level of coordination and cooperation among responders. Tucson-Pima County's Joint Hazardous Materials Emergency Response Team attacks the problem by sharing personnel, equipment, knowledge, training, and financial resources of five government agencies in responding to hazardous materials spills, leaks, and accidents within Pima County that threaten multiple political subdivisions or exceed the effective emergency response capability of any single jurisdiction.

Recognized by a 1981 National Association of Counties Achievement Award designed to give national recognition to progressive developments which demonstrate an improvement in county government's sevices to citizens; the response team draws members from three police, one fire, and one emergency services (Civil Defense) organizations. Sergeant Ed Scott and Patrolman Jim Richards of the Tucson Police Department are specialists in the handling of explosives. Dave Canterbury and Ron Huerta are fire inspectors and hazardous materials specialists with the Tucson Fire Department. Sergeant John Reagor and Patrolman Don Holliday work for the Pima County Sheriff's Department, while Hank Axtell and Chris Long are highway patrolmen with the Arizona Department of Public Safety. Richard Casanova and Bob Dean of the Tucson-Pima County Department of Emergency Services provide coordination and support services for the team but do not suit-up and respond on-site. All members of the team are trained radiological defense officers.

Who has command and control of a hazardous materials incident scene is one of the most vexing problems faced by responders. At a major incident recently in another part of the country, 31 different interests were represented; including agencies of federal/state/county/municipal government; carrier/shipper/manufacturer representatives; commercial response personnel; and various advisors and experts. One of the most often asked questions with regard to hazardous materials incidents is. . . "Who's in charge here?" The Tucson-Pima County Joint Hazardous Materials Emergency Response Team adheres to the following procedure. The Tucson Fire Department will command any incident within the city limits of Tucson except that the Tucson Police Department will have command authority within the city limits for incidents involving explosive devices or criminal conduct. Outside of the city, the Pima County Sheriff's Department will be in command of any incident within Pima County outside of any municipal jurisdiction where such authority may be superseded, except that the Arizona Department of Public Safety will command any incident occurring on federal or state highways.

Although a full team response is not automatically triggered by minor incidents that can be handled by the primary response authority, during those situations requiring expanded response involving resources from other political subdivisions the responsible jurisdiction gathers all pertinent information available from the responding police or fire units and notifies a team member. Such initial information includes, where possible, the exact location of the incident, type of material involved, time of occurrence, quantity of material involved, area endangered, personnel on-scene, actions already taken, identification of shipper/manufacturer/container type, and rail car or truck identification numbers. The hazardous materials team member calls out additional team members as he deems necessary and initiates notification of any other support personnel who may be needed. He also insures that a containment perimeter and command post are established; and assesses types of additional aid required, weather conditions, terrain considerations, population density, anticipated movement and run-off of materials, and bodies of water potentially involved.

The team concept involves not only the mutual sharing of knowledge and professional expertise but the joint use of equipment. Additional required support from city/county/state resources not available through existing mutual aid agreements is coordinated and obtained through the Tucson-Pima County Department of Emergency Services, an agency that also handles reports and actions required by state and federal authorities for disaster assistance.

Pima County covers an area of 9,241 square miles including the City of Tucson which has a population of more than 250,000. The Tucson metropolitan area contains an estimated half-million residents, double the figure of ten years ago, yet the current rate of growth is much faster than the former rate. As a result, Tucson-Pima County is one of the fastest growing areas in the country. As for local conditions that led to formation of the multijurisdictional response team, Robert O. Dean, Plans & Operations Coordinator for the Tucson-Pima County Department of Emergency Services says. . . "About two years ago we had an organization called American Atomic operating here in Tucson. The company got some national publicity it didn't want when cited by the state atomic energy commission for venting rather high concentrations of tritium into the atmosphere. The tritium situation became a major event here in Arizona, and it dawned on a lot of people that we really didn't know what we had to face. We did not have a handle on businesses using hazardous materials; nor on what they manufactured, stored, or shipped. The city council passed an ordinance making it mandatory that all companies obtaining or renewing business licenses declare a wide range of substances, basically those regulated in transportation by the U.S. Department of Transportation. Let me tell you, they were *deluged* with reports. People had no idea of the numbers, types, and variety of different chemicals being used in the community until reporting was made mandatory. It was an eye-opening experience. The fire department was assigned as the primary agency to pull the data together, and they were overwhelmed with material.

"Also, about eight months ago the Corporation Commission of the State of Arizona decided they were going to get a handle on hazardous materials being transported within and through the state," continues Bob Dean. "They stopped two out of every ten interstate trucks for a period of 72 hours or so. What they learned literally horrified them. The volume and variety of hazardous materials being shipped was quite high, and roughly 60 percent of the trucks were in violation of one regulation or another. Also, the volume of chlorine here is very heavy, and anhydrous ammonia goes through here a lot because Arizona is quite an agricultural area."

Once they developed an understanding of the volume and variety of hazardous materials present in the area, officials found that each of the local jurisdictions did not have the capability by themselves to address the problem of response. "We were faced with the hard, cold reality that we needed a hazardous materials response capability that just did not exist at that time," remembers Bob Dean. "The answer to the problem seemed so obvious, and really so simple, yet no one had previously addressed the bringing together of specialists from various local agencies and combining their equipment and expertise to form a multijurisdictional hazardous materials response team. Elected officials and department heads anywhere are likely to be a bit sensitive about combining their operations with that of anyone else. The biggest problem we ran into, and this was probably a bit of a surprise, was jurisdictional prerogatives. Everybody was immediately hesitant to contribute to a combined operation. They all thought of themselves as having full jurisdictional authority and control and were hesitant in the beginning to give up some of that for a combined operation. The concept of a multijurisdictional hazardous materials response team stands by itself if you look at it honestly. Perhaps it is the only real answer. Few jurisdictions can come up with the kind of bucks necessary to form this type of an operation within one department. An educational effort was required to get the community to support the concept, and to convince the local political people to recognize and support it. Eventually, through an educational process, and by pointing to incidents that had occurred all over the U.S. and Canada, the point was made that we could be confronted by similar incidents at any moment; and that living by Murphy's Law as we seem to do, we needed a capability we did not have."

"The team was formed in July of 1980 and has responded about 30 times in the last year-and-a-half," continues Bob Dean. "We have been fortunate; we have not had any major incidents although there have been a couple of close calls. A rail tank car carrying ethylene caught fire over on the Southern Pacific line. There was a leak in a valve, and evidently a hot-box had ignited the gas. Flame was impinging on the skin of the car, and although a BLEVE (Boiling Liquid Expanding Vapor Explosion) was a possibility the incident was successfully brought under control. When a local chemical company had an incident recently that killed one and injured two, the company made its initial call to the fire department. The fire department hazardous materials guys were there in about ten minutes; once they sized up the situation and realized they were probably going to have to evacuate the area, they immediately activated the whole team. The Tucson Fire Department had the site control. Chlorine vapors from a mixing accident were staying close to the ground and spreading out into the community, so it was necessary to evacuate about a ten square block area. Resolution of the problem involved diluting the product greatly and then flushing it down the sewer with the agreement of the environmental and sewer people who stated the system could handle it if brought down to a certain concentration. The team responded quite well; everything we had previously worked out on paper, they were able to put into effect there in 15 to 20 minutes. The evacuation was completed within roughly 30 minutes. Team members are working well together and are following procedures. They know and like one another, and we all have learned to depend on other team members. The more times we respond to an incident, the easier it seems to be."

Additional incidents responded to by the team have included a fire in a boxcar loaded with insulation that was generating formaldehyde fumes, an overturned and leaking tank-trailer of anhydrous ammonia on Interstate-10, a 200 gallon tank-trailer of propane overturned and leaking, an exploding 15 gallon carboy of waste chemicals, and back-to-back incidents at a truck terminal featuring phosphorus trichloride and organo phosphorus.

"The team is on-call 24-hours a day, seven days a week," adds Robert Dean. "Any team member from any participating jurisdiction can trigger the full team response. All of the men, who have regular duties within their own agencies of course, are available by pager at all times for a team response. By bringing together not only their skills and training but their equipment as well, we've got ourselves a little capability that is a heck of a lot better than we had before. At the moment, we are trying to get elected people at the county and city level to fund us. At the moment, we have no independent funding; each department provides what little it can to its two men who are team members, and they bring their equipment with them. Most of the departments have been agreeable about letting their men come out at any time day or night, and the individual departments have been picking up whatever expenses there have been for overtime.''

''We did get about $10,000 from the county as a special grant to go out and buy some desperately needed life-support equipment,'' continues Dean. ''We didn't have enough air packs or acid suits, so the county came through with ten grand. We have also been successful in digging into different funds; we were able to get enough funds from the Arizona Highway Department about a year ago to send some team members to Florida and Colorado for training. We are wandering around with our hand out; and we get help, support, and contributions from anyone willing to provide it. Abouth eight months ago we received a gift right out of the blue. The Tucson International Airport Authority donated to the team a complete set of eight acid suits as well as a couple of fire suits. The equipment was all brand new but surplus to their needs. The manager of the airport believed in what we were doing and wanted to make an effort at recognition, exactly the kind of community recognition and support that we have tried to achieve. It was a real suprise, and we were delighted to get the equipment. Currently, we are trying very hard to obtain enough funding so we can purchase a good-sized truck to carry the most essential equipment, and be able to send out the one big truck rather than having five or six separate vehicles show up at the scene. Presently, each department uses its own vehicles. Tucson fire has a vehicle for hazardous materials; the Sheriff's Department guys have a four wheel drive carry-all that they load down with what they have. The Tucson Police bomb squad has a vehicle of their own, really nothing more than a four door sedan, loaded with everything they carry. The two highway patrol officers are each assigned a cruiser; when they respond to an incident, they show up with whatever they have in the trunk.''

''The procedures for a multiagency response team such as ours are probably not much different than they would be if we all worked for one agency except that communications did give us a bit of a problem we have not completely resolved,'' notes Dean. ''The various police departments have their special radio frequencies, and they are not always able to cross-over and talk to each other. For a long time, we had a heck of a problem for the police people to be able to talk to the Sheriff's people. We are trying to get a separate net; we are thinking about using 'LERN' (Law Enforcement Radio Network) and attempting to allow everyone to use the same net so if there is a problem in an emergency we can switch over to the common network. Also, we are attempting to get an intergovernmental agreement signed by which we could call on any jurisdiction throughout the area for equipment and personnel needed on an emergency basis. We do have verbal, unwritten agreements with a number of people, including some of the local industries, that in the event of an incident of major proportions all we have to do is call and they will provide all the back-up equipment they can gather. One chemical company, for example, has agreed to provide us with testing equipment, patch kits, chlorine kits, and additional respirators and air packs in an emergency.''

Since the team has not as yet succeeded in obtaining funding on a yearly basis, things can sometimes get very tight financially. At the chlorine incident mentioned previously, the team expended all available chlorine testing tubes used with a hand-held air sampler and had to dip into petty cash to obtain additional units. Team member Sergeant John Reagor of the Pima County Sheriff's Department has been known to open a presentation to business and industry personnel who might provide donations of equipment with the classic remark. . . ''We figured out a list of our resources, and we put it on a 3 x 5 card.''

''For future 'wishes,' I would like to see the different jurisdictions designate their two people as fulltime,'' concludes Bob Dan, ''dedicate both the men and the equipment to this combined operation. I would hope for the community to recognize the value of the team and budget for it on an annual basis so we could send people to school and replace equipment periodically. The team needs a permanent funding source because as the city and county continue to grow the hazardous materials threat will continue to grow. Someday, we are going to be confronted with a major incident. Unless we have a dedicated organization that exists on a permanent basis and is properly equipped and trained; no matter how hard we might try to respond, we might be up the proverbial creek if we don't get the effort on a dedicated, permanent, fulltime basis. One of the things that bothers me most about hazardous materials—certainly it was evident here—is the tremendous lack of public awareness as to what we are really facing. I'm a great believer in public education. If we can educate the people to the threat, I believe an enlightened public will respond. We take a certain pride in the fact that we got this project off the ground, and that we are succeeding with it.''

Local Governments in Washington State Identify Regional Hazardous Materials Transportation Corridors

A broadscale regional study of problems relating to the transportation of hazardous materials in the central Puget Sound area of western Washington State, funded by the U.S. Department of Transportation and administered by the Puget Sound Council of Governments, is expected to encourage agencies, municipalities, and industry to coordinate and cooperate actively in the management of hazardous materials transportation; and may provide valuable feedback to fire service organizations in the area. Participation by various community interest groups, viewed as essential to effective completion of the ambitious study as well as for adoption of measures ultimately recommended as a result of the study, has been extensive with inputs from rural, city, county and state emergency services response personnel; state and federal regulatory agencies; manufacturers, shippers, carriers, and users of hazardous materials; professional associations; and the public at large.

Study administrators recognized that federal and state laws pertaining to hazardous materials often rely heavily on local governments for hazardous materials incident response and control yet gave little if any voice to local governments in the prevention of such incidents. Local governments that must respond to and attempt to control incidents often have extremely limited ability to regulate, or even become aware of, the movement of hazardous materials within their corporate limits. On one hand, the federal government dominates regulation and prevention in the area of hazardous materials transportation; on the other hand, local governments are left to assume responsibility for control and response with little involvement by federal agencies yet are given almost no say in the prevention of incidents.

Local government officials are rapidly becoming aware of the dangers presented by hazardous materials. Constantly increasing numbers and volumes of hazardous materials being shipped in ever-larger unit payloads over a relatively fixed highway and rail network, continued urbanization that results in industrial sites and transportation corridors being surrounded by large population centers, and continued population growth all result in increased exposure to hazardous materials; and have led local governments to attempt to improve the capacity of their emergency response organizations. Experiencing great difficulty in accurately identifying types and volumes of hazardous materials transported and stored within their borders, local governments have often had to prepare to meet all possible dangers. Acting alone, local governments are faced with an impossible task.

Out in the Puget Sound area of western Washington State, the Puget Sound Council of Governments engaged in investigating relationships and local capabilities that exist to determine whether greater involvement by local governments and the private sector in accident prevention, and by the federal government in accident response, can result in improved public safety while not unduly restricting the movement of hazardous materials in commerce. One measure of the success obtained by this study, and a primary reason for the national interest it has generated, is involvement in the study process by all levels of government as well as by carriers-shippers-manufacturers-users of hazardous materials and other private sector interests.

The study task force completed four major tasks. The first task called for an identification of varieties and volumes of hazardous materials being transported through King, Kitsap, Pierce, and Snohomish counties by highway, rail, air, water, and pipeline. Local communities were informed of materials that move through their areas so that public safety agencies could knowledgably plan their response and control efforts.

Using information obtained during the identification phase, an evaluation was undertaken of the roles, responsibilities, jurisdictions, and management capabilities of agencies having prevention or response mandates. The role played by private industry in hazardous materials transportation emergencies was also reviewed. This evaluation phase considered response capabilities in terms of funding, trained personnel, equipment and communication systems available—and the extent and degree to which there was coordination among various agencies.

In a third phase of the study, federal/state programs and projects around the country were surveyed to learn their applicability to a comprehensive hazardous materials transportation coordination system for the Puget Sound area.

In a fourth and final phase, study administrators used results from the first three phases to develop options for a "Regional Prevention And Response Plan" that considers the necessity, feasibility, cost effectiveness, and requirements of a central transportation information center; the advantages and costs of alternative prevention and response systems using resources and legal authorities of all levels of government and industry, and the training needs and location of personnel.

The broadscale, community-wide base of the study group enabled it to identify hazardous materials storage and flow within the Puget Sound region to an amazing degree, quite a departure from the lack of specific knowledge pertaining to hazardous materials movement faced by most communities. In order to develop specific training, organizational and equipment options for regional hazardous materials management, the Puget Sound Council of Governments Study Group identified the prominent hazardous materials transportation corridors as well as areas of industrial concentration where hazardous materials shipments originate or terminate. Identification and analysis of major hazardous materials commodity movements in the central Puget Sound region will permit individual communities to better train and equip their emergency response personnel, and allow for the development of communication and coordination strategies.

The primary means of identification of hazardous materials transportation corridors, as well as points of origin and termination for such shipments, was questionnaires secured from manufacturers-users-shippers-carriers. Fire departments and fire districts also provided feedback on hazardous materials storage and transportation within their respective districts.

Questionnaires concentrated on the shipment and storage of *bulk* quantities of hazardous materials, and provided standardized data as to the commodity name, annual throughput or production, and number/net weight/origin and destination of shipments. The resultant inventory of transportation and storage corridors and locations allowed an identification of key hazardous materials-related industries, an analysis of hazardous materials flows, and a summary of the major hazardous cargos transported and stored within the communities of the region. Much of this information is now being analyzed statistically to determine the degree of risk posed by rail, highway, and water transportation of hazardous materials within the region. One result will be a ranking of hazardous materials by degree of hazard and an identification of both the likelihood and potential consequences of incidents.

Work was conducted under the guidance of an advisory board composed of representatives of government, carriers, manufacturing, processing, and public safety organizations. Technical subcommittees for "Manufacture, Transportation, and Use," "Emergency Response," "Training and Prevention," "Recovery and Environmental Protection," and "Federal and State Agencies" insured that the overall effort generated input from a wide variety of interests. Industrial and transportation organizations involved included Burlington Northern Railroad, Cryogenics Northwest, Widing Trucking, Pennwalt, Western Natural Gas, Union Oil, Hooker Chemical, Union Pacific Railroad, Sea-Land, Flying Tigers, N.W. Pulp and Paper Associates, Chem-Nuclear Systems, Boeing, and Weyerhaeuser. Government and public safety agencies supplying subcommittee members included the State of Washington, Seattle Fire Department, U.S.—D.O.T., U.S. Coast Guard, the port of Everett, the Port of Tacoma, Tacoma Fire Department, METRO, Kitsap County Fire District No. 1, SEATAC, the Washington State Patrol, the Port of Seattle, Medic I, Marysville Fire Department, U.S.—FEMA, Bellevue Fire Department, and the Everett Fire Department.

Captain John Hadfield of the Seattle Fire Department reported there are a total of 175 fire service organizations in the four-county area. (69 fire departments and 106 fire districts) He noted that the emergency response survey, with extensive assistance and inputs from the private sector, assisted fire service and law enforcement agencies to determine the quantity, frequency, and nature of hazardous materials movement within the area. "Hopefully," added Captain Hadfield, "one outcome of the model we developed was a recognition of exactly what equipment and training public safety agencies need to respond effectively. For the guy on the tailboard, what type of training does he need? What type of training is needed for the Washington State Patrol, or the Seattle Police Department?

"Through the advisory committee process," continued Captain Hadfield, "we were placed in contact with some very key people from area industries. . . actually sitting down at a table with these gentlemen and listening to their side of the story, perhaps learning to empathize with them a bit and recognizing that they have a job to do as well. While we may not agree with them in all ways, we came to recognize that these guys were trying, that they were attempting in many ways to assist the local fire departments. They in turn became more aware of our side. . . where it may be unfair to ask us to risk our lives dealing with an unmarked container, or with something that is ultra-hazardous, if our people do not have the proper equipment and training. Industrial people volunteered to give certain training. They are supplying rail cars and container ship boxes that we can use in training.

"In the overall effort, we tried to gather as many ideas and opinions as possible and keep the representation over as wide a range of people and interests as possible" added Hadfield, "so that when we ultimately came up with the recommendations the people will truthfully feel such recommendations came from all over. . . from the smallest organization to the largest. We sought a solution that will help *all* of us within this four-county region. Our base had to be 'out there.'

"The level of participation by various fire departments, both paid and volunteer, was very good" said Hadfield. "We originally sent out 20 questionnaires to specific fire departments to learn their pre-fire planning. We sent out 26 questionnaires to learn their thinking on a specialized hazardous materials unit, how the chain-of-command should be set up, who should be in charge at an emergency, their receptivity to training, their desire to participate further in the study, and their thoughts on the direction the study should be taking. The City of _____ comes to mind because they did such detailed work; it is the most beautiful piece of work I have seen . . . such depth. They gave us everything off their pre-fire files."

Study organizers eventually plan to test how the system will work by having a series of mock incidents. An initial mock incident will be allowed to run its course so as to allow observation of the first arriving units, how the mutual aid pacts work in practice, how state and federal agency personnel fit in, etc. Then, according to plan, Rockwell International Corporation, using an expected grant from the federal government, will establish a total emergency response plan for two communities. Once this plan has been completed, additional mock incidents will be conducted to learn if the planning assistance has assisted communities to better respond to an incident.

Hazardous Materials Pre-Alert in Santa Barbara, California

Although Vandenberg Air Force Base north of Santa Barbara, California has been launching liquid fueled rockets since the late 1950s, planned initiation of the Space Shuttle program is expected to greatly increase the total volume of fuels and oxidizers (nitrogen tetroxide, inhibited red fuming nitric acid, liquid hydrogen, hydrazines, solid rocket propellants, etc.) being transported into the area. It has been estimated that a single Space Shuttle launch will require that an excess of 100 truckloads of hazardous fuels and oxidizers be delivered to the base within a three-day period prior to launching. That portion of the main highway transportation route lying between Carpinteria and Gaviota is composed of approximately 46 miles of heavily travelled, four-lane express-way passing thorugh an area inhabited by a half-million people. The massive Sierra Madre mountains to the east are nearly impassable and severely limit potential escape routes; with the Pacific Ocean hard-by to the west, evacuation routing would be limited pretty much to north/south highways.

After reviewing an Environmental Impact Statement (E.I.S.) related to the Space Shuttle program, fire service and other public safety officials from Santa Barbara City and Santa Barbara County worked closely with officials from Vandenberg Air Force Base to develop a four-hour prior notification to public safety personnel of hazardous materials shipments enroute to Vandenberg, a "Hazardous Material Information Guide" specifically keyed to commodities used at Vandenberg that has been distributed to city and county public safety people on a "need to know" basis, and training of area first responders in both the notification procedure and specific commodities.

"Carrier personnel are required to call Vandenberg A.F.B. four hours prior to arrival and inform them of the type of commodity enroute and the estimated time of arrival," reported Hank Howard, formerly Fire Chief at Vandenberg and currently Training Officer of the Santa Barbara City Fire Department under Chief Rich Peterson. "The Vandenberg command post then notifies the Santa Barbara County dispatcher who fans out the information to public safety agencies along the route.

"We have had nothing but excellent cooperation from the Air Force people," noted Howard. "The overall effort required a lot of hard work and numerous meetings and agreements between Air Force and City/County representatives, but in addition to outcomes such as the prior warning system/guide/training directly related to transport of hazardous materials into Vandenberg, the effort also spurred on new planning directions for our public emergency response agencies. The City of Santa Barbara is developing a new hazardous materials contingency plan. Santa Barbara City, Santa Barbara County, and Carpinteria are purchasing full acid suits. Agencies are undertaking advanced training in population evacuation—a major problem area due to lack of routes away from Santa Barbara. Since the Southern Pacific Railroad main line from Los Angeles to San Francisco parallels U.S. Highway 101 through much of the area, there have been briefings of people who live in the danger zone, and we have purchased a BioMarine one-hour SCBA for search and intelligence at train incidents. Renewed emphasis has been placed on management of freeway traffic accidents involving hazardous materials; Santa Barbara has the only stoplights on U.S. Highway 101 between Los Angeles and San Francisco, and there are approximately 500 traffic accidents and 50 traffic fatalities a year on 101 in the urban areas. There has been extensive interagency planning with the California Highway Patrol, Santa Barbara Police Department, the Sheriff's office, ambulance companies, and other first responders. In addition, there has been special training for police units who have vehicles with the public address systems necessary to announce removal and evacuation, and standardization of maps used by first responders has been undertaken."

Vermonters Stress State/Local Coordination and Communication To Maximize Incident Response Effectiveness

Hazardous materials incidents are often thought of as a problem of urban, industrialized areas. Vermont used to have more cows than people, and still has little or no heavy industry. Only 260 or so firms in the entire state employ 100 or more workers. Vermont is hardly an industrialized state.

Rather surprisingly, Vermont firefighters have recently been called upon to bring under control a significant number of hazardous materials incidents. Title 20, Section 2673, of the *Vermont Statues Annotated* spells out the powers and duties of fire officers during fires, *but also during threat of fires and explosions* as follows: "Where an emergency exists in a municipality and there is no fire, but there is an imminent threat of fire or explosion, the ranking member of the fire company responding will be in charge as long as the imminent threat continues." Thus, Vermont fire officers have clear authority in most hazardous materials incidents. One exception is in the area of explosive ordnance: "Where there is a threat of bombing, the fire department shall surrender responsibility to the police department having jurisdiction in the area."

In the past five years, and particularly within the past two years, Vermont firefighters have developed a new public image as hazardous materials incident response personnel. The image, and the reputation for competence that goes with it, has been earned through performance. Awareness of hazardous materials came early to Vermont firefighters. In the early 1970s, a tank truck dropped a load of L.N.G. on Interstate Highway 89 in Waterbury, the only incident of L.N.G. on the ground in New England known to this writer. A rather bizarre incident occurred when an unmaned F-111 bomber crashed in Londonderry, Vermont. In November of 1975, three railroad tank cars of liquid petroleum gas (L.P.G.) BLEVE'd at Fairlee, Vermont. In February of 1977, firefighters in East Ryegate responded to a tank truck incident to find an aluminum-bodied milk tanker with no placards. Tight discipline and a knowledgable 21-year old fire chief averted a potential disaster. Rather than milk, the liquid leaking from the holed tanker was 6,000 gallons of inhibited styrene a flammable liquid whose vapors form explosive mixtures with air, tend to react violently, and can cause permanent damage to the respiratory system. These early, isolated events proved merely a prelude of much more widespread hazardous materials incidents to come.

Vermont fire departments have recently responded to chlorine tanks spilled on I-89 in Waterbury, a 6,000 gallon gasoline spill in Morrisville, a wrecked and leaking liquid oxygen tank truck on I-89 in Montpelier, an overturned gasoline tank truck in minus 28 degree weather in Berlin, two separate derailments of propane tank cars in Colchester, a derailed propane tank car in Springfield, an overturned tractor-trailer spewing tanks of chlorine and ammonia off the Suzie Wilson Road in Essex, and an overturned propane tank truck in Westford. Vermont departments have also had their share of hair-raising incidents not directly related to transportation. At Colchester in 1978, firefighters had to bring under control a potential chemical reaction caused by leaking, deteriorating rocket fuel stored in an industrial shed with more than 15 other chemicals. In September of 1979, Colchester firefighters were summond when a motor vehicle inspector for the Vermont Agency of Transportation found a trailer truck parked in a vacant lot leaking chemicals. Wearing protective clothing and breathing apparatus, firemen separated and catalogued 144 different drums and barrels containing approximately 80 different chemicals.

You would need a large-scale map to locate Colchester Center, Vermont; yet in a recent 24 month period the Colchester Center Volunteer Fire Company has been responsible for handling four major hazardous materials incidents and providing assistance to a neighboring department for a fifth.

Of the 240-odd fire departments in Vermont, only 12 or so are career departments while 225 to 230 are volunteer. Vermont fire departments are overwhelmingly volunteer, and funding is extremely tight. Money to purchase acid-gas-entry-suits, explosimeters, chlorine kits, and other tools of the hazardous materials trade is virtually nonexistent for a single, small department. Yet Vermont has a number of very perceptive fire officers and state officials who are super-effective in what is sometimes called "inter-agency communications." They figured that *someplace* within the state was the equipment and expertise to allow each and every local fire chief to effectively respond to any hazardous materials threat *if* an effective communications system could be developed.

In late 1977 representatives of the Vermont State Firefighters' Association, the Vermont Agency of Transportation, and various other agencies of state and local government perceived a need to be able to handle hazardous materials incidents as a result of a recognized increase in such incidents. Beginning as an ad-hoc committee, initially a very informal body, they drafted a report including legislative recommendations which in-turn created a Vermont Hazardous Materials Committee about the same time that chemical disasters in Pensacola, Florida; Waverly, Tennessee; and Youngstown, Florida received a great deal of attention in the press. The committee identified two basic needs. One was to establish a *response plan* within state government having two main components: a communications system so knowledgable people with specific expertise could be alerted and moved to the scene of a problem, and up-to-date inventories of resources and equipment available to respond to incidents. The response plan included a "one-call" Hazardous Materials Emergency telephone number, 802-828-3100. During normal working hours, this number is answered at state civil defense headquarters in Montpelier; at all other times, it is answered by a Vermont State Police dispatcher at the Middlesex barracks located just off I-89 in central Vermont.

The second component of cooperative hazardous materials incident response in Vermont is the *Vermont Hazardous Materials On-Scene Action Guide,* basic information and guidance, including the one-call emergency telephone number, for people responding on-scene. When a fire chief in any Vermont town faced with a hazardous materials incident calls 802-828-3100 and provides the information requested by the "worksheet" section of the *Guide,* he can obtain assistance and/or equipment from one or more of the following state agencies as needed (the "worksheet" is purposely designed to provide essentially the same information requested by CHEMTREC so that once it is completed, either the local fire chief on-scene, the local dispatcher, *or* the statewide emergency dispatcher can call CHEMTREC if desired).

Vermont State Police (traffic control, security, air bottle compressor/filler and other equipment, and for all incidents where explosives are involved).

Vermont Office of Civil Defense (personnel, equipment, and for all incidents where evacuation is necessary).

Vermont Environmental Agency (personnel and equipment from the Department of Forests and Parks, game wardens from the Fish and Game Department, a hazardous materials spills specialist from the Water Resources Department for any incident of actual or threatened water pollution or contamination, etc.).

Vermont Department of Agriculture (pesticides specialist, etc.).

Vermont Department of Health (radiologists, health officers, safety and health inspectors, etc.).

Office of the Govenor.

Vermont Department of Labor and Industry (fire prevention specialists, etc.).

Vermont Agency of Transportation (trucks and heavy equipment from the Highway Department, Department of Motor Vehicles personnel for matters related to enforcement of CFR-49, etc.)

Vermont Public Service Board (gas specialist, pipeline specialist, and for incident concerns involving all public utilities).

Millitary Department of Vermont (National Guard personnel and equipment, etc.)

Vermont Agency of Human Services (when there is a need to relocate, feed, or house evacuees or victims, etc.)

Sixteen two-hour seminars, one in each Fire Mutual Aid District, were held to provide and explain the Vermont State-wide Hazardous Materials Incident Response Plan and the *Vermont Hazardous Materials On-Scene Action Guide* to 250 fire departments, 75-plus emergency medical squads, 60 police departments, and their respective dispatch personnel. Funded by the Vermont Agency of Transportation, and developed and conducted for the Vermont State Firefighters' Association by State Fire Instructors Norman Bird on the East and Walter Swarkowski on the West, the seminars were designed to "train the trainers" of each local emergency response organization.

In each seminar, instructors utilized six overhead transparencies and 130 color slides as building blocks of a visual program. In each of the 16 areas, they left local trainers with 12 color slides, six over-head transparencies, and a one-page lesson plan that contained the essence of the presentation.

A serious effort was made to maximize both the total number of trainees *and* the variety of emergency services personnel attending each session. Six weeks prior to the first session, a mass mailing of a covering letter and one copy of the *Vermont Hazardous Materials On-Scene Action Guide* went to 700 fire, police, emergency medical services, and dispatch personnel; as

VERMONT HAZARDOUS MATERIALS ON-SCENE ACTION GUIDE

GENERAL RECOMMENDATIONS

Never Under-estimate The Incident

Identify hazardous materials (look for placards, and if possible, labels and/or shipping documents).

Use worksheet on page 8 and 9.

ALL HAZARDOUS MATERIAL INCIDENTS
MUST BE REPORTED TO:

(802) 828-3100

and state **"THIS IS A HAZARDOUS MATERIALS EMERGENCY"**

Establish a command post as soon as possible.

Wear full protective clothing and breathing apparatus.

Allow only absolute minimum number of essential personnel within immediate danger area.

Consider all hazards when rescuing victims.

Evacuate 2,000 foot radius.

Stay upwind and upgrade.

Avoid/eliminate all ignition sources - (flares, combustible engines, smoking, electrical, etc.)

Avoid contact with spilled materials.

Avoid ends of tanks (heated cylinders and containers may explode).

Contain spillage (See Other Considerations).

Figure 2. Vermont Hazardous Materials On-Scene Action Guide.

well as to mayors and town managers. The mass mailing letter was a "teaser;" it told the recipient he could obtain more information and additional copies of the *Guide* at a local session to be announced later. The second notification provided more specific information and announced a time and place for the local meeting. The third notification was a phone call to each local emergency services organization seven days prior to the training session. Phone calls are time consuming and expensive but add a personal touch that helps to bring out attendees.

As a result of extensive state/local cooperation and communication in the small, rural state of Vermont; a fire chief or other emergency services responder on-scene is no longer on his own in obtaining expert assistance or specialized equipment.

The Hazardous Materials Unit of the Louisiana State Police: Encorcement, Investigation, and Emergency Response

Nearly a quarter of all the petrochemicals produced in the United States are produced within the State of Louisiana. In order to fulfill a legislative mandate given to the Secretary of the Louisiana Department of Public Safety to regulate the transportation of hazardous materials, the Louisiana State Police organized a Hazardous Materials Unit in mid-1978 to deal with hazardous materials enforcement, investigation, and emergency response. State law requires the Secretary to coordinate an emergency response system in conjunction with the Louisiana Department of Natural Resources and establish and equip emergency response teams at strategic locations throughout the state. Trained State Police officers are currently assigned to various locations in order to minimize their response time and maximize time spent during non-emergency hours coordinating various public and private agencies.

In early 1979, state troppers began to attend the U.S. Department of Transportation Safety Institute in Oklahoma City. Hazardous materials regulations adopted by the State of Louisiana are identical to those regulations adopted by D.O.T. Federal regulations that once applied to interstate transportation thereby became state regulations and applied to intrastate transportation where no regulations had previously existed. In addition to enforcement-related training, troopers assigned to the Hazardous Materials Unit have studied product identification, response management, spill control, protective equipment, and hostile environment rescue at institutes ranging from Emittsburg, Maryland to San Luis Obispo, California; with the bulk of response-related training undertaken at the Colorado Training Institute in Denver. Some team members have attended the two-week radiological emergency management course at Mercury, Nevada.

Hazardous Materials Unit troopers provided a good deal of enforcement-related training to other troopers and to municipal police departments throughout Louisiana. In conducting response-related training, the unit stresses coordination. New state regulations require the establishment of an on-scene command post at hazardous materials emergencies. Also covered are identification of the material involved by placards, container shape and markings, shipping documents, etc.; and the use of emergency action guides. Instruction covers the incompatibility of certain classes of chemicals and the by-products of their reaction. Trainees are familiarized with how the different classes of hazardous materials affect the environment in order that ecological damage can be minimized, potential physiological damage resulting from chemicals, and steps that can be taken to prevent or minimize injury.

Within the Hazardous Materials Unit, Pat Touchard is Director of Hazardous Materials Response. He is a Sergeant with ten years experience who has accumulated 400 hours of technical training in the hazardous materials field. "The overall Hazardous Materials Unit, currently staffed by 11 people with Lieutenant Goudeau in charge, has two primary responsibilities: enforcement and emergency response," says Touchard. "Personnel are organized into a five-man team at headquarters in Baton Rouge with two men responsible for inspection and follow-up investigations and two men responsible for education and training. Among the five-man field team, each trooper is responsible for all phases: education and training, follow-up investigation, inspection, emergency response, and routine enforcement. Right now we are spread pretty thin, but we are getting some good results. Basically, when there is a hazardous materials incident in Louisiana, state troopers who have been engaged in enforcing regulations governing transportation/manufacturing/storage then become the specialists involved in the police response to the emergency once hazardous materials have been released from their containers.

"The question of control of the scene. . . 'Who's in Charge?'. . . is the big question around the country," continues Sergeant Touchard. "It is important to maintain a unity of command. In Louisiana the state legislature gave the State Police the responsibility to 'control' a system of emergency response to hazardous materials emergencies. At our suggestion, 'control' was changed to 'coordinate.' For 'coordination' I envision a coordinator's eight arms reaching all areas as his brain acts as a clearinghouse for information. On the other hand, when I think of 'control' I think of one central figure who has the authority to say 'yea or nea, do it or don't do it.' We stress *coordination*, maximum possible involvement of all concerned agencies along horizontal lines, not just vertical, that may have a particular expertise to offer at the scene of an incident.''

In practice, control of the scene is spelled out in Louisiana's *Chemical Transportation Emergency Response Plan* as follows. In the event of an urban hazardous materials incident, either the fire department or the police department may be first on the scene. If the police department arrives first, they remain in charge of the scene until the arrival of a qualified fire department, at which time they may relinquish control of the incident and offer assistance. If the incident occurs in a rural area, the State Police will probably be dispatched. The trooper will control the scene until a qualified fire department or other agency capable of handling the incident arrives. The State Police may then relinquish control and offer whatever assistance is deemed necessary. If the incident occurs on a state or federal highway system, or at a railroad grade crossing of such highways outside incorporated municipalities, the State Police will be the primary controlling law enforcement agency. Troopers will control the scene until the local sheriff asks that control be relinquished to him.

In addition to the above, the state legislature passed a law that gave the District Fire Chief command of any and all fire personnel at an incident involving fire. ''We are still in a process of evolution,'' explains Sergeant Touchard, ''but generally any fire-related incident involving pure storage or manufacturing at a fixed-site remains the realm of the fire chief. He calls the shots: who gets evacuated, identification of potential chemical reactions, the size of area that may be affected, etc. He turns to us and says. . . 'Get me a chlorine specialist; get me a sulfuric acid specialist; locate a team that can handle hydrocyanic acid;'. . . or asks for whatever assistance he needs to fulfill his responsibilities. The fire chief pulls the strings.

''In other cases,'' adds Touchard, ''Specifically the non-fire-related transportation incident, we are likely to assume command. Historically in Louisiana, the State Police exercise jurisdiction over a transportation accident that occurs outside of city limits.

''With regard to emergency response, we rely heavily on industry,'' emphasizes Touchard. ''We take responsibility for a decision to bring in industry. If necessary, we may tell an industrial organization. . . 'Look, you are XYZ Chemical Company. ABC Chemical Company owns the product, but they are located 500 miles away and can't get here immediately. You are right across the street. You suggest to us what to do, and we will make the decisions and take the resonsibility.' In fact we get tremendous cooperation from industry. There *are* some exceptions; some industries are more willing and/or capable to participate than others. Overall though, we find that industry does not often ask 'Who?;' more often they merely ask. . . 'When and how did it happen?'. . . and they take care of the rest. That is, we are accepting responsibility for corrective actions taken; we feel state law gives us such responsibility, and we go ahead and ask for technical assistance from as many qualified people as possible before we make our decision. We want input. Companies have been willing as a public service to come in and assist us to control and contain, possibly even clean-up and dispose, even though until we contacted them they may have had no involvement or responsibility for the situation. Chlorine is a good example. For years the CHLOREP organization really has not cared who owns the cylinders, where they came from, where they are headed, or who manufactured the product. They provide all possible assistance whenever chlorine is involved. In many cases of gasoline spills, we contact the nearest refinery that is able to send competent assistance. In my experience, public safety concerns definitely come first with industry here in Louisiana.

''We have a fairly large number of incidents reported, some serious but most quite minor,'' reports Sergeant Touchard, ''both because of the heavy concentration of oil/chemical/gas industries in the area and because of some pretty strict laws. It can be a felony if you don't report an incident. Because of this we are getting a lot of precautionary reporting of releases.

We average one incident report a day. Many of these are very, very small releases; and we will take the reporting party's word if he says he is reporting it as a precautionary measure but has the situation completely under control. We ask exactly what occurred, and judging from his answers we get a feeling for whether or not a response is required from us. Eventually, some form of follow-up is done on even the very minor reports. We respond immediately to those incidents that obviously need emergency response. . . an average of two a week. . . but a very minor leak may result in a follow-up at some time in the future. We have to prioritize our response based on our knowledge of the seriousness involved.

"Officers in the field who are part of the team are presently in patrol cars; we have no immediate plans to change that," ways Sergeant Touchard. "They carry personal protective equipment, and we have on hand some very basic patching/plugging equipment and materials. Our philosophy is that if we can't stop the leak with common sense, we are going to control the situation and wait for qualified assistance. We do not want to get into the leak fixing business; that would be beyond our resources. We want to make clear that many times it is better to contain the situation then back-off and let industry come in. Remember, we use industry a great deal. We are not shy about saying. . . 'We need your help, equipment, and expertise.' Overall, the program has helped me to gain a healthy respect for industry."

In Louisiana, if you have enough hazardous materials on your property that a resultant spill or leak might exceed boundaries of your property, you must file a plan with the Department of Public Safety. Inspection/enforcement officers use these plans when inspecting trucking companies, terminals, storage complexes, refineries, chemistry laboratories, and similar facilities.

"For inspection and enforcement," adds Pat Touchard, "a team of five officers assigned to the field periodically selects an area that has a high volume of truck traffic and works steadily for a week or so, day and night, examining loads, manifests, etc. We invite federal D.O.T. people and state people to participate. We write a lot of violations, but do it with a cooperative attitude foremost in our minds. We are not bloodthirsty, but neither are we afraid to do our jobs. Many, many violations appear to be a direct result of ignorance of the regulations; a lack of knowledge."

The Louisiana Department of Public Safety has adopted a detailed *Chemical Transportation Emergency Response Plan* which includes a 24-hour "hotline" for required reporting of chemical releases. One component of the overall plan, a *Hazardous Material Emergency Response Plan Checklist*, is reproduced below.

   I.   **Establish and maintain communication with the scene.**
        1. Park the vehicle safely; this may later become the command post.
        2. Establish security perimeter.
  II.   **Identify hazardous materials involved.**
        1. Markings on container and container shape.
        2. Contact driver or conductor & attempt to obtain shipping documents.
        3. Contact CHEMTREC, shipper, or manufacturer.
 III.   **Determine potential danger.**
        1. What material is involved.
        2. How much material is involved.
        3. What other facilities may become involved.
        4. What is the wind direction.
  IV.   **Establish Command Post.**
        1. One member from each agency with portable radio.
        2. Policy level personnel at predetermined Emergency Operations Center.
   V.   **Notify appropriate agencies.**
        1. State of Louisiana Hazardous Materials Response Center.
        2. Sheriff's office and local police.
        3. Fire service.

Figure 3.

Annex Q of the Kentucky Natural Disaster Plan: State-Level Coordination of Inter-Agency Response to Hazardous Materials Incidents

In Kentucky, the Office of the State Fire Marshal is charged with the responsiblity of administering and enforcing regulations pertaining to the storage, transportation, handling and use of hazardous materials. Seventeen personnel who comprise the Hazardous Materials Section within the Office of the Fire Marshal have responded to more than 900 hazardous materials incidents since the office was initiated in 1974. In recent years, Kentucky has been recognized as a leader in the field of response to hazardous materials incidents due in large part to a unique method of statewide, interagency cooperation.

In an attempt to do away with bureaucratic duplication at an incident scene, Kentucky in 1978 held a series of meetings among all state agencies that might have a role to play at a hazardous materials incident. The result of these meetings was the Hazardous Materials Annex (Annex Q) to the Kentucky Natural Disaster Plan. Annex Q does not appoint an ultimate authority, but rather stresses a dual concept of coordination and operational control by recognizing areas of particular expertise.

Normally, the Kentucky State Police are the first to arrive at a disaster scene, and thus implement the applicable section of Kentucky's Natural Disaster Plan. If the incident features hazardous materials, Annex Q of the plan is put into operation. The State Fire Marshal will respond to the scene with a team to determine what needs to be done. If the operation is such as to require more than State Police and Fire Marshal involvement, a Kentucky Disaster Emergency Service Area Coordinator will respond to the scene to act as overall coordinator and to insure that all necessary resources are made available. A command post will be established on-scene at the vehicle of the senior law enforcement office present. Throughout the incident, the D.E.S. Coordinator functions as a "clearing house" for all information and activities while the agency representatives with the most pressing concerns retain operation control. For example, if fire or explosion is the greatest concern, the State Fire Marshal will direct all on-scene activities. If the main concern is a toxic cloud or chemicals entering a waterway or water supply, the Dept. of Natural Resources & Environmental Protection assumes overall operational control. In all cases, operating agencies share common data and information with the D.E.S. Coordinator who must often obtain resources required by various agencies.

A centralized State Emergency Operations Center in Frankfort, Kentucky provides back-up to on-scene agencies when activated at the request of the commander at the scene. All state agencies involved in the emergency response send to the E.O.C. in Frankfort a representative who is empowered to act for his agency. In addition, Annex Q spells out the primary responsibilities for eight state agencies for incidents involving hazardous materials.

Greensboro-Guilford County Radiological Assistance Team: (R.A.T.)

The purpose of the Greensboro-Guilford County, North Carolina Radiological Assistance Team (R.A.T.) is to provide requested on-site monitoring and advice on response to any incident involving radioactive materials within Guilford County. Local officials have put together an impressive inter-agency emergency response team for radioactive materials; an interesting feature is that nearly all training was done in-house by knowledgable persons within the community. This resulted in very good relations with those persons who serve as technical advisors as well as the industries and organizations they represent. The all-volunteer effort has pulled together firefighters, police officers, ambulance personnel, hospital staffs, and representatives of industry and citizens' groups who have worked together successfully to achieve a common objective: public safety.

The team operates under the direct control of the Radiation Protection Section of the North Carolina Department of Human Resources which has ultimate authority during radiological accidents within the state. At the scene of an incident the team cooperates with and advises the on-scene commander and participating agencies. Sergeant Roy Riggs of the Greensboro Police Department is the Team Leader while D. Jerold Stack, Deputy Chief of Operations for the Guilford County Fire Department; and Captain Dan Shumate of the Greensboro Fire Department; serve as Assistant Team Leaders. Overall, team members are drawn from six agencies of local government: Greensboro-Guilford County Emergency Management Assistance Agency, (EMAA) Greensboro City Police, Guilford County Fire Department, Greensboro City Fire Department, Guilford County Emergency Medical Service, and Guilford County Environmental Health.

Team members, who go on-site only if activiated by the North Carolina Radiation Protection Section or by local government, have been trained and equipped to undertake the following functions: *Identify the types of radiation present and the source type if possible. *Provide on-scene monitoring. *Provide personnel monitoring. *Assist in decontamination of personnel and equipment. *Record the incident from the time of the team's involvement to completion.

On-scene, the team may divide into three sub-teams. A Monitoring Group determines radiation levels and records readings in established proximities from the radiation source. A Personnel Monitoring Group monitors all injured and non-injured personnel at the site. Injured personnel are monitored by the team's medical officer who insures that radiation readings are furnished responding hospitals, and assists in determining if decontamination facilities should be activated. A Triage Team activated by the medical officer is sent to the decontamination center to monitor decontamination procedures.

The concept of R.A.T. team first arose in December of 1979 when the Greensboro-Guilford County Emergency Management Assistance Agency (EMAA) advised local governments that standard civil defense equipment and training had not provided adequate attention to peacetime radiation incidents. Basically, standard civil defense equipment measures the presence and amount of gamma radiation, and the presence of beta radiation, but does not register alpha radiation. A project involving the following components was proposed by local agencies. *Formation of a specialized response team. *Training of line personnel. *Procurement of adequate instrumentation. *Development of a medical response team.

EMAA researched equipment and advised procurement of instruments identical to those used by the North Carolina Radiation Protection Section's team. Initially, three sets were funded: one by the City of Greensboro, a second by Guilford County, and a third was funded jointly through the EMAA office. EMAA rebudgeted 1980 travel funds to cover expenses for an initial set of uniforms, and as of Fiscal year 1981 the EMAA budget includes a line item account for the team.

A special course was developed by EMAA for first responders. As of December of 1980, over 500 Guilford County firefighters had been trained with training for other departments continuing through 1981. With regard to development of a medical response system, Dr. John Krege, EMAA's Air Force medical advisor, worked with the Guilford County Emergency Medical Service to research and develop Standard-Operating-Procedures for peacetime radiation response. EMAA designed and filmed a video-tape for line personnel and arranged for special consultation with Dr. Ralph "Monty" Leonard of the Bowman Gray School of Medicine. Dr. Leonard's training was also videotaped. Concurrently, EMAA worked thorugh the Emergency Medical Service Regional Council to develop overall medical response in Guilford County.

*Dr. "Monty" Leonard and representatives from the North Carolina Radiation Protection Section trained personnel from every shift of each of the five hospitals in Guilford County.

*Representatives from four of the five hospitals in the county were trained in the use of civil defense instruments. Instruments were then provided to all five hospitals.

*Dr. John Krege/Dr. "Monty" Leonard/the N.C. Radiation Protection Section/EMAA compiled the five best prototype procedures for medical facilities then available in the United States and extended planning guidance to each facility so they could write their own plans. All five hospitals then proceeded to write their own.

In addition, personnel from the Guilford County Mental Health Department were trained in elementary health physics and general team procedures. Specialists in crisis control, they are available to assist on-site, or at any decontamination or medical facility requesting assistance.

Cost of equipment and the initial minimal outfitting of the team have been approximately $10,000. Team members are volunteers, all training and research was performed "in-kind," and all consultants donated their time to the effort. A particularly unusual aspect of the overall radiological preparedness in Guilford County is that from the very beginning representatives of diverse and disparate groups have worked together quite effectively. A number have already been mentioned. In addition, the American Friends Service Committee, the Cudzu Alliance, and the North Carolina Public Interest Research Group have continuously engaged in research, project coordination, and training to accomplish the overall objectives of the broad-based effort. As one EMAA official noted recently, "We calculated that the American Friends Service Committee alone saved us eight months of research." As further examples, workers designing and providing a special Core Team training week included representatives of the Duke Power Corporation's McGuire Nuclear Facility, the American Friends Service Committee, and N.C.-P.I.R.G., along with state and local government personnel; also, the Duke Power Company provided six health Physicists for a weekend to evaluate and offer advice during a simulation of a radiological emergency involving all five hospitals in the county.

1. Emergency Management Assistant Agency Office

6 Rolls 2" Masking Tape
1 E520 Geiger Counter
1 PIC-6 Ion Chamber
1 PAC4-G Alpha Meter
2 (ea) Dosimeters (low range, Eberline)
6 (ea) Propane Gas Cylinders
1 (Set) Radiation Check Sources
1 Box Surgical Gloves
4 (ea) CDV 777-1 Shelter Kits
1 Box Plastic Bags (Small)
1 Box Batteries (for instruments)
1 Box Marking Tape
6 (ea) 1 Gallon Plastic Bottles
1 Gallon Soap (Decontamination)

2. Fish Vehicle

8 (ea) CDV 777-1 Shelter Kits
1 (ea) Team Box with: forms/paper/pencils, etc.
1 Box 50 Dosimeters (high range) with charger
(For additional equipment carried on FISH vehicle for hazardous materials, see FISH Vehicle Equipment List)

3. County Fire Marshal's Office

1 (ea) E520 Geiger Counter
1 (ea) PIC-6 Ion Chamber
1 (ea) PAC4-G Alpha Meter
6 (ea) Propane Gas Cylinders

4. Team Members

1 Pair Coveralls (White
1 Pair Gloves (Brown)
2 Each, Hoods (White)
1 Pair Boots (Rubber)

Figure 4.

FISH (Fire Investigation Special Hazards)

1　10,000 lbs. blackhawk porta power and attachments
19　Body Stakes
20　Metal Tags
1　Hand light 7½ volts
1　Generator
2　Adaptors for drop cords
1　Entry Suit
3　Radiation Detection Kits (Civil Defense)
1　Exhaust fan
1　Explosimeter
1　Map ringer
1　Pair boots
2　1 gal. cans
1　20 lb. dry chemical extinguisher
8　Metal post
4　Nylon ropes (100 ft.)
1　Manila rope 5/8 (250 ft.)
2　Aluminum 10 qt. buckets
3　Packs ductseal
5　Books emergency medical tags
1　Box plaster of paris (5 lb.)
1　Flace
2　Corn forks
2　Long handle round tip shovels
1　Short handle round tip shovel
1　Flat head axe
1　Pitch fork
1　Roll fox wire
1　Bush axe
2　Fire axe
1　Yard stick
28　Pint cans
24　Mason pint jars
3　½ gal. cans—metal
6　Pint cans—metal
1　18 inch crow bar
4　Wood chisel ¼"–1"
1　12" crescent wrench
1　Combination wrench set 3/8–1¼ (14) piece
1　Adjustable pliers

1　14" pipe wrench
1　Short screw driver
1　½" chisel
1　Gas can (5 gal.)
3　Water extinguishers
1　Resuscitator
1　Telephone
1　Camera Kit 4x5
1　Strob. 500 flash
1　9 cup coffee pot
3　Plectrons
1　Box nylon bags
1　Film holder graflex (120 roll)
1　Camera instamatic-124
　　Assortment of pencils
5　Pack freezer bags (quart)
5　Cassette tapes
1　Box wastebasket liners (15 bags)
1　Hazardous material book
1　Pack "Keep Out" signs
1　Pack legal pads
1　Fire protection handbook
1　Flammable liquids gases book
1　Notebook highways & oil terminals
1　Chem card manual
1　Bulk oil storage manual
1　Poisoning book
1　Fiberglass stokes basket
1　24 ft. extension ladder
2　Hacksaws
20　No smoking signs
24　Mason quart jars
3　1 gal. cans—metal
6　Quart cans—metal
12　½ pint cans—metal
1　Tool box
2　Hacksaw blades
1　Long ¼" screw driver
1　Punch & chisel set (12 pcs.)
1　Small bolt cutter

Figure 4. Equipment and its location for the radiation team.

Virginia Hazardous Materials Emergency Response Program

　　　　The Virginia Office of Emergency And Energy Services (OEES) administers a statewide hazardous materials emergency response team program designed to provide local governments with critical skills and expertise from a variety of state agencies during any hazardous materials incident that exceeds local capabilities. OEES personnel provide a focal point for hazardous materials concerns within state government, and respond with others in a 25-foot Titan motorhome hazardous materials response vehicle equipped with the necessary tools/equipment/materials. Initial funding for the program was obtained in March of 1980 through a grant from the National Highway Transportation Safety Administration, with additional funding added by OEES in July of 1980.

4	sets firefighters clothing.	1	wet/dry vacuum.
18	sets vinyl spray suits.	1	butyl rubber acid suit.
6	Cairns Firefighter helmets	1	aluminized cover suit.
12	hardhats.	1	20 lbs. ABC extinguisher
6	sets gloves. (5 pr. set)	1	resuscitator.
	surgical gloves.	1	first aid station.
	neoprene fuel handling.	6	handlights.
	nitrile rubber.	10	equipment boxes.
	insulated glo-glove	1	microfiche reader.
	kynol-lined, leather.	1	telephone w/line tap.
12	sets Tyvek coveralls.	1	technical library kit.
6	pr. ¾ length boots.	2	1000-ft. extension cords.
6	pr. Leggin boots.	4	Onan generators.
12	sorbent pads.	1	CB radio.
25	lbs. dry sorbent.		
1	Matheson-Kitagawa Detector.	1	marine radio.
2	sets radiation equipment	1	scanner.
	Civil Defense: alpha, beta, gamma.	1	standard tool set.
	Ludlum: alpha, beta, gamma.	1	Gastechtor Model 1214.
	handling tongs.	1	law enforcement radio.
	dosimeters & chargers.	1	8-channel radio. (EMS/hospital, etc.)
6	Scott Air Pak II	1	aviation radio. (air-ground)
4	AO Full-face respirators	1	set nonsparking tools.

Figure 5. Equipment list: Va. OEES Response Vehicle.

Response team members with particular expertise drawn from more than a dozen Virginia state agencies in addition to OEES were in place and ready to respond by October of 1980. The team responds on a 24 hour basis when local government officials feel a situation will exceed their capabilities and/or resources. The state sponsored program is activated whenever a local official calls the OEES 24 hour number in Richmond, 804-323-2300, and defines the problem and the nature of assistance desired. Command resides with the local legal authority requesting assistance; the response team acts in an advisory role during a crisis, although active assistance will be provided when advisable or the situation warrants. There is no cost to a local government for a response by the state team.

A normal response team is composed of six members with capabilities ranging from communications to health physics. Although additional expertise is provided if a particular situation warrants, a basic response team is staffed by qualified volunteers available on short notice for assignments of from one-to-three days who are capable of performing the following functions.

Function 1: (Team Captain Forward) A person with direct knowledge of fire science, chemical formularies, and command/control techniques having necessary leadership qualities to enable him/her to work directly with local government officials at a forward command post.

Function 2: (Team Captain Rear) A person with background in management/logistics/communications plus knowledge of command-and-control and hazardous materials response equipment; responsible for setting up and managing the emergency response unit once it is on-scene.

Function 3: (Technical-Scientific Representative) A person with background as an industrial hygienist or health physicist, or with general laboratory exposure, who has a strong chemistry background and working knowledge of physical properties of various dangerous chemicals as well as the ability to critically assess threat to life in various situations in order to serve as a technical advisor.

Function 4: (Logistics Officer) A person familiar with heavy equipment and fire/EMS/pollution response apparatus, who has purchasing and acquisition experience in state government if possible; and is able to determine equipment requirements for teams as well as help to satisfy local government demands for equipment that exceeds their capability. Duties may include procurement of food and subsistence items for a team while deployed.

Function 5: (Public Information Officer) A person experienced in daily press contact, or with fulltime duties in the information services field, preferably with a background in emergency services and hazardous materials training who will be responsible for gathering and disseminating to the press vital information on any incident responded to. Also gathers data for team records.

Function 6: (Communications Officer) A person familiar with routine radio operations and having some background in emergency services communications who understands hazardous materials terminology and possesses the ability to maintain communications logs and thus is able to handle intrateam communications as well as incoming and outgoing traffic with other state agencies, local governments, and industry during the progress of an incident.

The state response effort operates out of five orbit areas. Area I in Richmond covers 19 towns; Area II covers 18 towns in the Norfolk/Newport News area; Area III centered in Roanoke includes 20 towns; Area IV covers 17 towns from Charlottesville; while Area V is considered to be the balance-of-state.

Multnomah County, Oregon Uses Multi Agency Approach to Haz Mat Response

In 1979 three separate agencies of county government in western Oregon —the Multnomah County Office of Emergency Management, Multnomah County Fire District 10, and the Multnomah County Division of Public Safety—submitted a joint proposal to the Defense Civil Preparedness Agency (later incorporated into the Federal Emergency Management Agency) and obtained $136,000 over two years to establish a hazardous materials management system. The grant required six basic tasks: completion of a hazard analysis; development of a resource inventory; design and implementation of a tactical information system; initiation of a multiagency response unit; provision of training; and implementation of a prevention program. There are many public safety agency hazardous materials response teams initiated and staffed by a single agency of local government such as Police, Fire, or Emergency Management; but the Multnomah County Hazardous Materials Response Unit is one of a limited number of multiagency response teams throughout the country initiated and staffed cooperatively by three separate agencies.

The eight-member response team that is a direct outgrowth of the Multnomah County effort is staffed by five firefighters, two deputy sheriffs, and one employee of the Office of Emergency Management. "The response unit will not respond into an area that has a recognized jurisdiction *unless* we are requested by the local fire agency," explains Team Captain, Len Malmquist, Assistant Fire Marshal with Multnomah County Fire District 10. "We have fire department back-up for everything we do; mainly because of the need for hoses and water for possible ignition, decontamination, and wash-down. When a local fire agency calls, we tell them what they can expect. We respond as advisors; we do not take over their scene. We advise the incident commander as to what would be the best course of action. If he does not have the necessary capabilities and equipment, we will do the job for him; but any equipment loss must be reimbursed. They pay for expended equipment only; they do not have to reimburse us for salaries of team personnel. As part of the grant, the fire department has control over operation of the response unit. My chief has decided that we will respond anywhere in the county, plus a 30-minute Code-3 perimeter around the county with no okay required; anything further than that, I have to obtain approval. Up to the present time, we have been able to respond anyplace we have been requested to go. We are currently working on agreements with the State of Washington that begins right across the river from us; it looks as though in the future we may be responding into Southwest Washington."

Team members initially did a hazard analysis of their area of coverage and used information obtained to establish requirements for their response unit. They identified all fixed facilities where hazardous materials are manufactured, stored, distributed, transferred, or sold; and determined the type and quantity of material involved. In addition, they examined the transportation of hazardous materials and the routes used in and through the area. Information obtained on major high-risk fixed facilities, major carriers of hazardous materials, main transportation routes, and types and quantities of materials was used in selecting and equipping a mobile response unit and in training personnel.

"We bought a 26-foot motorhome; Fire District 10 stripped it out, installed shelving, and generally finished it off to our specific needs," remembers team member Penny Roe of the Multnomah County Office of Emergency Management. "From the hazard analysis, we determined exactly what chemicals were in the area; Fire District 10 had already put together a manual system pertaining to what they had found during inspections, and we took that manual tactical information system to a consultant who did a computer software system for us. We installed a floppy-disc minicomputer. The actual computer language we use is called 'Syndex.' We have a screen, a keyboard, and a printer at the Office of Emergency Management; and a terminal inside the motorhome that is hooked to the computer by mobile telephone. We can access the computer by product name, by synonym name, and by location/address within the county. Each category is cross-referenced to the assigned Department of Transportation number or Chemical Abstract Service (CAS) number. There is also a notation to indicate whether or not the chemical can be found in the OHM-TADS system. (Oil and Hazardous Materials Technical Assistance Data System) We have a listing of industrial organizations we can call for additional information about specific products. Also, there is a 'comment section' within which there are several different items: how the chemical is classified under the National Fire Protection Association '704 System;' (rating of health, fire, reactivity and other related hazards created by short-term exposure as might be encountered under fire or related emergency conditions. Applies to industrial and institutional facilities only; does not apply to transportation or to use by the general public.) how it is placarded according to the D.O.T. system; what fire protection is needed; and a number of other notations that would be of importance to response personnel—flashpoint, specific gravity, vapor density, reactivity toxicity, and flammability. Our 'synonym section' has the D.O.T. number, the CAS number, (Chemical Abstract Service) and the synonym name."

"When we go to 'location,' we have the street number, the street name and the city;" adds Penny Roe, "as differentiated from the entire City of Portland, we used the number of the fire engine closest to that particular facility so we know the area of the city being referred to. We have two emergency phone numbers for each particular facility, plus another 'comment section' where we list how the building is placarded according to the 704 system; where the product can be found on the property; other chemicals at this particular company that we should be aware of; how the product(s) is stored within the building; and so forth. Such information can be accessed either from the Office of Emergency Management or by the response team operating the motorhome. The system contains those chemicals we *know* are to be found within the county. We also maintain an extensive, 92-volume reference library in case we come up against something that is not normally found within the county so we can research a particular product and add data on it to the system if desired. The system can be continuously updated."

"Basically, we are using the system in three major ways," explains Len Malmquist. "We are using it to inventory chemicals that are permanently stored within our area; we are using it for an inventory of chemicals that move through the area; and we are using it as a reference source for any particular chemical at any time. Although information put into the system initially was basically that the fire department had put together, over-time we have greatly expanded the database."

As to what would be a normal query to the system, Len Malmquist says, "If we were at a location that had a '704' placard and we were not familiar with what was inside the building, we could check the database by address and obtain a printout that tells the products and STCC code(s) (Standard Transportation Commodity Code) of the products that are stored in that particular location. Then by turning around and using that information, we obtain a printout of all information for a particular product." (See Figure 7a-c, Chapter 12, page 363).

Equipment

"We chose a motorhome response vehicle as differentiated from some of the other styles available because we determined that there was no other such vehicle or team in the State of Oregon that could respond with the level of capability we can. Also, we might have to go long distances and remain on-site for some period of time," notes Malmquist. "The motorhome works best for us because it is fully selfcontained with its own power source, water supply, and sleeping facilities. We changed the piping system in the motorhome so it is chemical-resistent; we can do our own decontamination on-board the unit and maintain the contamination for later neutralization and/or disposal. The equipment purchased to stock the motorhome was based strictly on the risk analysis that we did. We knew what chemicals we had in the area, and we knew what equipment would be required to deal with such chemicals, then we bought equipment to handle the previously identified chemicals. The motorhome is a 26-foot 'Cruise-Aire' manufactured by Georgie Boy. We selected this particular vehicle because it has a roof capacity of 3,000 pounds; removing internal partitions did not decrease the stability of the vehicle."

The motorhome is stocked with an extensive array of response gear, and has all-round communications capability. In addition to a mobile telephone, there is a preprogrammed 50-channel scanner and a citizen's band radio. Since the motorhome is an assigned police vehicle, it carries a police department base set with all metropolitan police frequencies. There is also a tie-in to local state fire and radio networks.

Team Personnel:

Duty as a multiagency response team member is over-and-above normal duties personnel are assigned within their own agencies. Team Captain, Len Malmquist of Fire District 10, is Deputy Fire Marshal and a fire investigator during the day, as well as Hazardous Materials Officer for the fire district. Back-up Team Captain, Brian Reynolds, is a Deupty Sheriff with the Multnomah County Division of Public Safety, and operates as a Truck Inspector on the freeways and in the industrial areas of Portland insuring that safety equipment on trucks is accessible by Oregon law. Dave Rouse is a Deputy Sheriff, Chemist, and Truck Inspector. Dave Harms, Jim Gallagher, Vern Hamilton, and Mike Hendrix are Fire Officers/Fire Inspectors with Fire District 10. Penny Roe is Hazardus Materials Coordinator within the Office of Emergency Management and has additional responsibilities related to disaster preparedness and response to natural disasters.

"In the motorhome, my main function is to run what we call the 'resource center,' " notes Penny Roe. "This consists of operating the computer, answering any telephone calls, and answering calls on the fire radio. I also monitor the television set we have on-board in order that we know what information is being broadcast; if information is incorrect, we can call the station and make a correction so the public is not unnecessarily alarmed. We also have a reference library on-board, and if we need to look up information that is not in the computer I would do it as well as maintain all reports. I also handle all necessary notifications to other agencies/organizations as we are enroute to the scene. As I obtain information about the chemical or product we are dealing with, I call the agencies or individuals who need such information so they can do their jobs. In addition, I am in phone contact with Myra Lee, Manager of the Office of Emergency Management, regarding additional resources we need. She in turn arranges to obtain what we need and lets me know when it will arrive on-site."

"The three agencies supplying members to the team all use different radio frequencies, so we utilized the fire department dispatch center and provided each team member with an all-call pager," says Len Malmquist. "When the dispatch center wants to call out the hazardous materials team, they merely push a button that tones all the members pagers at the same time. They say whether they want one, three, or all of us; and we respond based on that page."

One of the most often asked questions related to hazardous materials incidents is. . . "Who's in charge here?" How does a multiagency response team handle the question of command? "We just sat down and talked about it," replies Len Malmquist. "If it is a fire-type function, the fire department will be in charge. If it is more a police function; such as a truck accident on the highway, then the police department will be in charge. It depends on the type of function the incident represents. There is an internal agreement among the three agencies as to who is in overall charge at any type of incident."

"As far as we are concerned, a hazardous materials incident is a fire *and* police department problem," adds Deputy Sheriff Brian Reynolds. "If you evacuate people, or get involved in a number of other efforts common to hazardous materials incidents, and don't have a policeman *on the team* who is capable of communicating with fellow policemen—you can really have a problem. By having two qualified police representatives on this team, we don't have firemen telling policemen what to do, or vice-versa. We can appreciate the communications problem, and inform our respective command structures what is coming down the pike, what we are doing and why we are doing it; as long as they have faith in what we are telling them, we do not have a problem."

Training:

"We have used a two-part program with regard to training," notes Penny Roe. "All team members attended a course in hazardous materials put on by Dr. Woods whom we brought out from Nashville where he did the training program for the State of Tennessee. Len and Brian have been to San Luis Obispo for the course put on by the California Specialized Traninng Institute. The three of us have attended the radiological course at the Nevada Test Site, as well as other courses. Also, we have drill every Thursday afternoon for three hours. During that time period we take specialized training, bring in people to give classes, and do all of our equipment tests to insure that everyone is familiar with operation of the extensive equipment we carry. Our drills include taking the motorhome to each of the facilities in the area that handle hazardous materials and having the people at such facilities go through the unit. In turn, we tour their facility to see what types of products are being made and where they are stored. At the same time, we attempt to get a copy of their emergency plan that we file on-board the motorhome. During such visits, we attempt to initiate a good working relationship so they know what to expect from us, and we know what to expect from them."

"Also, such companies supply a lot of the chemicals we use for our schools," adds Lem Malmquist, "as well as manpower, assistance, and money in some cases. Without support from industry, I'm not sure we could carry on the level of effort we currently maintain."

Incidents:

During the first 10½ months of calendar year 1981, the team responded to 39 incidents ranging from acids to corrosives to poisons, and flammable liquids. "Intentional dumping or purposeful abandonment and transportation are the major sources of our incidents," says Malmquist. "We have had a few incidents at fixed sites, but only a very few. For example, there was a spill of pesticides on a roadway near I-84. We had some explosives dumped in the river; a poison, chloropicrin, dumped along the freeway; and a number of chemicals dumped in fields. We handle purposeful abandonment of hazardous materials quite often. Of course, flammable liquids are the most frequently transported commodity, but we have not had much problem with them. Farming is a large industry out here, so pesticides tend to be one of our most frequent problems; the incidents we respond to often involve pesticides."

Lessons Learned:

"We would recommend to any group that they first perform a complete hazard analysis of their area before they get too far into establishing a response program," emphasizes Len Malmquist. "An initial hazard analysis is crucial. Too many people seem to have bought equipment, learned it was not the correct equipment for what they later found to be in their area, and then had to buy additional equipment. It is critical that you understand what you will be dealing with *before* you buy equipment. Another important task is to contact everybody you can possibly think of who would conceivably be involved in any incident you might be asked to respond to. Sit down and find out what their concerns are; make clear what your concerns are so each agency has a good idea of what the other expects. Iron-out any difficulties well in advance because it just does not work if you don't. A third tip is that the media will kill you if they can. We've had to modify our procedures so that we have an inside-perimeter and an outside-perimeter beyond which we keep the news media and all others not directly involved with controlling an incident. We feed them only information that is relative to the incident—good, factual information—and we keep somebody there talking to them so they don't gather hypotheses and put it out on the news. As for general procedures developed through our experience, our contract required development of a manual that allows anyone to use any portion of the entire system we have devised. They can obtain information on the resource inventory only, the computer system only, the response unit and equipment, or everything we have done. Such data is outlined in the contract document that's available to anybody who writes for it.''

''You need to get the right team members, people who are able to work together under a wide variety of conditions,'' adds Penny Roe. ''We turn around once we have an incident and critique our actions very thoroughly, making sure that we say something to each team member so that feelings are not hurt and no one gets upset by the actions of the others. You really have to work to keep the lines of communication open, not only among government agencies and industrial personnel but among team members as well. Sometimes such matters must be handled in a very delicate fashion. We try to insure in the way we state comments that we are not criticizing personalities but rather just making sure that we do not repeat mistakes. Hopefully, that way we will not have problems reoccurring that we have already dealt with.''

10 TOOLS OF THE HAZARDOUS TRADE: EQUIPMENT, MATERIALS, AND REFERENCES USED BY INCIDENT RESPONSE TEAMS

Because hazardous materials normally become extremely dangerous only when released from some type of containerized environment, the ability to patch/plug leaks in a wide variety of containers has assumed critical importance. A number of hazardous materials emergency reponse teams around the country have assembled necessary equipment and materials that allow them to control numerous types of leaks in literally hundreds of different types of containers. Although critical support equipment . . . particularly vehicles, communications equipment, detection and monitoring devices, and personal protective gear . . . is often complex and costly; much of the equipment and materials actually used for patching and plugging is quite ordinary, inexpensive, and readily available.

Nearly every team the writer has contact with owns or has access to Chlorine Emergency Kits manufactured to specifications of the Chlorine Institute (342 Madison Avenue, New York, New York 10017. Tel. 212-682-4324) that are produced by and available from Indian Springs Manufacturing Co., Inc. (P.O. Box 112, Baldwinsville, New York 13027. Tel. 315-635-6101). Separate kits are available for chlorine cylinders, (Kit A) chlorine ton containers, (Kit B) and chlorine tank cars/tank trucks (Kit C). The kits operate on the principal of capping off leaking valves and/or fusible plugs and, in the case of cylinders and ton containers, of sealing off a leak in the container wall itself. The kits may be ordered directly from Indian Springs Manufacturing Company with current prices F.O.B. Baldwinsville, New York as follows:

Kit A: $658
Kit B: $780
Kit C: $775

A number of response teams arrange to borrow Chlorine Emergency Kits from organizations in their local area as needed, or arrange custody on permanent loan. The St. Johns County, Florida Hazardous Materials Team, for instance, obtained A and B chlorine kits from the local water department. Instead of the water department keeping the kits on its premises, the St. Johns County team keeps them on their response vehicle.

By obtaining Publication #35, *Location of Chlorine Emergency Kits*, (60 pages, current price: $5.00) from the Chlorine Institute, response organizations can learn the location and type (A,B, or C) of Chlorine Emergency Kits in all states. This is the easiest way to determine what industrial organizations, government agencies, and business/trade/technical associations in a particular area have kits available.

If you purchase or gain access to a Chlorine Emergency Kit (A,B, or C) you can obtain from the Chlorine Institute a training package for *each* kit composed of newly developed slides, a cassette recording synchronized to the slides, plus a complete script keyed to the slides. Each kit (specify A,B, or C) has from 59 to 72 slides and currently costs $100. Also, the Chlorine Institute can supply instructional booklets for each Chlorine Emergency Kit (specify A,B, or C) at a current price of $6.00 each.

It may be helpful to review the actual tools/equipment/materials carried by selected hazardous materials response teams around the country. The Memphis, Tennessee Division of Fire Services in February of 1978 formed two, five-man hazardous materials REACT teams that are staffed round-the-clock by a total of 55 men on all shifts. The primary purpose of these teams is to handle normal fire suppression calls, but either team can be reassigned on a moment's notice to respond to a chemical emergency. Each REACT team is equipped with a snorkel truck while two hazardous materials specialists assigned to the Office of the Fire Marshal each have four-wheel-drive Suburbans. All vehicles carry the equipment described below.

Communications equipment includes a fire department portable radio with recharger and detachable speaker mike, a fire department vehicle radio, and a civil defense vehicle radio. For protective clothing there are two chemical suits, two self-contained air masks with extra bottles, two proximity suits, a Nomex jumpsuit, rubber gloves, and six disposable rainsuits with boots. For "sniffers" and detection/monitoring equipment, Memphis REACT squads carry a gas indicator, one chemical detector with tubes, a pack of pH papers, and a radiological monitoring device consisting of a dosimeter/Geiger counter/ion chamber.

With regard to general tools/equipment/materials, REACT squads carry fieldglasses, a set of sparkproof tools, maps of hydrants/watermains accessible from the Interstate highway system, a handheld spotlight, spades and shovels, a 35mm camera, wooden plugs tapered from one inch to six inches, a roll of laboratory film, hoseclamps, a plumber's inflatable pipe plug, two rolls of duct tape, a resuscitator, and grounding cables. In addition, the two REACT snorkel trucks carry two flexible hatch funnels, four dome lid clamps, A/B/C chlorine kits, an air chisel, two lifelines, and four airbags.

Due to an increase in illegal disposal and accidental spills of hazardous materials in San Bernadino County, California; the Board of Supervisors and the county Environmental Health Services Department developed a hazardous materials response team. The team is somewhat unique in that it operates under the direction of the Health Officer. Its role is to perform field identification, site assessment, and initial containment of spills. The fire service agencies in the county are reasonably well prepared to handle incidents involving flammables and explosives, but the hazardous materials team is equipped with the specialized protective equipment needed to deal with many toxic, corrosive, sensitizing, carcinogenic, mutagenic, teratogenic, or irritating substances. The team is on-call at all times and will respond to any point in the county; geographically, the largest county in the United States.

Most of the team's work involves off-highway incidents. The two most common types of incidents are illegal disposals, and spills or releases in any of the three rail classification yards in the county. As duly authorized agents of the Health Officer, team members are charged with enforcement of the California hazardous waste control laws. The team provides on-scene experts in the proper and legal disposal of spilled materials, and team personnel are equipped to gather needed evidence in incidents involving illegal activity prior to cleanup operations (See Figure 1 for a listing of tools/equipment/materials carried aboard the County of San Bernadino's Environmental Health Services response vehicle).

From its inception in January of 1977 through June of 1981, the Hazardous Materials Team of the Jacksonville, Florida Fire Protection Division responded to 976 hazardous materials incidents, an average of 217 a year. The unit handles ordinary fire and rescue calls within its stationhouse area, and responds to hazardous materials emergencies anywhere within the City of Jacksonville's 822 square mile area. The 12-man team covers rotating shifts of 24-hours on duty and 48-hours off with four men assigned to each shift. The team operates two apparatus that carry acid/gas suits, oxygen rebreather masks, positive pressure air masks, radiological monitors, explosimeters/combustible gas detectors, heavyduty tools for rail cars and tank trucks, neutralizers, sorbent materials, chlorine kits, gas and liquid container repair kits, 300 gallons of foam and special foam applicators, an extensive hazardous materials reference

Equipment

1. Air packs with spare bottles
2. "LCD Combo" O_2 and combustible gas detector
3. Detector kit with polytest, phenol, chlorine, acrylonitrile tubes
4. Electronic footmeter
5. Toolbox with: slip joint pliers, crescent wrench, crowbar, screwdriver, duct tape, spark-proof bung wrench
6. Fence posts, barricade tape, and warning signs
7. 10 lb. sledge
8. Shovel
9. First aid kit and eye care kit
10. Sampling kit: scoops, plastic and glass jars, turkey basters, pipettes, pH paper
11. Explosion-proof flashlights
12. 35 mm camera with databack, autowinder, 500 mm telefoto
13. Disposal bags
14. Hazardous waste labels
15. Organic vapor/acid gas respirators
16. Binoculars
17. 4 gallons fresh water
18. Paper towels and pre-moistened towelettes
19. 20 lb. A-B-C dry chemical fire extinguisher
20. 85 gallon P.E. recovery bags
21. Safety line and belt
22. Rad meter

1. County road map
2. Thomas Bros. guide
3. Field Operations Manual: instructions for equipment, technical data on protective clothing, codes and guidelines, etc.
4. Reference books:
   a. Sax, I. Dangerous Properties of Industrial Materials
   b. Int'l Tech. Info. Inst.: Toxic & Hazardous Industrial Chemicals Safety Manual
   c. Gosselin, et.al.: Clinical Toxicology of Commercial Products
   d. Aldrich Chemical Co. Catalog
   e. Proctor & Hughes: Chemical Hazards of the Workplace
   f. The Merck Index
   g. D.O.T. Hazardous Materials Emergency Response Guidebook
   h. NIOSH/OSHA Pocket Guide to Chemical Hazards
   i. Robinson: Hazardous Chemical Spill Cleanup

Clothing

1. Gloves: Neoprene, Nitrile, PVA
2. Splash goggles
3. Tyvek coveralls
4. Nomex coveralls
5. P.E. coated Tyvek coveralls
6. Neoprene boots
7. Latex booties
8. PVC booties
9. Hardhats
10. Acid hood and suit
11. P.E. coated Tyvek suit with hood and pouch for air pack
12. Saranex coated suits

Communications

1. Mobile radios (2)
2. Personal communicators (4)
3. 50-channel scanner

Figure 1. San Bernadino HAZ MAT Team Vehicle Inventory.

The two fire apparatus used by the Jacksonville Fire Division's Hazardous Materials Team are standard type fire engines (pumpers) that have been modified to carry special tools and equipment.

1980 1500 6 PM Pumper with a 750 gallon water/booster tank:

One explosimeter
One flammable gas/vapor and oxygen detector with hydrocarbon ppm component
One air/vapor sampler kit—ppm
Two radiological monitors
One pH meter and pH paper
Four acid/gas suits
Four 30-minute positive pressure air masks
Two 25' and two 50' air hose extensions for air masks
Two 1-hour oxygen rebreather masks
Two 4-hour oxygen rebreather masks
Two voice amplifiers (Vox)
One portable radio
One pair binoculars (10 x 50 power)
One gas repair kit (plugs, adapters, caps, etc.)
One rubber gasket and clamp kit
One tool kit which includes nonspark type tools
Various steel plates for tank car patching, etc.
100' of various lengths and sizes of chain
One "A" Chlorine Patch Kit (100 and 150 lb. cylinders)
One "B" Chlorine Patch Kit (one ton containers)
100 gallons of AFFF Light Water (200 extra gallons kept at station)
20 gallons of Hi X foam (100 extra gallons kept at station)
Several foam eductors and nozzles, 1½" and 2½", 44 GPM to 250 GPM
One 250 GPM forcing foam maker
50 lbs. of soda ash—neutralizer (1000 lbs. kept at station)
Absorbents (granular and blankets) for picking up hydrocarbon spills, etc.
Hazardous materials reference library
One 20-lb. Purple K fire extinguisher, one CO_2 fire extinguisher and two foam (AFFF)
   fire extinguishers
One 45-lb. container of Class D fire extinguihsing powder
One wind sock
One product depth measuring device (DART)
Rain gear, extra clothing and 12 pair rubber gloves
12 corrosive vapor filter masks
5 Nomex hoods
1 heavy-duty ¾" drive socket tool kit for rail cars, ships, etc.
1 universal 2½" adapter for ships, etc.

1963 750 6 PM Pumper with a 1000 gallon water/booster tank:

One flammable gas/vapor and oxygen detector
Two radiological monitors
One pH meter
Two acid/gas suits
Two 30-minute positive pressure air masks
One "C" Chlorine Patch Kit (rail car and tank truck)
150 gallons AFFF Light Water
50 gallons Hi X foam
Several foam eductors and nozzles 1½" and 2½", 44 GPM and 250 GPM
Absorbent pads and blankets for picking up hydrocarbon spills
Hazardous materials reference library
1200' 5" hose
United Nations/universal ship fittings/adapters, etc.

Figure 2. Continued.

One rail car valve T-wrench
One manhole bar/hook/opener
Wooden plugs and shims
One 20-lb. Purple K fire extinguisher
One heavy-duty hand truck for carrying heavy equipment
One portable radio
One wind sock
One portable loud speaker
Two portable foam (AFFF) fire extinguishers

Figure 2. Special tools and equipment carried by the Jacksonville, Florida, Fire Division's Hazardous Materials Team.

library, and many additional items. (See Figure 2 for list of response equipment carried on-apparatus by the Jacksonville team.).

Separated by hazard class, the 976 incidents handled by the Jacksonville team from January, 1977 through June of 1981 can be broken out as follows:

1. Flammable/Combustible Liquids accounted for 26.8 percent of all incidents. (n = 262)
2. Flammable Gases accountered for 25.2 percent of all incidents. (n = 246)
3. Miscellaneous hazardous materials accounted for 19.6 percent of all incidents. (n = 191)
4. Nonflammable Gases accounted for 8.7 percent (tie) of all incidents. (n = 85)
5. Poisons accounted for 8.7 percent (tie) of all incidents. (n = 85)
6. Corrosives accounted for 7.8 percent of all incidents. (n = 76)
7. Oxidizers accounted for 1.4 percent of all incidents. (n = 14)
8. Flammable Solids accounted for 1.0 percent of all incidents. (n = 10)
9. Water Reactive Materials accounted for 0.3 percent of all incidents. (n = 3)
10. Radioactives accounted for 0.2 percent (tie) of all incidents. (n = 2)
11. Explosives accounted for 0.2 percent (tie) of all incidents. (n = 2)

Of all hazardous materials incidents responded to by the Jacksonville team during its first four-and-one-half years of operations, 59.3 percent (n = 579) were related to transportation while 40.7 percent (n = 397) involved fixed locations and facilities. Of those 579 incidents related to transportation:

42.7 percent (n = 247) involved the highway mode;
21.8 percent (n = 126) involved the pipeline mode;
18.1 percent (n = 105) involved the air mode;
10.2 percent (n = 59) involved the rail mode;
7.2 percent (n = 42) involved the water mode.

In 1977 the City of Tukwila, Washington initiated one of the first fire department hazardous materials response teams on the west coast. Through experience gained over a five year period responding to incidents involving gasoline, sulfuric acid, hydrofluoric acid, parathion, crude oil, alcohol, chlorine, propane, and a host of other chemicals within an area that contains two major Interstate highways, four mainline railroads, three major state highways, and the flightpaths of two commercial airports; the Tukwila Fire Department Team selected and evaluated an extensive assortment of tools/equipment/materials. (See Figure 3 for a complete listing of Tukwila equipment.)

The Orange County District Fire Department that covers 1,038 square miles of county land around the City of Orlando, Florida—one of the fastest growing areas in the country—is a paid department of over 500 personnel encompassing what was until recently 16 separate fire control districts.

All firefighters stationed at each of three firehouses spaced throughout the county have volunteered for and received indepth training in hazardous materials response and control. Personnel at other firehouses receive basic training in hazardous materials awareness and recognition. The three hazardous materials stationhouses are actually dual purpose units

Unit 56:

4 x 4¾ ton towing vehicle, 16,000 double line winch w/200 feet of 3/8'' cable, on-board foam tank containing 105 gallons of 3 percent AFFF. 1 250 GPM foam nozzle w/eductor and 200 feet of 2½'' pre-connected hose. 30 gallons of 6 percent AFFF in cans, 30 gallons of high expansion foam. Mobile radio 4 frequency, 35 watt. Portable radio 4 frequency, 5 watt. 4 traffic cones. DOT—P 5800.2 Emergency Response Guidebook. Tukwila Fire Department Pre-Fire Book. First Aid Kit. 5 lb. dry chemical extinguisher.

Trailer A:

  50 Gallons of 6/9 percent Alcohol AFFF
  20 Gallons of Protein Foam
200 Lbs. of Soda Ash
100 feet of 1½'' chemical hose
   3 Bales of absorbent pads (hydrocarbon)
   1 Bale of absorbent pads (acid)
   3 Fully self-contained entry suits (chemturion)
   1 Smoke ejector
   1 Plastic/Nylon ground tarp 18' x 12', yellow
   4 four-hour rebreathers
   2 thirty-minute air packs
   3 Household brooms
   1 Roll black visqueen 12' x 50'
   3 Scoop shovels
   1 Contractors wheel barrow
   1 Pointed shovel
   1 Fire eater high expansion foam nozzle
   2 one ton come-a-longs
   1 Movable chalk board
   2 50 lb. cans of Met-L-X powder (class D fires)
   1 Mobile radio Fire 2 only (telephone patch frequency)
   3 Hard Hats
   1 Hand truck
   1 Shower set-up
   2 Boxes of chain (100 ft. each)
   1 Electrical Cord Reel w/100 feet of cord
   1 100 ft. section of 5/8'' garden hose
   1 Set of Recovery Drums (1–85 gallon, 1–55 gallon, 1–12 gallon)
   8 One-gallon containers of Industrial strength bleach
   1 One-gallon ammonia

Lockers on Trailer A:

   1 Radiological Monitoring Kit
     1 CDV 715 Monitor
     1 CDV 717 Monitor
     1 CDV 700 Monitor
     6 Dosimeters
     1 Headset
     1 CDV 750 Dosimeter Charger
     8 D Cell Batteries
  11 Rolls of bright surveyors tape (10 min.)
   2 Nucle-Clean Suits w/decontamination kits
  24 Packaged disposable, Yellow Hooded, splash suits
   2 Containers of pH paper (0–11), (0–14)
   1 Gastechtor Model 1238 w/extensions, (0–500 ppm)
   1 Draeger Instrument w/selected vials

Figure 3. Continued.

1 ammonia
1 chlorine
1 methane
1 vinyl chloride
1 hydrocarbon
1 carbon dioxide
1 nitric acid
1 acetone
1 toluene
1 carbon monoxide
1 alcohol
1 Portable pump, 150 GPM, w/hard suction hose
5 Mechanical sealers
2 30 lb. Nitrogen cartridges for Ansul extinguishers
1 battery charger, for on-board electrical supply
2 cans anti-fog spray
5 plugs, Redwood
   Assortment of zip-lock bags
2 buckets, plastic, 5 gallon
1 bucket, plastic, 2½ gallon
4 Quarts of 4x Fire water
100 30'' steel rods with marker flags
1 first aid kit
2 hold-downs, rubber
24 55 gallon plastic can liners
4 blankets, disposable
16 flares, red, (10 min.)
4 Nomex flash hoods
2 rolls of duct tape (2'' x 60 yards)
10 Cyalume 30 minute high intensity light sticks
48 disposable respirators
12 Pair safety Goggles (green lense)
3 Cans industrial sealer (one pound each)
10 Bars Ivory soap (10 min.)
10 One-minute smoke bombs
1 Bottle baby powder
1 Pair of Binoculars

Reference Material Trailer A:

N.F.P.A. Hazardous Materials 5th edition Pacific Northwest Pest Control Handbook Guidelines—Handling Hazardous Materials (FST) Fire Fighters Handbook of Hazardous Materials, 3rd edition; Emergency Handling of Hazardous Materials (Bureau of Explosives) 1980; DOT-P 5800.2 Emergency Response Guidebook; Handbook of Reactive Chemical Hazards, 2nd edition; Code of Federal Regulation 49 Dangerous Properties of Industrial Materials, Sax, 5th edition; United States Coast Guard, CHRIS Manual; GATX Tank Car Manual Chemical Company Index (Tukwila Fire Department); Chemturion Manual Industrial Control Bulletins.

Exterior Trailer. A:

1 30 lb. ABC Extinguisher
2 30 lb. Class ''D'' Extinguishers
1 30 lb. BC Extinguishers
2 2½ Gallon Pressurized Water Extinguishers
2 Side Mounted Flood Lamps (500 watt each)
1 Portable typewriter

Figure 3. Continued.

Unit 53:

½ ton towing vehicle. Covered back. Mobile radio 4 frequency 35 watt. Portable radio 4 frequency, 5 watt. DOT—P5800.2 Emergency Response Guidebook. Tukwila Fire Department Pre-Fire Book. First Aid Kit. 5 lb. dry chemical extinguisher.

Trailer B:

 75 Bags of Sand (Approximate weight @ 48 lbs. 3600 lbs.)
  2 Cubic yards of saw dust
  4 Bales of straw
     Assorted lengths of 2 x 4's
     Assorted sizes of plywood
  1 Set of Recovery Drums (1–85 gallon, 1–55 gallon, 1–12 gallon)

Figure 3. Tukwila Fire Department Hazardous Material Team Equipment

handling both fire suppression calls and hazardous materials response; one hazardous material response vehicle at each of these three stations carries the types of specialized tools/equipment/materials listed in Figure 4. In addition, each of the three hazardous materials response vehicles is stocked with the following reference materials.

1. *Fire Protection Guide on Hazardous Materials.* (National Fire Protection Association)
2. *Chemical Hazards Response Information System.* (U.S. Coast Guard)
3. *Hazardous Materials Emergency Response Guidebook.* (U.S. Department of Transportation)
4. *Emergency Handling of Hazardous Materials In Surface Transportation.* (Bureau of Explosives/Association of American Railroads)
5. *Agricultural Chemical Handbook.*
6. *Hazardous Material Handbook* by James H. Meidl. (Glencoe Press)
7. *Flammable Hazardous Materials* by James H. Meidl. (Glencoe Press)
8. *Explosive & Toxic Hazardous Materials* by James H. Meidl. (Glencoe Press)
9. *Hazardous Materials Reference Manual.* (Labelmaster)
10. *Emergency Evacuation Procedures.* (Amtrack)
11. *Fire Officer's Guide to Dangerous Chemicals* by Charles W. Bahme. (National Fire Protection Association)
12. *Detector Tube Handbook.* (National Draeger)
13. *Emergency Action Guide For Selected Hazardous Materials.* (U.S. Department of Transportation)

It should be understood that all fire service hazardous materials teams carry foam, foam, and more foam for extinguishing flammable liquid fires and suppressing flammable liquid and other types of vapors. The Jacksonville, Florida two-truck unit, for example, keeps 300 gallons of various types of foams readily available. Four quadrant teams that make up the overall St. Johns County, Florida group each maintain 50 to 100 gallons of foam, while an additional 200 gallons is available at the office of the county fire coordinator. Houston's on-apparatus foam supply includes 25 gallons high expansion, 25 gallons protein, and 30 gallons hydrocarbon emulsifier. All teams carry foam educators although some rely on backup units for water and hoses. Also, all teams tend to carry Purple-K, AFFF, (Aqueous Film Forming Foam) Met-L-X, and CO-2 extinguishers.

Response personnel repeatedly stress how often they use common, ordinary tools and materials in controlling leaks of hazardous commodities. They point out that they do not try for "the ultimate patch," but rather work to temporarily stop the flow. . . or at least reduce the flow. Team members often work under frightening conditions, but their basic stock-in-trade is their ability to temporarily patch any leak in any vessel, tank, cylinder or drum. "We are a first aid company for leaks involving hazardous materials just as a Rescue company is to victims," notes one experienced responder. "Our primary objective is to control the situation."

Quantity	Description
1	Vice Grips-6
1	Vice Grips-7
1	Vice Grips-8
1	Vice Grips-10
1	Vice Grips-11
1	8" "C" Clamp
1	3" "C" Clamp
1	4" "C" Clamp
1	#708 Crescent Wrench
1	#712 Crescent Wrench
1	Bolt Cutter
	Sheet Metal Screws (various sizes)
1	Ammonia Spray & Swab
1	Pipe-Tube Cutter (small)
	Wooden Plugs (various sizes and shapes)
	Rubber Patches (various sizes)
50	Lb. Soda Ash
100	Ft. Chain
1	Crowbar
1	Key Hole Saw
2	Civil Defense Radiation Detection Units (Geiger Counter)
55	Gallon Foam-Angus 3% Flouroprotein
3	Proximity Suit-Bunker Coat
3	Proximity Suit-Bunker Pants
1	Hydraulic Jack-10 ton
1	Hoist Puller
1	Rubber Mallet
1	Screw Driver Set
1	Nut Driver Set
1	Spray Bottle
1	Brush
3	Pair Rubber Gloves
1	Caulking Gun
1	Tube Silicone Caulking
	Assorted Glues
1	Roll Masking Tape
1	Saw (Wood-hand)
1	Whisk Broom
1	Bottle Liquid Wrench
1	Wrench Set (various sizes)
1	Igloo Cooler (5 gal.)
2	Jumbo Sponges
1	Funnel
100	Ft. Rope
1	Roll Electrical Tape
2	Chlorine Patch Kit—"A"
2	Chlorine Patch Kit—"B"
1	64 Dodge-750 GPM Pumper (B)
1	76 Dodge-Squad Truck (C)
1	53 Seagrave/International-500 GPM Pumper (A)
2	2½" Foam Eductors
4	1½" Foam Eductors
4	1½" Foam Nozzles
1	1700 Watt APU

Figure 4. Continued.

2	60 Min. Bio-Paks
2	Pair Binoculars
6	Visor Assembly for Acid Suits (Stock)
1	Bio-Marine Gas Analyzer w/Chg. & Calibrator
2	2½'' x 1½'' Gated Wye-Elkhart
1	12' Fiberglass Trailer (D)
1	Ctn. PB-35 Plug & Dike
8	Pair Wolverine Firewall Gloves
4	CP-2000 Acid Suits
2	Sets Chris Manuals
3	NFPA Hazardous Chemical Handbook
30	Gallon Foam-Alcohol
20	Gallon Foam-A.F.F.F. 6%
3	Roll Plumbing Tape
1	Set Measuring Cups
1	Socket Set (15 pc.)
1	Pair Shears—Metal Type
1	Pipe Wrench—Large
1	Shovel—Square Type
1	Shovel—Round Type
1	Grub Hoe
	Assorted "C" Clamps—small
	Assorted Hose Clamps
4	Hack Saw Blades
1	Measuring Tape
2	Klein Pliers
1	Pouch-Carpenter's Type
1	Belt-Carpenter's Type
2	Rolls Duct Tape (Silver)
1	Hammer—Claw Type
1	Hack Saw
1	Council Axe
4	3/8'' Screw Pin Anchors & Schackles
4	½'' Screw Pin Anchors & Schackles
2	5/8'' Screw Pin Anchors & Schackles
2	½'' Crosby Cable Clamps
2	¼ Slip Hooks
2	5/10 Slip Hooks
1	1½'' Foam/Water Nozzle (Navy-type)
1	Wind Sock (with pole)
1	Explosimeter
1	Briefcase
1	Flashlight
	Assorted Manuals & Handbooks
1	Orlando-Central Florida Area Atlas
1	Alkacid Test Kit
1	Box Epoxy Putty
1	Jar Vaseline
1	Battery Pliers
1	Gas Wrench
4	Road Cones
1	5 Gallon Container Vermiculite
1	5 Gallon Container Soda Ash
1	Spray Bottle Ammonia
1	Replacement Filter/Biopak (Soda Sorb)
1	Block & Tackle

Figure 4. Continued.

1	Road Flag
2	Road Reflectors
2	Hydrant Wrenches
2	Barway Spanners
2	2½'' Spanners
1	Brush
1	Rubber Mallet
1	Meter Puller (electrical metters)
1	Bottle Ammonia (with swab)
	Assorted Gloves
1	Hux Bar
1	Water Tee Shut-off
1	Pickhead Axe
1	Flathead Axe
1	Handsaw
1	Boltcutters
1	Sledgehammer
1	Metal Plate
2	Wheel Chocks
1	Tool Box (with assorted handtools)
	Assorted Pike Poles & Ladders
1	5 Gallon Container Diking Material
1	Acid Suit/Hard Hat
1	15 Minute Scott Pak
2	30 Minute Scott Paks
1	20' 3/8'' Nylon Rope
1	Container Talcum Powder (for donning acid suit)
1	60 Minute Bio-Pak
	Assorted Cardboard Splints
1	Portable Acetylene/Oxygen Torch & Goggles
2	First Aid Kits
1	Multitrauma Kit
1	Box Plastic Bags
1	Box ''Banner Guard'' (scene protection)
2	Radiological Personnel Dosimeters
1	Box Sodium Chloride (for irrigation usage)
	Assorted of Foam & Rubber Pads (for patching)
25	Absorbing Blankets
1	Oxygen Resuscitator
	Assorted Cribbing Materials
2	Salvage Covers
	Assorted Leather/Rubber Patching Materials
1	50' Braided ½'' Rope
1	50' Manila ½'' Rope
1	50' Manila ¾'' Rope
2	500 Watt ''Circle D'' Lights
1	AC/DC Smoke Ejector (with hanger)
	Assorted Electrical Cords & Plugs
1	15 lb. CO-2 Extinguisher
1	20 lb. Dry Chemical Extinguisher
1	Onan Portable 2.5 kw Generator
1	''Porta-Power'' (10 ton)

The Orange County team also carries a selection of hoses, couplings, nozzles, and other assorted equipment.

Figure 4. Hazardous Materials Team Inventory (Orange County Fire Dept., Orlando, Florida)

Pop-rivet tools, tubeless tire/plug patch kits, assorted rubber patches/cement/gasket material, assorted O-rings/washers/nuts, various tapes, (duct, Teflon, electrician's, etc.) combinations of square/conical/wedgeshaped wooden plugs wrapped with felt or cloth, T-bolts, C-clamps, pipe plugs and caps, and similar materials are used repeatedly. It may require a $100,000 combination of vehicle/detection and monitoring equipment/communications gear/personal protective equipment to get a responder into position; but to actually plug the leak causing all the trouble he may use a T-bolt/gasket/thin metal sheeting/concentric washers/wingnut patch, or possibly even a togglebolt/rubber ball/washer/wingnut patch. He may even place an innertube over a leaking drum and tighten the tube with a stick to form a fairly good seal. He may jam a bicycle tube into a leaking pipe and then inflate it. For bigger pipes, if no plumber's inflatable pipe plug is available, he may use the bladder from a boxing bag. So much of the work done down in the mud requires improvisation and innovation. Basically, circumstances dictate procedures. As a result, teams tend to carry precut rubber patches/backing/gaskets of various sizes and shapes, pointed wooden dowels and plugs that can be hammered into ruptured pipes and other leaks, hydraulic jacks and chains for clamping patches to tanks of all dimensions, and assorted pipe fittings, sleeves, brackets, compression plates, valves, elbows and joints.

Professional response personnel point out repeatedly that the bulk of success on their assignment comes from being able to temporarily patch/plug a leak. "We are glorified plumbers," notes one, "if we can't stem the flow of any fluid known to man or God, we are in the wrong business." Various teams and individuals have cited the following very rough groundrules.

Rule One: Shut Off The Flow

By shut-off valves, electrical circuits, or other means built into the system that someone just forgot to activate. Too many times people have tried to patch leaks when they could have walked 20 feet and thrown a switch or turned a valve.

Rule Two: Rotate The Container

For simple nonpressurized containers such as drums, the simplest way to stop flow from a leak is to turn the container so the leak is at the topmost point of the container. For some pressurized containers, (Chlorine cylinders, for example) if the cylinder is inverted you may get liquid flowing from a valve leak. It may be possible to turn the cylinder so that gas rather than liquid finds its way through the leak.

Rule Three: Attempt To Decrease The Pressure

Decreasing pressure on many containers can substantially reduce the rate of flow from a leak. For a nonpressurized container, the goal is to restrict entry of outside air into the leaking container. For a pressurized container, the goal is to cool the tank and its contents through application of water, or possibly CO_2 from a fire extinguisher.

Rule Four: If Rule One, Two, And Three Didn't Work; (and your relief is nowhere in sight) Start Patching And Plugging.

Most hazardous materials response teams stock wooden dowels and plugs, rubber materials of various sizes and shapes, pipe plugs and caps, chains and hydraulic jacks, and patching compounds. A knowledgable responder once estimated that there are from 50 to 75 different types of patching compounds. Some are made specifically for hazardous materials usage; most are not. "Plug N' Dike" (TM) is a nontoxic, granular material that absorbs over 300 times its weight in water to produce a sealant that stops penetration of fuels and chemicals. A handful of "Plug N' Dike" kneaded with water for 30 seconds forms a sticky paste for patching and plugging. Larger quantities can be used to form dikes for containment or to divert spills away from sewers and other exposures. A common usage is to control truck fuel tank leaks and spills. "Plug N' Dike" is used by a number of fire departments, has an unlimited shelf life, and costs roughly $40 for a 50-pound box. It is available from Construction Aids Technology, Inc. 812 102nd N.E., Bellevue, Washington 98004. Tel. 206-454-2034.

Additional commonly used patching compounds and materials are detailed below. In any discussion of patching compounds/materials, the reader should understand that such compounds/materials are often made for uses other than the sealing of hazardous materials leaks; and neither the manufacturer nor the author makes any warranty, implied or otherwise, for their use with hazardous materials. Responders who use such materials normally confine their application to nonpressurized containers, select only compounds/materials that will not react with the leaking product, and practice application through drills and excercises in a safe atmosphere before ever attempting to patch/plug a leak of hazardous materials. This section is not intended to recommend or endorse the use of such materials/compounds with hazardous commodities, but merely to report such usage.

•Laboratory film is a stretchy film somewhat like very heavy Saran-Wrap. It is commonly used in laboratories for covering test tubes, beakers, etc., to provide a moisture barrier that allows gases to pass through. Response personnel sometimes use it to wrap threads or valves. Laboratory film is available in various widths from laboratory supply firms. A 50-foot role of film 20-inches wide costs about $25.

•Duct Seal compound is an oil-based putty available in blocks and is normally used by electricians for sealing around service entries, junction boxes, flashings and similar applications. A permanently soft but strong adhesive used to seal out moisture, duct seal compound is generally available from electrical supply houses. A one-pound block costs about $1.39

•Certain products available through hobby shops that sell model railroader supplies are normally used to make "mountains" on model railroad layouts. Once activated, such products expand both greatly and rapidly. For a large hole, a four-mil plastic bag can be inserted half-in/half-out of the hole with the closed end of the bag toward the inside of the hole. The compound is then mixed in the closed end of the bag and allowed to expand to a hard plastic-like material within the bag thus sealing the hole. The open end of the bag is kept open until expansion is complete, both to control the direction and area of expansion and to keep from breaking the bag.

•Closed-cell polyurethane foam (not open cell) will expand when placed in contact with a number of petroleum products and will not allow the petroleum product to pass. Jammed into a ruptured diesel or gasoline tank, the portion inside the tank will expand to block the hole.

•There are a number of leak sealers available through auto supply stores designed to repair fuel tanks without removing the fuel or dismantling the tank. These compounds tend to be fast acting, most work at temperatures as low as 30° F, and some are toxic or flammable and require due care when using.

The following materials have been used, sometimes under extremely adverse conditions, to temporarily but successfully patch/plug leaks of hazardous materials. It should be noted that not all will work under water, some are temperature sensitive, and a few may require ultraviolet light for setup and thus can be used only in sunlight. Normal sources of supply include auto body shops, hardware stores, lumberyards, auto supply stores, and marine supply houses.

Epoxy-type products	Duct tape
Marine Sealers	Liquid adhesive sealers
Plaster-of-paris	Autobody repair tape
Dental Impression material	Autobody sealants
Clay	Channel bonding adhesive
Gasket seating compounds	Urethane foam

With regard to tools, a number of teams make their own or modify existing tools to meet their special needs. T-bars, tool extensions, and crow's feet" (for applications such as turning valves in rail tank car domes from outside the dome) are common examples. Jacksonville team members have a variety of visegrips with a welded concave extension on each jaw they used for clamping off hoselines and other applications. Jacksonville personnel also make their own windsocks and erect them at any incident involving fire, vapor, toxic fumes or gases. A number of teams make their own dome lid clamps for tank trucks.

One fire officer noted recently, "The first thing a team has to do is beg, borrow, and . . . ahhh . . . not quite steal the necessary tools, equipment and training. The St. Johns County team in northeast Florida obtained two acid suits on permanent loan from industry. The Houston Fire Department team obtains equipment and materials from various industrial organizations, and has had training and travel expenses of team members paid by industry. A local rubber dealer gives Jacksonville and St. Johns County team members free-run of his scrap pile. Many teams gather the bulk of their frangible discs, fusible plugs, backcheck valves, pipe caps, couplings and other equipment related to specific industries during preplanning and training visits to these industries.

Many public safety agency response teams look to other agencies for assistance with start-up efforts. Memphis received more than $20,000 for its fire department team. The Virginia Office of Emergency & Energy Services obtained a motorhome response vehicle and equipment through a grant from the National Highway Taffic Safety Administration. Fire/police/emergency management agencies in Multnomah County, Oregon received $136,000 from the Federal Emergency Management Agency to develop and implement a hazardous materials response capability. The Denver Fire Department team obtained its response van because the Denver Office of Emergency Preparedness provided funding from contingency monies reserved for purchases of pressing emergency need items. The Jacksonville Fire Department team obtained a $20,000 Federal grant through the State of Florida Governor's Highway Safety Commission to assist with necessary tools, equipment and materials. A number of teams have obtained radiological monitoring devices on permanent loan from other agencies, and most at one time or another have borrowed equipment from the military.

Regarding tools/equipment/materials, personnel of a number of teams repeatedly offered the same advice: "If you don't have it; know where you can get it." All teams maintain a listing or cardfile with the name, address, contact person, and telephone number of organizations that can supply heavy equipment, vacuum trucks, wreckers, sorbent materials, booming and diking equipment, and other types of specialized hazardous materials response gear. In all such cases, team personnel attempt to obtain definite prior commitments from each potential provider.

One of the most important "tools" available to a hazardous materials responder is an attitude rather than a piece of equipment. Experienced responders make every possible effort to contain spilled hazardous materials and limit contamination to the smallest possible area. Containment is necessary to insure health and safety, protect property, avoid unnecessary environmental damage, and minimize potential liability. Containment includes not only the immediate control of the spill but continuous control of manpower and equipment as well.

Appropriate use of water is a prime consideration when attempting containment of a hazardous materials spill. Washing a toxic chemical into a storm drain, sewer, water source, or wildlife habitat can create a secondary emergency far more serious than that posed by the original spill. Water may dilute the chemical, but it will spread it as well.

Commonly used containment techniques include dikes, dams, sumps, booms, and diversionary waterways. A dike is an obstacle or barrier, most commonly a bank of earth, used to enclose, restrain, or protect. Shovels can be used to construct a dike of earth or sand around a limited spill area in a matter of minutes; larger spills or wider areas will require earthmoving equipment. If it is not possible to completely dike the spill itself, or if too much time would be required; limited diking can be constructed around catchbasins, sewer intakes, or drains that might allow the product to spread. A dam is a barrier built to obstruct the flow of water. Depending upon the rate of flow, dams can be constructed to completely stop the flow of water, allow only the surface portion to pass, or allow only the bottom portion to pass. For products having a specific gravity of less-than-one that are insoluble in water, including most petroleum products, an inverted siphon dam (sometimes called a pipe skimming dam) is commonly used to separate product from flowing water so the product can be obtained and picked up while clean water is released. Inverted siphon dams are not well able to handle large volumes of water such as rivers or large streams but are quite successful in dealing with limited volumes of flow, particularly in containing/separating contaminants from run-off water. A dam of earth or sandbags can be constructed over and around pipes that pass through the bottom area of the dam. The pipes are inclined slightly so the low end accepts the stream flow and the high end discharges it, and blocked until the dam has pooled water/contaminant to above the intake level of the pipes. Insoluble contaminants with a specific gravity of less-than-one will rise to the surface of the pooled water, the pipes can be unplugged, and uncontaminated water allowed to flow through the pipes. Contaminated water remains behind the dam and can be picked up with pumps, skimmers, or sorbent materials.

A material that is insoluble but has a specific gravity of greater-than-one, will tend to sink in water. If total flow is limited, a dam submerged just slightly below the surface will contain the sunken contaminant while allowing excess surface water to flow over the dam.

A sump is a pit, well, trench or hole where liquid is collected. A backhoe can dig a sump around a tank truck or tank car in a short period of time.

A boom is an obstruction that floats on the surface of water to contain or direct a floating contaminant. Since commercially made booms are often not available; makeshift booms can be erected of wire fencing, snowfence, netting or other restraining devices. For example, in the eastern states responders sometimes use tobacco netting available from farm supply stores. A 33-foot by 15-foot section costs about $7.50 and is easily stored. Straw, hay, or sorbent materials that float on water are then placed on the upstream side of the restraining device to entrap floating contaminants that are then picked up with skimmers or sorbent materials.

Dams are of no use with soluble contaminants unless it is feasible to dam the total waterflow. That is generally a very short-term expediency at best. However, if contaminant has entered a natural waterflow, it may be possible to dam the natural watercourse both above and below the spill area, then either dig a diversionary channel or use pumps and hoses to temporarily direct flow around the contaminated area. This method might be called for if the contaminated ground area is expected to leach contaminants for some time that without a diversionary waterway would enter the natural waterflow overtime.

If contaminated material has already entered stormdrain piping, inflatable plumber's pipe plugs are available that can close off a pipe ahead of contaminated drainage. Because plumber's pipe plugs are not always available, response crews have used inflated truck tubes, air bags, (normally used for lifting/positioning/leveling heavy objects) and even inflated beach balls to get the job done.

When dealing with flow rates, pipe capacities, volumes, and areas a little figuring is often required. With time often of the essence, first responders try not to let rusty math skills slow them down. Figure 5 provides an example of a conversion table that can be found in many response vehicles.

Cleanup can be very, very expensive. Containment can often be done quickly, easily, and cheaply. In some instances even intricate, time-consuming, expensive containment efforts can be cost-effective. It is almost always cost-effective to minimize the spread of a hazardous material and limit the area of contamination.

The containment procedures discussed previously related to spread of contamination by water flow or run-off from firefighting or hazard control efforts. Recognize that the spilled hazardous material could be a powder, granular material, or dust that could be spread by wind or carried away on clothing, footwear or equipment used by personnel on-scene. Salvage covers or tarps offer some protection from wind, and personnel and equipment can be con-

CONVERSION TABLES

Multiply	By	To Obtain
Acres	43,560	Square feet
Acres	4047	Square meters
Acres	1.562×10^{-3}	Square miles
Acres	4840	Square yards
Acre--feet	43,560	Cubic feet
Acre--feet	325,851	Gallons
Acre--feet	1233.48	Cubic meters
Atmospheres	76.0	Cms. of mercury
Atmospheres	29.92	Inches of mercury
Atmospheres	33.90	Feet of water
Atmospheres	10,332	Kgs./sq. meter
Atmospheres	14.70	Lbs./sq. inch
Atmospheres	1.058	Tons/sq. ft.
Barrels--oil	42	Gallons--oil
Barrels--cement	376	Pounds--cement
Bags or sacks--cement	94	Pounds--cement
Board feet	144 sq. in. X 1 in.	Cubic inches
British Thermal Units	0.2520	Kilogram--calories
British Thermal Units	777.6	Foot-lbs.
British Thermal Units	3.927×10^{-4}	Horsepower--hrs.
British Thermal Units	107.5	Kilogram--meters
British Thermal Units	2.928×10^{-4}	Kilowatt--hrs.
B.T.U./min.	12.96	Foot--lbs./sec.
B.T.U./min.	0.02356	Horsepower
B.T.U./min.	0.01757	Kilowatts
B.T.U./min.	17.57	Watts
Centares (Centiares)	1	Square meters
Centigrams	0.01	Grams
Centiliters	0.01	Liters
Centimeters	0.3937	Inches
Centimeters	0.01	Meters
Centimeters	10	Millimeters
Centimtrs. of mercury	0.01316	Atmospheres
Centimtrs. of mercury	0.4461	Feet of water
Centimtrs. of mercury	136.0	Kgs./sq. meter
Centimtrs. of mercury	27.85	Lbs./sq. ft.
Centimtrs. of mercury	0.1934	Lbs./sq. inch
Centimeters/sec.	1.969	Feet/min.
Centimeters/sec.	0.03281	Feet/sec.
Centimeters/sec.	0.036	Kilometers/hr.
Centimeters/sec.	0.6	Meters/min.
Centimeters/sec.	0.02237	Miles/hr.
Centimeters/sec.	3.728×10^{-4}	Miles/min.
Cms./sec./sec.	0.03281	Feet/sec./sec.
Cubic centimeters	3.531×10^{-5}	Cubic feet
Cubic centimeters	6.102×10^{-2}	Cubic inches
Cubic centimeters	10^{-6}	Cubic meters
Cubic centimeters	1.308×10^{-6}	Cubic yards
Cubic centimeters	2.642×10^{-4}	Gallons
Cubic centimeters	9.999×10^{-4}	Liters
Cubic centimeters	2.113×10^{-3}	Pints (liq.)
Cubic centimeters	1.057×10^{-3}	Quarts (liq.)
Cubic feet	2.832×10^{-4}	Cubic cms.
Cubic feet	1728	Cubic inches
Cubic feet	0.02832	Cubic meters
Cubic feet	0.03704	Cubic yards
Cubic feet	7.48052	Gallons
Cubic feet	28.32	Liters

Figure 5. Continued.

Multiply	By	To Obtain
Cubic feet	59.84	Pints (liq.)
Cubic feet	29.92	Quarts (liq.)
Cubic feet/min.	472.0	Cubic cms./sec.
Cubic feet/min.	0.1247	Gallons/sec.
Cubic feet/min.	0.4719	Liters/sec.
Cubic feet/min.	62.43	Pounds of water/min.
Cubic feet/sec.	0.646317	Millions gals./day
Cubic feet/sec.	448.831	Gallons/min.
Cubic inches	16.39	Cubic centimeters
Cubic inches	5.787×10^{-4}	Cubic feet
Cubic inches	1.639×10^{-5}	Cubic meters
Cubic inches	2.143×10^{-5}	Cubic yards
Cubic inches	4.329×10^{-3}	Gallons
Cubic inches	1.639×10^{-2}	Liters
Cubic inches	0.03463	Pints (liq.)
Cubic inches	0.01732	Quarts (liq.)
Cubic meters	10^6	Cubic centimeters
Cubic meters	35.31	Cubic feet
Cubic meters	61023.	Cubic inches
Cubic meters	1.308	Cubic yards
Cubic meters	264.2	Gallons
Cubic meters	999.97	Liters
Cubic meters	2113	Pints (liq.)
Cubic meters	1057	Quarts (liq.)
Cubic yards	764,554.86	Cubic centimeters
Cubic yards	27	Cubic feet
Cubic yards	46,656	Cubic inches
Cubic yards	0.7646	Cubic meters
Cubic yards	202.0	Gallons
Cubic yards	764.5	Liters
Cubic yards	1616	Pints (liq.)
Cubic yards	807.9	Quarts (liq.)
Cubic yards/min.	0.45	Cubic feet/sec.
Cubic yards/min.	3.366	Gallons/sec.
Cubic yards/min.	12.74	Liters/sec.
Decigrams	0.1	Grams
Deciliters	0.1	Liters
Decimeters	0.1	Meters
Degrees (angle)	60	Minutes
Degrees (angle)	0.01745	Radians
Degrees (angle)	3600	Seconds
Degrees/sec.	0.01745	Radians/sec.
Degrees/sec.	0.1667	Revolutions/min.
Degrees/sec.	0.002778	Revolutions/sec.
Dekagrams	10	Grams
Dekaliters	10	Liters
Dekameters	10	Meters
Drams	27.34375	Grains
Drams	0.0625	Ounces
Drams	1.771845	Grams
Fathoms	6	Feet
Feet	30.48	Centimeters
Feet	12	Inches
Feet	0.3048	Meters
Feet	1/3	Yards
Feet of water	0.0295	Atmospheres
Feet of water	0.8826	Inches of mercury
Feet of water	304.8	Kgs./sq. meter
Feet of water	62.43	Lbs./sq. ft.
Feet of water	0.4335	Lbs./sq. inch
Feet/min.	0.5080	Centimeters/sec.
Feet/min.	0.01667	Feet/sec.

Figure 5. Continued.

Multiply	By	To Obtain
Feet/min.	0.01829	Kilometers/hr.
Feet/min.	0.3048	Meters/min.
Feet/min.	0.01136	Miles/hr.
Feet/sec.	30.48	Centimeters/sec.
Feet/sec.	1.097	Kilometers/hr.
Feet/sec.	0.5924	Knots
Feet/sec.	18.29	Meters/min.
Feet/sec.	0.6818	Miles/hr.
Feet/sec.	0.01136	Miles/min.
Feet/sec./sec.	30.48	Cms./sec./sec.
Feet/sec./sec.	0.3048	Meters/sec./sec.
Foot--pounds	1.286×10^{-3}	British Thermal Units
Foot--pounds	5.050×10^{-7}	Horsepower--hrs.
Foot--pounds	3.240×10^{-4}	Kilogram--calories
Foot--pounds	0.1383	Kilogram--meters
Foot--pounds	3.766×10^{-7}	Kilowatt--hours
Foot--pounds/min.	2.140×10^{-5}	B.T.U./sec.
Foot--pounds/min.	0.01667	Foot--pounds/sec.
Foot--pounds/min.	3.030×10^{-5}	Horsepower
Foot--pounds/min.	5.393×10^{-3}	Gm.--calories/sec.
Foot--pounds/min.	2.260×10^{-5}	Kilowatts
Foot--pounds/sec.	7.704×10^{-2}	B.T.U./min.
Foot--pounds/sec.	1.818×10^{-3}	Horsepower
Foot--pounds/sec.	1.941×10^{-2}	Kg.--calories/min.
Foot--pounds/sec.	1.356×10^{-3}	Kilowatts
Gallons	3785	Cubic centimeters
Gallons	0.1337	Cubic feet
Gallons	231	Cubic inches
Gallons	3.785×10^{-3}	Cubic meters
Gallons	4.951×10^{-3}	Cubic yards
Gallons	3.785	Liters
Gallons	8	Pints (liq.)
Gallons	4	Quarts (liq.)
Gallons--Imperial	1.20095	U.S. Gallons
Gallons--U.S.	0.83267	Imperial gallons
Gallons water	8.345	Pounds of water
Gallons/min.	2.228×10^{-3}	Cubic feet/sec.
Gallons/min.	0.06308	Liters/sec.
Gallons/min.	8.0208	Cu. ft./hr.
Grains (troy)	0.06480	Grams
Grains (troy)	0.04167	Pennyweights (troy)
Grains (troy)	2.0833×10^{-3}	Ounces (troy)
Grains/U.S. gal.	17.118	Parts/million
Grains/U.S. gal.	142.86	Lbs./million gal.
Grains/Imp. gal.	14.254	Parts/million
Grams	980.7	Dynes
Grams	15.43	Grains
Grams	.001	Kilograms
Grams	1000	Milligrams
Grams	0.03527	Ounces
Grams	0.03215	Ounces (troy)
Grams	2.205×10^{-3}	Pounds
Grams/cm.	5.600×10^{-3}	Pounds/inch
Grams/cu. cm.	62.43	Pounds/cubic foot
Grams/cu. cm.	0.03613	Pounds/cubic inch
Grams/liter	58.416	Grains/gal.
Grams/liter	8.345	Pounds/1000 gals.
Grams/liter	0.06242	Pounds/cubic foot
Grams/liter	1000	Parts/million
Hectares	2.471	Acres
Hectares	1.076×10^{-5}	Square feet
Hectograms	100	Grams

Figure 5. Continued.

Multiply	By	To Obtain
Hectoliters	100	Liters
Hectometers	100	Meters
Hectowatts	100	Watts
Horsepower	42.44	B.T.U./min.
Horsepower	33,000	Foot--lbs./min.
Horsepower	550	Foot--lbs./sec.
Horsepower	1.014	Horsepower (metric)
Horsepower	10.547	Kg.--calories/min.
Horsepower	0.7457	Kilowatts
Horsepower	745.7	Watts
Horsepower (boiler)	33,493	B.T.U./hr.
Horsepower (boiler)	9.809	Kilowatts
Horsepower--hours	2546	B.T.U.
Horsepower--hours	1.98×10^6	Foot--lbs.
Horsepower--hours	641.6	Kilogram--calories
Horsepower--hours	2.737×10^5	Kilogram--meters
Horsepower--hours	0.7457	Kilowatt--hours
Inches	2.540	Centimeters
Inches of mercury	0.03342	Atmospheres
Inches of mercury	1.133	Feet of water
Inches of mercury	345.3	Kgs./sq. meter
Inches of mercury	70.73	Lbs./sq. ft.
Inches of mercury (32°F)	0.491	Lbs./sq. inch
Inches of water	0.002458	Atmospheres
Inches of water	0.07355	Inches of mercury
Inches of water	25.40	Kgs./sq. meter
Inches of water	0.578	Ounces/sq. inch
Inches of water	5.202	Lbs./sq. foot
Inches of water	0.03613	Lbs./sq. inch
Kilograms	980,665	Dynes
Kilograms	2.205	Lbs.
Kilograms	1.102×10^{-3}	Tons (short)
Kilograms	10^3	Grams
Kilograms--cal./sec.	3.968	B.T.U./sec.
Kilograms--cal./sec.	3086	Foot--lbs./sec.
Kilograms--cal./sec.	5.6145	Horsepower
Kilograms--cal./sec.	4186.7	Watts
Kilogram--cal./min.	3085.9	Foot--lbs./min.
Kilogram--cal./min.	0.09351	Horsepower
Kilogram--cal./min.	69.733	Watts
Kgs./meter	0.6720	Lbs./foot
Kgs./sq. meter	9.678×10^{-5}	Atmospheres
Kgs./sq. meter	3.281×10^{-3}	Feet of water
Kgs./sq. meter	2.896×10^{-3}	Inches of mercury
Kgs./sq. meter	0.2048	Lbs./sq. foot
Kgs./sq. meter	1.422×10^{-3}	Lbs./sq. inch
Kgs./sq. millimeter	10^6	Kgs./sq. meter
Kiloliters	10^3	Liters
Kilometers	10^5	Centimeters
Kilometers	3281	Feet
Kilometers	10^3	Meters
Kilometers	0.6214	Miles
Kilometers	1094	Yards
Kilometers/hr.	27.78	Centimeters/sec.
Kilometers/hr.	54.68	Feet/min.
Kilometers/hr.	0.9113	Feet/sec.
Kilometers/hr.	.5399	Knots
Kilometers/hr.	16.67	Meters/min.
Kilometers/hr.	0.6214	Miles/hr.
Kms./hr./sec.	27.78	Cms./sec./sec.
Kms./hr./sec.	0.9113	Ft./sec./sec.
Kms./hr./sec.	0.2778	Meters/sec./sec.

Figure 5. Continued.

Multiply	By	To Obtain
Kilowatts	56.907	B.T.U./min.
Kilowatts	4.425×10^4	Foot--lbs./min.
Kilowatts	737.6	Foot--lbs./sec.
Kilowatts	1.341	Horsepower
Kilowatts	14.34	Kg.--calories/min.
Kilowatts	10^3	Watts
Kilowatt--hours	3414.4	B.T.U.
Kilowatt--hours	2.655×10^6	Foot--lbs.
Kilowatt--hours	1.341	Horsepower--hrs.
Kilowatt--hours	860.4	Kilogram--calories
Kilowatt--hours	3.671×10^5	Kilogram--meters
Liters	10^3	Cubic centimeters
Liters	0.03531	Cubic feet
Liters	61.02	Cubic inches
Liters	10^{-3}	Cubic meters
Liters	1.308×10^{-3}	Cubic yards
Liters	0.2642	Gallons
Liters	2.113	Pints (liq.)
Liters	1.057	Quarts (liq.)
Liters/min.	5.886×10^{-4}	Cubic ft./sec.
Liters/min.	4.403×10^{-3}	Gals./sec.
Lumber Width (in.) X Thickness (in.) / 12	Length (ft.)	Board feet
Meters	100	Centimeters
Meters	3.281	Feet
Meters	39.37	Inches
Meters	10^{-3}	Kilometers
Meters	10^3	Millimeters
Meters	1.094	Yards
Meters/min.	1.667	Centimeters/sec.
Meters/min.	3.281	Feet/min.
Meters/min.	0.05468	Feet/sec.
Meters/min.	0.06	Kilometers/hr.
Meters/min.	0.03728	Miles/hr.
Meters/sec.	196.8	Feet/min.
Meters/sec.	3.281	Feet/sec.
Meters/sec.	3.6	Kilometers/hr.
Meters/sec.	0.06	Kilometers/min.
Meters/sec.	2.237	Miles/hr.
Meters/sec.	0.03728	Miles/min.
Microns	10^{-6}	Meters
Miles	1.609×10^5	Centimeters
Miles	5280	Feet
Miles	1.609	Kilometers
Miles	1760	Yards
Miles/hr.	44.70	Centimeters/sec.
Miles/hr.	88	Feet/min.
Miles/hr.	1.467	Feet/sec.
Miles/hr.	1.609	Kilometers/hr.
Miles/hr.	0.8689	Knots
Miles/hr.	26.82	Meters/min.
Miles/min.	2682	Centimeters/sec.
Miles/min.	88	Feet/sec.
Miles/min.	1.609	Kilometers/min.
Miles/min.	60	Miles/hr.
Milliers	10^3	Kilograms
Milligrams	10^{-3}	Grams
Milliliters	10^{-3}	Liters
Millimeters	0.1	Centimeters
Millimeters	0.03937	Inches
Milligrams/liter	1	Parts/million

Figure 5. Continued.

Multiply	By	To Obtain
Million gals./day	1.54723	Cubic ft./sec.
Miner's inches	1.5	Cubic ft./min.
Minutes (angle)	2.909×10^{-4}	Radians
Ounces	16	Drams
Ounces	437.5	Grains
Ounces	0.0625	Pounds
Ounces	28.3495	Grams
Ounces	0.9115	Ounces (troy)
Ounces	2.790×10^{-5}	Tons (long)
Ounces	2.835×10^{-5}	Tons (metric)
Ounces (troy)	480	Grains
Ounces (troy)	20	Pennyweights (troy)
Ounces (troy	0.08333	Pounds (troy)
Ounces (troy)	31.10348	Grams
Ounces (troy)	1.09714	Ounces (avoir)
Ounces (fluid)	1.805	Cubic inches
Ounces (fluid)	0.02957	Liters
Ounces/sq. inch	0.0625	Lbs./sq. inch
Parts/million	0.0584	Grains/U.S. gal.
Parts/million	0.07015	Grains/Imp. gal.
Parts/million	8.345	Lbs./million gal.
Pennyweights (troy)	24	Grains
Pennyweights (troy)	1.55517	Grams
Pennyweights (troy)	0.05	Ounces (troy)
Pennyweights (troy)	4.1667×10^{-3}	Pounds (troy)
Pounds	16	Ounces
Pounds	256	Drams
Pounds	7000	Grains
Pounds	0.0005	Tons (short)
Pounds	453.5924	Grams
Pounds	1.21528	Pounds (troy)
Pounds	14.5833	Ounces (troy)
Pounds (troy)	5760	Grains
Pounds (troy)	240	Pennyweights (troy)
Pounds (troy)	12	Ounces (troy)
Pounds (troy)	373.2417	Grams
Pounds (troy)	0.822857	Pounds (avoir.)
Pounds (troy)	13.1657	Ounces (avoir.)
Pounds (troy)	3.6735×10^{-4}	Tons (long)
Pounds (troy)	4.1143×10^{-4}	Tons (short)
Pounds (troy)	3.7324×10^{-4}	Tons (metric)
Pounds of water	0.01602	Cubic feet
Pounds of water	27.68	Cubic inches
Pounds of water	0.1198	Gallons
Pounds of water/min.	2.670×10^{-4}	Cubic ft./sec.
Pounds/cubic foot	0.01602	Grams/cubic cm.
Pounds/cubic foot	16.02	Kgs./cubic meters
Pounds/cubic foot	5.787×10^{-4}	Lbs./cubic inch
Pounds/cubic inch	27.68	Grams/cubic cm.
Pounds/cubic inch	2.768×10^{4}	Kgs./cubic meter
Pounds/cubic inch	1728	Lbs./cubic foot
Pounds/foot	1.488	Kgs./meter
Pounds/inch	1152	Grams/cm.
Pounds/sq. foot	0.01602	Feet of water
Pounds/sq. foot	4.882	Kgs./sq. meter
Pounds/sq. foot	6.944×10^{-3}	Pounds/sq. inch
Pounds/sq. inch	0.06804	Atmospheres
Pounds/sq. inch	2.307	Feet of water
Pounds/sq. inch	2.036	Inches of mercury
Pounds/sq. inch	703.1	Kgs./sq. meter
Quadrants (angle)	90	Degrees
Quadrants (angle)	5400	Minutes

Figure 5. Continued.

Multiply	By	To Obtain
Quadrants (angle)	1.571	Radians
Quarts (dry)	67.20	Cubic inches
Quarts (liq.)	57.75	Cubic inches
Quintal, Argentine	101.28	Pounds
Quintal, Brazil	129.54	Pounds
Quintal, Castile, Peru	101.43	Pounds
Quintal, Chile	101.41	Pounds
Quintal, Mexico	101.47	Pounds
Quintal, Metric	220.46	Pounds
Quires	25	Sheets
Radians	57.30	Degrees
Radians	3438	Minutes
Radians	0.637	Quadrants
Radians/sec.	57.30	Degrees/sec.
Radians/sec.	0.1592	Revolutions/sec.
Radians/sec.	9.549	Revolutions/min.
Radians/sec./sec.	573.0	Revs./min./min.
Radians/sec./sec.	0.1592	Revs./sec./sec.
Reams	500	Sheets
Revolutions	360	Degrees
Revolutions	4	Quadrants
Revolutions	6.283	Radians
Revolutions/min.	6	Degrees/sec.
Revolutions/min.	0.1047	Radians/sec.
Revolutions/min.	0.01667	Revolutions/sec.
Revolutions/min./min.	1.745×10^{-3}	Rads./sec./sec.
Revolutions/min./min.	2.778×10^{-4}	Revs./sec./sec.
Revolutions/sec.	360	Degrees/sec.
Revolutions/sec.	6.283	Radians/sec.
Revolutions/sec.	60	Revolutions/min.
Revolutions/sec./sec.	6.283	Radians/sec./sec.
Revolutions/sec./sec.	3600	Revs./min./min.
Seconds (angle)	4.848×10^{-6}	Radians
Square centimeters	1.076×10^{-3}	Square feet
Square centimeters	0.1550	Square inches
Square centimeters	10^{-4}	Square meters
Square centimeters	100	Square millimeters
Square feet	2.296×10^{-5}	Acres
Square feet	929.0	Square centimeters
Square feet	144	Square inches
Square feet	0.09290	Square meters
Square feet	3.587×10^{-8}	Square miles
Square feet	1/9	Square yards
$\dfrac{1}{\text{Sq. ft./gal./min.}}$	8.0208	Overflow rate (ft./hr.)
Square inches	6.452	Square centimeters
Square inches	6.944×10^{-3}	Square feet
Square inches	645.2	Square millimeters
Square kilometers	247.1	Acres
Square kilometers	10.76×10^{6}	Square feet
Square kilometers	10^{6}	Square meters
Square kilometers	0.3861	Square miles
Square kilometers	1.196×10^{6}	Square yards
Square meters	2.471×10^{-4}	Acres
Square meters	10.76	Square feet
Square meters	3.861×10^{-7}	Square miles
Square meters	1.196	Square yards
Square miles	640	Acres
Square miles	27.88×10^{6}	Square feet
Square miles	2.590	Square kilometers
Square miles	3.098×10^{6}	Square yards

Figure 5. Continued.

Multiply	By	To Obtain
Square millimeters	0.01	Square centimeters
Square millimeters	1.550×10^{-3}	Square inches
Square yards	2.066×10^{-4}	Acres
Square yards	9	Square feet
Square yards	0.8361	Square meters
Square yards	3.228×10^{-7}	Square miles
Temp. ($^{\circ}$C.) + 273	1	Abs. temp. ($^{\circ}$C.)
Temp. ($^{\circ}$C.) + 17.78	1.8	Temp. ($^{\circ}$F.)
Temp. ($^{\circ}$F.) +460	1	Abs. temp. ($^{\circ}$F.)
Temp. ($^{\circ}$F.) - 32	5/9	Temp. ($^{\circ}$C.)
Tons (long)	1016	Kilograms
Tons (long)	2240	Pounds
Tons (long)	1.12000	Tons (short)
Tons (metric)	10^3	Kilograms
Tons (metric)	2205	Pounds
Tons (short)	2000	Pounds
Tons (short)	32,000	Ounces
Tons (short)	907.1848	Kilograms
Tons (short)	2430.56	Pounds (troy)
Tons (short)	0.89287	Tons (long)
Tons (short)	29166.66	Ounces (troy)
Tons (short)	0.90718	Tons (metric)
Tons of water/24 hrs.	83.333	Pounds water/hr.
Tons of water/24 hrs.	0.16643	Gallons/min.
Tons of water/24 hrs.	1.3349	Cu. ft./hr.
Watts	0.05686	B.T.U./min.
Watts	44.25	Foot--lbs./min.
Watts	0.7376	Foot--lbs./sec.
Watts	1.341×10^{-3}	Horsepower
Watts	0.0143	Kg.--calories/min.
Watts	10^{-3}	Kilowatts
Watt--hours	3.414	B.T.U.
Watt--hours	2655	Foot--lbs.
Watt--hours	1.341×10^{-3}	Horsepower--hrs.
Watt--hours	0.8604	Kilogram--calories
Watt--hours	367.1	Kilogram--meters
Watt--hours	10^{-3}	Kilowatt--hours
Yards	91.44	Centimeters
Yards	3	Feet
Yards	36	Inches
Yards	0.9144	Meters

Approximate Conversions:

Material	Barrels per Ton (long)
Crude oils	6.7-8.1
Aviation gasolines	8.3-9.2
Motor gasolines	8.2-9.1
Kerosenes	7.7-8.3
Gas Oils	7.2-7.9
Diesel Oils	7.0-7.9
Lubricating oils	6.8-7.6
Fuel oils	6.6-7.0
Asphaltic bitumens	5.9-6.5

(As a general rule-of-thumb use 6.5 barrels or 250 gallons per ton of oil).

Figure 5. Conversion Tables.

trolled. Site control, zonal deliniation, and decontamination are used to prevent unnecessary distribution of contaminants by personnel and equipment. Site control insures that only personnel and equipment absolutely necessary to control the situation are allowed into a contaminated area. Limiting traffic of persons, vehicles, and equipment into and out of a contaminated area is both a health and safety measure and an effort to limit transfer of contamination. Zonal delineation is a component of site control; a danger area or "hot" area immediately surrounding the spill or contamination area is clearly marked, an entry/exit checkpoint is initiated, and all personnel and equipment deemed necessary to enter/exit the central area pass through the checkpoint. An outer circle around the hot zone contains the decontamination area; all entry/exit is through a control point, and the area is considered to be potentially contaminated. All decontamination is handled within this area, and no personnel or equipment is allowed to exit until being checked for contamination and decontaminated if necessary. Decontamination can include washing/rinsing of personnel, personal protective equipment, and general equipment and materials—but it also may include when necessary proper disposal of equipment/clothing/apparatus that cannot be successfully decontaminated. A third outer circle is considered the staging area—a noncontaminated area for support activities, operation of a command post, preparation of equipment and personnel needed within either the decontamination or hot areas, and provision of briefings to officials or members of the media. Stringent zonal deliniation requires extensive planning, personnel, and equipment; and is normally applied to situations where the volume or toxicity of a hazardous material warrants extreme care.

Hazardous materials response personnel encounter hazards where serious injury or death can result from inhalation, ingestion, skin absorption, burns, frostbite, and even injection (if a responder happens to step on a nail, or tear his skin on jagged metal, that has been coated with pesticide). Levels of protection for protective equipment and clothing are often required to insure that responders are protected adequately in what could conceivably be a "worst case" situation. What may appear as overkill in protective equipment and clothing to inexperienced observers, is prudent judgement for a hazardous materials team and often a requirement for participation. Figure 6 details levels of protection demanded by U.S. Environmental Protection Agency protocols.

Corrosives often require neutralization before they can be safely handled or disposed of. Adding acid to an alkaline solution or an alkali to an acid solution to adjust the pH can render the mixture neutral although it must be understood that mixing some acids with some bases can release dangerous amounts of heat, and that some acids are poisonous as well as corrosive. Also, earth contaminated by corrosives may have to be decontaminated to avoid longterm environmental damage. Bases often used to neutralize acids include agricultural lime, (slaked lime) crushed limestone, or sodium bicarbonate. Vinegar or other dilute acids can be used to neutralize certain bases.

Sorbent materials used to pick up hazardous fluids are available in great variety. Commercial suppliers can provide pads, pillows, sheets and booms (capable of absorbing both floating and pooled substances) or granular sorbent materials. "Kitty Litter" available in any supermarket is routinely used to pick up liquid spills on dry land as is rice-hull or other varieties of "grease sweep" available through service station and garage suppliers. When a large volume of sorbent material is required, fly ash and cement powder can be used with certain products. The problem with sorbent materials is what to do with them after you use them. Depending on the chemical involved, disposal of saturated sorbent material may have to be done very carefully and permission and guidance may have to be obtained from E.P.A. or a state agency.

Dispersants are used to break up floating petroleum products by decreasing cohesion among individual droplets, thus allowing the petroleum product to enter the water base. Dispersants can be spread on an oil slick in a large body of water, and wave action will mix the dispersant/water/petroleum, or can be injected into a fire hose water stream by use of an inductor. Since many dispersants are a combination of solvents/surfacants/stabilizers and may pose an environmental threat to aquatic life and drinking water supplies, some state and

FACT SHEET   United States
Environmental Protection
Agency

April 1982

LEVELS OF PROTECTION

When response activities are conducted where atmospheric contamination is known or suspected to exist, personnel protective equipment must be worn. Personnel protective equipment is designed to prevent/ reduce skin and eye contact as well as inhalation or ingestion of the chemical substance.

Personnel equipment to protect the body against contact with known or anticipated chemical hazards has been divided into four categories:

1. **Level A** protection should be worn when the highest level of respiratory, skin, eye, and mucous membrane protection is needed.

    a. Personal Protective Equipment

        — Positive-pressure (pressure demand), self contained breathing apparatus (MSHA/ NIOSH approved).
        — Fully-encapsulating chemical resistant suit.
        — Gloves, inner, chemical resistant.
        — Gloves, outer, chemical resistant.
        — Boots, chemical resistant, steel toe and shank; (depending on suit boot construction, worn over or under suit boot.)
        — Underwear, cotton, long-john type.*
        — Hard hat (under suit).*
        — Coveralls (under suit).*
        — Two-way radio communications (intrinsically safe).

        * Optional

2. **Level B** protection should be selected when the highest level of respiratory protection is needed, but a lesser level of skin and eye protection. Level B protection is the minimum level recommended on initial site entries until the hazards have been further identified and defined by monitoring. sampling, and other reliable methods of analysis. and personnel equipment corresponding with those findings utilized.

    a. Personal Protective Equipment

        — Positive-pressure (pressure-demand), self contained breathing apparatus (MSHA/ NIOSH approved).
        — Chemical resistant clothing (overalls and long sleeved jacket, coveralls, hooded two

piece chemical splash suit, disposable chemical resistant coveralls.)
— Coveralls (under splash suit).*
— Gloves, outer, chemical resistant.
— Gloves, inner, chemical resistant.
— Boots, outer, chemical resistant, steel toe and shank.
— Boots, outer, chemical resistant.*
— Two-way radio communications (intrinsically safe).*
— Hard hat.*

* Optional

3. **Level C** protection should be selected when the type of airborne substance is known, concentration measured, criteria for using air-purifying respirators met, and skin and eye exposure is unlikely. Periodic monitoring of the air must be performed.

    a. Personal Protective Equipment

        — Full-face, air-purifying respirator (MSHA/ NIOSH approved).
        — Chemical resistant clothing (one piece coverall, hooded two piece chemical splash suit, chemical resistant hood and apron, disposable chemical resistant coveralls.)
        — Gloves, outer, chemical resistant.
        — Gloves, inner, chemical resistant.*
        — Boots, steel toe and shank, chemical resistant.
        — Boots, outer, chemical resistant.*
        — Cloth coveralls (inside chemical protective clothing.)*
        — Two-way radio communications (intrinsically safe).*
        — Hard hat.*
        — Escape mask.*

        * Optional

4. **Level D** is primarily a work uniform. It should not be worn on any site where respiratory or skin hazards exist.

    Refer to the Office of Emergency and Remedial Response. Environmental Response Division. **Interim Standard Operating Safety Procedures** for full details.

Figure 6.

federal agencies discourage their use. They would normally be used only when a flammable vapor hazard from a product such as gasoline presents a clear threat to life or property, and permission for their use may be required from the local environmental agency. It is important to recognize that a dispersant merely allows a petroleum product to enter a water base; without water, a dispersant cannot do a thing.

No hazardous materials response team is ever without an on-apparatus reference library of manuals, guidebooks, schematic diagrams, and other reference materials that allow personnel to identify chemicals, potential reactions, and proper control and containment methods. The four most commonly used hazardous materials response manuals that are listed below all vary as to content. Many times responders will refer to two or three different manuals at a single incident. All are designed to provide *basic* information necessary to handle an immediate emergency situation and may not detail all potential hazards for a given chemical, particularly in situations where chemicals have become mixed. Each emergency response manual recommends that expert advice and assistance be obtained as quickly as possible to supplement the basic information provided.

- *Chemical Hazards Response Information System.* (U.S. Coast Guard)
- *Emergency Handling Of Hazardous Materials In Surface Transportation.* (Bureau of Explosives/Association of American Railroads)
- *Fire Protection Guide On Hazardous Materials.* (National Fire Protection Association)
- *Hazardous Materials Emergency Response Guidebook.* (U.S. Department of Transportation)

Pages 225 through 228 of Chapter 8 provide more detailed information on the contents of specific manuals.

Within one or more of the manuals mentioned above a responder can obtain chemical-specific information for most hazardous materials transported in commerce. Such information may consider. . .

- The hazard class of the commodity.
- Actions to be taken if material is involved in fire.
- Actions to be taken if material is not involved in fire.
- Personnel protection.
- Evacuation guidelines.
- Environmental considerations — land/air/water.
- Health hazards.
- First aid measures.
- Assistance in identification of unknown materials.
- Chemical properties of liquids/gases/solids (flash point, ignition temperature, flammable limits, specific gravity, vapor density, boiling point, etc.).
- Methods of extinguishment.
- General chemical descriptions.
- Neutralization.
- Cleanup.
- Potential chemical reactions.
- Usual shipping containers.
- Proper storage methods.

Fire departments and other public safety agency emergency response organizations locally develop preprinted forms that help to insure necessary information is obtained for use in preplanning, on-scene operations, reporting, and evaluation of experience. Figure 7 shows a *Hazardous Materials Incident Data* form filled out by the Memphis Division of Fire Services fire alarm office when a hazardous materials call is received. The form is a guide to securing specific, pertinent information required by first responders.

Figure 8 depicts a *Hazardous Materials Spills* report form also used by the Memphis Division of Fire Services to allow evaluation of hazardous materials incidents overtime as an aid to continuous planning.

The Denver Fire Department uses the *Hazardous Materials Storage* form in Figure 9 when making preplanning visits to facilities that store hazardous materials so that such information will be available to first responders when and if the facility has an incident.

Figure 10 depicts a *Hazardous Materials Report* form used by the Denver Fire Department

HAZARDOUS MATERIAL INCIDENT DATA

DIV. FIRE SERVICES

CITY OF MEMPHIS

LOCATION OF INCIDENT:

VEHICLE I.D. #	TIME OF INCIDENT :

WEATHER CONDITIONS :	WIND DIRECTION	WIND SPEED	RELATIVE HUMIDITY	TEMP. (°F)	BARO-METER

CHEMICAL NAME:

SYNONYM (S):

TRADE NAMES:

STCC NO.:	CLASSIFICATION:

HAZARDOUS :

A. HEALTH

B. REACTIVITY

C. FLAMMABILITY

D. SPECIAL HAZARDS

E. OTHER

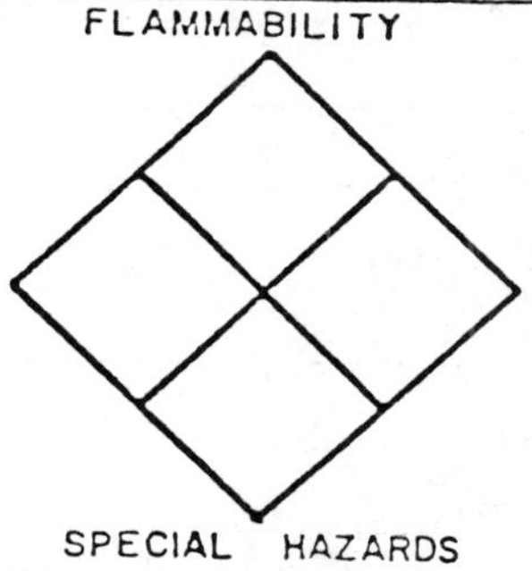

FLAMMABILITY

HEALTH

REACTIVITY

SPECIAL HAZARDS

CHARACTERISTICS:

A.	BOILING POINT (°F)	FLASH POINT (°F)	IGNITION TEMP (°F)	FLAMM. LIMITS	UPPER	LOWER
B.	SPECIFIC GRAVITY	VAPOR DENSITY	WATER SOLUBLE :	SLIGHT	YES	NO

AGENTS REQ'D:	WATER FOG	WATER STRAIGHT STREAM	AFFF	ATC	PROTEIN FOAM	UNOX	HI-EXPANSION FOAM	DRY CHEMICAL

TYPE CONTAINERS INVOLVED :	PRESSURIZED	NON-PRESSURIZED	INSULATED	NON-INSULATED

AMOUNT FUEL (PRODUCT) INVOLVED :	GALS, TON ETC.

SHIPPER :	TELEPHONE:

MANUFACTURER :	TELEPHONE:

REMARKS :

TIME DATA REC'D:	DATE:	SHEET of

FORM NO. F2000 - .186

CHEMTREC 1-800-424-9300

Figure 7.

CITY of MEMPHIS

DIVISION OF FIRE SERVICES

HAZARDOUS MATERIALS SPILLS

LOCATION OF INCIDENT _____

DATE AND TIME OF INCIDENT _____

CARRIER _____

NAME OF OPERATOR _____

MANUFACTURER _____

MATERIAL INVOLVED _____ GAS ____ LIQUID ____ SOLID ____

PROPERTIES OF MATERIAL _____

_____

RESOURCES USED _____

QUANTITY INVOLVED _____

TYPE OF VEHICLE _____ ID NO. _____

POINT OF DEPARTURE _____

POINT OF DESTINATION _____

WIND DIRECTION AND SPEED _____ AMOUNT OF SPILLAGE _____

PROXIMITY TO POPULATED AREA _____ STREAM OR SEWER SYSTEM _____

EXTINGUISHING AGENT _____

CO. RESPONDENT _____

CONDITION AT SCENE AND ACTION TAKEN: _____

_____

_____

_____

_____

_____

_____

                            HAZARDOUS MATERIAL REP.
_____

Figure 8.

DENVER FIRE DEPARTMENT
HAZARDOUS MATERIALS STORAGE

Station: _____ Chief: _____

Name: _____ Sprinklered: Yes ☐ No ☐ Date: _____

Address: _____ Basement: Yes ☐ No ☐

Phone: _____ Stories: _____ Order: Yes ☐ No ☐

Owner: _____ Manager: _____

Type of Business: _____
_____

CHEMICAL AND HAZARD RATING

	Chemical	Health	Flamm.	React.	Other	Amount
1						
2						
3						
4						
5						
6						

COMMENTS

_____

_____

_____

_____

_____

_____

_____

_____

Figure 9.

when actually responding to a hazardous materials incident. This type of form is used both to secure information that may be needed to bring the incident under control, and information that will provide guidance for future planning, evaluation and training.

Another example of a form used when making preplanning visits to facilities that store hazardous materials is the *Emergency Guide to Hazardous Materials Storage* form utilized by

HAZARDOUS MATERIALS REPORT

DATE:_____　　TIME:_____　　INCIDENT:_____ _____

TEMPERATURE:_____　WIND DIRECTION:_____　WIND VELOCITY:_____

(1-800-424-9300)
WEATHER:_____　CHEMTREC CALLED:_____　TIME:_____

LOCATION OF INCIDENT:_____

TRANSPORTATION:　　　　YES:☐　　　　NO:☐

RAIL:☐　　　　　　　　MOTOR CARRIER:☐　　　　　　AIR:☐

MANUFACTURE:_____　CARRIER NO._____

ADDRESS:_____　PLACARDED:_____

CITY AND STATE:_____　MATERIAL:_____

PHONE:_____　HAZARD:_____

SHIPPER:_____　SIZE OF RELEASE:_____

ADDRESS:_____　TYPE OF CONTAINER:_____

CITY AND STATE:_____　FLASH POINT:_____

PHONE:_____　BOILING POINT:_____

CONSIGNEE:_____　AUTO IGNITION:_____

ADDRESS:_____　SPECIFIC GRAVITY:_____

CITY AND STATE:_____　WATER SOLUABLE:_____

PHONE:_____　TOXICITY:_____

CARRIER:_____　AGENCIES CALLED:

ADDRESS:_____　1._____

CITY AND STATE:_____　2._____

PHONE:_____　3._____

　　　　　　　　　　　　　　　　　　4._____

　　　　COMMENTS:

_____

_____

_____

Figure 10.

the Philadelphia Fire Department (See Figure 11). Many fire departments use similar forms that are completed and made available to the fire unit(s) that would respond when and if there was a fire or hazardous materials incident at the particular firm covered by the form.

Figures 12 and 13 show an example of a *Preplan Information* form that includes a map or layout diagram of the facility in question with particular emphasis on the types of information firefighters would need to know. A further example of a preplan information form is the *Pre-Fire Planning Hazardous Materials Survey Form* (Figure 14).

There are also information accumulation/preplanning forms used to obtain information on special types of facilities that handle specific chemicals. One example is the *Pool Survey* form developed by the Baltimore County Fire Department (See Figure 15).

A number of worksheets or checklists are in use around the country that were adopted to assist responders in acquiring pertinent information while on-scene at an actual incident, or to insure that necessary actions are taken. Figure 16 depicts the *Worksheet—Hazardous Materials Emergency Report* that is part of a *Vermont Hazardous Materials On-Scene Action Guide* distributed to all emergency response vehicles (fire, police, civil defense, ambulance) in that state.

Figure 17 provides an example of a *Hazardous Materials Incident Control Checklist* utilized by responders/trainers affiliated with Safety Systems, Inc. in Jacksonville, Florida.

A variation of the on-scene checklist is the *Incident Commander Checklist* developed for instructional purposes by Al Mozingo, an instructor/responder on the west coast Figure 18).

The forms reviewed here have merely provided a sampling of the types of data accumulation devices actually used by incident response personnel around the country. Reliable, specific information is a critical factor in response to hazardous materials incidents. There are so many chemicals, interests, conditions and variables potentially involved in an incident that a standard format for acquiring information is considered a necessity by many responders. Worksheets/checklists are one way in which experienced response personnel seek to. . .

- Gather the maximum possible amount of preplanning information.
- Insure the right questions are asked and answered.
- Obtain and record information from actual experience to aid in training, evaluation of experience, and future planning as a result of that experience.
- Maintain incident vigilance and discipline.
- Refrain from underestimating the potential seriousness of an incident.
- Maximize safety of response personnel.
- Secure reliable data for reports and possible testimony.
- Quickly determine the identity and characteristics of the product(s) involved.
- Insure thoroughness in responding to incidents.

Hazardous materials response personnel are often called upon to work in hazardous atmospheric conditions that cannot easily be detected by normal human senses such as sight, smell, hearing or touch. Three quite common dangers are oxygen deficient atmospheres, combustible gases and vapors, and toxic gases and vapors. It is important to recognize that some quite deadly gases and vapors have absolutely no odor (carbon monoxide, for example). Response personnel are conditioned to check for oxygen deficiency, combustible gases and vapors, and toxic gases and vapors and carry a variety of meters and "sniffers" for this purpose. Oxygen deficiency meters provide a warning when an atmosphere contains less than 19.5 percent oxygen; normal breathing air required to sustain life contains roughly 21 percent oxygen. So-called "continuous reading" meters provide a constant indication of the oxygen level, and some are equipped with a bell or buzzer that is activated whenever the oxygen level drops below 19.5 percent.

Any source of ignition can cause a fire or explosion in an area where flammable or combustible gases are present in sufficient concentration. There are many different types of instruments for measuring combustible gases and vapors, and some units combine the capabilities of an oxygen deficiency meter/combustible gas indicator. Combustible gas in-

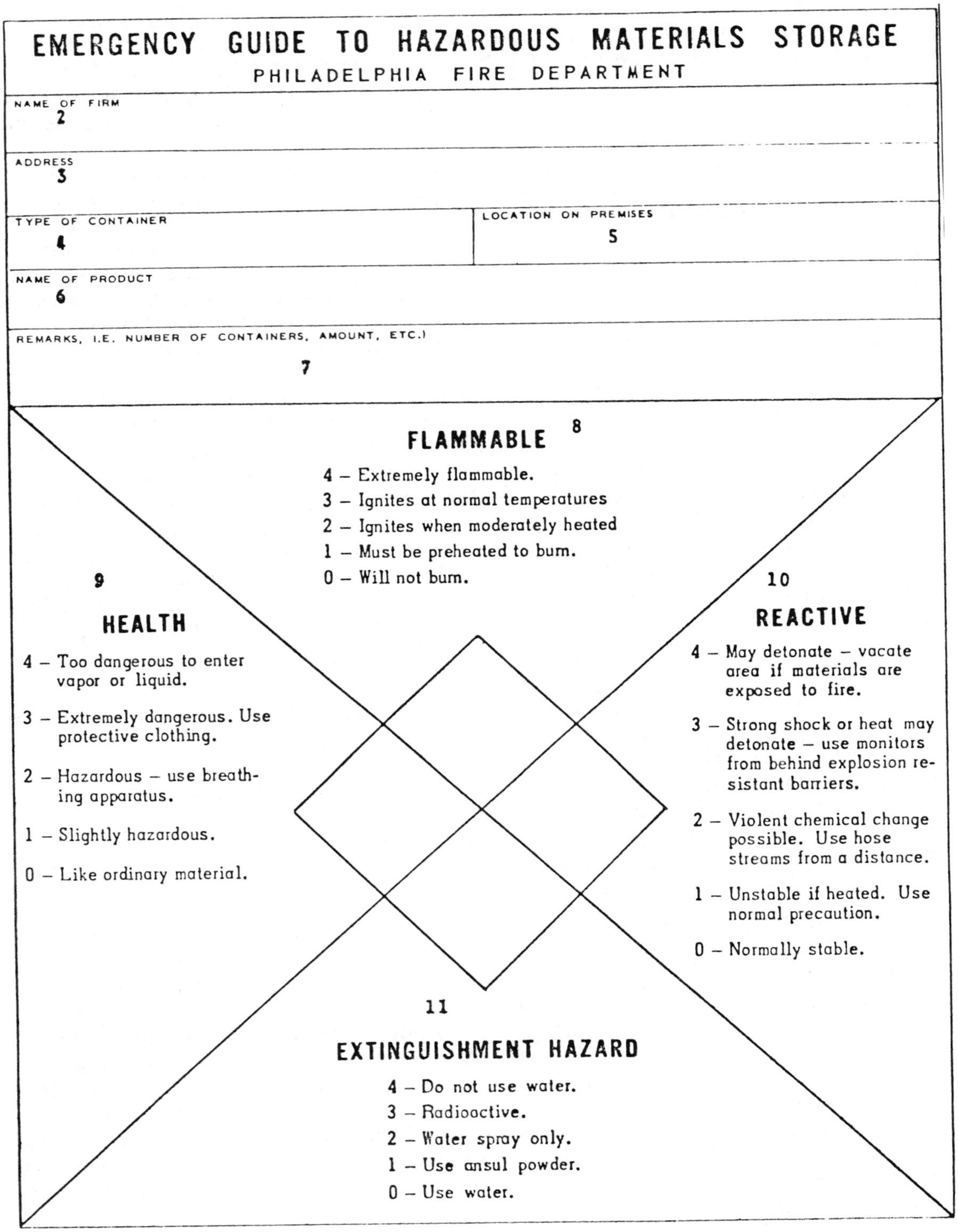

EMERGENCY GUIDE TO HAZARDOUS MATERIALS STORAGE
PHILADELPHIA FIRE DEPARTMENT

NAME OF FIRM
2

ADDRESS
3

TYPE OF CONTAINER
4

LOCATION ON PREMISES
5

NAME OF PRODUCT
6

REMARKS, I.E. NUMBER OF CONTAINERS, AMOUNT, ETC.)
7

FLAMMABLE 8

4 – Extremely flammable.
3 – Ignites at normal temperatures
2 – Ignites when moderately heated
1 – Must be preheated to burn.
0 – Will not burn.

9
HEALTH

4 – Too dangerous to enter
vapor or liquid.

3 – Extremely dangerous. Use
protective clothing.

2 – Hazardous – use breathing apparatus.

1 – Slightly hazardous.

0 – Like ordinary material.

10
REACTIVE

4 – May detonate – vacate
area if materials are
exposed to fire.

3 – Strong shock or heat may
detonate – use monitors
from behind explosion re-
sistant barriers.

2 – Violent chemical change
possible. Use hose
streams from a distance.

1 – Unstable if heated. Use
normal precaution.

0 – Normally stable.

11
EXTINGUISHMENT HAZARD

4 – Do not use water.
3 – Radioactive.
2 – Water spray only.
1 – Use ansul powder.
0 – Use water.

76-112

Figure 11.

PREPLAN INFORMATION

OCCUPANT __XYZ CORP.__ ADDRESS - __1647 ABC ST.__

GENERAL DESCRIPTION - __· MANUFACTURING PLANT.__

__117,600 SQ. FT.__   __ADT #000__

HYDRANTS/DRAFTING LOCATIONS - CORNER OF ABC + DEF STREETS. CORNER OF _DEF + GHI_ ST. +. TWO ON _G H I_ + ST. SIDE OF PLANT. *ALL HYDRANTS ARE PUBLIC HYDRANTS.

SPRINKLERS [YES] # OF RISERS [5] F.D. CONNECTION(S) - THERE ARE 4 - SEE MAP. CONTROL VALVE(S) - PIV & OSY FOR RISERS AT CONNECTIONS. OSY FOR FLAM. LIQUIDS STO? PRIVATE WATER SUPPLY INFO. - ALL SUPPLY IS FROM 8 INCH PUBLIC MAINS.

STANDPIPE [YES] TYPE [WET] # OF RISERS [2] F.D. CONNECTION(S) - CONNECTIONS C & D. HOSE LOCATIONS - 12 LOCATIONS THROUGHOUT PLANT. OUTLETS - ALL ARE 1½ INCH. CONTROL VALVES - AT EACH HOSE STATION. PRIVATE WATER SUPPLY INFO. - SAME AS SPRINKLERS.

SPECIAL PROTECTION - CO$_2$ SYSTEM IN PRINTING PRESS AREA - HEAT ACTIVATED.

EXPOSURES - ONE STORY WOODEN WORLD WAR II BUILDING ON ABC St.

UTILITIES -
ELECTRICAL CUTOFF(S) - INSIDE BOILER ROOM ON NORTH WALL - SEE MAP.
GAS [YES] TYPE [NATURAL] CUTOFF(S) - OUTSIDE BOILER ROOM DOORS AND AT BOILERS.
NOTES - GAS USED FOR BOILERS - THERE IS A 3 INCH GAS LINE ON THE ROOF THAT GOES TO THE BOILERS. — (SEE NOTES FOR LP GAS INFORMATION)

CONSTRUCTION - NO. OF STORIES - ONE (26 FT. HIGH)   SEE NOTES!
FRAME - METAL/BRICK/STEEL EXTERIOR - BRICK/ASBESTOS/GLASS, ROOF - FLAT COMPOSITION. NOTES!
ROOF OPENINGS - PRINTING ROOM HAS HATCH TO ROOF WITH LADDER.
HEAVY OBJECTS ON ROOF - ELECTRIC MOTORS - 25 HP / 40 HP
STAIRWELLS - PRINTING ROOM TO ROOF /OFFICE TO WAREHOUSE / LOBBY TO 2ND FLOOR
ELEVATORS - NONE   MAKE - N/A
FIRE WALLS - PARTITION AROUND PRINTING PRESS — SEE MAP.
FIRE DOORS - PRINTING ROOM — HEAT ACTIVATED.
OTHER CONSTRUCTION - METAL SHED AT REAR OF BUILDING - FLAMMABLE LIQUIDS STORAGE
— DRAFT CURTAINS IN MAIN PLANT AREA.

ROOF IS A LIGHT COMPOSITION CONSTRUCTION - WILL NOT HOLD HEAVY LOAD!

*1000 CU. FT. PROPANE TANK AT SE SIDE OF PLANT USED TO FILL FORK LIFT TRUCK TANKS.

Figure 12.

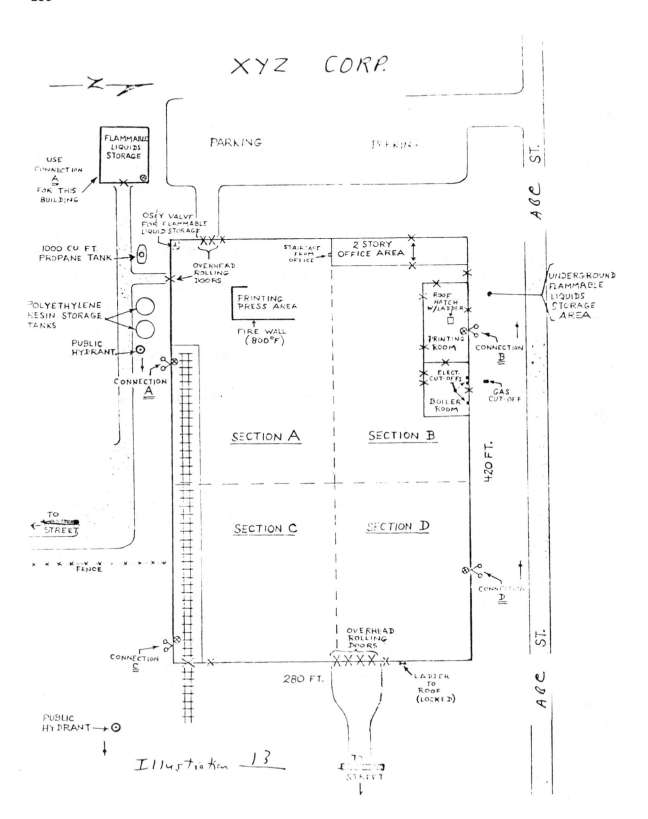

Figure 13.

PRE-FIRE PLANNING HAZARDOUS MATERIAL SURVEY FORM

ADDRESS _____ DATE _____

BRAND NAME _____

CHEMICAL NAME _____

TYPE OF MATERIAL _____

CLASSIFICATION

1. ☐	FLAMMABLE LIQUID	5. ☐	COMPRESSED GAS	9. ☐	CORROSIVE
2. ☐	FLAMMABLE GAS	6. ☐	LIQUIFIED GAS	10. ☐	POISON
3. ☐	FLAMMABLE SOLID	7. ☐	WATER REACTIVE	11. ☐	RADIOACTIVE
4. ☐	UNSTABLE MATERIAL	8. ☐	COMBUSTIBLE METAL	12. ☐	OXIDIZING AGENT
13.	OTHER				

_____

STORAGE

AMOUNT (Lb., Gal., Bbls.) _____
LOCATION AND STORAGE _____
SEGREGATED BY _____
PROTECTED BY _____

REACTIVITY: How will this material react when

	EXPLODE	BURN RAPIDLY	GIVE OFF HEAT	TOXIC VAPORS
IN CONTACT WITH WATER				
IN CONTACT WITH HEAT				
IN CONTACT WITH AIR				
IN CONTACT WITH ORDINARY MATERIAL				
IN CONTACT WITH OXIDIZING MATERIAL				

TOXICITY: What effect this material or vapors have on the human body?

☐	TOXIC	☐	CHEMICAL BURNS ON SKIN	☐	ASPHYXIATING
☐	CHOKING	☐	IRRITATION OF SKIN	☐	TEAR PRODUCING
☐	NAUSEA	☐	HEADACHE	☐	IRRITATION OF EYES
	OTHER				AND NOSE

OTHER _____

FIRE SCENE INSTRUCTIONS

BREATHING APPARATUS REQUIRE   SKIN PROTECTION REQUIRED

☐ YES   ☐ YES
☐ NO   ☐ NO

FULL PROTECTIVE CLOTHING

☐ YES
☐ NO

IF MATERIAL IS ON FIRE OR INVOLVED IN FIRE: _____
_____
_____
_____
_____

Figure 14.

dicators can be either "detectors" requiring repeated sampling to provide successive readings, or "monitors" that provide continuous readings and are often equipped with alarms that are activated when a combustible concentration of gas is detected. Since many combustible gases are heavier-than-air (having a vapor density of greater-than-one) and will tend to seep into low spots or travel at low levels for a considerable distance to a source of ignition and flash back,

BALTIMORE COUNTY FIRE DEPARTMENT

POOL SURVEY

1. NAME OF POOL: _____

2. LOCATION: _____

3. CAPACITY: _____ F.D. USE: _____

4. TYPE OF SYSTEM: CYLINDER____ LIQUID____ CRYSTAL____ PELLET____ OTHER_____

5. OWNER: _____ MANAGER: _____

6. TRAINING COURSES: _____

7. MAINTENANCE CONTRACTOR: _____

8. CYLINDERS CHAINED_____ "HTH" STORED PROPERLY____ LIQUID TANKS COVERED_____

9. ADEQUATE EXITS____ CONDITION OF PUMP ROOM____ STORAGE ROOM____

10. CONDITION OF WATER HEATER_____ ELECTRICAL SYSTEM_____

11. VENTILATION SYSTEM_____

12. PHONE NUMBER AT POOL:_____

    Does pool manager and/or life guard know what to do in case of a chlorine

    problem? _____

    COMMENTS: _____

    _____

    DATE_____ OFFICER_____ STATION_____ CYCLE_____

Figure 15.

NOTES

WORKSHEET
HAZARDOUS MATERIALS EMERGENCY REPORT

Time: _____   Date: _____

Caller's Name/Organization: _____

Call Back Number: _____

CHEMTREC: Called ☐   to be called by: Local ☐   State ☐   Not Needed ☐

PROBLEM: Type of Accident: _____

Time of Accident: _____

LOCATION: Town/City: _____

Highway/Road/Other: _____

Mile Marker/Other Locator: _____

DETAILS: Injuries/Fatalities/Observed Ill Effects: _____

_____

Weather: _____   Populated/Open Area: _____

On-Scene Status: _____

_____

PRODUCT(S) INVOLVED (Spell It Out):   Chemical Name(s): _____

_____

Trade Name(s): _____

_____

Characteristics of material (Vapor, Liquid, Solid - Other Observations): _____

_____

Quantity: _____   Type of Container: _____

☐ Single Load   ☐ Mixed Load

CARRIER/FACILITY: _____

Railroad Car No.: _____   Truck/Trailer No.: _____

Origin/Shipper: _____

Destination/Consignee: _____

Bill-Lading/Waybill No.: _____

ASSISTANCE: What are you Requesting? _____

_____

_____

Figure 16.

HAZARDOUS MATERIALS INCIDENT CONTROL CHECK LIST...

INCIDENT LOCATION AND TIME_____

HAZARDOUS MATERIALS INVOLVED_____

DANGERS OF INVOLVED MATERIALS_____

WHAT HAPPENED...WHAT CAUSED THE INCIDENT...TALK TO ALL PERSONS INVOLVED AND ESPECIALLY THE
PERSON WHO WAS INITIALLY INVOLVED, EVEN IF YOU HAVE TO TALK TO THE PERSON AT THE HOSPITAL,ETC

_____

_____

_____

CIVILIAN COORDINATOR_____INCIDENT COMMANDER_____

WEATHER CONDITIONS AT TIME OF INCIDENT_____

WEATHER FORECAST_____

SHIPPER & SHIPPING POINT_____

CARRIER & VEHICLE OPERATOR_____

CONSIGNEE & DESTINATION_____

VEHICLE TYPE & CHARACTERISTICS_____

VEHICLE #_____VEHICLE DOT CLASSIFICATION_____BILL OF LADING #_____

CONTACT SHIPPER, CARRIER AND CONSIGNEE AND LOG TO WHOM YOU SPOKE, NATURE OF CONVERSATION AND
TIME OF CALL...

_____

_____

_____

CONTACT OTHER AGENCIES THAT MAY ASSIST YOU...SUCH AS CHEMTREC, EPA, COAST GUARD, EARTH MOVERS
PRODUCT TRANSFERRERS, ETC...AND LOG TO WHOM YOU SPOKE, NATURE OF CONVERSATION AND TIME OF CAL

_____

_____

_____

AIR MONITOR READING_____WATER MONITOR READING_____

OTHER INFORMATION_____

ON OPPOSITE SIDE DRAW SKETCH OF INCIDENT...SHOW EMERGENCY RESPONSE UNITS AND THEIR ACTIVITIES
EXPOSURES, WATER/PRODUCT RUN OFF, DIKE CONSTRUCTION, VAPOR CLOUD MOVEMENT, EVACUATION AREAS,
ETC...IS YOUR OPERATION PLAN WORKING...REEVALUATE AND UPDATE YOUR OPERATION PLAN PERIODICALLY
KEEP INCIDENT VIGILANCE AND DISCIPLINE

CHEMTREC.................... -800-424-9300
NATIONAL RESPONSE CENTER.... -800-424-8802

Figure 17.

HAZARDOUS MATERIALS INSTRUCTION

AL MC ...
...
(714 447 ....

Incident Commander Checklist

HAZARDOUS MATERIAL:

Name _____

Check Resource Material _____

Chemical Properties _____

_____

Physical Properties _____

_____

_____

Called:	YES	NO
CHEMTREC 1 800 424 9200	O	O
MANUFACTURE	O	O
SHIPPER	O	O
HAZ MAT TEAM	O	O
MUTUAL AID	O	O

OBJECTIVES: UNIT(S) IN PRO. COM.

		IN PRO.	COM.
Rescue		O	O
Evacuation		O	O
Exposures		O	O
Extinguishment		O	O
Ventilate		O	O
Salvage		O	O
Contain		O	O
Control		O	O
Neutralize & Remove		O	O

UTILITIES: UNIT(S) IN PRO. COM.

		IN PRO.	COM.
Electric		O	O
Gas		O	O
Water		O	O
Other(s)		O	O

SUPPORT CONSIDERATIONS:

STAFF

Incident Commander _____

Suppression & Rescue _____

Resource Officer _____

Water Supply Officer _____

Safety Officer _____

Public Information _____

Planning Officer _____

ADDITIONAL

Refuge Area _____

Staging Area _____

Relief Personnel _____

Fire Investigator _____

Traffic Control _____

Ambulances _____

EPA _____

UNITS ASSIGNED

_____

_____

_____

DISPOSITION

_____

_____

_____

Figure 18.

HMT TRANSPORTATION VEHICLE INCIDENT CHECKLIST

Incident Info - (Date, Location, Circumstances, Etc.)_____

_____

_____

Railroad Car/Truck Trailer Initial & Number _____

Vehicle Type (RR Tank Car, Truck Trailer, Etc.)_____

Hazardous Material in Vehicle Container _____

Form of Hazardous Material (Liq., Gas, Liq.Gas, etc.)_____

Shipper's Name_____

Name of Driver,Engr.,Etc._____

Shipping Point(City & State, & Chemical Co., Etc.)_____

_____

Shipping Destination(City & State, & Chemical Co.,Etc.)_____

_____

Details of Vehicle's Container (Leaking, Spilled, Fire, Etc.)_____

_____

Action Taken by HMT to Remedy Vehicle's Problem _____

_____

REMARKS:

Figure 19.

the detection of combustible gases requires a continuous checking at various locations around a hazardous materials scene. Standing in one area and taking a few samples will not provide the necessary protection. When combustible vapors are potentially present, many response teams assign one person to continuously move throughout the area taking repeated readings with particular attention to low spots such as cellar holes, ditches, and slopes.

Instrumentation normally available to hazardous materials responders to detect toxic gases and vapors consists of chemical-specific glass tubes which change color to indicate gas concentration by the length of stain or color change. That is, as a sample is taken, chemical in the tube is discolored if the suspected gas or vapor is present in the atmosphere; the longer the stain, the greater the concentration. Many tubes have calibrated markings to indicate the existing concentration in Parts-Per-Million (PPM). To suck an air sample into the glass tube, either a hand-operated bellows pump or a slide-type pump is used depending on the make of a particular unit. A specific tube is required to test for a specific type of gas. To test for ammonia, you need an ammonia tube; to test for chlorine, you nee a chlorine tube, etc. Several hundred tubes are available to test for corresponding toxic vapors and gases, and a single tube may cost from $1.50 to $2.50 depending on the gas it is designed to detect. Limitations of detector tubes are that you must have the proper tube for a specific gas, they provide a one-time sampling capability rather than a continuous one, and their accuracy can be limited when more than one toxic vapor or gas may be present. There are over-riding positive features, however; detector tubes are quick, easy, reliable, and precise even at low concentrations.

Although most hazardous materials response teams are not hesitant to point out that if an incident involves radiation they will immediately pull out, evacuate the area, and call in a specialty team that handles only radiation incidents; many teams do carry basic radiation detection devices. Equipment generally available to response teams measures the presence and amount of Gamma radiation, the presence but not the amount of Beta radiation, and does not measure Alpha radiation.

In addition to meters and "sniffers," some response teams carry hand-held infrared sensors that can detect heat patterns as small as a lighted cigarette at 18 feet. Some units are capable of detecting heat patterns through wooden walls, metal bulkheads, or glass windows; and may emit an audible alarm when pointed at a heat source—the hotter the source, the higher the pitch of the alarm.

Nearly all teams carry inexpensive pH papers to determine a relative measure of the degree to which a substance is an acid or a base. The pH scale runs from zero (very acid) to 14 (very basic) with 7 being neutral.

pH Value   0 1 2 3 4 5 6 7 8 9 10 11 12 13 14
           Acidic   Neutral   Basic

The pH scale is logarithmic so a pH of 2.5 is far more acidic than a pH of 3.0. A pH paper is dipped in the substance to be tested, and the resultant color change that becomes apparent on the test strip is matched against a color code chart to provide the pH reading. pH meters are also available, but pH papers are inexpensive, (strips cost two to three cents apiece) quick, and reasonably accurate.

A gas chromatograph is a rather substantial and expensive piece of equipment not needed by nor available to first responders, but certain highly trained response teams do use portable units on-site rather than put up with the delay of sending samples to a laboratory for identification. Because of its great dynamic range, gas chromatography can be applied to a wide variety of organic compounds and offers a near universal tool when mixtures are involved. The process allows separation, identification, and measurement of individual unknown components in a mixture down to levels as fine, for example, as one Part-Per-Billion of P.C.B. Using glass or stainless steel "columns" of known length, a liquefied sample is placed into a hot injection port and becomes a gas, then is flushed through the column using nitrogen as a carrier. The carrier gas will move through the column unhindered, but various components of the sample will undergo various levels of partitioning. Components are thus retarded depending on their individual chemistry. Rate of retardation or "retention time," within the column is thus the identification or quantification factor. As the sample comes off the column, a detector makes a line on a data printer as soon as one organic molecule is sensed. The magnitude of the resulting "peaks" is in direct proportion to the amount of component present. By then comparing through the column process a known quantity of a known material, chemists complete the process of both quantifying and qualifying unknown components within a mixture. Any material that can be volatized can be measured by gas chromatograph.

11 HAZARDOUS MATERIALS EMERGENCIES—CASE HISTORIES

At 12:49 A.M. on Sunday July 26, 1981 the St. Johnsbury, Vermont Fire Department under Chief Jerald Fournier responded to an explosion and fire at an LPG mixing plant operated by the Gas Company of Vermont. Approximately four minutes before, St. Johnsbury city police had been alerted to a possible problem at the plant by a direct alarm between the mixing plant and police headquarters. Officer June Kelly was enroute to investigate when an explosion shook the surrounding area, shattering two-thirds of the four-brick thick building housing the mixing plant, showering adjacent South Main Street with brick and debris, and blowing in windows of 8–12 residences. A combination of ultra-thick brick walls and thin wooden roofing was thought to contain concussion, perhaps sparing immediately adjacent residences.

At the Vermont State Police barracks less than a mile away, a trooper was just returning from patrol when a brilliant flash lit his rearview mirror. In one motion, he floored the cruiser in a U-turn and radioed his dispatcher. As a result, three state police officers were reportedly on the scene blockading access roads within four minutes of the explosion.

"Our initial response was a 750 gpm pumper, a 75-foot aerial, and four men," says Jerald Fournier, Chief of the St. Johnsbury Fire Department staffed by 15 paid and 14 call firefighters. "We found the large brick building that had housed the mixing plant two-thirds destroyed by the blast, and a broken two-inch pipe releasing LPG that immediately vaporized and ignited sending flames 70–80 feet in the air. We had preplanned the building and were aware of a number of critical exposures in addition to immediately adjacent residences that state troopers and sheriff's deputies were already evacuating."

Exposures of first concern to firefighters included a 30,000 gallon rail tank car filled with LPG sitting on a siding 60 feet from the roaring blaze, a 30,000 gallon stationary above-ground tank approximately 40 percent loaded with LPG that sat 75 feet away, a 16,000 gallon LPG tank estimated to be 40 percent full situated within 100 feet, and a 16,000 gallon tank of propane gas located within 20 feet of the fire. Additional exposures included a second building containing 100 pound, household-type cylinders; and a third building used to store company vehicles.

"Our initial action—after calling the railroad that runs a 24-hour switching operation in the adjacent yards and asking them to immediately remove the loaded LPG tank car—was to get water on the tanks to keep them cool," says Chief Fournier. "There did not appear to be too much danger to these tanks at this time even though they were quite close to the fire. Flames fed by vaporized LPG escaping from a ruptured two-inch pipe were spectacular, but they were going straight into the air as there was almost no wind. We purposely refrained from any attempt to extinguish the burning, leaking gas so we would not have to contend with an accumulation of explosive vapors while gas company technicians were enroute to shut-down the valves.''

"We immediately got a 2½'' line on the 30,000 gallon stationary tank, and established a deluge set-up to cover the railroad tank car," adds Fournier, "then positioned additional lines for coverage of a 16,000 gallon propane tank and other exposures. Within five minutes of arrival, we called for two additional pumpers, a 1,000 gpm and a 750 gpm. The first of these additional units laid approximately 1200 feet of 2½'' line to the scene, then went back and hitched to a hydrant to service one handline and one line to the deluge set-up. We then laid two more lines off the 1,000 gpm pumper, one for a handline and a second to the deluge apparatus. A 750 gpm pumper made two 1600 foot lays to a hydrant, and then pumped in tandem with the original 750 gpm pumper located 1200 feet nearer the scene.''

"A railway crew moved the 30,000 gallon tank car to safety, and gas company technicians by then had shut-off the flow of escaping LPG," adds Chief Fournier. "As soon as the gas was shut-off, the remaining fire was not serious. The mixing plant had been nearly all brick and steel except for a light wooden roof, and combustibles within the building were limited to a few stored tires and some bags of coal left over from the days when the plant had manufactured gas from coal. Once the gas was shut-off, we were able to bring the remaining fire under control within an hour. Basically, we were able to get good water, and there was almost no wind. Also, there was tremendous, instantaneous response from state and local police and sheriff's deputies who immediately closed-off access to the area and began house-by-house evacuation.''

Although 8–12 frame residences lining the road across from the gas plant had many of their windows and a few doors blown, there was no fire damage other than to the exploded mixing building itself. LPG and propane tanks as close as 20 feet to the fire later showed little or no readily apparent heat or flame damage, although one tank buckled slightly from the weight of falling debris and necessitated precautionary repairs later in the day.

The Vermont State Hazardous Materials one-call system was activated by state police, primarily to secure additional gas detectors plus the services of a gas systems specialist and the state civil defense services coordinator. The only reported injury was to a resident who suffered minor cuts from flying glass.

Area residents who fled in vehicles south on South Main Street so as not to pass by the flaming plant hard-by the roadside, were stopped a half mile down the road by a chainlink fence that deadended the road when the new Interstate highway was constructed recently. Roland Duprey, a wrecker service operator with cutting torches, cut a roadway through the fence allowing state and local police to drive evacuee's vehicles onto the Interstate so they could continue on to homes of friends or to a temporary shelter established by town officials.

*   *   *   *   *

At 10:00 A.M. Tuesday, January 30, 1979 a tank truck driver for Good Housekeeping Gas Co., Inc., of Jacksonville, Florida unloaded 9,000 gallons of liquid propane gas into aboveground storage tanks at the Brunswick Gas Fuel Company located on U.S. Highway 17 in downtown Darien, Georgia across the road from 630-student Todd-Grant School. The driver moved the truck without first disconnecting a hoseline, causing the hose to break away from a valve on one of two stationary, above-ground storage tanks. The driver stepped down from the truck to find himself standing in a waist-deep vapor cloud flowing from a 3-inch main pipeline.

The vapor reached a point of ignition in approximately eight seconds. Flames from the broken connector hose flooded the truck cab, bouncing back to impinge on stationary storage tanks containing 16,000 gallons of L.P.G.

Vince Lanasa, Assistant Chief of the Darien-McIntosh Volunteer Fire Department, and Fireman David Bluestein, who were working across the street; gave the alarm and brought the first equipment to the scene. Fire Chief Melvin Amerson, who is also Assistant Manager of the gas company in Darien, was contacted by radio while driving in nearby Brunswick and was on the scene in ten minutes.

"Coming across the bridge into Darien," remembered Amerson, "I saw black smoke billowing above town. I assumed the tar-coated roof of the office building was on fire, but come to find out it was the tires on the tractor-trailer that were burning and causing the black smoke. A BLEVE was always a 100 percent possibility in my mind. Flame from the broken pipe under the storage tank was throwing flame into the tractor cab and off the cab back onto the closest of our above-ground tanks. The truck-tank had beem emptied before ignition took place and was thus filled with vapor and very dangerous. When I went through Georgia Fire Academy, they taught us a gas fire attack using seven men. I put three men each on two 2½" hoses with a master-man in the middle. They put out a water fog to keep the fire rolling so few or no flames were hitting me directly as I went in wearing protective clothing to shut-off two valves."

"The gas company also sends men to service school each year where instruction includes what to do in case of a fire of this nature," added Chief Amerson. "Our experience here was pretty much like they teach in the classroom. The basic difference was that here I was the 'MAIN MAN' required to take a real giant step: instead of being there to watch it, I was there to *DO IT*. Thankfully, I had been well-trained. I just let my mind go blank because I knew what had to be done—and went ahead and did it."

"Once the gas flow was shut-off, and we had everything cooled-down as much as possible, we separated the tractor and trailer and had the trailer towed to a desolate spot four miles outside of town. Georgia Forestry is most outstanding with us," noted Amerson, "they had a dozer here in 20 minutes to move the burning tractor away from the trailer. Then a fellow here in town who has a lease tractor put it under the empty trailer and towed it out of the downtown area."

"We never did have an explosion," concluded Amerson. "There was no tank rupture. The gas that was in the air ignited and burned out of control for the time it took us to get in close and shut-off two valves, about 35 minutes."

Due to immediate response and aggressive assault by the Darien—McIntosh Volunteer Fire Department, no one was injured in the vapor-ignition and resultant fire. Property damage was estimated to be in the area of $50–60,000.

To the extent that any selected incidents can be "typical," those in St. Johnsbury, Vermont and Darien, Georgia described above are representative of hazardous materials emergencies that occur annually throughout the United States. The commodity involved (propane) is widely used in sizable volume; there existed a serious potential for death and destruction yet the situation was brought under control without loss of life or serious injury; there was little or no prior warning the incident would occur; initial response was primarily a function of local government through an organized fire department, either paid or volunteer; fire officials were prepared, either through pre-planning or specialized training, to respond to the incident even though prior to its occurrence the potential necessity for such action may have appeared remote; and a fire officer on-scene as the incident commander requested and obtained assistance from other government/industrial/commercial organizations in controlling a recognized threat to public safety.

The intent of this chapter is not to downplay or disregard the serious destructive potential of hazardous materials emergencies, nor to minimize the various complex threats faced by response personnel on-site; but rather to emphasize that the great bulk of hazardous materials emergencies are successfully contained and controlled.

Certainly, there have been catastrophes. At 2:39 P.M. on October 24, 1944 the East Ohio Gas Company works in Cleveland was 20 years ahead of its time, a facility capable of liquefying 600 cubic feet of natural gas to one cubic foot of liquid thus saving on transportation and storage costs. At 2:40 P.M. the plant exploded killing 130 persons. The Cleveland disaster signaled the elimination of most LNG (Liquefied Natural Gas) usage in the United States for two decades. Not until the fuel shortages of the mid-1960s were additional LNG facilities constructed in this country.

On April 16, 1947 at Texas City, Texas outside Galveston the French freighter, Grandcamp, loaded with 1400 tons of ammonium nitrate fertilizer caught fire. As authorities initiated action to tow the Grandcamp out to sea, it disintegrated in a blast felt 150 miles away. Subsequent explosions destroyed the nearby Monsanto chemical plant and a second freighter loaded with nitrates. Although exact death and injury figures were never recorded due to the massive destruction, there were 468 known dead and 1000 seriously injured.

Railroads carry an estimated 40 percent of the hazardous materials transported within this country. Due to the sheer volume of hazardous materials carried in a single tank car, the greatest potential for a transportation-related hazardous materials catastrophe within a populated area may lie with the nation's railroads. On January 25, 1969 in Laurel, Mississippi one broken wheel resulted in an LPG explosion that killed two, injured 33, destroyed 54 residences, damaged 1,350 residences plus six schools and five churches, and graphically illustrated the problems encountered by small town fire departments in responding to such incidents. On February 18, 1969 nine died and 53 were injured when a toxic cloud of anhydrous ammonia seeped from a derailed tank car and smothered Crete, Nebraska. On June 21, 1970 multpile BLEVEs, rocketing tank cars, and flaming propane destroyed the business district of Crescent City, Illinois injuring 64 townspeople. On October 19, 1971 at Houston, Texas approximately 50 million B.T.U.s of energy were released from a ruptured tank car of vinyl chloride in less than one second, killing one and injuring 54. More than 230 were injured by an explosion of liquefied petroleum gas (propylene) at a railyard in East St. Louis, Illinois on January 22, 1972. On July 5, 1973 BLEVEs and rocketing tank cars of propane killed 13 and injured 95 at Kingman, Arizona. On February 12, 1974 multiple BLEVEs and rocketing of propane-laden tank cars injured 54 at Oneonta, New York. In one three-day period in late February of 1978 one tank car of LPG claimed 16 lives and decimated both the fire and police departments at Waverly, Tennessee while a single car of chlorine killed eight in Youngstown, Florida. One November 10, 1979 chlorine from a ruptured tank car at Mississauga near Toronto caused the largest peacetime evacuation in North American history when a quarter million people were forced to leave their homes.

However, roughly 85 percent of all transportation-related hazardous materials incidents occur on the nation's highways and involve motor vehicles. In June of 1964 six persons died at Marshall's Creek in Pennsylvania when a truckload of explosives caught fire and detonated. On May 30, 1970 two died and 30 were injured in Brooklyn, New York when a truckload of liquefied oxygen exploded in a hospital parking lot. Six died in Berwick, Maine on April 2, 1971 when a delivery truck pumped sodium hydrosulfide into the wrong storage tank causing a chemical reaction that sent deadly fumes through an industrial plant. On June 4, 1971 a truckload of dynamite exploded near Waco, Georgia killing six and injuring 33. Methyl bromide containers involved in a highway accident outside Gretna, Florida on August 8, 1971 sustained punctures and valve failures; exposure to the spilled cargo resulted in the death of four and injury to 12. On March 9, 1972 at Lynchburg, Virginia a tractor-trailer carrying liquid propane overturned, ruptured, and its leaking cargo ignited killing two and injuring five. The explosion of a tractor-trailer load of propylene 25 minutes after a traffic accident on the New Jersey Turnpike injured 28 persons on September 21, 1972. A propane tank truck overturn, explosion, and fire near Eagle Pass, Texas killed 16 and injured 51 on April 29, 1975. On May 11, 1976 a highway incident involving anhydrous ammonia killed six in Houston, Texas. In Beattyville, Kentucky on September 24, 1977 a truck carrying 8,255 gallons of gasoline went out of control while descending a hill and overturned in the downtown area. Spilled gasoline ignited killing seven; six buildings and 16 automobiles and other vehicles were destroyed. In the Detroit, Michigan area during 1977 there were 13 serious accidents involving tandem-trailer tank trucks ("double-bottoms") that killed five persons and injured 16. Dynamite cargos injured 13 at Keystone, West Virginia on April 5, 1979 and nine at Lancaster, New York on April 23. On June 9, 1979 a tractor-trailer crashed into a gasoline station causing an explosion and fire that killed four at Polar Bluff, Missouri.

Catastrophies there have been and will continue to be, yet to deal only with that percentage of incidents resulting in tradgedy would be misleading. The pages that follow describe 379 incidents that occurred during the two-year period April, 1980 to April, 1982. By no means are these 379 incidents all that occurred during that period, but merely those for which the author was able to obtain basic information. Some were tragedies; most were successfully controlled. The incidents are divided among six categories; stationary facilities, (206 incidents) highway transportation, (77 incidents) rail transportation, (48 incidents) marine transportation, (32 incidents) pipeline transportation, (14 incidents) and air transportation (2 incidents). Within each of the six categories, incidents are listed in reverse chronological order.

Stationary

March 18, 1982: Petersburg, Virginia. A fire touched off a series of explosions in downtown Petersburg, killing an elderly woman and a city firefighter and injuring at least 26. Damage was estimated at more than $3 million.

March 11, 1982: Forest Hills, Texas. Forest Hills Fire Department and the Fort Worth Fire Department hazardous materials squad responded to a chlorine leak at Wright Manufacturing where chlorine bleach is produced for retail sale. A 1¼ inch liquid line from a R.R. tank car on a siding to a chlorinator had ruptured, and a vapor cloud traveled north routing patrons from a cafe and motorcycle shop. Eleven people were injured, none seriously. Firefighters were able to shut-off the piping from the tank car, and industrial personnel handled clean-up.

March 5, 1982: Fort Worth, Texas. A citizen found four one-gallon jugs of an unknown chemical in a shed at the rear of her property. Firefighters and a commercial response contractor identified the liquid as hydrofluoric acid and sulfuric acid. The contractor found a local community college chemistry lab willing to accept the material in an innovative bit of disposal.

March 5, 1982: Fort Worth, Texas. When Fort Worth police broke up a drug manufacturing ring, they confiscated some eight pound cylinders of anhydrous hydrogen chloride. A commercial response crew found the cylinders to be sound and shipped them back to the manufacturer.

March 1, 1982: East Los Angeles, California. At least 11 people were injured and 100 persons evacuated from a one-square mile area after 100-gallons of chlorine overflowed from a storage tank. The leak was stopped within ten minutes of the time it started, but a westerly wind spread chlorine fumes.

February 26, 1982: Richmond, Virginia. Eighteen people were injured, with two in critical condition and one listed in serious condition, by fumes from a toxic gas leak at an industrial park north of Richmond. Workers at Environmental Laboratories, Inc. were converting an acidic boron gas to harmless hydrogen gas when the accident occurred.

February 15, 1982: Malone, New York. A 12-square block area was evacuated after a spill from a 30,000-gallon storage tank at Purdy Oil. No fire.

February 14, 1982: Athens, Georgia. Approximately 40,000 gallons of gasoline flooded a tank farm storage area, apparently released by a faulty gasket. Area residents were temporarily evacuated; no fire; no injuries.

February 11, 1982: Newark, New Jersey. A Rutgers University biology building, Smith Hall, was closed indefinitely after a study done in September was released indicating airborne levels of estradiol benzoate, a hormone believed to cause cancer, in five of seven classrooms sampled. A state health department report submitted last October stated that eight cases of cancer among 436 faculty members and other personnel interviewed between May, 1980 and April, 1981 was an excessive number. Estradiol benzoate has been used by the Institute of Animal Behavior on the fourth floor of Smith Hall since the modernistic building opened in 1968. A safe exposure level for workers is thought to be on the order of 50 nanograms per cubic meter of air. Samples taken in Smith Hall ranged from 170 nanograms per cubic meter of air in a third floor classroom to 1,070 nanograms per cubic meter of air near a breaker used to mix the hormone.

February 8, 1982: Point Tupper, Nova Scotia. An explosion in an evaporator used in a pulp making process at Nova Scotia Forest Industries, Ltd. tore through a steam plant and released poisonous gases. At least five died.

February 5, 1982: Essex Junction, Vermont. Xylene, a chemical solvent that ignites when exposed to air leaked at an I.B.M. plant and ignited. Sorbent material was used to clean the area. Earlier in the evening, a small spill of another solvent had occurred at the same plant.

February 5, 1982: Portland, Oregon. A 25,000 ton fixed-site tank with a malfunctioning valve dropped three to five tons of anhydrous ammonia. Response personnel evacuated an area three miles downwind, used water fog to knock-down vapors, and had the spill vacuumed.

February 4, 1982: Covert, Michigan. A hydrogen gas explosion in the turbine exciter of an electric generator closed the Palisades nuclear power plant run by Consumer's Power. One employee was injured seriously. Emergency plans had been approved the previous week, and an NRC member was on-site for a scheduled exercise. There was no release of radiation. The plant had been shut-down for an earlier, unrelated accident when a valve in the cooling tower malfunctioned. It was not immediately known what caused the hydrogen leak or what ignited the gas.

February 4, 1982: Oswego, New York. Radioactive sludge overflowed from a tank at the Nine Mile nuclear power plant operated by Niagara Mohawk when a valve on a 500-gallon holding tank allowed sludge to ooze onto the floor contaminating one building. Radioactive contamination was discovered on the outer clothing of three workers in what was classified as an "operational event" or non-emergency situation.

February 3, 1982: Omaha, Nebraska. A leak at the Fort Calhoun nuclear power plant released radiation into the environment and forced the evacuation of several workers before the leak was isolated.

February 2, 1982: Manchester, New Hampshire. An explosion and fire in a West High School chemical storage room injured a chemistry teacher and three firefighters. The teacher was reportedly putting away chemicals when the explosion occurred. Officials theorize a spark from faulty wiring touched off highly flammable benzene. The teacher and one firefighter were hospitalized for inhalation of toxic chemical fumes; two others were treated and released.

February 1, 1982: Lower Alloways Creek, New Jersey. The Salem I nuclear power plant had a spill of 23,000 gallons of radioactive water when a coupling in a temporary line gave way while water used to cool fuel in the spent fuel pond was being transferred. Sixteen workers had their shoes, socks and trousers wet by the slightly radioactive water, and had to be decontaminated with soap and water solution. Salem I was temporarily out-of-service for scheduled maintenance.

January 31, 1982: Oak Hill, West Virginia. An estimated 100 homes and businesses were damaged when an explosion authorities say was apparently caused by leaking natural gas in a downtown printing shop blasted the area at 2 A.M. Total damage was estimated at $1 million, but no injuries were reported.

January 20, 1982: Elizabeth, New Jersey. A father and son died in a flash fire triggered by oxygen leaking from a tank. Five oxygen tanks were stored in the house for use by another son being treated for a respiratory ailment. Rugs and furniture were reportedly saturated with oxygen, believed to have ignited when the father struck a match to light a cigarette.

January 25, 1982: Berkeley, California. Two University of California students were exposed to low levels of radiation when they accidently broke the protective covering on a source of Iron-55 material.

January 25, 1982: Ontario, New York. A tube ruptured at the Ginna nuclear power plant releasing radioactive steam into the atmosphere and causing a "site emergency." Following the rupture, the reactor was shut down and flooded with water in what has been called the worst nuclear scare since the accident at T.M.I.

January 25, 1982: Camden, New Jersey. A blast in basement classroom at Camden County College injured 48 people, three critically. The county fire marshal described the incident as a "gas" explosion, but was investigating to determine if it was methane gas from a broken sewer line or natural gas.

January 25, 1982: Worthington, Pennsylvania. A fire at Delta Chemical, Inc. where chemicals used by other industries are reprocessed, caused an estimated $400,000 damage. Six firemen suffered minor frostbite, but no injuries were reported from chemical exposure.

January 18, 1982: Los Angeles, California. The L.A. Fire Department responded to the Marquardt Co., an aerospace manufacturer in Reseda, where a vat used to heat trichloroethane overheated causing residue to ignite. Fumes given off caused a toxic cloud in the area, forcing the evacuation of approximately 500 persons. Trichloroethane, often used as a degreaser, is toxic and can be absorbed through the skin.

January 14, 1982: Poughkeepsie, New York. An explosion and fire destroyed the Berncolors Poughkeepsie, Inc. dye works. Two were listed as dead, two missing, and at least six were injured. Officials speculated the explosion may have been caused by the mixing of chlorine and muriatic acid (hydrochloric acid). All city buses were taken off their normal routes to aid in the evacuation of the 18-story Rip Van Winkle apartments next door to the scene due to smoke and toxic fumes.

January 14, 1982: Boston, Massachusetts. Two thousand persons were evacuated from the historic Quincy Market area's shops and restaurants when a noon hour leak from an electrical transformer sent ammonia-like fumes from insulating oil through the area. Four persons were treated for exposure to toxic fumes. The insulating oil contained PCBs, a suspected cancer-causing agent.

January 13, 1982: Shelby Township, Michigan. Two workers died after inhaling toxic fumes, possibly hydrogen sulfide, as employees at Liquid Disposal, Inc. were pumping sodium hydroxide from a tanker truck into a holding tank. A chemical reaction of unknown origin caused toxic fumes to develop around a pumping connection. Nine additional persons were injured.

January 12, 1982: Catlettsburg, Kentucky. Two workers were burned when an asphalt-separating unit exploded at an Ashland Oil, Inc. refinery while frozen pipes leading to processing tanks were being cleared.

January 11, 1982: Chicago, Illinois. A faulty furnace apparently released carbon monoxide fumes into a two-story apartment building, requiring hospitalization of an estimated 13 persons and treatment of 14 more. Several patients were reported to have blood gas levels of carbon monoxide near 30 percent (normal = less than five percent).

January 11, 1982: Athens, Alabama. Two workers repairing a gas leak caused when the water level in a radioactive waste sump became too low, were slightly contaminated at T.V.A.'s Brown's Ferry nuclear power plant.

January 8, 1982: Harrisburg, Pennsylvania. Workers inside the Three Mile Island nuclear power plant, scene of the nation's worst commercial nuclear accident in March of 1979, were clearing water from a pneumatic air line into a drainage system when contaminated dust particles were blown about a fuel handling building. Twelve workers were evacuated from two buildings during a 1½ hour "unusual event" alert, while 2500 workers in other buildings remained at work. One person was slightly contaminated.

December 26, 1981: Fraser, Michigan. A burning truck inside a building in an industrial complex reportedly released chlorine gas and caused police to evacuate a one square mile area, later extended to two miles.

December 23, 1981: Fort Worth, Texas. Liquid propane being pumped from a truck into an underground tank at the Homestyle Restaurant ignited/exploded, killing two and injuring four.

December 21, 1981: Danville, Illinois. Sixteen workers were injured when two explosions hit the Lauhoff Grain Co. grain processing plant located in downtown Danville. The cause of the blast was not immediately known. At the time of the blast, corn oil was being extracted, a process that requires the use of hexane, a volatile liquid.

December 21, 1981: Sylacauga, Alabama. An explosion centered in a coin-operated laundry caused structural damage to 42 buidings and caused a fire that destroyed half a city block, yet only three persons were reported injured, none seriously.

December 15, 1981: Danbury, Connecticut. Up to 15 propane tanks exploded in a garage at Union Carbide's under-construction world headquarters, forcing the evacuation of 2,000 people. The cause of the initial explosion was under investigation, particularly the possibility of a gas leak from a faulty 100 pound cylinder being unloaded from a truck. Reportedly, the bottom of the cylinder fell out, allowing gas to flow on an open-flame heater, ignite, and cause a chain-reaction explosion of up to 14 additional cylinders.

December 9, 1981: Port Arthur, Texas. Twenty-one persons were injured, two seriously, when a catalytic cracking unit exploded at a Gulf Oil Co. refinery.

December 1, 1981: Haddam, Connecticut. Approximately 4,000 gallons of sodium hypochlorite, a bleach-line substance, was accidently released into the Connecticut River from the Connecticut Yankee nuclear power plant. The discharge was described as about ten percent chemical and 90 percent water, and the chemical as water soluble. Sodium hypochlorite, often used in swimming pools, is used to clean the plant's reactor of algae.

November 14, 1981: Soddy Daisy, Tennessee. An electrical malfunction at T.V.A.'s Sequoyah Nuclear Plant caused a control rod to stick, and officials declared a first-level emergency. Reportedly, a boron-coated rod used in regulating the nuclear chain reaction became stuck in an "up" position during low-power tests. The problem was described as minor, yet either by design or unfortunate coincidence, two days later 15 personnel from the Tennessee Dept. of Health began going door-to-door in the Soddy Daisy community to distribute vials containing 14 potassium iodide pills designed to prevent thyroid cancer if taken 30 minutes prior to exposure to airborne radiation. Pills were given to 7,000 families.

November 6, 1981: Ventura, California. An explosion at a Getty Oil Company oil field compressor plant in the Ventura oilfields ignited five natural gas fires and injured one worker. Explosive forces scattered debris throughout the facility and twisted steel girders of the 100-foot long compressor plant. Portions of the compressors were melted, and nearby storage tanks were scorched.

November 6, 1981: Saugus, California. Waste rocket fuel exploded during a burn-off operation at the Bermite Div. of Whittaker Corporation. No injuries.

October 31, 1981: Weatherford, Texas. Extremely heavy rains and resultant flooding that hit parts of Texas in recent months caused a spill of 5,000–7,000 gallons of isopropyl alcohol and naptha at Power Service Products, Inc.; a manufacturer of diesel fuel additive. Flood waters floated a 12,000 capacity tank, broke the piping, and left the tank floating free. A nearby creek rose a total of six feet, sometimes at a rate of three inches every 10 minutes. Response crews from Western Emergency Services of Keller, Texas had to fight flood waters between Fort Worth and Weatherford to reach the scene. Inside a broken dike, they found four horizontal tanks; three contained enough product that they did not float, but were leaking and had to be patched. Six vertical tanks did not present a problem. Response personnel wearing lifejackets were able to secure the floating tank with ropes only by working thigh-deep in rushing waters too fast for a boat.

October 30, 1981: Fort Worth, Texas. Police and fire officers responded *very carefully* to a cache of badly deteriorated dynamite found in an abandoned well-house. Approximately three inches of nitro had pooled in the bottom of the box. Responders used kerosene to stabilize the dynamite, then moved it 15 sticks at a time to a drillfield for burning.

October 27, 1981: St. Marys, West Virginia. Five workmen were injured when an explosion at an oil drilling rig sent flaming oil surging over them. The well was being drilled by Adams Well Service for Berg Petroleum on the grounds of Colin Anderson State Hospital, a state-operated school for the retarded.

October 22, 1981: Holbrook, Massachusetts. An explosion destroyed Aerosol Research Laboratories killing three and hospitalizing 25 to 30. The cause of the blast at the facility where windshield washer solvents and aerosol cans were manufactured, is unknown. Three weeks after the explosion, 14 persons were still hospitalized.

October 18, 1981: Ventura, California. Four workers were burned, two seriously, when leaking vapors from a portable tanker carrying 20,000 gallons of crude oil ignited near an oil well in the Conoco oilfields in Ventura County.

October 14, 1981: Los Alamos, New Mexico. Eleven persons were exposed to plutonium radiation at the Los Alamos National Laboratory. One worker was possibly seriously contaminated while ten others were described as receiving minor contamination.

October 13, 1981: Justin, Texas. Three 30,000 gallon capacity tanks (two gasoline and one diesel fuel) floated, broke their piping, and rolled over during heavy flooding at a truckstop. Approximately 10,000 gallons of product was spilled. After flood waters receded, Justin Fire Department personnel used foam to seal remaining vapors, and brought in vacuum trucks to pick-up residue.

October 13, 1981: Freeport, Texas. Six workers died at Dow Chemical's Freeport plant from an explosion and fire in a chemical container that had been shut-down for maintenance. The victims had been replacing equipment in Dow's polyethelene unit when an explosion occurred in an outlet line at the bottom of a separator vessel in Plant B located near the center of the 4,500 acre complex. The polyethelene unit produces plastics, chlorine, and solvents. The primary bases used are brine and petroleum which are combined to produce about 100 chemicals.

October 12, 1981: South Portland Maine. A faulty gauge was blamed as the cause of a 10,000 gallon storage tank overflow of gasoline at a Gulf Oil Corp. facility. The gas was contained within a protective dike. Gasoline was being pumped into the tank from the barge Gulf Pennsylvania when the overflow occurred. Firefighters pumped water into the earthen floor dike in order to "float" the gasoline and keep it from seeping through the ground into Portland Harbor, then used protein foam to prevent the gasoline from evaporating into explosive vapors.

October 9, 1981: Richland, Washington. A potentially dangerous malfunction occured at the Hanford nuclear reactor during a routine maintenance operation while the reactor was shut-down. The ventilation system lost its vacuum capabilities temporarily, requiring that 50 employees be checked for contamination.

October 1, 1981: Long Beach, California. Pressure from a mixture of sulfuric/nitric/hydrochloric acid blew a plate off a storage tank at a waste disposal company operated by Chancellor & Ogden, Inc.-Div. of BKK Corporation; allowing toxic vapors to escape. Approximately 300 people were evacuated for up to 4 hours.

September 30, 1981: Toledo, Ohio. An uncontrolled leak of radioactive krypton and xenon gas at Toledo Edison's Davis-Besse nuclear plant caused the temporary evacuation of 75 workers. The amount of gas that escaped was described as relatively insignificant.

September 30, 1981: Amherst, Massachusetts. The 17-story Graduate Research Tower at the University of Massachusetts was evacuated after two graduate students in the polymer chemistry lab inadvertently formed what was believed to be thallium acetylide, a very unstable compound described as a shock-sensitive, potentially toxic explosive.

September 28, 1981: Norwalk, Connecticut. At least 16 persons were treated at Norwalk area hospitals after a chlorine leak at a city sewage plant sent a toxic cloud over a waterfront business/residential area. Those treated included at least four firemen and five plant workers involved in stopping the leak.

September 21, 1981: Cambridge, Massachusetts. Chlorine gas leaking from a Harvard University athletic building injured 34 persons and caused the evacuation of seven nearby buildings. An employee was reportedly changing chlorine tanks near a swimming pool when gas began to escape under pressure. Firefighters closed-off tanks, then used water and ventilating fans to clear the building.

September 17, 1981: Good Hope Louisiana. At least 12 persons were injured at Good Hope Refinery when a fire in a vacuum heater flashed out of control forcing evacuation of nearby homes and a school.

September 11, 1981: San Diego, California. San Diego firefighters called to an alley on 4th

Avenue found a couple of leaking drums discarded by persons unknown. After identifying the contents as hydrochloric acid and formaldehyde, the drums were removed and disposed of.

September 10, 1981: Chattanooga, Tennessee. Approximately 50 employees were contaminated by radioactive gas at T.V.A.'s Sequoyah nuclear power plant. Exposure levels were described as within limits established by the N.R.C.

August 27, 1981: San Diego, California. The San Diego Fire Department responded to a spill of methyl isobutyl ketone (flammable liquid) reportedly resulting from a clean-up operation. Firefighters diked approximately 60 feet of a parking lot area, and used sand to absorb the liquid.

August 21, 1981: South San Francisco, California. A truck backed into a pipe on a 2,000 gallon tank of silicon tetrachloride at the M&T Chemical Co. located between Candlestick Park and San Francisco Airport, releasing approximately 1,000 gallons of the corrosive/acidic material that immediately formed a vapor cloud. Approximately 6,000 workers in nearby industrial plants were evacuated as the Coast Guard ordered boats to keep out of an eight square-mile area of San Francisco Bay and the FAA barred air traffic from the area. At least 28 people required medical care, although injuries were described as "not serious." Much of the spilled material was caught in an underground holding tank. Crews from IT Corp. used rubber-lined vacuum trucks to suck-up the chemical, a colorless, fuming liquid with a pungent odor that is decomposed by water to hydrochloric acid.

August 21, 1981: San Diego, California. Two children brought a container to a San Diego fire station. The box, once used to carry a radioactive material used for medical purposes and having a half-life of 66 hours, indicated a zero reading on the outside and proved to be empty.

August 16, 1981: Mettler, California. Two pumping station employees were injured critically when an 80,000 barrel crude oil tank exploded at the Emidio Pump Station. The two men had finished draining the tank, but as they removed a manhole cover, vapors from oil residue remaining in the tank found a source of ignition.

August 13, 1981: Seattle, Washington. Employees alerted by the sound and smell of escaping gas vacated the Wienker Carpet Service, Inc. block-long building minutes before an explosion destroyed the structure. A fire department spokesman stated a forklift driver had apparently broken a gas meter, either by hitting it or by dropping a carpet roll. Only two minor injuries were reported.

August 11, 1981: South Burlington, Vermont. A leaking 150-pound cylinder of sulfur dioxide gas at a Merriman-Graves welding supply company warehouse brought firefighters to remove the cylinder and hose down the remaining vapors. Lime was spread to neutralize any remaining acid. Sulfur dioxide is a non-flammable gas/corrosive. It is soluble in water to form sulfurous acid, a corrosive. Water spray is used only on resultant vapors, not on the material itself.

August 10, 1981: Keyser, West Virginia. An explosion at the Navy Ordnance System Command's Alleganey Ballistics Laboratory operated by Hercules, Inc. under a Navy contract completely destroyed a building where rocket fuel components are made and liquid explosives for rocket propulsion systems are prepared. No trace remained of two workers in the building. The 1,000-square-foot building was surrounded by timber and earth to direct explosive forces upward, and 100 similar buildings evenly spaced throughout the complex were not effected.

August 10, 1981: San Diego, California. Firefighters were called to a business on Harney Street to assist in controlling a leaking container of corrosive material. They removed the container from the building, diked the spill, and called IT Corp. for clean-up.

August 10, 1981: The San Diego Fire Department responded to a recreation center at 2909 Marcy Street to assist in stopping the flow from a chlorine cylinder.

August 9, 1981: Tustin, California. 1500 persons were evacuated for 14 hours after a 3,800 gallon leak of phosphoric acid, apparently resulting from a broken gas gauge on the side of a storage tank at Larry Fricker Co., a fertilizer plant.

August 5, 1981: Lordstown, Ohio. Approximately 100 gallons of cancer-causing polychlorinated biphenyls (PCBs) spilled on the floor of a powerhouse at General Motors Lordstown assembly plant when a 2,000 kilowatt transformer exploded releasing about one-third of its PCB-laden oil.

July 31, 1981: Moab, Utah. One died and nine were critically injured when a lightning bolt reportedly knocked down a power pole severing a gas line at the Buckeye Gas Products storage facility, sending a wall of flame through a nearby campground and forcing 3,000 people to flee their homes. Gas company employee, Doug Farnsworth, made a valiant attempt to shut-down an emergency valve after the line burst, but the gas ignited just as he reached the valve, and Farnsworth was blown across the yard into a water tank. Rescue workers were worried about 50,000 gallons of liquid propane in an underground salt cavern beneath the campground, but this large resevoir did not ignite. Authorities theorize the source of ignition was campfires in the campground.

July 30, 1981: Monticello, Minnesota. Up to 2,000 gallons of radioactive water leaked from the Northern States Power Co. nuclear reactor into the Mississippi River from which Minneapolis and St. Paul take their drinking water, but authorities stated natural dillution in the river would keep the pollution from posing a health problem. Reportedly, an employee hooked a hose to a service water pipe that leads from storage tanks holding contaminated water in order to use the water to mix cement. The hose later came loose, and contaminated water flowed a block and a half into the river.

July 30, 1981: Grantsville, Utah. Three powerful explosions leveled the remote Mining Services International explosives manufacturing facility, killing all five graveyard shift workers. Major explosions occurred at 4:20 A.M., 6:30 A.M., and 7:00 A.M., completely leveling the five acre site.

July 28, 1981: Peoria, Illinois. Search crews aided by a helicopter found 43 of 51 cannisters of deadly methyl bromide gas buried at a construction site. The cannisters had earlier been taken from an Orkin Exterminating Co. facility. Still missing were eight cannisters of methyl bromide and a five pound container of potentially deadly cyanogas. A heavy equipment operator told police he had buried containers similar to those mentioned in a news report, thus leading authorities to the construction site. Methyl bromide in its various forms is a Poison B.

July 26, 1981: St. Johnsbury, Vermont. An early morning explosion of unknown cause destroyed the Gas Company of Vermont mixing plant, shattering two-thirds of the 75 year old building. Although windows were blown in 8–12 nearby homes, the four-brick thick walls of the old facility were thought to have contained most of the shock effect of the blast. Primary exposures found by fire-fighters who had preplanned the facility included a 30,000 gallon rail tank car on a siding sixty feet from the blast site, a 30,000 gallon stationary tank 75 feet away, as well as a 16,000 gallon tank 100 feet from the blast—all either fully or partially loaded with LPG or propane. A smaller tank of propane was 20 feet from the fire, and a second building used to fill household propane cylinders was nearby. While firefighters used handlines and a deluge set-up to cool exposed tanks, a rail crew removed the tank car and Vermont Gas Co. technicians shut valves to stop the flow of LPG that was vaporizing and igniting as it hit the atmosphere. Once the flow of LPG was stopped, firefighters were able to bring the remaining blaze under control within an hour. The only reported injury was one person cut slightly by flying glass.

July 23, 1981: Norco, Louisiana. An explosion and fire ripped apart a 25,000 barrel gasoline storage tank at the General American Transportation Co. (GATX) gasoline tank terminal, critically injuring two workers. Flaming gasoline was hurled onto La. Hwy. 48.

July 23, 1981: Martinez California. More than 100 employees were evacuated from the Shell Oil Co. refinery when an explosion and fire caused minor injuries. The cause of the blast was said to be oil line obstruction.

July 18, 1981: San Onofre, California. A radioactive steam leak to the atmosphere was reported at the San Onofre nuclear power plant. No injuries.

July 11, 1981: Santa Fe Springs, California. Burning drums of hydrocarbons at a chemical storage yard, including several drums that BLEVE'd, eventually required three alarms to control.

July 11, 1981: Orangeburg, New York. Approximately 25 persons were treated at a local hospital for inhalation of chlorine fumes after a leak at the Orangetown sewage treatment facility on Rte. 303 forced 80 people from a drive-in theatre 500 yards away that was showing "Escape From New York." The chlorine was stored in pressurized cylinders and officials indicated a faulty valve as the probable cause.

July 10, 1981: San Diego, California. A civilian walked into San Diego Fire Station #15 with a glass vile containing a radioactive material identified as a beta emitter. Radiological testing equipment was used to determine that there was no measurable outside reading, and the device was taken to a disposal facility.

July 7, 1981: Duncannon, Pennsylvania. Officials gradually discharged a half-million gallons of herbicide-contaminated (2,4D) water from the Duncannon resevoir into a tributary of the Susquehanna River prior to flushing the system after tests showed .92 PPM of the weedkiller in the water supply. (.1 PPM = federal safe drinking water level.) Officials fear sabotage as the cause.

July 7, 1981: St. Johnsbury Center, Vermont. Approximately 200 gallons of liquid propane escaped from a tank at Adirondack Gas Co., releasing an estimated 5,400 cubic feet of vapor, apparently through a missing or malfunctioning-backcheck valve.

July 3, 1981: Louisa, Virginia. An electrical transformer fire that burned part of the wall of a turbine building adjacent to the reactor containment structure at the Virginia Electric & Power Co.'s North Anna nuclear power plant sent large clouds of black smoke billowing from the plant, but reportedly resulted in no injuries or release of radioactivity, and did not require evacuation.

June 26, 1981: Barberton, Ohio. An explosion in a four-by 25 foot tank on the ground level of a seven story tower in the hydrogen peroxide area at the PPG Industries plant resulted in a fire that blackened part of the tower structure. PPG fire brigade personnel and Barberton city firefighters wearing SCBA were able to bring the blaze under control without injury. Hydrogen peroxide, used as a bleaching agent in the textile and paper industires, has a flammability level similar to that of kerosene. A corrosive and oxidizer, it can cause skin and lung irritation.

June 23, 1981: Pasadena, Texas. Explosions and fire in an area of the Crown Central Petroleum Corp. refinery near where hydrochloride acid is stored injured five plant workers and one firefighter before being brought under control. The fire along the Houston Ship Channel forced the closing of nearby roads and the Washburn Tunnel that links highways on both sides of the channel.

June 22, 1981: Rocklin, California. A malfunction in the overflow alarm allowed approximately 40,000 gallons of gasoline to overflow a storage tank at a Southern Pacific Transportation Co. tank farm. More than 100 firefighters responded, and mutual aid calls were made to nearby communities. A layer of foam was put down, and the gasoline did not ignite.

June 19, 1981: Lyndonville, Vermont. Approximately 250 tons of liquid nitrogen fertilizer spilled on the ground at the Old Fox Chemical Co. plant, reportedly the result of vandalism, caused the temporary closing of a village water well near the site. Old Fox is being requested to speed excavation and replacement of the saturated soil, and a geologist will be hired to determine if and where the nitrogen is moving through the ground.

June 13, 1981: Lancaster, Pennsylvania. Two EMTs and a volunteer fireman died one after another in attempting to save an eight year old boy who had climed down a 14-inch pipe to retrieve a lawnmower grasscatcher he had dropped into an abandoned septic tank used over the years for disposal of grass clippings. The boy was overcome by methane gas produced by decomposition of the clippings over time. A Medic entered the tank but was overcome. A second Medic wearing an air pack entered the tank and managed to secure a rope to the boy, but apparently removed his mask in an effort to save the first Medic and was overcome, the mask only a few feet from his face. Two volunteer firemen crawled down the pipe, and one was overcome. The boy was pulled to safety and survived, but the two Medics and one firefighter died.

June 12, 1981: San Diego, California. San Diego firefighters responded to a minor chlorine spill; someone had poured approximately 10 gallons of liquid chlorine into a stormdrain. Since the chlorine was already in the stormdrain in a limited quantity, it was flushed with water.

June 11, 1981: San Diego, California. The San Diego Fire Department used water and sand to dilute/absorb approximately 50 gallons of sulfuric acid leaked from a 3,000 gallon tank, then had the tank off-loaded.

June 10, 1981: Barnesville, Minnesota. Thirty persons were injured, eight seriously, when 40–50 tons of anhydrous ammonia escaped from the Farmland Industries storage terminal during a period estimated at 25 minutes. A toxic cloud floated across a nearby highway, and many of the injured were motorists who drove into the cloud, stalled or crashed their vehicles, and stumbled through the cloud on foot coughing and wretching. A number suffered skin burns. The leak was reportedly caused by failure in a heating system that allowed pressure to build triggering a release valve. The Minnesota Pollution Control Authority reportedly planned to cite Farmland Industries both for the release and for a possible delay in notifying authorities.

June 1, 1981: Janesville, Wisconsin. A natural gas main, apparently ruptured during excavation work, caused a series of explosions that ripped through a downtown office and apartment building, set fire to three adjoining buildings, and injured 12 persons, four seriously.

May 27, 1981: Oak Ridge, Tennessee. A highly toxic radioactive gas, (uranium hexafluroide gas) used in production of fuel for nuclear power plants, was accidently released from a government uranium processing center. Officials evacuated 125 workers from the plant and buildings downwind. No injuries were immediately reported, but 17 workers were tested by urinalysis to determine how much radiation they may have ingested. An unknown amount of gas, described as "a white puff," apparently escaped and hung over the facility for a short period of time after a small hose was cut during routine maintenance. The gas is both radioactive and a fluoride and thus can cause problems with the respiratory system. Uranium hexafluoride is a solid at room temperature but becomes a gas when slightly heated.

May 26, 1981: San Diego, California. IT Corp. of Wilmington, California was contracted to handle clean-up after a survey at the University of California at San Diego indicated that of all transformers on campus, 100 contained P.C.B. (polychlorinated biphenyls). Of the 100, three were found to be leaking P.C.B.

May 26, 1981: Artesia, New Mexico. Leaking propane and butane gases ignited and exploded at the Navajo Refining Co. facility injuring 17, two seriously. The explosion reportedly occurred when gases escaped and ignited as employees were working on a pumping unit of a new catalytic converter that had been shut-down two days earlier because of a leak.

May 26, 1981: Groton, Connecticut. Local health authorities imposed a ban on shellfishing in the Poquonnock River after a four-alarm fire destroyed Johnson's Home & Hardware store. Chemicals leaking into the waterway left hundreds of fish dead, either from a lack of oxygen caused by the chemicals or from direct contact with the chemicals themselves.

May 22, 1981: Athens, Alabama. A leak in a reactor cooling system triggered a site alert and the shutdown of one of three units at the Brown's Ferry Nuclear Plant as an estimated 10,000 gallons of radioactive water leaked from the stem of a discharge valve on one of two recirculation pumps in the drywell surrounding one of the reactors. T.V.A. officials stated that all cooling water was contained within the drywell and recirculated by a drainage system, and that no radiation was released.

May 22, 1981: Chicago, Illinois. A 55-gallon drum of cresylic acid exploded releasing toxic fumes that injured at least 28 persons in a commercial/industrial area on the west side of the city.

May 20, 1981: Childersburg, Alabama. 900 people were evacuated from within a ten mile radius of Alpine Laboratories after the pesticide plant was leveled by an explosion that released toxic fumes. Continuing explosions over a period of hours kept firemen at bay. Attention was directed toward protecting exposures until hours later when firefighters were eventually able to approach the ruins and use light water to mop up burning residues.

May 19, 1981: Trujillo Alto, Puerto Rico. A spill of nearly four tons of liquid chlorine at a water purification plant in a suburb of San Juan spawned a toxic cloud that forced the evacuation of 1,500 persons by city buses pulled off regular routes and sent to the scene. 200 persons received emergency treatment at six San Juan hospitals. The contents of three tanks and most of the fourth escaped, with the liquid evaporating quickly but forming a toxic cloud that spread through two nearby housing projects.

May 12, 1981: DeWitt, Iowa. Twenty-seven people attending a 4-H meeting at a private home were injured, eight critically, by an explosion and fire believed caused when the homeowner attempted to light a propane gas furnace in the basement.

May 11, 1981: Phoenix, Arizona. Approximately 110 fourth graders at Sunburst Elementary School were treated at local hospitals for nausea and related symptoms. Officials reported workmen had been using an industrial cleaner to remove graffitti from a wall, and that some cleaner residue apparently settled on drinking fountain spouts.

May 7, 1981: San Diego, California. Approximately ten gallons of an unknown substance was found on the ground at a construction site at Pacific Hwy. and Taylor. The substance was believed to be oil from a military-type transformer spilled when an effort was made to salvage copper inside the transformer. The area was isolated and IT Corporation was called in to make a positive identification by laboratory examination and perform clean-up. The substance was determined to contain 50PPM polychlorinated biphenyls (PCBs).

May 6, 1981: Radford, Virginia. Four workers were burned and the roof was blown off a building by an explosion during a test of a new powder making process involving water-wet nitrocellulose at the Army ammunition plant operated by Hercules, Inc.

May 3, 1981: Sacramento, California. Leaking radioactive steam delayed plans to restart the nuclear power plant at Rancho Seco. The leak, described as five gallons-per-minute and "insignificant" by a utility spokesman, was inside the plant's reactor building. No workers were thought to be exposed to the radioactive steam, and there was no immediate evidence that steam escaped from the building.

May 2, 1981: Shippingport, Pennsylvania. A two-minute release of radioactivity occurred at the Beaver Valley nuclear power plant No. 1. Low-level radioactive material used as a tracer for water droplets in steam generators during efficiency tests was released to the atmosphere when a feedwater control valve malfunctioned causing a safety valve on a steam generator to open. The release was described as "less than ten percent of the level normally allowed under N.R.C. regulations."

April 29, 1981: Newington, New Hampshire. An empty, underground jet fuel storage tank exploded in flames while being cleaned by a crew from Jet Line Pollution Control of Massachusetts. Two men received second and third degree burns and two died of asphyxiation when the 100 foot by 24 foot deep concrete tank collapsed during the first day of routine cleaning operations. The U.S. Defense Logistics Agency owns the tanks and pipes jet and motor fuel from them to nearby Pease Air Force Base. New England Tank Industries of N.H. operates the tank farm. The cause of the explosion was not immediately known.

April 29, 1981: Santee, California. The Santee Fire Department reponded when a couple of pallets at Price Club, a department store, were tipped over and chlorine and ammonia became mixed. Approximately ten persons had been trying to clean-up the spill when they began to show signs of respiratory distress and headaches. Roughly 30 employees were evacuated and six treated.

April 23, 1981: Bethesda, Maryland. A small package of highly radioactive cobalt at the Armed Forces Radio-Biological Research Institute posed a tricky problem for officials. The cobalt was exposed by the jamming of a small elevator-like device designed to carry the container from a storage location in 14 feet of water to the surface for use in weapons research. The cobalt was said to pose no immediate threat as it was contained within a heavily insulated experiment room. The room is designed to be flooded so workers can enter the room in a rubber boat and use manual tools to free the cobalt.

April 23, 1981: Highland City, Florida. A blast that severely damaged a chemical plant injuring at least one workman caused a plume of smoke and gas 1,000 feet high as officials evacuated a one square mile area and blocked off nearby Hwy. 98. The plume drifted toward the southwest and was reported to be dissipating rapidly. Fire trucks and ambulances from Highland City, Lakeland, and five other nearby towns responded.

April 20, 1981: Wiscasset, Maine. Radioactive gas leaked to the atmosphere for several days at Maine Yankee's nuclear power plant. The leak occurred when a quarter-inch valve was not shut and air from a containment building was released. N.R.C. personnel reported the release was 14 percent of what could be considered within limits during a purge of all air in the plant's containment building.

April 15, 1981: Erie, Pennsylvania. Erie County officials reported cyanide, chloroform, toluene, and napthalene . . . some in dangerously high levels . . . at an old dump site near a home for the retarded by Lake Erie. The chemicals were thought to be coal gas products. Cyanide was said to be present at 100 PPM. State drinking water standards set the allowable levels of cyanide at .01 PPM.

April 14, 1981: Epping, New Hampshire. The E.P.A. declared an emergency at Keefe Environmental Services and declared the site must be cleaned-up at once. Heavy rains that had earlier threatened to overflow a large waste holding lagoon, and extreme cold that ruptured several drums stacked on the ground added to a long series of problems at the site. A number of barrels bulged or were distorted, and red and green liquids flowed on the ground in some areas.

April 13, 1981: Golden, Colorado. A fire at the Colorado Specialties plant burned for five hours, forced the evacuation of an estimated 250 persons, and allowed leaking flammable styrene to enter the town's sewer system. Frequent explosions were heard throughout the afternoon. Chemicals in the plant included toluene, heptane, styrene, butadine, sodium disposium, lycon and zilon. All nearby manhole covers were raised to aerate the sewer system but preliminary tests indicated the styrene was present only in smaller amounts not likely to cause further explosion or fire (Styrene has a flashpoint of 90°F. Vapors are irritating to eyes and mucus membranes. When contaminated or subjected to heat, it may polymerize. Polymerization inside a container may subject the container to violent rupture).

April 13, 1981: National City, California. The National City Fire Department responded to an acid spill of unknown origin on a city street. Firefighters in breathing apparatus traced the flow to a nearby facility where an open valve on a hydrochloric acid tank was located. The facility business manager was notified, the police department was contacted to block-off the street, and a waste disposal company called to handle clean-up.

April 13, 1981: Coventry, Rhode Island. The State of Rhode Island resumed clean-up work on the state's biggest illegal toxic dump, the Picillo Pig Farm, where at least 10,000 barrels of PCBs, organic solvents, cyanide, and other chemicals are expected to be unearthed. Work resumed when state officials decided to finance excavation with a $1 million bond originally scheduled to cover expenses for disposal of the waste, rather than wait for delayed federal funds. Peabody Clean Industry, Inc. has been hired as clean-up contractor.

April 9, 1981: Detroit, Michigan. Employees of Delta Resins & Chemical Refractories were attempting to cool a vat of phenol formaldehyde/carbolic acid when a safety valve ruptured emitting a plume of steam that carried a corrosive mist into the air. An acid cloud dripping ''white rain'' covered a square mile of Detroit's east side, sending 19 persons to local hospitals for treatment of burns, and eye and respiratory problems. The total amount of chemical involved was estimated at 2,000 pounds. There were reports of dead animals and paint stripped from buildings and vehicles. No evacuation was ordered.

April 8, 1981: Buchanan, New York. An estimated 120 gallons of water in puddles near the Consolidated Edison Indian Point No. 2 nuclear power plant was found to contain low levels of radioactivity during a routine inspection. An unknown amount of contaminated water entered the Hudson River through a storm drain leading to a discharge canal. The plant, which had been out-of-service due to a water leak, is located 35 miles outside New York City.

April 4, 1981: Middle Granville, New York. A fire at a dump containing toxic foam products burned for 12 hours, letting off toxic fumes that high winds carried toward residences forcing the temporary evacuation of 14 families. The dump is used by Norton Sealant Co. and Telescope Folding Furniture Co. Among the chemicals stored at the Deopt Street site were polyvinyl chloride and urethane material. Eventually, public works crews covered the dump with sand from the town sandpit to prevent additional fumes from escaping.

April 3, 1981: Lawrenceville, Illinois. For unknown reasons, three separate sources of electricity at a Texaco refinery went out, allowing petro-wastes to back-up in a burn-off tower and catch fire within the area of a containment dike. Pump power was eventually restored, averting a potentially serious situation.

April 2, 1981: Newport, Kentucky. Two were killed and 25 injured in what police described as the explosion of a bootleg fireworks manufacturing plant. The explosion shattered windows in about 100 buildings within a four-block area, and leveled the concrete block garage where a Newport man was believed engaged in the illegal manufacture of fireworks. An estimated 100 persons were displaced temporarily from their homes pending fire inspections.

March 27, 1981: Aiken, South Carolina. Approximately 33,000 curies of radiation (tritium) escaped at the Savannah River nuclear plant after a broken pipe began leaking. A representative of the state E.P.A. termed the leak "a moderate size release that is not significant in terms of a major health impact, that we are aware of right now."

March 26, 1981: Arnettsville, West Virginia. Two men were injured, one seriously when a propane tank exploded above ground at Consolidation Coal Company's No. 93 Arkwright Mine as they were installing a pump at an air shaft at the unused mine.

March 25, 1981: El Segundo, California. Leaking fumes from a large compressor were cited as the probable cause of a 2½ hour blaze at a Chevron refinery that consumed up to 42,000 gallons of oil. Company officials cut-off the flow of oil into the area, allowing the blaze to eventually burn itself out. No injuries were reported.

March 23, 1981: Appleton, Wisconsin. After a Fox River Paper Co. warehouse was cleaned out prior to leasing, a truck carried a load of debris to a local landfill where a broken container began to release purple fumes and an acid odor. An unbroken container marked "Boron Tri Fluoride Phenol Complex" was located and a call put in to the Institute of Paper Chem. The Institute informed local response personnel the substance was water reactive and hydrofluoric acid forming. The landfill was closed off and Environmental Emergency Services of St. Louis, Missouri was contracted to handle clean-up. E.E.S. crews flew into Green Bay the next morning, and entered the contaminated area at 11:30 A.M. clad in full protective clothing; they found three unbroken bottles of Boron Tri Fluroide Phenol Complex, two bottles of duelene, and one small box of unknown white powder. The amount of spilled product was not positively identified. Clean-up and decontamination followed. Cost of clean-up was estimated to be in the area of $30,000.

March 11, 1981: Lower Alloways Creek, New Jersey. Ten workers were evacuated from an auxiliary building at the Salem I nuclear power plant when low levels of radioactive material leaked from a waste-gas compressor pump. Radiation levels were calculated at .02 millirems.

February 27, 1981: Peru, Indiana. One died and 13 were injured by an explosion at an Olin Corp. plant where railroad safety flares are made.

February 25, 1981: Fort Lauderdale, Florida. Forty-five persons went to the hospital when two 500 gallon tanks of chlorine and one 300 gallon tank of muriatic acid began leaking at a swimming pool company. Toxic by themselves, chlorine and muriatic acid combine to form deadly hydrochloric acid. Firefighters wearing SCBA were able to plug the leaks after the area was evacuated. Since the valves on all three tanks were broken between the tank and the control valve, authorities cited vandalism as the cause. Victims were treated, then released.

February 24, 1981: San Diego, California. A sewage treatment plant accidently released an unknown quantity of ferric chloride, estimated by the U.S.S Coast Guard to be less than 1,000 gallons, into the ocean near Point Loma. Ferric chloride is heavier than water, and the spill could not be contained or removed. The California Dept. of Fish & Game is investigating the incident and checking shellfish in the area for biological damage. Local officials say less than 1,000 gallons is unlikely to damage marine life.

February 13, 1981: Louisville, Kentucky. A before dawn series of explosions in about two miles of Louisville's sewer system caused an estimated $42 million in damage to pavements and sewers within a 15 by 14 block area, and caused the evacuation of 100 families. The blasts, which engineers estimated to carry the force of 100 tons of TNT, left 23 craters in the streets and blew manhole covers along 13.5 miles of streets. The blasts were apparently caused by an estimated 150 gallons of hexane, an industrial solvent, that reportedly spilled into the sewer system from a Ralston-Purina Co. Soybean Division mill. Amazingly, only four persons were injured, none seriously.

February 12, 1981: Stockton, California. Three petroleum storage tanks exploded in a ball of fire at an Arco Supply terminal in the Port of Stockton, requiring a four-alarm response.

February 11, 1981: Chicago Heights, Illinois. A powerful explosion at the DeSoto, Inc. chemical plant killed one and injured 24 in a blast felt up to 20 miles away. The cause of the blast, believed centered in a resin processor unit, was not immediately known.

February 11, 1981: Chattanooga, Tennessee. A valve at the Sequoyah nuclear power plant opened accidently, spraying 14 T.V.A. maintenance workers with 100,000 gallons of slightly radioactive water. The contaminated workers were rushed to scrub-down showers, and then checked with radiation detection equipment.

February 10, 1981: Forked River, New Jersey. An accident of unknown cause allowed 25,000 gallons of radioactive waste to spill onto a building floor at Oyster Creek nuclear generating station. Approximately ten gallons seeped to an area around the plant building that houses the plant's radioactive waste treatment system.

February 10, 1981: Bennington, Vermont. A drain hose from a chemical tank detached from its fitting leaking a mixture of methanol, hydrochloric acid, and alcohol on an electric pump motor causing a fire that destroyed 50 percent of the Transister Electronics plant. Three firemen were treated for inhalation of toxic fumes or minor acid burns.

February 9, 1981: Zionsville, Indiana. One worker died after falling into chemical wastes while trying to fix a leak on a tanker trailer at the Enviro-Chem Corp. plant. Fire others were injured while attempting rescue efforts.

February 8. 1981: Plateville, Colorado. A third release of radioactive helium coolant in as many days was reported by Fort St. Vrain nuclear power plant personnel and Public Service of Colorado. Tests of monitoring equipment were suspended until the cause could be learned.

February 5, 1981: Binghampton, New York. A pre-dawn explosion and fire damaged an electrical transformer containing more than 1,000 gallons of pyranol containing polychlorinated biphenyls, (PCBs) causing contaminated smoke to seep through the ventialation system of an 18 story state office building. The County Health Commissioner noted: "Every hidden space on all 18 floors is contaminated—every desk, every light fixture, in the air conditioning system and the air ducts—just everything." Officials fear the building may be closed for months, even years, during cleanup operations. Governor Carey reportedly offered to drink a tumblerful of PCBs from the building, and implied the cleanup would be a snap stating: "If I had a couple of willing hands and a few vacuum cleaners, I'd clean that building myself." Government regulations allow up to five parts-per-million of PCBs in edible fish; the soot in the building in Binghampton is reported to contain roughly 100,000 ppm. The PCBs may not be the main danger; the soot also contains dioxin, the chief hazard in Agent Orange.

January 30, 1981: Portland, Oregon. Approximately 6,000 gallons of gasoline seeped from an underground storage tank at a Flying A service station in the Raleigh Hills area just west of Portland. Authorities shut off electricity and natural gas to the area and diverted vehicle and pedestrian traffic. No explosion; no fire.

January 23, 1981: West Deptford, New Jersey. Two workers were killed and five injured by fumes they inhaled while cleaning inside an eight-foot by 20-foot chemical mixing vat at Poly-Rez Chemical Company. Chemical solutions used in the cleaning process gave off phenol and unital fumes, causing two men to die of heart failure.

January 17, 1981: Plymouth, Massachusetts. Approximately 150 gallons of water and 100 cubic feet of resins, (filtering material) both slightly radioactive, seeped under a door and into the yard at Boston Edison's Pilgrim atomic power plant. The spill occurred during routine operations as workers were transferring resins from one demineralizer to another, causing a three hour alert. An open valve was cited as the cause of the release.

January 12, 1981: Ipswich, Massachusetts. Five members of one family were injured, two critically, when their home exploded at 1:30 A.M. Family members had previously smelled gas and had just decided to evacuate when the explosion occurred. The home was reduced to rubble.

December 29, 1980: Clarkton, North Carolina. A fire at an agricultural chemicals warehouse operated by Clarkton Farmers Exchange, Inc. led police to evacuate all 750 residents of the small tobacco farming town for several hours. There were no serious injuries, but ten persons were treated for smoke inhalation. Commodities in the warehouse included a variety of pesticides and cylinders of methylbromide.

December 28, 1980: El Dorado, Kansas. A fire at a Getty oil refinery spread to three tanks of petroleum solvent causing a series of explosions injuring two.

December 22, 1980: Wethersfield, Connecticut. A 15 mile long slick of fuel oil and gasoline fouled the Connecticut River after pressure of ice against an earthen supporting dock used to unload fuel barges to onshore tanks reportedly ruptured a pipeline at Mercury Oil Company. Additional pipes cracked, allowing 50–60,000 gallons of petroleum product, mainly No. 2 fuel oil, to enter the river. Possibly 50 percent of the spilled fuel was retained by booms and recovered by vacuum truck.

December 18, 1980: Soddy-Daisy, Tennessee. An open valve on a pipeline at the Sequoyah Nuclear Plant released small amounts of Ribidium 88 and Cesium 137 slightly contaminating two workers—the third time in a month workers have been affected.

December 17, 1980: Guilford, New Hampshire. A fire at the Urethane Molding, Inc. building, believed caused by a short-circuited fuse, was brought under control by 40 firefighters from nearby towns using foam to knock down the hot chemical fire. A dozen 55 gallon drums of polyurethane and other chemicals used for making water pipe insulation were in the building. No injuries.

December 17, 1980: Concord, New Hampshire. Toxic fumes, apparently caused when an employee mixed chlorine and another chemical, forced evacuation of the downtown Y.M.C.A. with firefighters in air masks leading inhabitants to safety. Thirty-eight adults and children were treated at the local hospital. The nearby Concord fire station was used as an evacuation depot to assemble evacuees in frigid weather.

December 11, 1980: Henderson, Nevada. A propane tank explosion at State Industries, a water heater manufacturing plant, injured ten persons, one critically.

November 26, 1980: Indian Trail, North Carolina. A series of explosions believed triggered by electrical sparks on an automative fluids assembly line at the Radiator Specialty Co. plant destroyed the plant building and injured 47 workers. Injuries ranged from smoke inhalation to burns of all degrees, and a number of the injured were listed in serious condition.

November 16, 1980: York County, Pennsylvania. Radioactive gases escaped for approximately 20 minutes when Unit #2 of the Peach Bottom Atomic Power Station was put back into operation after a valve was repaired.

November 16, 1980: Lusby, Maryland. Radioactive gas leaked into the atmosphere at the Calvert Cliffs nuclear power plant was the result of a faulty valve mechanism in the plant's water purification system. This was the fifth release of radioactive gas from this plant this year.

November 15, 1980: Benson, Arizona. Toxic oxidizer propellant escaped into the atmosphere at a Titan II missle site. Workmen were using a flame to burn-off toxic fumes from a missle propellant tank when the flame went out allowing fumes to escape.

November 11, 1980: Mississauga, Ontario. One year and one day after an incident involving propane and chlorine caused the evacuation of a quarter million people, a tractor trailer truck rolled into three tanks of liquid propane while unloading corn at Forest Lawn Farmers, Ltd. Wind carried the leaking propane into nearby fields, and there were no injuries reported.

October 21, 1980: New Castle, Delaware. An explosion at an Amoco Chemical plant while propylene was being unloaded from railroad tank cars took at least five lives and injured 29 more. State police evacuated the nearby area, and firefighters from Delaware, New Jersey, Pennsylvania, and Maryland battled the resultant blaze for 11 hours. Windows were blown from buildings two miles away.

October 17, 1980: Bridgman, Michigan. Radioactive gas leaked from a nuclear reactor at the Donald C. Cook power plant, contaminating one worker and forcing the evacuation of others.

October 17, 1980: Buchanan, New York. Fifty workers were exposed to approximately 10 millirems of radiation at the Indian Point No. 2 power plant when a water leak of several days duration allowed 100,000 gallons of water to leak onto the floor of the containment building.

October 16, 1980: Neal, West Virginia. Two workers were killed and another seriously burned when an explosion ripped through an Ashland Oil Co. chemical plant that produces maleic anhydride used to make polyester resins.

October 15, 1980: Cincinnati, Ohio. An explosion and fire leveled the four-story Hill and Griffith Foundry Supply Co. where coal is pulverized and bagged for resale to foundries. At least eight employees were injured. The explosion was believed caused by ignition of coal dust particles.

October 14, 1980: Olpe, Kansas. A tank containing 1500 gallons of diesel fuel exploded causing the evacuation of all 530 residents of the Town of Olpe. No injuries were reported.

October 14, 1980: Pekin, Illinois. Two explosions that rocked Commonwealth Edison's Powertown generating plant injuring an estimated 15 persons and causing possibly as much as $100 million in damage was believed touched-off by ignition of coal dust particles in the air.

October 10, 1980: Franklin, Louisiana. An Exxon natural gas well that burned out-of-control for three days was brought under control with no injuries. The well was being plugged as part of a close-down operation at the time of the incident.

October 9, 1980: Richland, Washington. Eight employees of the Rockwell-Hanford Operations group at D.O.E.'s processing site were contaminated when plutonium scrap caught fire. Two workers ingested a small amount of plutonium oxide powder. Unlike metallic plutonium, plutonium oxide does not normally flash to flame, and officials are at a loss to explain the cause of this incident.

October 7, 1980: Haddam, Connecticut. For the second time in two weeks, the Connecticut Yankee nuclear power plant experienced an unplanned release-to-the atmosphere of radioactive gas. The most recent release lasted six minutes; the earlier release lasted three minutes and was reportedly caused by a technician opening the wrong valve.

September 19, 1980: Fitchburg, Massachusetts. One of three tanks of vinyl chloride exploded at the Great American Chemical Corp. causing the evacuation of nearly 3,000 persons. Firefighters poured water on two nearby tanks to keep them cool as a cloud of toxic gas rose from the ruptured tank and slowly dissipated. A total of 18 persons were reported injured, two seriously.

September 19, 1980: Hamden, Connecticut. About 80 persons were treated in unrelated incidents when caustic fumes at two separate restaurants within a few blocks of each other caused patrons to become ill and suffer fainting spells. Exhaust fumes from a gas-fired hot water heater were the apparent cause at the Sleigh House Restaurant, while a backup of sewer gas was cited at Jimmies Restaurant nearby. Rescue crews from Hamden and nearby towns had just brought the situation at the Sleigh House under control when they were called to Jimmies.

September 18, 1980: Damascus, Arkansas. A fuel leak and subsequent explosion in an underground Tital II missle silo killed one and injured approximately 20. Nearly 1,400 people were evacuated from three small towns nearby when the fuel leak, reportedly caused by a 3 lb. socket dropped from a height of 70 feet, first occurred. An explosion eight hours later wrecked the silo, blowing debris including a nuclear warhead a couple of hundred feet over the nearby area. A 1978 leak from an above-ground fuel tank had previously required evacuation of the Damascus complex. In that incident, several people were hospitalized for inhalation of toxic fumes. Officials stated there was no leakage of radiation during either incident.

September 17, 1980: Edmond, Oklahoma. A stolen car crashed into a natural gas well knocking off its cap allowing thousands of cubic feet of gas to enter the atmosphere.

Firefighters arrived to find the auto, its engine still running, sitting atop the well. A crane was used to gingerly lift the car off the wellhead, and a crew from Halliburton Services Co. was called in to recap the well. Two lanes of I-35 were closed for a time, and about 100 families were evacuated.

September 14, 1980: Muscatine, Iowa. A concrete, vertical storage tank filled to approximately one-fourth its 164,000 gallon capacity with styrene monomer ruptured at an industrial facility. Commodity was contained within a diked area, but there was a reported evacuation of an industrial plant across the road and some cottages along the nearby Mississippi River. Local public safety officials were reportedly not informed of the spill until seven hours later.

September 8, 1980: Beaver, Utah. Farmer Sheldon Roberts, his son Steven, and Beaver County Sheriff, Dale Nelson, died inside a concrete manure pit after being overcome by methane gas as each tried unsuccessfully in-turn to rescue the other.

September 7, 1980: Albany, New York. A fire and explosions in two 1.5 million gallon capacity storage tanks, one containing 30,000 and the other 6,000 gallons, threatened to spread to other full tanks at Mobil Oil Co. storage facility in the Port of Albany. Two explosions 30 minutes apart occurred three hours after the fire started. The fire began about 10 A.M., reportedly in a separate tank where a vacuum truck was siphoning a foot or so of residue from unleaded gasoline previously stored in the tank. A fire of unknown origin started under the truck and spread to the nearby tanks. Eight persons, mostly firemen, were injured as nearby residents were evacuated.

September 1, 1980: Mansfield, Louisiana. Four men evacuated a nearly completed natural gas drilling rig just before it blew out, sending a surge of gas to the surface that ignited and melted the steel framework of the derrick. C. L. Morris Drilling Co. personnel bulldozed a trench around the rig to prevent spread of the fire to nearby woodlands.

August 26, 1980: Chicago, Illinois. A major fire at a chemical warehouse located along the Chicago River destroyed more than 2,000 55-gallon drums of organic chemicals and required 350 Chicago firefighters to bring it under control. Sewers were blocked to prevent runoff of contaminated water, although later runoff was allowed to enter the sewer system in controlled amounts over a period of a week. The Chicago Fire Dept. lost one piece of apparatus when a wall collapsed, and had seven more damaged by intense heat.

August 26, 1980: Tampa, Florida. A regulator being replaced on a one-ton cylinder of chlorine at Seabrook International Foods, Inc., allowed escape of chlorine gas that caused the evacuation of 200 persons. A worker from the Tampa office of Thompson-Hayward Chemicals Co., supplier of the chlorine, arrived in less than an hour and stopped the leak without the protection of SCBA.

August 15, 1980: Military firefighters from Hickam Air Force Base joined civilian firefighters in a prolonged foam attack on a burning 55,000-barrel gasoline storage tank along Honolulu waterfront as two tanker ships exited the area with Coast Guard help in the fastest such maneuver since December 7, 1941. The Cielio Diroma and the Santa Paula were able to slip transfer connections and lines in a matter of minutes and clear the pier area. Five injured, two critically.

August 14, 1980: Morgan City, Louisiana. An apparent overflow and ignition at a Texaco gasoline tank truck loading rack caused more than $1 million in damages and forced the evacuation of seven blocks in downtown Morgan City. Twenty-three fire depts. joined forces to end the blaze.

August 12, 1980: Jackson, Michigan. The second leak of radioactive gas in two weeks from Consumer Power Company's Palisades nuclear power plant near South Haven lasted about three minutes and released approximately 11 percent of the limit allowed by N.R.C. The facility was under investigation by N.R.C. for a third incident that occurred July 25, 1980 when "human error" allowed a valve in the plant's reactor back-up cooling system to remain open for 36 hours.

August 9, 1980: Cabool, Missouri. A 20,000 gallon spill of liquid fertilizer (ammonium phosphate) from the Ozark Fertilizer Co. plant entered the Big Piney River. Ozark officials said the spill was caused by the actions of vandals on company property.

August 8, 1980: Bridgewater, New Jersey. A package dropped from a conveyor system at the United Parcel Service facility resulted in six small bottles of acetic anhydride releasing their contents and forced emergency room treatment for 26 U.P.S. personnel. Local firefighters cleaned-up.

August 5, 1980: Sandwich, Massachusetts. A nearly full 7.7 million gallon fuel oil tank on the banks of Cape Cod Canal owned by New England Gas & Electric Systems/Canal Electric Co. suffered an explosion and burst into flames as two workmen installing insulation at the top of the tank fled for their lives. Firefighters were able to confine the blaze to the destroyed tank.

July 30, 1980: New Orleans, Louisiana. An explosion and fire at a pesticide storage warehouse operated by the New Orleans Mosquito Control Program consumed approximately 50 gallons of malathion and damaged containers of Dibrom-14. A dozen persons were examined-on-site.

July 29, 1980: San Diego, California. Three soldiers, including one female sergeant, assigned to the Army's 70th Explosive Ordnance Team at the Navy Submarine Support Facility on Point Loma were killed when Mexican fireworks, confiscated from tourists returning to the U.S. prior to the 4th of July, exploded while being transferred from Battery McGrath, a World War II concrete bunker on San Diego Bay. The fireworks, mainly Roman candles and cherry bombs, were being carried to a pick-up for transfer to a desert storage site. Additional fireworks in the bunker did not explode. The cause of the explosion was not known.

July 24, 1980: Jamaica Queens, New York. Sparks from a welder's torch apparently set off volatile lacquer fumes at the Vogue Metal Craft plant causing a flash explosion that killed nine and injured 30. Sixteen firefighters and four paramedics were treated for smoke and toxic fume inhalation.

July 21, 1980: Borger, Texas. The third major explosion at Phillips Petroleum Co.'s Borger, Texas plant in nine months occurred when two five-gallon cylinders containing ethylene oxide gas exploded injuring nine workers.

July 19, 1980: Michigan City, Indiana. Tanks of liquid propane gas apparently exploded inside the Michiana Gas Service plant causing a series of fires and explosions that destroyed the plant and a nearby animal hospital and forced the evacuation of nearby residents.

July 16, 1980: Murray Hill, Staten Island, New York. An explosion believed caused by gas service restored prematurely ripped through the Parker Crescent hi-rise injuring 20 residents. Con Edison had earlier shut-off all gas valves to search for a leak after residents had complained of a strong odor of gas. It was not known who had turned the gas back on.

July 12, 1980: Bayonne, New Jersey. Approximately 350,000 gallons of ethylene glycol flowed into a Kill Van Kull waterway between Bayonne, N.J. and Staten Island, N.Y. when a tank split at Rollins Terminals plant in Bayonne. Although the tank was diked, a hole in the dike allowed the bulk of the product to escape.

July 9, 1980: Red Wing, Minnesota. The source of a radioactive leak of two-weeks duration was located at the Northern States Power Company's Prairie Island nuclear power plant in Red Wing.

July 3, 1980: Rapid City, South Dakota. A two to three gallon spill of PCB occurred at the South Dakota Cement Plant, shutting down part of the plant for a week. Workers built a dike around the leak source, an electrical transformer, barricaded the building containing the spill, posted a 24-hour guard, and called in a crew of chemical experts from General Electric Corp. in Denver, Colorado.

June 4, 1980: Laramie, Wyoming. A fireworks stand exploded killing one adult and three juveniles who were loading fireworks into trucks for transfer to retail outlets. Assistant Fire Chief, Don Bradley, hooked a tractor to one burning trailer and pulled it to where Laramie firefighters could get water on it. County tanker trucks had to be called into supply enough water to battle the hot fire.

June 4–5, 1980: Elizabeth New Jersey. Two additional fires broke out June 4 and 5 at the Chemical Control Co. hazardous waste site, scene of a disastrous explosion and fire on April 21, 1980. On June 4 a drum of phosphorous material ignited, but was quickly smothered. On

June 5 an exothermic reaction ignited 100 barrels. Total cleanup costs for this longterm dump site may exceed $10 million.

June 3, 1980: Buffalo, New York. Three workers entering a 55-foot deep sewer construction shaft in a bucket-lift were overcome by hydrogen sulfide fumes and fell. The three workers recognized they had entered a pocket of heavier-than-air gas and reentered the bucket-lift but fell back down the shaft.

May 29, 1980: Seattle, Washington. Eight hundred employees were evacuated for five hours at a Boeing plant in Seattle when nitric acid vapor was released during transfer of 800 gallons from a holding tank to a tank truck and drawn into the plant through the ventilation system. Boeing personnel used water mist to control the vapor cloud. Cause of the accident was felt to be either improper lining within the truck tank or a reactive residue remaining within the tank.

May 21, 1980: Aiken, South Carolina. A leak in a pipeline carrying hydrogen sulfide gas at the Savannah River nuclear plant closed highways and forced a small-scale evacuation.

May 15, 1980: DePue, Illinois. Schools closed and scores of families were evacuated after a woman died from hydrogen sulfide gas leaking from a village sewer system. Gas may have been formed in the sewer system after a spill of sulfuric acid at nearby Mobil Chemical Company that may have seeped into a sewer from a storm drain.

May 13, 1980: New Castle, Delaware. Eleven one pint glass jars of pesticide broken in a fall from a department store shelf caused toxic fumes that sent 60 persons to area hospitals. Shenango Township firemen were called in after a check with the manufacturer disclosed the solution contained malathion.

May 6, 1980: Spangler, Pennsylvania. A series of explosions at Terrizzi Fireworks killed at least one and injured at least nine.

May 3, 1980: Idaho Falls, Idaho. A problem with a high-pressure air line at Idaho National Engineering Laboratory where uranium is recovered from spent reactor fuel caused a smokey fire and a small evacuation of workers. No radiation was released.

April 27, 1980: White Springs, Florida. A 75 foot pollution control tower exploded at Occidental Chemical Corporation's Suwanee River plant blasting out four workers who were inside cleaning, and injuring seven others. The tower is used to removed emmisions from a reactor that turns raw phosphate into animal feed supplements.

April 25, 1980: New Orleans, Louisiana. A series of underground explosions sending flames 12 feet high from manhole covers ripped through a three-block area of downtown New Orleans during Friday rush hour causing slight injuries to nine.

April 22, 1980: Potwin, Kansas. A leak of liquid fuel oxidizer from a drain valve assembly at an Air Force Titan II missle silo lasted eight hours and released a small cloud of poisonous gas.

April 22, 1980: Fort Hall, Idaho. A fire at Russett Chemical Company caused a several-day evacuation of 700 residents during Earth Day environmental celebrations. Airborne toxic fumes from herbicides and pesticides were kept from towns of Blackfoot and Pocatello by a light breeze. Thirty firefighters were treated for inhalation of fumes.

April 21, 1980: Elizabeth, New Jersey. Fire at a chemical storage facility on the waterfront owned by Chemical Control Corporation belched poisonous smoke causing officials to warn 400,000 area residents to remain indoors. Up to 20,000 barrels of explosives, solvents, pesticides, acids and mercury were consumed in 1400 degree heat. Eight firefighters were injured.

April 1, 1980: Laurel, Montana. Two construction workers were killed when an explosion shattered a storage tank of liquid asphalt at the Cenex Refinery.

Highway

April 7, 1982: Contra Costa, California. A bus hit a gasoline-laden tank truck carrying 8,800 gallons in the westbound bore of the Caldercott Tunnel between Contra Costa County and San Francisco causing a fire that killed at least seven.

April 1, 1982: Dallas, Texas. When three drums tipped over inside a truck causing one puncture and two loose bungs, approximately 70 gallons of ammonium hydroxide containing a heavy concentration of copper in suspension spilled. (waste from copper etching.) The Dallas Fire Department secured the area and called in a commercial contractor, Western Emergency Services, to handle clean-up. Since copper is an environmentally hazardous substance, responders used additional ammonia to soften the copper, then picked it up with absorbent clay.

March 29, 1982: Houston, Texas. A gravel truck struck a gasoline tank truck stopped due to an earlier traffic accident up ahead, set off an explosion and fire, then ran over a police officer. Heat from the resultant fire melted parts of an overpass. It was initially unknown whether the police office died from impact or was burned to death.

March 1, 1982: Bridgewater, Maine. A Bangor & Aroostock train hit a gasoline tank truck resulting in an explosion and fire that killed the truck driver, setting fire to the train and a nearby building.

February 18, 1982: Kamloops, British Columbia. In a potential disaster feared by all responders, a tractor trailer carrying 190 45-gallon drums of sodium cyanide briquets lost its brakes on Columbia Street hill in downtown Kamloops. One died and five were injured in the resultant traffic accident. Approximately 20 drums were damaged and some contents spilled in the road hard-by the 1,000 bed Royal Inland Hospital, the 200-bed Chronic-Care Hospital, a Ponderosa Lodge and other motels. The sodium cyanide briquets, used in the process of gold and silver recovery, were enroute to a mine in Pine Point, North West Territories. A command post was established at the Travelodge Motel by Provincial Emergency Program officials. Immediate fears were that the sodium cyanide would react with water and form deadly hydrogen cyanide gas, or that even worse . . . battery acid might leak and mix with the chemical which would cause immediate release of lethal gas. Contents of ruptured barrels were swept from the highway by firefighters in SCBA and protective clothing working in seven-man teams for 45-minute shifts, sealed in special chemical containers secured from Weyerhaeuser Canada, Ltd., then stored in a Department of Defense bunker. Officials planned to spray the area with sodium hypochlorite, a bleach, to neutralize sodium cyanide residue but were initially unable to obtain a sufficient supply. A response team of specialists from C.I.L., manufacturer of the sodium cyanide, arrived next morning to continue clean-up. A frontend loader was used to scrape snow that was then placed in waterproof containers and taken to the bunker. The accident occurred even though trucks are restricted from carrying dangerous goods (haz. mat.) in the City of Kamloops, and all trucks are required to use a highway by-pass to avoid the steep Columbia Street hill. A few years ago, sodium cyanide was spilled in the Kamloops area when a truck went off Hwy. 5 just south of Rayleigh.

February 9, 1982: Littleton, New Hampshire. A tank truck carrying 9,000 gal. of propane overturned near two stationary propane tanks causing the evacuation of 1,500 people. No leak; no fire.

February 5, 1982: Fort Worth, Texas. A tank truck similar to those used to transport L.P.G. skidded on icy pavement, hit a bridge abutment, rolled and burst into flames on Highway 287. Firefighters were able to extinguish the blaze, visible for 20 miles, using a combination of water and foam. Although a smell like natural gas was tremendously strong throughout the area, commercial response personnel from Western Emergency Services taking readings with a combustible gas indicator got only slight indication of readings in the flammable range. The natural gas-like smell was so overpowering, however, that responders soon realized they had a truck load of mercaptan. The truck had just left Baytown, Texas enroute to Edmonton, Alberta and was carrying 5,800 gallons of the odorent used to identify natural gas. Approximately 4,800 gallons burned; the tank ruptured but not violently. Dikes were constructed to contain run-off using two truckloads of sand obtained from the highway department. The tank, still containing an estimated 1,000 gallons of product, was patched, the area foamed, and airbags used to allow slings to be slid under the tank in order to lift it onto a lowboy trailer for transfer to a Class I dump. A vacuum truck was used to pick up concentrated commodity, then HTH and water used to neutralize residue and combat extremely strong odor. Response personnel expect to eventually excavate six to eight inches of soil and some asphalt to combat the lasting odor.

February 3, 1982: Stroudsburg, Pennsylvania. A flatbed truck dropped a load of aluminum chloride on I-80. Aluminum chloride reacts with water to form hydrochloric acid and aluminum hydroxide with release of heat. Eleven persons were treated and released, and 300 evacuated from the area.

January 29, 1982: El Cajon, California. A hazardous waste tank truck parked on the street in front of the residence of its driver leaked a gallon or less of hydrofluoric acid from a loose coupling at the rear of the truck. A block-long stretch of South Mollison Avenue was closed for two hours, and 150 residents downwind were evacuated. Firefighters eventually ascertained that the truck was empty except for residue, and the small spill was sucked up using a vacuum hose attached to the tank truck; then the spill site was washed down.

January 21, 1982: Barnet, Vermont. A tank truck carrying sodium aluminate, a corrosive, overturned on I-91 and rolled down an embankment. Response crews emptied the tank and transferred the product into another vehicle.

January 18, 1982: Boston, Massachusetts. An automobile carrying 12 packages of radioactive medical supplies, including radioactive iodine, was stolen while the driver made a phone call. NRC officials were not immediately successful in locating the hot car.

January 16, 1982: Holbrook, Arizona. A portion of I-40 was closed for seven hours as a precautionary measure when fire broke out in a truck carrying 13 pounds of methyl alcohol and ten pounds of radioactive tritium in four containers. The fire was quickly extinguished, and no radioactivity escaped.

January 16, 1982: Oil City, Pennsylvania. A motorist who found a canister lying in a roadway intersection later called police after hearing news reports that the cylinder could produce a fatal dose of radiation with an hour's exposure. Police responded and found the container intact.

December 20, 1981: Texarkana, Texas. Propenol, a toxic/flammable allylalcohol, leaked from an overturned tanker truck after a collision between the tanker and a pick-up truck. Officials expected to be able to contain the spill and off-load the chemical.

December 17, 1981: Vista, California. A truck carrying 1,000 gallons of propane flipped into a ditch while backing out of a dirt driveway, causing residents in 30 homes within a one square mile area to be temporarily evacuated as a precautionary measure. The truck was righted with no leakage.

December 8, 1981: West Rutland, Vermont. Traffic was diverted for seven hours while response personnel transferred 2,100 gallons of propane from an overturned truck. No fire.

November 14, 1981: Texas Creek, Colorado. Seven died, three were critically injured, when a 9,000 gallon tank truck carrying gasoline sideswiped a flatbed trailer on a twisting mountain highway and burst into flames leaving a wall of flames for 1,000 feet down the road.

November 5, 1981: Castaic, California. Approximately 2,000 gallons of propylene dichloride, a degreasing agent, leaked from a tank truck at a truckstop on I-5, sending 58 persons to the hospital with nausea and dizziness. Eleven persons admitted to the hospital included three masked hospital emergency room workers who breathed fumes from the patients. Cleanup operations were delayed when several hours after the incident it was learned the truck also contained a 2,000 gallon tank of tetrahydrofuran, a flammable liquid that gives off toxic vapors. However, this tank did not leak.

November 4, 1981: Buskirk, New York. A motorist drove through a stop sign into the path of a tank truck loaded with 8,500 gallons of gasoline causing a spectacular fire. The occupants of the car received only minor injuries.

September 17, 1981: San Diego, California. After a Caltrans worker located two leaking drums beside a highway, IT Corp. was contacted to test and remove the liquid, identified as toluene and xylene.

September 15, 1981: Huntsville, Alabama. A gasoline tank truck was struck in the rear portion by a Southern Railways freight at a grade crossing, dousing waiting automobiles with flaming gasoline. Five persons died immediately, and two additional persons, including the truck driver, died the next day. Erroneous reports that LPG was the commodity involved apparently stemmed from the fact that response and railroad personnel had to remove an endangered LPG car from the area. Firefighters cooled the LPG car with water while rail crews separated the train and pulled the uninvolved portion to safety. The National Transportation Safety Board is expected to issue an investigative report on the overall incident.

September 14, 1981: Pueblo, Colorado. Six persons were injured at the Pinon Truck Stop ten miles north of Pueblo on I-25 when an unoccupied semi-tractor rig left in the parking lot rolled into a stationary propane tank, rupturing it. The resulting cloud of gas eventually ignited in a ball of flame that could be seen for miles, completely destroying the truck stop's restaurant, gift shop, and service station.

September 8, 1981: Concord, California. A truck owned by Erickson, Inc. of Richmond, California carrying 1,000 gallons of a toxic acid-metal mixture leaked part of its load on I-680 sending a cloud of noxious fumes over San Ramon Valley neighborhoods. Twenty-one persons received medical treatment, and approximately 3,000 students were evacuated.

September 4, 1981: St. Marys, Georgia. Two men were killed at a Gilman Paper Co. woodyard when an explosion occurred while a liquid oxygen tank truck owned by Air Products and Chemicals, Inc. was filling a storage tank. Neither the Office of the State Fire Marshal nor investigators from Air Products have been able to identify the cause. There was no evidence of fire prior to the explosion, although fire soon spread to an office building, the LOX tanker, and a Gilman Co. fuel truck. The shell of the storage tank that ruptured at both ends has been taken to Air Products headquarters in Pennsylvania for additional examination.

August 13, 1981: Winsted, Connecticut. A tanker truck enroute from a Union Carbide battery products division plant in Bennington, Vermont to the Connecticut Treatment Corp. in Bristol leaked nearly all of its 5,000 gallon load of highly alkaline toxic waste along public roads, apparently through a defective inlet valve.

July 22, 1981: Blythe, California. Blythe, California and Ehrenberg, Arizona were evacuated of 15,000 people when a tanker truck carrying 26,000 pounds of fuming red nitric acid sprung a leak on I-10 just outside of town releasing a toxic cloud. The leak, reportedly located at a gasket between the tank and the elbow going down to the valve, posed a tricky problem for initial responders that could provide a possible lesson for others. Leaking nitric acid pooled on the ground around the truck tires, eventually causing them to ignite repeatedly. Hindsight might indicate a need in future, similar incidents to bank, dike, or divert the flow of acid away from the truck tires to lessen the danger of reaction causing repeated ignition. A repair crew from W.S. Hatch Co. in L.A. was called in to plug the leak, and IT Corp. of Wilmington, California handled clean-up. The nitric acid, used as an oxidizer for some space launch efforts, was enroute from Vandenberg AFB near Santa Barbara, California to Holloman AFB in New Mexico.

July 21, 1981: Akron, Ohio. A tractor-trailer loaded with photographic chemicals from Eastman Kodak Co. in Rochester, New York and operated by C&L Trucking of Bald Knob, Arkansas crashed off an expressway access in Akron, Ohio while enroute to San Ramon, California. CHEMTREC was called with a list of the reactive chemicals on board. Although the load contained 1,200 pounds of sulfuric acid along with small quantities of formic acid, phenyl-2 thiourea, methylmercaptan, monochloroacetone, alcanesulfonic acid, allyl alcohol, and other miscellaneous corrosive/solid items; and thus posed a serious problem due to potential reactions between chemicals, only a small amount of sulfuric acid actually leaked. Responders established a perimeter 500 yards from the crash and secured technical information from CHEMTREC, E.P.A., and Kodak. Tarpaulins were brought to the scene to cover the truck in case of rain due to a fear of reaction between the sulfuric acid and water. City workers blocked the entrance to a nearby storm sewer, and workers from PPG Industries in Barberton and Coastal Industries of Richfield wore chemical suits to remove chemicals from the wrecked truck and place them in another.

June 21, 1981: Homer, Michigan. A hose that burst while a tanker truck from Aero Liquid of Grand Rapids was pumping anhydrous ammonia into a farm supply company's truck allowed 7,000 gallons to spill, causing a toxic cloud to spread through the area. Approximately 300 residents were evacuated for two hours.

June 16, 1981: Atlanta, Georgia. A 150 pound chlorine cylinder that fell from a truck at a loading dock in the garage of the 25-story Hilton Hotel sent 33 persons to the hospital with six in critical condition and 12 in poor condition. The garage was evacuated, as well as the Hilton lobby and one of the hotel's convention areas, along with the 22-story Trust Company building next door. The injured who suffered severe pneumonia-like reactions, included six firemen and four police.

June 10, 1981: Stark County, Ohio. A truck owned by Shellwell Services Co. hauling sealed containers of radioactive americum beryllium crashed off US-30 near Massillon, Ohio and came to rest on a Chessie System railroad track. Rte. 30 was closed for four hours, and there were conflicting reports whether radioactivity could or could not be detected at the site, and whether radioactivity if present came from the lead-shielded cylinders or from longterm contamination of the truck body. Shellwell is an oil well surveying company and uses the radioactive material in its work.

May 27, 1981: Wasta, South Dakota. A semitrailer loaded with 71 barrels of low-level nuclear waste oil enroute from a nuclear plant in New York state to a nuclear waste dump near Richland, Washington was cordoned off under guard at a rest stop after the driver discovered a small leak. Approximately a pint of radioactive waste, measuring .15 millirems on one side of the truck and .12 on the other side, leaked during a period of hours.

May 26, 1981: Randolph, Vermont. State Police operating a routine truck weighing check at a rest area on I-89 found xylene, an industrial cleaner, seeping from a faulty weld on a tanker truck carrying 7,200 gallons of the product, described as mildly flammable and potentially explosive. State Civil Defense officials, hazardous materials specialists, and two local fire departments responded. A second tanker sent by the shipper, Hall Chemical Co. of Montreal, failed to meet state safety requirements and a third tanker was ordered. Transfer of the product was successfully completed after 18 hours. The shipper was charged with seven violations, including improper placement of valves and inadequate documents.

May 25, 1981: Essex, Vermont. An estimated 500 gallons of sulfuric acid reportedly leaked from a tractor-trailer tanker at Folino Industries. Town authorities said they learned of the spill when they were called by a reporter who had received an anonymous telephone tip. Quick-dry material and lime, plus evening rains, helped to neutralize the product. Expectations call for approximately 5 tons of the chemical/earth mixture to be picked-up, placed in drums, and transported to a dump near Buffalo, New York.

May 18, 1981: San Diego, California. A 90-pound container fell off the rear of a chemical company truck. A motorist who stopped and rolled the container off the road received second degree burns and had to be transported to the hospital by paramedics. IT Corporation responded, picked-up the container, and did minor clean-up.

May 20, 1981: San Diego, California. About two gallons of a pesticide was deposited by a driver washing out his pesticide truck. The driver was cited for illegal dumping of a pesticide product. The product evaporated quickly and no subsequent clean-up operation was required.

May 19, 1981: San Diego, California. At Pacific & Harbor, the rear trailer of a tanker truck tandem carrying Jet A 50, a flammable liquid, turned over when the driver failed to negotiate a curve. A limited amount, possibly 200 gallons of total load of 1,600 gallons was released and entered a storm drain 200 yards from the oceanfront. The U.S. Coast Guard responded to the spill, and Chevron Chemical Co. contracted for a spill clean-up company.

May 11, 1981: Newburgh, New York. A loading rack blast at the Agway Petroleum Co. depot killed one and injured 12 when a tanker truck exploded during loading or unloading operations and touched off a small fire.

April 30, 1981: Hoover Dam, Arizona. Approximately 100 pounds of sodium chlorate, an oxidizer that yields toxic fumes if involved in fire, spilled on U.S.-93 a half mile south of

Hoover Dam when a truck under tow overturned. No explosion; no fire. Traffic backed up for 2½ miles during a 5½ hour neutralization and clean-up.

April 25, 1981: Malibu, Claifornia. A gasoline tank truck carrying 9,000 gallons hit an embankment and burst into flames. Flaming gasoline flowed across a street and into an art gallery setting it afire. A motel behind the art gallery suffered smoke damage before the fire was brought under control. Brake failure was reportedly the cause of the accident. The driver suffered scrapes and bruises when he jumped from the truck just before it crashed.

April 16, 1981: Oakland, California. Burning gasoline spilled across both lanes of Hwy.-17 following a collision between a gasoline tank truck and an automobile. The truck driver was presumed dead, and the highway remained closed to traffic for some hours after the fire due to roadway damage from the heat.

April 14, 1981: Seekonk, Massachusetts. Leaking chemical drums in a 40-foot trailer at the Red Star Express truck terminal gave off toxic vapors forcing up to 100 residents from within a half mile and the closing of heavily-traveled Route 6. Seekonk Fire Chief, John Shaw, reported the commodity was liquid thionyl chloride, and noted that it could explode on contact with water. Employees of Jet Line Services of Stoughton, Massachusetts dressed in protective suits and helmets, were able to neutralize the spill and proceed with clean-up. (Thionyl chloride is corrosive, acidic, poisonous; a colorless fuming liquid with a suffocating pungent odor. It reacts with water to form sulfer dioxide and hydrochloric acid. Corrosive to metals and tissues.)

April 11, 1981: New Brunswick, New Jersey. New Jersey State Police hippitty-hopped after an errent Easter bunny when dozens of vehicles were dyed Easter-egg purple, and a 15-mile stretch of the New Jersey Turnpike was closed for six hours, after a tractor trailer spilled a dye, apparently not hazardous, that was activated by rainfall. Two State Police officers followed the purple brick road down Route 287 ten miles to the American Cyanamid Co. plant in Bridgewater Township where they found a BIG Easter bunny with egg on his face.

March 26, 1981: Williams, Arizona. Eleven people became ill after being exposed to nitrogen-tetraoxide, an oxidizing agent used as a component in Titan II rocket fuel. A state hazardous materials inspector was apparently unaware his clothing had become contaminated while inspecting a truck, and exposed the ten other persons over a five-hour period while going about routine duties. His condition became known only after he became violently ill five hours after exposure. He was later reported in good condition at a Flagstaff Hospital.

March 19, 1981: Weathersfield, Vermont. A tractor-trailer loaded with newsprint crashed into a disabled truck loaded with materials destined for a local hardware store, setting off explosions and sending both rigs down an embankment on Interstate 91. Small liquid propane cylinders were believed to have caused the explosions and fed the fire that was fueled by cases of motor oil, pesticides, and aerosol cans of paint stripper. Firemen wore SCBA and used foam to extinguish the blaze. Sand and hay bales were used to keep burning chemicals from entering a small tributary of the Connecticut River. New England Marine Contractors handled clean-up.

March 7, 1981: Bolton Flats, Vermont. A tractor-trailer loaded with 43,000 pounds of carbon dioxide in a pressurized tank caught fire when two rear tires ignited, burning a four-by-eight foot hole in the outer layer of the double-lined tank. Carbon dioxide is nonflammable, but intense heat could have caused the pressurized gas to explode. No injuries; cargo transferred.

March 2, 1981: Sunshine Key, Florida. One person died when the bucket on a backhoe being transported over Seven Mile Bridge on flatbed trailer swung up, hitting a steel girder of the bridge then slammed down into a propane tank next to a bridgetender's shack. The bridgetender died as additional propane tanks exploded in a chain-reaction.

February 26, 1981: San Diego, California. The San Diego Fire Department responded to a small mercury spill at the University of California/San Diego involving a leaking truck tank. IT Corporation was called in to remove 20–25 pounds of mercury, a job that required 15 hours. The release caused no injuries and required no evacuation.

February 19, 1981: Princeton, British Columbia. A semitrailer loaded with 45,000 pounds of zinc sulfate fertilizer in bags crashed on Highway 3. Numerous rescue workers suffered ill effects of vomiting, nausea, respiratory distress, and skin irritation. Stan Thompson of Model Transfer vacuumed the spilled material and put it into containment barrels. Emergency personnel burned the paper bags and washed down the area.

February 18, 1981: National City, California. The National City Fire Department responded to a 25–30 gallon spill of pesticide (Chlordane-72%) spread over a one block area by a pesticide company truck. The area was blocked off, access limited, and a call put into Chemtrec. The D.P.W. provided a load of sand for diking, as a command post was initiated. Using information from D.O.T.'s *Emergency Response Guide* and from Chemtrec, the Incident Commander directed efforts to absorb the chemical and have it hauled to a waste disposal site. The street was then decontaminated by scrubbing with detergent.

February 10, 1981: Hamilton, Ontario. A truck hauling two tanks of 62 percent liquid chlorine collided with a second truck on the Burlington Skyway Bridge over the mouth of Hamilton Harbor, then crashed into a guardrail. Several hundred cars were abandoned on the bridge and its approaches, and nearby residents were evacuated. The total amount of chlorine involved was estimated to be 20,000 pounds.

January 29, 1981: Shelbyville, Kentucky. A truck loaded with explosives overturned on rural Route 53, caught fire, then exploded with a blast felt more than 25 miles away . . . leaving a 30 foot crater in the roadway. The driver escaped before the explosion. Three firefighters approximately 200 yards from the exploding truck were injured slightly.

January 22, 1981: Williston, Vermont. A Coastal Trucking tank truck carrying 6,000 gallons of resin plasticizer (used in making fibreglass) enroute from New Haven, Connecticut to Valleyfield, Quebec, Canada overturned at 5 A.M. on I-89. The product is not extremely hazardous, but is flammable and can cause respiratory problems after long exposure. State officials called CHEMTREC for identification and handling information. Sand was spread on the spilled liquid by personnel wearing special rubber gloves and boots. An attempt to off-load the cargo while the tank was lying on its side was unsuccessful, so three wreckers were used to upright the tanker.

January 14, 1981: El Cajon, California. The El Cajon Fire Department responded to a spill of unknown origin. The area was cordoned off until the product was identified by staff of the county agricultural department as formaldehyde. Firemen wearing turnout gear and SCBA used absorbent material (Kitty Litter) to clean-up the commodity, then had it removed to a waste disposal site. One fireman visited the hospital for observation but was released the same day.

January 8, 1981: Boulevard, California. A truckload of smoldering sulfur on I-8 raised an interesting issue for response personnel. Chemtrec, the shipper, and the manufacturer reportedly each gave the dispatcher somewhat different recommendations for extinguishing the sulfur. Chemtrec recommended using a CO-2 fire extinguisher or a flammable metal type of extinguishing agent; the shipper recommended covering the sulfur with additional sulfur in order to exclude oxygen; the manufacturer recommended using water, saying that a garden hose would do.

January 3, 1981: Santee, California. The Santee Fire Department responded to a vehicular accident involving a pesticide company truck rearended by a speeding auto. The impact ruptured a container of Lindane (100:1) releasing approximately eight gallons onto the highway. The San Diego County Highway Department delivered a load of sand for diking, and the California Highway Patrol called in IT Corporation for clean-up.

December 30, 1980: Birch Run, Michigan. Two died and three were injured when a Mobil Oil Corporation filling station exploded and erupted in flames while a twin-tank trailer was unloading fuel.

December 27, 1980: Ashland, New Hampshire. A tank truck loaded with 9,000 gallons of unleaded gasoline crashed on the median strip of I-93 and burst into flames. Although the truck was destroyed, the driver and a passenger escaped.

December 24, 1980: LacLaHache, B.C., Canada. Ice and snow conditions caused an accident on Hwy. 97 seven miles north of town involving a C.P.R. tanker transport and an automobile. The tanker rolled over, spilling a considerable amount of 11,000 gallons of -35 degree heating oil it carried. No fire. One dead, two critically injured due to the vehicle accident; no injuries caused by the spill.

December 23, 1980: Deer Isle, Maine. A Petrolane propane gas truck overturned on an icy stretch of Route 15 in a rural area. There were no leaks, and the product was safely transferred to another truck the following day. No fire.

December 22, 1980: Marcy, New York. A tank-trailer truck carrying fuel oil swerved to avoid a snowplow, hit a car, then crashed through a house injuring the elderly owner, the truck driver, and the car driver. No fire.

December 22, 1980: Oklahoma City, Oklahoma. An explosive device being transported to an oilfield in a Midco, Inc. pick-up truck exploded on an elevated section of I-40 near downtown, scattering additional explosive devices over the highway and into city streets below. The Oklahoma City bomb squad was called in to recover the charges and Primacord. No injuries.

December 17, 1980: San Diego, California. An engine company responding to a vehicle accident for an extrication call involving two persons injured and one dead were providing first aid treatment when they noticed containers thrown about from a truck involved in the accident. One of the five gallon containers was used to make an identification using normal reference sources . . . Organic Phosphate, Pesticide. Later it was found that paramedics and injured persons were becoming ill from delayed effects of pesticide poisoning. A waste disposal firm was called to the scene, and soil in the area was scrapped and removed to a disposal site. The pesticide was said to permeate leather products; therefore, police officers and firefighters were instructed to dispose of all leather products taken into the site on their person.

December 16, 1980: 70 Mile House, B.C., Canada. Ronald Osmond was driving a Trimac tanker truck out of Burnaby, B.C. when about 23 kilometers north of Clinton on Hwy. 97 the rear pup trailer flipped on its side and ruptured, igniting 49,800 litres of gasoline. Realizing a ball of fire was following him, Osmond continued down a hill to keep ahead of the flaming trailer, shut-off the engine, jumped and ran. The blazing tanker damaged overhead power lines cutting off power to Clinton for a day.

December 7, 1980: Houston, Texas. A 16-inch square carton containing radioactive molybdenum and technetium enclosed within lead shielding was found to be missing from a truck shipment originating at Mallinckrodt, Inc. in St. Louis. The vial of radioactive material could emit significant amounts of radiation over time if removed from its lead-shielded container. A radioactive drug shipment from Mallinckrodt, Inc. in St. Louis to Boston was reported missing November 5th but found intact November 24th at a Boston airport.

November 29, 1980: Charlotte, North Carolina. Three persons were seriously burned when a gasoline tank truck, reportedly held-up by traffic as it attempted to cross the tracks, was hit by a Seabord Coast Line freight train causing an explosion and fire.

November 25, 1980: Kenner, Louisiana. Negligent homicide charges are expected to be filed against the driver of a gasoline tank truck that failed to beat a freight train to a grade crossing causing a fire that killed at least seven persons. Resultant explosions blew down utility poles and set additional fires as approximately 300 persons were evacuated. Five of the dead were in a bar near the rail crossing in downtown Kenner, a suburb of New Orleans.

November 25, 1980: King of Prussia, Pennsylvania. A pick-up truck containing a sealed container of Iridium-192 was stolen from U.S. Testing Co. offices but located three blocks away. Iridium-192 emits significant amounts of gamma radiation if not continuously housed in a special container. When located, the safety container was undamaged and no injuries were reported.

November 5, 1980: Richmond, California. A tanker truck being loaded with nitrous oxide ("laughing gas") exploded blowing out three sides of Puritan Bennett Chemical Corporation plant. The blast could reportedly be heard up to 50 miles away, but no injuries were reported. A sizable vapor cloud of nitrous oxide was not viewed as a serious threat.

September 22, 1980: New Lebanon, New York. A Consolidated Freightways tractor-trailer hit a bridge abutment while rounding a curve, sending the vehicle into a dry riverbed. Two low-yield radioactive materials, including Iridium 192, were located amid spilled house paint and reloaded on another truck. Initial reports indicated no leakage of radioactivity.

September 16, 1980: Bakersfield, California. An 11-man chemical emergency response crew from IT Corporation in Wilmington, California was called in to clean-up a major spill of toxic liquid defoliant (hydrooxydimethy-arsineoxide) that shut down the main north-south highway (I-5) in a mountainous stretch known as "the Grapevine" for 21 hours. Wearing SCBA and protective clothing, clean-up workers spread bleach over the spill to oxidize the liquid defoliant leaving a solid, powdery residue that was then picked up using special vacuum trucks. The accident was reportedly caused when a tank truck hit a parked California Dept. of Highways truck. The arsenic and acid-based defoliant oozed across all four lanes of northbound I-5, covering a total area of one-quarter mile in length. A California Dept. of Highways crew, the tank truck driver, and four highway patrolmen were taken to Bakersfield area hospitals for observation. Their contaminated clothing was burned, and their vehicles and personal belongings quarantined.

September 15, 1980: Branford, Connecticut. A small leak in a tank truck containing 4,000 gallons of hydrofluoric acid and operated by Chemical Leaman Tank Lines, Inc. necessitated bringing the truck into a Leaman terminal for patching and offloading. Branford fire officials ordered a stand-by alert for residents of the nearby area until offloading was completed. The cause of the leak was thought to be hydrofluoric acid corrosion of a welded area. A hazardous materials response team from the Connecticut Dept. of Environmental Protection applied a patch to the leaking tank prior to offloading.

August 29, 1980: Baldwin, Pennsylvania. A garbage truck operator, finding his load on fire, dumped part of the load on a residential driveway. Mixed chemicals discarded by a high school science department released cyanide fumes that sent nearly 100 persons to a local hospital. Chemicals involved included quart bottles of asbestos, arsenic, sodium cyanide, potassium cyanide, hydrofluoric acid, mercury, phosphorus, and a quarter-pound bottle of uranyl acetate that did not break. The probable cause of the fire was thought to be either spontaneous combustion or the action of the truck's compactor.

August 26, 1980: Bayonne, New Jersey. A dump truck/gasoline tanker collision on the New Jersey Turnpike caused an explosion and fire that critically injured the tank truck driver and seriously injured the dump truck operator. Rush hour traffic was blocked for a number of hours.

August 19, 1980: Grayslake, Illinois. A truck that failed to clear a rail crossing was clipped by a freight train, rupturing a number of 85 drums of white house paint and spilling thousands of gallons of paint into an intersection. Two persons were treated for minor injuries.

August 7, 1980: New York, New York. A leaking propane tank truck owned by Ritter Transportation Co. of Rahway, New Jersey shut down the George Washington Bridge for eight hours and caused a monumental traffic jam that backed-up traffic as far as the Tappan Zee Bridge and the Connecticut state line 15 miles away. Alternate routes through the Lincoln and Holland Tunnels quickly became jammed bringing traffic between New York City and New Jersey to a near standstill. Approximately 2,000 residents were evacuated from the bridge area on the Manhattan side as firefighters used a cherrypicker to hose down the leaking tanker. Attempts to offload the cargo were hampered by the defective pressure release valve that caused the original leak. Eventually, Police Office Chris Drauer of the Emergency Services Squad #2, a former plumber, suggested use of a simple plumber's plug that was obtained from a Bronx hardware store and brought to the bridge by police motorcycle escort where a Ritter Safety Coordinator was able to successfully attach it to the tanker. The propane was offloaded to another truck and the bridge was reopened in the early evening, eight hours after being closed at 10 A.M. New York City later sued Ritter Transportation Co. for $600,000 to cover city costs in responding to the incident.

July 24, 1980: Verona Isle, Maine. Seven thousand pounds of lobster were contaminated by Sevin-4, the pesticide used in Maine's highly controversial spruce budworm spraying program.

A Merrill Transport tank truck reportedly delivered 6,000 gallons of saltwater, later found to contain Sevin-4 oil, to Jeff's Lobster Pool. It was reported the truck had previously been used supplying Sevin-4 to aircraft used in the budworm spraying program. Sevin-4 oil contains cabaryl, a toxic chemical whose longterm effects are thought to be dangerous to human health. No contaminated lobsters were sold, and the water was isolated.

June 27, 1980: Oakland, California. Highway workers in Oakland had a problem when a tank truck transporting molten sulfer from a refinery overturned and burst open in flames. Firefighters had the fire under control in 45 minutes, but once the smoke had cleared tons of hardened sulfer lay across four southbound and one northbound lane of Highway 17. High-pressure water hoses proved ineffective in removing the mess, and officials were considering using a huge grinder.

May 28, 1980: Swift Current, Saskatchewan, Canada. A fuel tank truck was in collision with a bus carrying C.P. Rail crew, killing 22 and injuring 11. Fire destroyed both vehicles.

Rail

March 26, 1982: Ukiah, California. Toxic formaldehyde spilled from a Georgia Pacific Co. tank car sitting on a siding, apparently after vandals knocked-off the tank car valves, entered the Russian River, source of drinking water for much of Sonoma County. County water agency officials immediately shut-down intake pumps from the contaminated river. Thousands of dead fish lined the river bank downstream from the spill. Roughly 21,000 gallons was thought to have spilled, but perhaps about 60 percent of the product was trapped by an emergency dam thrown across a drainage ditch near the leaking tank car.

March 16, 1982: Norton Shores, Michigan. Twelve Chesapeake & Ohio rail cars carrying caustic soda and chlorine gas derailed, forcing evacuation of approximately 600 residents for 12 hours. Caustic soda spilled from two cars but no chlorine was released. Cleanup crews from Hooker Chemical and C & O R.R. spread lime on the caustic soda.

February 28, 1982: Orillia, Ontario. Thirty-seven cars of a CP train carrying anhydrous hydrogenfluoride, isopropanol, hydrochloric acid, vinyl acetate, butane and resin solution derailed outside Orillia, reportedly causing evacuation of 3,000 persons from within a wide area. The Canadian Transport Commission is currently conducting an investigation of the incident.

January 31, 1982: Lockbourne, Ohio. A 33-car Norfolk & Western Railway Co. train derailed and tumbled off a bridge. A tank car carrying isobutylaldehyde, a flammable liquid used in the rubber and chemical industry, exploded in flames. Approximately 350 area residents were temporarily evacuated due to fear of toxic fumes. No injuries.

January 20, 1982: Haleyville, Alabama. A tank car of chlorosulfonic acid sprung a leak that caused a toxic cloud and initiated a precautionary evacuation of half the 5,000 residents of Haleyville. No injuries. Reportedly, the car had been loaded in frigid Chicago, possibly resulting in the cargo expanding until it ruptured a seal in the 50 degree temperature of Alabama. An emergency team from DuPont was called in to fix the leak.

January 7, 1982: Thermal, California. Fourteen cars of a 55-car Southern Pacific freight derailed. State health officials quarentined the desert area nearby for six hours to search through the debris until a five gallon drum containing radioactive Americium-241 was found intact.

December 6, 1981: Orange, Texas. Residents of a 12 square block area were evacuated for five hours after an early afternoon derailment of 16 cars and two locomotives of a 100-car Southern Pacific train left a car of butyl alcohol punctured and leaking. No injuries were reported. Response personnel constructed sand dikes in roadside ditches to contain the flammable liquid.

November 19, 1981: Portsmouth, New Hampshire. If you are going to have an incident, you might as well be prepared. Three Boston & Main Railroad cars (an "empty" containing toxic methyl methacrylate vapors, an L.P.G. tanker; and an empty boxcar) derailed near the

National Gypsum Company plant; only the boxcar overturned. By coincidence, just three miles away at Pease A.F.B. hazardous materials/wastes experts from throughout the northeast were staging a spill simulation for trainees in a University of New Hampshire/C.E.T.A. training program for Haz. Mat./Haz. Waste Technicians. Hazardous waste removal experts from state and local government agencies and a number of commercial/industrial firms forgot about the simulated incident and went to work on the real one. No fire, no leak, no injury. "Hands-on experience is probably the best way to train," said Dave McIntyre of the E.P.A. Regional Response Team based in Lexington, Massachusetts.

October 31, 1981: Hamburg, New York. A 97-car Chessie System train derailed nine boxcars plus a tank car loaded with 16,000 gallons of methyl chloride that ruptured forcing the evacuation of approximately 4,500 people for eight hours in a suburb of Buffalo. Some boxcars caught fire, endangering the methyl chloride gas that is shipped as a liquefied gas under its vapor pressure. It is easily ignited, and under fire conditions the cylinder or tank car may violently rupture and rocket. However, no rupture occurred.

October 6, 1981: Everett, Washington. More than 1,000 people were evacuated after a freight train derailed, overturning five tanks of liquid chlorine and two of butane. No leak, no spill, no fire.

August 7, 1981: Bridgman, Michigan. More than 1500 people had to be evacuated from a 25 square mile area of southwestern Michigan when 14 cars of a Chessie System freight derailed at 5:30 A.M. about a block from the main business district of Bridgman. A tank car of fluosulfonic (i.e. "fluorosulfonic") acid was punctured forming a toxic cloud that moved in a northeasterly direction. Response personnel were able to temporarily patch the leak seven hours after the derailment. Clean-up efforts were hampered because water cannot be used on fluorosulfonic acid as the chemical is decomposed by water to hydrofluoric acid and sulfuric acid with release of heat. Fourteen persons were treated for burning eyes and throats.

July 20, 1981: Dupo, Illinois. A rail car leaking nitric acid caused a large toxic cloud that forced 1,400 residents from the area as emergency response crews worked to plug the leak. Authorities closed off a seven mile stretch of Illinois-3, used town sirens in both Dupo and Cahokia, and went door-to-door to awaken sleeping residents after the leak was reported at 3 A.M. Lime was used on the spilled acid to decrease the amount of fumes being let off.

July 3, 1981: Thorp, Wisconsin. A wheel problem reportedly caused a Soo Line freight to derail at 2 A.M. spilling about 17,000 gallons of acetic anhydride, a corrosive used in the manufacture of cellulose acetate, or as a dehydrating agent in the handling of fatty and volatile oils. A slow moving cloud headed north over Taylor County forcing the evacuation of an estimated 700 people from Thorp and Aurora. A Deputy Sheriff and a train crew member were examined at a hospital.

June 27, 1981: Newark, New Jersey. A state-of-emergency lasting 40 hours required a major evacuation of an area around Newark Airport when a damaged rail tank car leaked ethylene oxide and became involved in fire. A flammable liquid and corrosive subject to possible polymerization if contaminated, ethylene oxide is a clear, colorless, volatile liquid used in the manufacture of antifreeze, cosmetics, pharmaceuticals, detergents, and other consumer products. Responders purposely refrained from extinguishing the fire since they were uncertain they could stop the flow, but initially concentrated on cooling the tank with water fog rather than direct streams. The strategy was to allow escaping ethylene oxide to burn-off while nitrogen was was pumped into the tank car to displace oxygen that could fuel an explosion. Later, firefighters attached a pipe to a valve on the tank allowing speedier discharge of the fluid which was also ignited. The fire eventually burned itself out with no reported injuries.

June 13, 1981: Deerfield, Massachusetts. A Boston & Maine Railroad freight derailed at least seven cars. Toxic freon gas escaping from a refrigerated car that fell approximately 60 feet from a bridge into the Connecticut River caused the evacuation of an estimated 200 people. Freon, often used as a refrigerant, is nonflammable but toxic. The freon eventually dispersed by itself.

June 11, 1981: Greensburg, Pennsylvania. Chemical vapors trailing from a leaking tank car carrying octyl alcohol, a perfume base described as nontoxic, sickened at least 63 passengers aboard Amtrak's Broadway Limited enroute from Chicago to New York. Passengers suffering from dizziness and nausea were treated at three area hospitals, with two admitted to intensive care.

May 11, 1981: Barberton, Ohio. A derailed tank car containing 11,000 gallons of butadiene, a flammable gas, was safely rerailed with no injuries.

March 30, 1981: Flagstaff, Arizona. Three minor leaks occurred in propane laden cars after a total of 13 cars derailed. Gov. Babbitt declared a state of emergency and called up approximately 50 National Guardsmen to handle evacuation of 100 area residents and to guard property in the closed-off area. Assistant Fire Chief, Leon Wilder, stated that authorities did not consider any of the three leaks as serious.

March 25, 1981: Enos, Indiana. A flatcar on a Milwaukee Railroad freight train being operated over conrail track derailed striking rubber-tired, portable 1,000 gallon tanks of anhydrous ammonia. The freight train conductor died when he walked into the resultant toxic cloud.

March 6, 1981: Gibbstown, New Jersey. Twelve derailed cars of a Conrail freight, including five containing chlorine, one carrying vinyl chloride, and one filled with hydrogen peroxide; remained upright after leaving the tracks, and no leaks were reported.

January 26, 1981: San Diego, California. The San Diego Fire Department implemented its Chemical Disaster Plan after responding to a railroad siding spill of phostoxic, a pesticide used in railroad boxcars for fumigation of grains. With assistance from representatives of the Agriculture Department, State Office of Disaster Preparedness, the railroad, and other agencies; firefighters wearing full turnout gear and SCBA initiated cleanup that consisted of taking the chemical in the form of pellets and submerging it in a mixture of water and liquid detergent before proceeding with normal disposal procedures.

January 8, 1981: Point Pleasant, West Virginia. Five cars of a Chessie System train derailed about 440 yards from a deep well that supplies drinking water to Point Pleasant, allowing at least one car to release toxic and explosive vinyl chloride forcing a small-scale evacuation in a rural area. A crew worked for a day to plug a leak in a 90-ton tank car, but halted operations at dark fearing artificial lighting devices might ignite the gas.

December 29, 1980: Salem, Virginia. Two Norfolk and Western freight trains collided in dense fog after one had left the tracks. A total of 27 cars and nine locomotives derailed, injuring five. Hazardous materials cars in the consist did not derail, although approximately 20,000 gallons of diesel fuel spilled.

December 22, 1980: Vancouver, B.C., Canada. A collision between a Vancouver Wharves yard—train and two B.C. Railway tank cars in the B.C. Rail yards just east of Lion's Gate Bridge caused an explosion and spectacular fire fueled by volatile and poisonous liquid methanol. 17,000 gallons of methanol spilled igniting at least six boxcars of newsprint just a short distance from stored propane and chlorine. The boxcars looked "like a string of burning Presto Logs" according to one West Vancouver firefighter. The incident occurred the same day a North Vancouver council meeting was to discuss a proposal by Vancouver Wharves to locate a multimillion dollar methanol storage facility on the North Shore. North Vancouver and West Vancouver firefighters arrived to find the rail siding on fire, a burning tank car, a railyard loaded with cars of unknown contents, and six flaming telephone polls carrying high tension wires of 12,000 to 60,000 volts. Nearby tank cars containing propane and chlorine were shunted to safety. Pools of methanol were diked to prevent spread, and product remaining in the extinguished tank car was off-loaded into two tank trucks.

December 14, 1980: Jefferson County, Kentucky. A 33,662 gallon tank car of butadiene derailed in a populated area of Jefferson County near the large Rubbertown chemical complex. Jefferson County HAZ-MAT, a mutual assistance hazardous materials team of public/private interests, responded to the incident. Hulcher Emergency Services personnel righted the car. No deaths, no injuries, no property damage except R.R.

November 1, 1980: Rolla, Missouri. Fourteen cars of a Frisco freight train derailed and caught fire causing at least two explosions. A tank car of naptha did not catch fire, and no one was injured.

September 28, 1980: Bainville, Montana. Ten families were evacuated following a 19-car Burlington & Northern derailment in which a tank car leaked the insecticide, propylene dichloride, following collision with an automobile on the tracks. No fire, no reported injuries.

September 20, 1980: Bradford, Pennsylvania. One person was killed and ten injured by a flash explosion as crude oil was being pumped out of a tank car at the site of a Chessie System freight train derailment. The explosion occurred shortly after firemen used ropes to rescue two persons overcome by fumes inside the car.

September 17, 1980: Custer City, Pennsylvania. A train hauling 15 tank cars of crude oil and an unknown number of caustic soda cars derailed about 10 P.M. causing an explosion and oil fire that destroyed three homes and resulted in the evacuation of nearby residents. No injuries reported.

September 4, 1980: Cook's Point, Texas. Approximately 15,000 gallons of carbon tetrachloride leaked from two tank cars when a Southern Pacific train derailed shearing the bottom outlet pipe from one car and damaging the valve on another. A vacuum truck was used to reclaim product, and contaminated soil was scrapped and trucked to a landfill site.

August 18, 1980: Sutton, Nebraska. A derailed Burlington-Northern hopper car hit two 18,000 gallon propane tanks causing a fire that forced the evacuation of more than 1,000 persons.

August 4, 1980: Beaumont, California. A five gallon container of white phosphorus immersed in water may have evaporated as a Southern Pacific Railroad train crossed a desert area, burst into flames igniting a total of 12 containers. A California Dept. of Forestry fire crew used water to control the fire and restabilize the white phosphorus, then had two cars of the cargo moved outside of town where the commodity was packed in ice until proper repacking materials became available three days later.

July 26, 1980: Muldraugh, Kentucky. An Illinois Gulf Central train derailed and six of ten derailed tank cars burned causing a 3,000 foot column of toxic smoke visible over a ten mile area that forced the evacuation of approximately 7,500 area residents. Burning and/or derailed tank cars included six of vinyl chloride; other derailed cars contained chlorine, acrylonitrile, and toluene. Police wearing protective gear patrolled the streets of Muldraugh to prevent looting. Evacuees included personnel from nearby Fort Knox where the nation's gold reserves are stored. The cause of the derailment was not immediately known. The derailment occurred 100 yards from the main street of Muldraugh. The train, a daily from Memphis to Louisville, was reported to have derailed eariler this month near W. Paducah, Kentucky. In 1979, the train was reported to have derailed three times within Kentucky, forcing two evacuations. Response personnel were planning to use plastic explosives to blast the still-burning cars of vinyl chloride. On 7/28 an E.P.A. monitoring team detected traces of hydrochloric acid, vinyl chloride and phosgene present near the derailed cars.

July 26, 1980: Gardner, Massachusetts. Ten cars of a 107 car train derailed blocking a main track. The outer skin of an empty propane tank car was pierced, but the inner shell remained intact and there was no leak or fire.

July 21, 1980: Truckee, California. An "empty" tank car of phosphoric acid caused a short-term evacuation of 2,000 people when a Southern Pacific train stopped to let off train crew members complaining of respiratory problems and nausea. The Truckee Fire Dept. called for an evacuation and had the train moved out of town. A loose manway cover was alleged to have allowed toxic fumes to escape from approximately 200 gallons of product remaining in the car.

June 21, 1980: Muscatine, Iowa. A leaking tank car of inhibited butadine was safely moved to a rural location and the leak plugged through combined efforts of Muscatine Fire & Police departments, County Civil Defense, Monsanto, the Milwaukee Railroad, and Fruitland Fire Department.

June 17, 1980: Middleboro, Massachusetts. A tank car leaking propane was off-loaded.

June 17, 1980: Hammond, Louisiana. Six Illinois Gulf Central rail cars derailed in downtown Hammond causing a leak in one derailed car of 23,500 gallons of styrene monomer. Response crews were able to tighten a vent connection to stem the leak. Approximately 2,600 people were evacuated. Hammond Fire Department crews built an earthen dike to contain spillage which was recovered by Spill Control Services of Hammond using sorbent materials. There were no injuries.

June 12, 1980: Delhi, Louisiana. An Illinois Gulf Central Railroad tank car containing chlorine derailed, but no leakage was reported. 1500 people were evacuated.

June 9, 1980: London, Ontario, Canada. Thirty-four cars of a 102-car Canadian Pacific train jumped the tracks at the northwest limits of London, tossing some cars hundreds of yards from the tracks to mow down power lines and poles. The initial explosion sent flames roaring 60 feet in the area and caused a shock wave that blew out windows in area homes. 300 persons were forced to flee the area. Only seven of the derailed tank cars were loaded; products included sulfer dioxide, coal tar, and light oil.

June 6, 1980: Garland, Texas. Nine cars of a Kansas City Southern Railway Co. train derailed at 2:15 A.M. spilling 5,000 gallons of styrene monomer. Approximately 6,500 persons were evacuated, and five persons were injured slightly. Firefighters used water to dissipate toxic vapors, and AFFF foam to lessen the potential for combustion. Two siphon dams were constructed on a nearby creek by local Civil Defense personnel, and entrapped water and styrene monomer were pumped into tank trucks for disposal at a sanitary landfill. One car of propane was reported leaking as well. No injuries.

May 12, 1980: Barstow, California. A Sante Fe freight slammed into the rear of another train, killing the conductor, injuring other rail workers and setting 15 cars ablaze including one 10,000 gallon car of ethelene glycolether, a combustible liquid.

May 6, 1980: Mobile, Alabama. A rail tank car ruptured releasing hydrogen chloride forcing near total evacuation of a wide area centered at the port facility, and closing all roads leading into the city.

April 19, 1980: Madison, Georgia. Tank cars carrying L.P.G. and sulfuric acid derailed causing an explosion and fire. No injuries were reported.

April 13, 1980: McNeill, Mississippi. Several hundred people fled their homes when 18 rail cars derailed and three overturned, one spilling explosive and toxic styrene.

April 3, 1980: Kearny, New Jersey. Four tank cars derailed with one spilling highly flammable methyl methacrylate. Workers were evacuated from a nearby industrial area.

April 3, 1980: Somerville, Massachusetts. A locomotive sideswiped a chemical tank train three miles from downtown Boston releasing clouds of hydrochloric and phosphoric acid causing evacuation of more than 10,000 people and the closing of I-93, area businesses and schools as a toxic cloud floated over the city. There was no fire or explosion but approximately 300 were injured by fumes.

Marine

April 1, 1982: Montz, Louisiana. After a tanker carrying crude oil collided with a tug and three barges setting off a spectacular blaze, crude oil leaked for 15 hours until the fire could be put out and the leaks plugged. Coast Guard crews ringed the tanker with containment booms and placed additional booms around drinking water intakes near the collision site. Patches of oil flowed 25 miles downriver past New Orleans but waterworks personnel there said city drinking water intakes were submerged below the level of the floating oil.

March 6, 1982: 700 Miles East of Bermuda. A boiler room explosion aboard the empty tanker, Golden Dolphin, left nine crewmen missing and feared dead. Sixteen survivors were picked up by the Swedish merchant vessel, The Norrland.

February 26, 1982: New York, New York. An explosion aboard an empty gasoline barge in the East River killed one and threw seven others into the water. Windows were blown in nearby buildings, and the barge sank one hour after the initial explosion near the Williamsburg Bridge.

February 15, 1982: Kalama, Washington. A ship fire aboard the West German grain freighter, Protector Alpha, believed started when bunker fuel spilled into a generator room beneath the fuel tanks, injured six persons, three seriously. The freighter, longer than two football fields and carrying 2,000 tons of grain, was towed upriver from a North Pacific Grain Growers elevator in the Port of Kalama and beached on the Oregon side of the Columbia River away from river traffic.

January 26, 1982: San Diego, California. Three sailors died aboard the guided missile cruiser USS Bainbridge, and seven were injured from freon gas escaping from a ruptured pipe in the forward air conditioning plant. Six of the injured were crewmen while the seventh was a corpsman from the naval hospital who became ill after administering mouth-to-mouth resuscitation to the injured.

January 3, 1982: Galveston, Texas. A Monsanto Co. barge carrying 400,000 gallons of acrylonitrile (flammable liquid; poisonous; polymerizable) valued at one million dollars struck the Galveston Causeway railroad bridge and caught fire. Resultant explosions caused one end of the barge to sink, spilling an undertermined amount of acrylonitrile into the 12-foot deep waterway.

December 25, 1981: Matane, Quebec, Canada. Four crewmen were believed dead and up to four others were missing after the 4,000 ton/460 foot Hudson Transport oil tanker caught fire in the St. Lawrence River about 250 miles northeast of Quebec City.

December 22, 1981: Buzzards Bay, Massachusetts. One person was killed and at least seven injured aboard the Massachusetts Maritime Academy training ship, Bay State, when spilled oil ignited. Engineers were reportedly changing oil strainers in the engine compartment of the 524-foot ship when a valve broke. Hot oil apparently burst out, hit a casing, and ignited.

December 9, 1981: Louisville, Kentucky. Two towboats pushing barges collided in the Ohio River about 75 miles from Louisville. A third barge damaged in the accident spilled 42,000 gallons of furnace oil into the river. An industrial response team from Olin-Matheson Chemical Corp. of Brandenburg, Kentucky was hired by the barge owner to combat fire on the first two barges with foam.

July 17, 1981: Akron, Ohio. A reported billion gallons of pungent, bacteria-laden chemicals tinted 20 miles of the Cuyohoga River red, seriously affecting ability of the Botzum sewage treatment plant to purify water before it is released into the river. Investigators believe the unidentified chemicals were illegally dumped into the city's sewer system, possibly by a plastics manufacturer. Bacteria counts taken at the Botzum facility were 300 times higher than the federal standard for safe swimming.

June 9, 1981: Eddystone, Pennsylvania. Three workers died and three were injured when a carbon dioxide firefighting system on a ship under repair at Sun Ship, Inc. discharged for unknown reasons, displacing oxygen in the engine room as fire doors automatically closed trapping five workers. A sixth person was injured during rescue efforts. Rescue workers wearing SCBA removed the dead and injured from the engine room of the 680 foot SS Lash Atlantico. Carbon dioxide, widely used in extinguishers, is a colorless, odorless, incombustible gas one-and-a-half-times as dense as air. When released in an enclosed area, it will replace available oxygen.

February 6, 1981: San Diego, California. An anonymous telephone call citing "an intended discharge of 2600 pounds of ammonia" into San Diego Bay caused Coast Guard officials to board a ship of Costa Rican registry. It was determined that a small amount, approximately 20 gallons, had actually been released. Officials obtained a written statement, and the shipowner was cited for violation of the Federal Water Pollution Control Act. Civil action was also recommended.

February 1, 1981: Portsmouth, Virginia. An engine room fire aboard the empty Greek tanker, Aikaterini, was extinguished after 13 days only to have it reignite the following day. Specialists in oil rig fires who were flown in from Houston repeated the procedures they had used previously by filling the engine room with foam, dispersing accumulated smoke, and ventilating the entire tanker.

January 12, 1981: Lake Charles, Louisiana. A natural gas blowout aboard an off-shore drilling rig forced an emergency evacuation of 31 workers. Two men were reported in serious condition from exposure after 30 minutes in the chilly water.

January 6, 1981: Alameda, California. Approximately 1,000 gallons of oil seeped into San Francisco Bay near the Alameda Naval Air Station and fouled the west end of Robert W. Crown Memorial State Beach. The spill is thought to have come from either the aircraft carrier, Coral Sea, or the supply ship, USS Wabash which can carry up to 175,000 barrels of petroleum. Booms and absorbent materials were used to deal with the spill.

December 20, 1980: Bayonne, New Jersey. The 180 foot chemical barge, "High Grade 42," was damaged by two explosions. The first, believed caused by naptha gas leaking from a valve, critically injured a crewmember. The second, believed caused by a build-up of gas in the engine room, injured three firefighters, knocking one into the waterway encumbered by SCBA. Two firefighters who went to the rescue of the first were slightly injured.

November 21, 1980: Rockland, Maine. The coastal supply tanker, Christian R. Reinauer, carrying 714,000 gallons of gasoline and fuel oil received an eight-foot gash in the hull and lost approximately 10,000 gallons of commodity when it ran aground on Metinic Ledge four miles off the Maine coast. Jet Line Pollution Control, Inc. boomed coastal areas while Boston Fuel Transportation Co., owner of the tanker, chartered a barge to remove the remaining cargo.

October 11, 1980: Corpus Christi, Texas. An engine room fire on the Ocean Traveler drilling platform 50 miles southeast of Corpus Christi in the Gulf of Mexico caused 59 crewmen to evacuate using floating escape capsules. Although the fire burned for hours, the well itself did not catch fire.

October 10, 1980: Stockton, California. Between 200,000 and 300,000 gallons of JP-4 fuel entered the Stockton Channel and the Coast Guard closed the port to all traffic due to the danger of fire from an inch of more of the fuel that covered an area of a half mile by 200 yards. The spill occurred while the Navy tanker, Shosone, was unloading.

September 6, 1980: Paducah, Kentucky. A barge carrying 675,000 gallons of methanol exploded in the Ohio River and burned out of control for more than a week. Additional explosions occurred up to ten days later. No injuries were reported.

August 31, 1980: Sabine, Texas. A well drilling rig off the Texas coast blew wild and caught fire, injuring five persons.

August 30, 1980: Corpus Christi, Texas. An unmanned oil barge containing 100,000 gallons of crude oil was located about 45 miles off the Texas coast and found to be awash and leaking. A 150 yard wide slick extended five miles from the barge as Coast Guard personnel began offloading.

August 30, 1980: Port O'Conner, Texas. An Ocean Drilling & Exploration Co. "Ocean King" jack-up drilling rig 18 miles off the Texas coast burst into flames after a pressure surge released natural gas and left the well burning out of control. Two were known dead, six injured, and three missing. The well, owned by Cities Service Oil Co., was located in 100 feet of water. No attempt was made to fight the out-of-control fire, but 24 hours later material collapsing from the side of the drill hole extinguished the flames and plugged the well.

August 25, 1980: Intracoastal City, Louisiana. A Mesa Petroleum Co. gas well suffered a blow-out as the rig's first well was completed at a depth of 5,700 feet. Four of 37 personnel on the rig located in 240 feet of water were injured slightly while evacuating the burning platform.

August 21, 1980: New Orleans, Louisiana. The 565-foot tanker, Texaco North Dakota, steaming at 16–18 knots with 2 million gallons of gasoline aboard rammed the unlighted, abandoned base of an oil drilling rig in the Gulf of Mexico setting the tanker afire as 39 crewmen abandoned ship.

August 7, 1980: Houston, Texas. A collision in the Houston Ship Channel between a U.S.-owned tug and barges and a British grain ship resulted in an explosion and fire of one 33,000 gallon tank of butadiene. Two burning barges were towed out to sea just ahead of Hurricane Allen where they burned for five days.

July 26, 1980: Off New Orleans, Louisiana. Personnel from the U.S. Coast Guard's Gulf Strike Team from Bay St. Louis, Mississippi supervised clean-up of 4,000 barrels of light heating oil after the Exxon Houston, bound from Baton Rouge, Louisiana to Everett, Massachusetts struck a submerged object at 4 A.M. The tanker was immediately anchored and the flow of oil was stopped.

July 22, 1980: Shell Beach, Louisiana. Up to 41,000 pounds of hydrobromic acid fouled part of the Mississippi waterway after the German M.V. Testbank collided with the Panamarian-flag vessel Sea Danial, a bulk carrier. When coming in contact with water, hydrobromic acid causes a voilent reaction and can burn like frostbite. 3,000 persons were evacuated from the adjacent shore area. More serious was the involvement of crystalline pentachlorophenal. (PCP, a wood preservative not to be confused with PCP known as "Angel Dust.") Response personnel used sonar gear in an effort to locate a large metal container of PCP that had dropped into the water. Six days later, it was announced the 32 Taiwanese crewmen of the Sea Danial had been exposed to 300 pounds of PCP spilled in the main deck and tracked about the ship. Divers in hard hats continued to search the muddy waters attempting to locate 12.5 tons of PCP contained in a metal container. Blood and urine samples from clean-up workers were being checked repeatedly. PCP spread into Lake Borgne, and a close watch is being kept on Lake Pontchartrain. The Sea Danial has been quarentined, and the rich shrimping and fishing area 25 miles east of New Orleans has been closed to all boats. On July 27, a Navy Minesweeper was brought in to aid in the search for the PCP. On August 2nd, the previously lost container was raised but found to be empty of PCP.

June 27, 1980: Tampa, Florida. A leak in a barge carrying 85,000 gallons of No. 6 fuel oil spilled 1,600 gallons into Tampa Bay. The leak was plugged 11 hours later.

June 8, 1980: Off New Orleans, Louisiana. A Marathon Oil Co. oil and gas platform 200 miles southwest of New Orleans exploded and caught fire causing 42 workers to scramble for "flying saucer" escape devices or jump 45 feet to the Gulf. Seven persons were injured. The offshore platform had been in operation for five years; the cause of the blast is unknown. Emergency equipment functioned properly to close-off flow from the 24 oil and gas wells served by the platform.

May 24, 1980: Cape Hatteras, North Carolina. The U.S. Coast Guard Atlantic Strike Team from Elizabeth City, No. Carolina boarded a 644 foot oil tanker loaded with 300,000 barrels of crude oil taking on water from a faulty 24-inch valve.

April 2, 1980: Galveston, Texas. An empty tanker and inbound freighter collided setting fires on both vessels. One creman treated for inhalation of fumes.

Pipeline

January 28, 1982: Centrailia, Missouri. There were conflicting opinions as to cause and process when gas lines in this town of 3,500 people sent flaming gas under pressure shooting out of furnaces, stoves, and heaters. Four houses were destroyed and several businesses damaged by fires that broke out simultaneously in different areas of town. Only one person was reported injured.

January 10, 1982: Sterling Heights, Michigan. A fire of unknown cause erupted in a natural gas line at a Consumers Power Co. regulating station in suburban Detroit, causing a fireball that could be seen two miles away.

December 30, 1981: Huffman, Texas. Two were injured and 150 evacuated when a backhoe reportedly ruptured a natural gas line and escaping gas ignited.

December 8, 1981: Santee, California. The Santee Fire Department blocked access to the area of an underground natural gas line break as workers from San Diego Gas and Electric wearing totally-encapsulating suits/airlines/safety harnesses and lines fixed the break.

August 25, 1981: San Francisco, California. As many as 40,000 people may have been exposed to cancer-causing polychlorinated biphenyls (PCB) when a contractor's drill cut a hole in a 16-inch gas main sending a shower of PCB-laden oil and gas through San Francisco's financial district. Escaping gas laden with residual PCB-containing oil from compressors at pumping stations along the pipeline coated trees, streets, buildings, and automobiles. Although there was extreme danger of a major fire until the broken pipeline was capped, the escaped gas did not ignite. More longlasting, and less understood, is longterm danger posed by PCB-laden oil coating the area. Clean-up crews were still using steam scrubbers and powerful vacuums, as well as large amounts of "Kitty Litter," to recover dripping oil when controversy broke out over how dangerous the oil really was, whether or not persons being allowed back into the area were seriously endangered, and what should be done with contaminated clothing.

August 1, 1981: Austin, Texas. A pipeline break in a heavily residential area of southwest Austin brought the evacuation of 1,000 families as highly explosive propane/butane gas mixture was carried by wind over parts of the city. The propane/butane gas, mixed for transportation, was scheduled to be refined later into separate gaseous mixtures. As firefighters stood by in special protective clothing, equipped with fog nozzles for diffusing the gas, repair workers used a lighted welding torch in the midst of an extremely heavy concentration of vapors to repair the break, hoping the mixture would be too rich to explode.

April 16, 1981: San Luis Obispo, California. An eight-inch petroleum pipeline ruptured during excavation work by a construction crew, allowed semi-refined product to be released into a creek. A citizen called the San Luis Obispo City 911 Emergency Dispatch Center to report a strong odor. Police and fire units were dispatched to determine the degree of hazard and source of the release. They found petroleum flowing in the creek and traced it to its source. Appropriate coordinating agencies were notified and requested to meet at a field command post established at the local waste water treatment plant. No significant fire hazard appeared imminent although a significant pollution problem existed. Approximately 12 different agencies and organizations cooperated in diking and skimming the creek. After a post-incident critique the following morning, further clean-up and monitoring activities were assumed by Union Oil Company, the U.S. Coast Guard, and the County Emergency Services Coordinator.

January 15, 1981: Allentown, Pennsylvania. An apparent natural gas explosion destroyed one home in a 4:30 A.M. blast and set fire to others forcing 15 families to flee. An infant was presumed dead and two children hospitalized for burns. A large crater was left in the street in front of the destroyed home.

January 1, 1981: Northern Alaska. An estimated 210,000 gallons (5,000 barrels) of oil spilled on the frozen Arctic tundra when a valve broke in a small (1.5 inch) by-pass line section of the Trans-Alaska Pipeline about 115 miles south of Prudhoe Bay causing a 13 hour shut-down. There was no leak from the main 48-inch line. Crews from Alyeska Pipeline Service Company repaired the by-pass line working in the constant darkness of the Arctic winter. Darkness made it difficult to assess the exact amount of product spilled that soon congealed in frigid temperatures. If the 210,000 gallon estimate is correct, it would be the largest spill to date on the pipeline that runs from Prudhoe Bay to Valdez. The latest leak may be the fifth since the 800 mile pipeline opened in 1977 to carry 1.5 million barrels a day.

December 1, 1980: Long Beach, California. Petroleum pouring from a ruptured 10-inch pipeline ignited, destroying nine homes and injuring four persons. Rivers of flaming fuel oil flowed through the 2700 block of Gale Avenue as residents fled for their lives. Long Beach city and L.A. County firefighters put down a three-foot deep blanket of foam throughout the area and had the fire under control in two hours.

November 19, 1980: Port Arthur, Texas. A pipeline owned by Independent Refining Corp. ruptured, allowing 29,000 gallons of gasoline to flow into a residential area forcing the evacuation of 240 families. No explosion, no fire, no injuries.

May 9, 1980: Medfield, Massachusetts. An estimated 15,000 gallons of gasoline spilled from an underground pipeline causing environmental damage. No fire.

April 16, 1980: Roseville, Minnesota. An explosion and fire at the Williams Pipeline station in a St. Paul suburb killed one and injured three.

April 11, 1980: Columbus, Georgia. A natural gas explosion damaged the Oglethorp office complex in Rankin Square.

Air

March 19, 1982: Wonder Lake, Illinois. An Air Force tanker with 20 people aboard exploded in the air while on an approach to O'Hare International Airport and fell near a residental area and elementary school. No survivors in the aircraft; no injuries reported on the ground.

November 17, 1981: San Diego, California. Firefighters wearing acid suits separated airline baggage after a corrosive, tentatively identified as acetic acid, was reported by baggage handlers.

12 STRATEGIES, TECHNIQUES AND METHODS OF HAZARDOUS MATERIALS RESPONSE

Hazardous materials by their very nature are extremely unpredictable when released from their normal containerized environment by leak, spill, rupture or carelessness. Combustion, detonation, simple temperature and atmospheric changes, and reactions between or among two or more hazardous materials; can increase the variable threats to people and property by geometric progression. Fire and smoke can definitely kill, but many hazardous materials can kill or injure by ingestion, absorption, inhalation or radiation. The expansion in volume and variety of hazardous materials in commerce has been so rapid and broad that few have been able to keep fully informed on the potential multiple threats faced by personnel responding to incidents. For these and other reasons, a hazardous materials incident requires extremely cautious and deliberate on-scene evaluation by response personnel.

Someone once said that hazardous materials incident response, at least conceptually, is quite simple. It involves getting the right people to the right place at the right time with the right equipment, the right knowledge, and the right training. In the real world, initiating and maintaining an effective response capability is far more complicated. Hazardous materials do not recognize jurisdictional boundaries, yet response personnel are often restricted by such boundaries. Although regulation and control of hazardous materials has been predominently a federal responsibility, response to hazardous materials incidents has remained almost solely a function of local government, industry, and commercial organizations. Prior knowledge of potential variables, and an understanding of proper methods of response, has assumed critical proportions for response personnel. The "judgement calls" required of responders on-site can be overwhelming: identification of the commodity, identification of the specific hazards involved, assessment of potential reactions between two or more materials, protection of primary and secondary exposures, evaluation of containment and run-off considerations, and determination whether or not evacuation is necessary. Simultaneously, response personnel may have to consider air monitoring needs, decide whether to patch or plug a leak, boom or dike a spill, and 101 other concerns. They may have to plan and carry out an attack, review neutralization and pH considerations, or assess and evaluate the results of their actions as they direct their own personnel and those of other agencies in seeking to obtain control.

To respond effectively to such concerns requires trained and experienced personnel, specialized tools/equipment/materials, and a high level of coordination and cooperation among responders. The pages that follow provide a review of some of the specific tasks required of any organization responding to hazardous materials incidents.

Planning

Throughout this book the focus has been on *first responders;* the tools, equipment, methods, strategies and "lessons learned" of persons who actually respond on-scene to contain and control accidental releases of hazardous materials. Every response organization the author has dealt with has had to engage in detailed prior planning in order to develop and implement a systematic approach to maximize effective utilization of available resources. Such prior planning serves to identify unknowns, establish productive relationships, minimize variables, and identify responsibilities. Planning attempts to identify what needs to be done and establish a framework for getting the job completed; it does not insure the job will be done properly.

Hazardous materials preincident planning efforts often consider the following specific components: Development of a Hazard Analysis; Creation of a Resource Inventory; Accumulation of Product Information; Implementation of a Response Capability; Provision of Training; (Including exercises and simulations in order to test planning assumptions) Preplanning of Specific Facilities; and Development of Standard Operating Procedures to guide response personnel on-scene. Such efforts are normally integrated with, and part of, a *Contingency Plan* developed by an organization responsible for the safety of people and property within a specific geographical area. A contingency plan generally seeks to define and achieve mutual acceptance of organizational procedures, responsibilities and required actions of involved parties or organizations; identify necessary and available resources for dealing with contingencies; and provide coordinating mechanisms.

A number of fire departments and other primary first response agencies use written response standards (Standard Operating Procedures) developed specifically to provide general guidance and instruction for response to hazardous materials emergencies. The *Hazardous Materials Response Standard* used by the North Charleston District Fire Department in Charleston, South Carolina (See Figure I) is one example of such a written standard.

Objective:

The objective of this standard is to provide a basic plan for Hazardous Materials Responses. This plan is written to insure that all District Fire Department personnel will have access to a standard operating procedure that will apply to most known hazardous material incidents. This standard operating procedure also provides for the maximum safety of personnel while operating at hazardous material incidents. Personnel should observe all safety precautions and wear full protective clothing and self contained breathing apparatus (where appropriate) when dealing with hazardous materials incidents.

Definition:

A *potential hazardous material incident* encompasses a wide variety of potential situations including fires, spills, transportation accidents, chemical reactions, leaks, explosions and similar events. Hazards involved may include toxicity, flammability, radiological hazards, corrosives, explosives, health hazards, chemical reactions and combinations of factors. This standard provides a general framework for handling a hazardous materials incident but *does not* address the special tactics or control measures for particular incidents.

Dispatchers:

The dispatcher will attempt to obtain any and all information from the person reporting a Hazardous Materials incident. The information should, if possible, include material name and/or type, amount and size of container(s), problem (leak, spill, fire, etc.) and dangerous properties of the materials. If possible, the dispatcher should stay on the phone with the caller to gain additional information after dispatching the initial response.

Any additional information shall be relayed to responding apparatus after they have been dispatched.

Figure 1. Continued.

When a potential hazardous material incident is reported, a *full* response including Squad 10 will be dispatched. Whenever possible, Squads 10 and 12 should *always* be dispatched together since some of our hazardous materials equipment is carried on both trucks.

If a call comes from a person with particular knowledge of the hazardous situation, have that person meet and assist the responding apparatus.

First Arriving Unit:

The first arriving officer will assume *Command* and begin a size-up. *The first arriving unit must avoid committing itself to a dangerous situation.* When approaching, slow down or stop to assess any visible activity taking place. Evaluate effects of wind, lay of the land and location of the situation.

The first arriving unit will advise *all other units to stage* until instructed to take specific action. Units must *stage* in a safe location, taking into account wind, spill flow, explosion potential and similar factors in any situation. The duty chief will assume command after conferring with first arriving officer. It may be necessary for Command to designate *staging* area for other responding apparatus.

Size-up

Command must make a careful size-up before deciding on a commitment. It may be necessary to take immediate action to make a rescue or evacuate an area, but this should be done with an awareness of the risk to Fire Department personnel, and taking advantage of available protective equipment.

The objective of the size-up is to identify the nature and severity of the immediate problem and gather sufficient information to formulate a valid action plan. *A Hazardous Materials incident requires a more cautions and deliberate size-up than most fire situations.*

Avoid premature commitment of companies and personnel to potentially hazardous locations. Proceed with caution in evaluating risks before formulating a plan and *keep uncommitted companies at a safe distance.*

Identify a hazardous area based on potential danger, taking into account materials involved, time of day, wind and weather conditions, location of the incident and degree of risk to unprotected personnel. Take immediate action to evacuate and/or rescue persons in critical danger if possible, providing for safety of rescuers.

The major problem in most cases is to identify the type of materials involved in a situation, and the hazards presented, before formulating a plan of action. Look for labels, markers, and shipping papers and ask personnel at the scene (plant management, responsible party, truck drivers, fire department specialist). Utilize reference materials carried on apparatus and have dispatcher contact other sources for assistance in sizing up the problem (Chemtrec, other agencies, fire department Hazardous Materials Team, manufacturers of materials, etc.).

Action Plan

Based on the initial size-up and any information available Command will have to formulate an action plan to deal with the situation.

> Most hazardous materials are intended to be maintained in a safe condition for handling and use through confinement in a container or protective system. The emergency is usually related to the material escaping from the protective container or system and creating a hazard on the exterior. The strategic plan must include a method to get the hazardous material back into a safe container, dispose of it, neutralize it, or allow it to dissipate safely.

> The specific action plan must identify the method of hazard control and identity the resources available and/or required to accomplish this goal. It may be necessary to select one method over another due to the unavailability of a particular resource or to adopt a "holding action" to wait for needed equipment or supplies.

Figure 1. Continued.

348   JOHN R. CASHMAN

The Action Plan Must Provide For:

1. Safety of Citizens
2. Safety of Fire Fighters
3. Evacuation of Endangered Area if Necessary
4. Control of Situation
5. Stabilization of Hazardous Materials, and/or
6. Disposal or Removal of Hazardous Material.

Avoid committing personnel and equipment prematurely or "experimenting" with techniques and tactics. Many times it is necessary to evacuate and wait for special equipment or expert help.

Control of Hazardous Area

A hazardous material incident has two zones associated with the scene. They are the *HAZARD ZONE* and the *EVACUATION ZONE*.

Hazard Zone

The Hazard Zone is the area in which personnel are potentially in immediate danger from the hazardous condition. This is established by Command and controlled by the Fire Department. Access to this area will be *rigidly* controlled and only personnel with proper protective equipment and an assigned activity will enter. All companies will remain intact in designated staging areas until assigned. Personnel will be assigned to monitor entry and exit of all personnel from the Hazard Zone. The Hazard Zone should be geographicall described to all responding units if possible.

Responsibility for control of personnel in this zone includes not only Fire Department personnel, but any others who may wish to enter the Hazard Zone (Police, press, employees, tow truck drivers, ambulance personnel, etc.). *Command is responsible for everyone's safety.*

Evacuation Zone

The Evacuation Zone is the larger area surrounding the Hazard Zone in which a lesser degree of risk to personnel exists. All civilians would be removed from this area. The limits of this zone will be enforced by the Police Department based on distances and directions established in consultation with Command. The area to be evacuated depends on the nature and amount of the material and type of risk it presents to unprotected personnel (toxic, explosive, etc.).

In some cases it is necessary to completely evacuate a radius around a site for a certain distance (i.e. potential explosion).

In other cases it may be advisable to evacuate a path downwind where toxic or flammable vapors may be carried (and control ignition sources in case of flammable vapors).

NOTE: When toxic or irritant vapors are being carried downwind it may be most effective to keep everyone indoors with windows and doors closed to prevent contact with the material instead of evacuating the area. In these cases Police and EMS would be assigned to patrol the area and evacuate persons with susceptibility to respiratory problems.

In all cases the responsibility for safety of *all* potentially endangered citizens rests with *Command*.

Use of Non-Fire Department Personnel

In some cases it may be advantageous to use non-Fire Department personnel to evaluate hazards and perform certain functions for which they would have particular experience or ability. When such personnel are outfitted with breathing apparatus, Chemical suits, etc., they must be made aware of the functions, limitations and safety precautions necessary in their use. Fire Department personnel with the necessary protective equipment must closely monitor and/or accompany such personnel for safety.

Figure 1. Continued.

BE AWARE THAT COMMAND IS RESPONSIBLE FOR THE SAFETY OF ALL PERSONNEL INVOLVED IN ANY ACCIDENT.

General Factors to Consider

Due to the wide variety of situations Fire Department personnel may encounter in dealing with hazardous materials, these considerations will not attempt to provide specific guidelines on any one individual chemical or situation and are not listed in any priority.

It is *imperative* that the first arriving Fire Department unit determine what hazardous material(s) is involved, and how much, prior to taking action to stabilize the incident.

Entering the scene to make positive identification may be a considerable risk. The danger of explosions, leaking gas and poisoning may be great.

Action taken prior to determining the product involved may be totally wrong and may severely compound the problem.

Transportation emergencies are often more difficult than those at fixed locations. The materials involved may be unknown, warning signs may not be visible, or obscured by smoke and debris, the driver may be killed or missing. D.O.T. hazardous materials marking systems are inadequate because some hazardous materials in quantities up to 1000 lbs., do not require a placard and there may be combinations of products involved with only a "dangerous" label showing. Sometimes only the most evident hazard is identified, while additional hazards are not labelled.

The following items may be significant to consider at any Hazardous Materials incident. (Not all will be significant at any particular incident.)

1. Cooling Containers
   a. Use adequate water supply
   b. Apply heavy streams to vapor space
   c. Use unmanned streams
   d. Use natural barriers to protect personnel

2. Remove Uninvolved Materials
   a. Move individual containers
   b. Move tank cars away from flame
   c. Cool containers before moving

3. Stop the Leak
   a. Close valves
   b. Place plug in openings
   c. Place container in upright position
   d. Use water spray to approach leak

4. Apply Diluting Spray or Neutralizing Agent
   a. Dilute water-soluble liquids
   b. Flush corrosives to reduce vapor
   c. Use fog streams to absorb vapor
   d. *Use water with caution on some materials*

5. Construct Dams, Dikes or Channels
   a. Direct running liquid away from exposures
   b. Control run off from corrosive materials
   c. Use sand or dirt

6. Remove Ignition Sources
   a. Start down wind
   b. Eliminate all sources of heat, spark, friction

7. *Call for Additional Resources* when their need is only anticipated. The *actions* taken by Command in the first few minutes of an incident affect the outcome more than any other single factor.

Figure 1. Continued.

The dispatcher has a reference list of personnel and organizations which may be helpful during a Hazardous Materials Emergency.

These include:

1. Fire Department Hazardous Materials Team.
2. Authorities in charge of landfills and dumps where Hazardous Materials may be disposed.
3. Commercial chemical experts with experience in handling and disposing of most common chemicals.
4. Pesticide consultants and disposal teams with equipment to clean-up agricultural chemical spills.
5. Personnel from State and Federal Regulatory Agencies. These personnel should be contacted for incidents involving transportation of Hazardous Materials.
6. Railroad information numbers.
7. Tank Truck Companies with defueling capability (in case carrier involved in incident has none).
8. Radioactivity and Military Weapons emergency contacts.

Conclusion

This standard is written to insure safe, professional action by the District Fire Department while operating at Hazardous Materials incidents.

Figure 1. North Charleston District Fire Department Hazardous Materials Response Standards

Written response standards vary widely but all represent an effort to provide a general framework for resonding to hazardous materials incidents without dictating specific tactics or control measures. A response standard may be written by and for a specific agency or be an interagency agreement detailing general guidelines agreed upon by two or more organizations. A critical consideration is that all personnel must be thoroughly trained in the use of a written response standard so that questions can be identified and amplifications provided when and where necessary. The standard must be available to all persons affected by it; a response standard is not a headquarters document but rather an all-hands protocol routinely carried on all vehicles used for response.

Written hazardous materials response standards utilized by various agencies tend to recognize incidents as a somewhat unique type of emergency by focusing attention on some or all of the following areas.
   • Preplanning of Industrial Facilities and Transportation Routes.
   • Definition of Terms.
   • Cautious Evaluation and Size-Up.
   • Safety of Personnel.
   • Thoroughness of Response.
   • Control of the Scene.
   • Communications.
   • Duties and Responsibilities of Command Personnel.
   • Command Post Operation.
   • Utilization of a Staging Area.
   • Use of Specialized Equipment.
   • Correct Reporting of Critical Information.
   • Relationships On-Scene With Other Agencies or Individuals.
   • Evacuation Considerations.
   • Assistance Available from Government, Industry, and Commercial Contractors.
   • Public and Media Relations.
The Philadelphia Fire Department uses a series of written response standards related to hazardous materials. One example is the standard for *Hazardous Chemicals and Materials*.

(See Figure 2) In addition, Philadelphia provides its first responders with individual operational procedures in the following areas.

- *Disposal of Hazardous Materials*
- *Petroleum Properties and Chemical Plants.*
- *Hazardous Materials Identification On The Baltimore & Ohio Railroad.*

- LNG/LPG Emergencies.
- *Conrail Hazardous Materials Identification System.*
- *Radiological Incident Procedure.*

Subject: Hazardous Chemicals and Materials

I. Purpose

To provide guidelines for Philadelphia Fire Department operations at incidents involving hazardous chemicals and/or materials.

II. Responsibility

It will be the responsibility of each member to exercise the appropriate control dictated by his rank in the implementation of this Operational Procedure.

III. Definitions

A. Hazardous Chemicals Task Force (HCTF)

A task force consists of an Engine, Ladder, Chemical Unit and a Foam Unit housed together and manned with specially trained personnel, designed to respond as a single entity on all incidents involving hazardous chemicals or materials. Designated HCTF's are (E.7—L.10—E.107—Chem.3), (E.19—L.8—E.119—Chem.2), E.60—L.19—E.160—Chem.1).

In those situations when all required components are not available from the same station, the Operations Officer, FCC, will consider the most expedient dispatch, based on circumstances, to insure that the four required components necessary to complete an HCTF are dispatched.

NOTE: Chemical 3 will be dispatched for all chlorine related incidents.

Typical responses would include, but not be limited to:

1. Chlorine or ammonia leaks.
2. Refinery fires.
3. Radiological incidents.
4. Carbon Monoxide incidents.
5. Chemical spills and/or explosions.
6. Petroleum products transportation fires.

B. Staging Area

A designated location outside the perimeter of the incident. First aid equipment, standby manpower and logistical support will be marshalled here.

C. Chemtrec

An agency that can be reached through the FCC where hazardous material information is available on a 24 hour a day basis.

IV. Procedures

A. General

1. Pre-planning

Hazardous material locations will be pre-planned by the local company. Required forms will be updated on an annual basis and station exercises will be conducted on all platoons to familiarize the members with conditions and to discuss specific fire fighting operations

Figure 2. Continued.

that may be encountered. Where necessary, due to the size or complexity of the facility, pre-planning tours will be coordinated through the Assistant Chief, FFF. When a facility is large enough to require pre-planning, copies of the pre-plans will be forwarded to all three (3) HCTF engine companies who will maintain a separate book containing this information.

2. Communications

Communications will be maintained at all times between operating units and the FCC.

As wind direction and velocity are extremely important in relation to chemical spills or leaks, the following format will be used by the FCC when giving wind conditions.

"Wind from the North to South at 12 MPH" or "Wind from East to West at 6 MPH", etc.

The FCC Supervisor will insure that when obtaining weather information the actual readings at time of request are recorded. The fireground commander will be appraised of any changes that might affect fireground operations.

The FCC Operations Officer will closely monitor the situation and if in his opinion, the number of casualties warrant PREMDOP Hospital Alert and three (3) alarms have not been struck, he will notify the Municipal Switchboard to initiate a PREMDOP alert.

B. Standard Operating Procedures

1. First arriving units will size up the situation and give a complete report to the FCC. If it is determined that a hazardous chemical or material is involved, the FCC will dispatch the nearest available HCTF.

2. Until proper identification of the product or material has been made, it should be considered toxic and explosive.

3. Members should *ANTICIPATE* and not delay in calling for assistance as a limited situation can quickly become a major problem if not handled expeditiously. If evacuation is deemed necessary, it should be started immediately, moving those closest to the problem first and working away from the incident.

4. All protective clothing, including breathing apparatus, will be worn in handling these incidents. If initial dispatch indicates a hazardous chemical, Scott Paks will be donned before entering the contaminated area.

5. The first arriving chief will assume command of the operation until properly relieved. He will also designate the staging area. Points to be considered in selecting a staging area would include wind direction and velocity, topography and accessibility.

6. The Fireground Commander will coordinate the establishment of a secure perimeter and the control of site access with the Police Commander on the scene.

7. Subsequent arriving units will, in the absence of specific instructions, report to the staging area.

8. The second arriving chief will assume command of the staging area.

9. Chief officers will be cognizant of available monitoring equipment (radiological, vapo-tester, etc.) and utilize them to best advantage.

10. Use of hose streams for flushing, cooling or absorption should be considered and stretched where indicated.

11. Fires in flammable gases should not be extinguished unless the flow of gas can be stooped.

C. Special Conditions

1. Underground Garages—When the FCC is notified of a Carbon Monoxide problem in an underground garage, a full box will be struck including a HCTF, rescue squad and a Deputy Chief with a CO analyzer. After taking CO readings, the garage will be im-

Figure 2. Continued.

mediately evacuated if a reading of 100 PPM (parts per million) or greater registers. The time of the evacuation order will be noted by the commanding officer, as will the "Fire Under Control" time. These times will be placed in the Remarks Sections of the Fire Report by the 1st in Engine Company.

Fire and Police personnel will be utilized in evacuating all persons from the garage.

When the evacuation order is given, all cash collections will stop and entrance lanes will be converted to exit lanes. Emergency lights, activated by intermediate CO detectors, will start flashing in the garage area before high concentrations of CO build up.

2. Gases

  a. Chlorine—A toxic, pungent, irritating greeinsh-yellow gas that is non-combustible in air. While non-flammable, it can react with many organic materials corrosively and, in some instances, explosively particularly with acetylene, turpentine, ether, gaseous ammonia, hydrocarbons, most fuel gases and finely divided metals. It is toxic enough to be considered as a poison and in its liquid state can burn skin. It is approximately 2½ times heavier than air. Stop leak if without risk. Use water spray to reduce vapors. Isolate area until gas has dispersed.

  b. Ammonia—A toxic, pungent, irritating, corrosive and flammable gas. (ER 16 to 25%) It is lighter than air (.60) and the presence of oil or other combustible materials increase the fire hazard. Stop flow of gas. Use water spray to keep fire exposed containers cool, to reduce vapors, and to protect men effecting the shut-off. In the event of an explosion, the minimum safe distance from flying fragments according to the Department of Transportation is 2000 ft. in all directions. Ammonia is water soluble. Runoff to storm sewers is acceptable if water deluge or flooding is possible.

  c. Carbon Monoxide (CO)—Colorless, odorless gas (Vapor Density = 1.0) produced by incomplete combustion at fires or as an exhaust product of internal combustion engines. It is highly flammable with limits ranging from 12.5 to 74% and is toxic.

3. Equipment

  a. Vapo-tester—Carried by Battalion Chiefs. A portable instrument designed to give fast and accurate indications of safe or unsafe concentrations of flammable gases and/or vapors, when mixed with air, in terms of their explosibility. (See F.T.M. for detailed description and operating instructions.)

  b. Vaportron—A hand held instrument carried by Fire Marshals and used to detect the presence of dangerous vapor accumulations. It can be used to detect virtually any hydrocarbon vapor as well as select inorganic vapors. This unit gives off a tone which increases in pitch when there is an increase in gas concentrations.

  c. Chlorine Capping Device—There are three types of chlorine capping devices carried on Chemical 3—#1 Kit "A" for 100 and 150 lb. chlorine cylinders, #2 Kit "B" for 1 ton chlorine containers, #3 Kit "C" for 16, 30 and 50 ton chlorine tank cars. These units are used to cap containers or to plug leaks in chlorine containers. (See F.T.M. for detailed description and operating instructions.)

  NOTE: Repairs to items a, b and c will be made via normal repair procedures.

  d. Carbon Monoxide Analyzer—Carried by Deputy Chiefs. This instrument is a direct-reading, portable, AC or battery operated gas analyzer. It functions as a survey or spot-check monitor of the working environment. (See F.T.M. for detailed description and operating instructions.) Analyzers will be calibrated by the City Air Management Laboratory once a month.

  Deputy Chiefs will see that their CO analyzer is delivered to Battalion 10 on the day prior to the scheduled calibrating day. Battalion 10 will deliver the CO analyzer to Air Management as early in the morning as possible, and pick up the unit the next day, seeing that the calibrated unit is then returned to the Deputy Chief as soon as possible.

Figure 2. Continued.

Deputy Chief 1—First Tuesday of each month.
Deputy Chief 2—Second Tuesday of each month.
Deputh Chief 3—Third Tuesday of each month.

The above schedule is for calibrating only. Battery replacement or repairs will be made as needed by delivering the unit to Batallion 10.

V. Guidelines

A. Training

The HCTF's will be trained annually by Fire Training College personnel.

B. Record Keeping

1. HCTF's will maintain an individual record of activities that will include:

   a. Date

   b. Time

   c. Location

   d. Units Responding

   e. Service Time

   f. Summary of Activities

VI. References

For more detailed information on this subject the following texts are recommended.

A. Fire Protection Guide on Hazardous Materials—NFPA.

B. Fire Protection Handbook—NFPA.

C. Hazardous Materials—Emergency Action Guide—Department of Transportation.

D. Training Manual—Philadelphia Fire Department.

E. Operational Procedure 12—Railroads—Philadelphia Fire Department.

Figure 2. Philadelphia Fire Department Operation Procedure 2. February 1978.

A written standard used by the Memphis Fire Department (See Figure 3) seeks to minimize exposure of responding personnel until the materials involved and general incident conditions have been evaluated. The first arriving company is expected to establish the primary command post; secure the area; and attempt to identify the materials involved. Unless life hazard or fire potential exists which would require immediate action on the part of the first arriving company, additional action is postponed until the arrival of specially teamed and trained fire department hazardous materials "REACT" squads that specialize in response to and control of hazardous materials incidents.

BASIC CONSIDERATIONS ON-SCENE:

The initial size-up or evaluation of a hazardous materials incident is a crucial component of effective response. Size-up includes detection and identification of hazardous materials; evaluation of fire, explosion, reaction, health and environmental hazards and impacts; immediate action; (action taken prior to initiation of supression and control activities—rescue, request for outside assistance, estimation of danger areas, definition of potential secondary emergencies, identification of critical exposures, initiation of evacuation efforts, etc.) and containment/control efforts.

I. Fire Alarm Office

    A. Who is to be dispatched?

        1. Pumper Company
        2. District Chief
        3. REACT Fire Squad
        NOTE: Notify Deputy Chief. He is dispatched at the discretion of the Fire Alarm Office.

    B. Information given to responding units.

        1. Location (the exact number related to street corner, intersections, cross streets, or other physical land markers, such as large buildings, signs, etc.)
        2. Disposition (if known)

            a. Type of hazardous material involved
            b. Visible activity showing (smoke, fire, etc.)
            c. Type of vehicle involved (truck, train, etc.)
            d. Type of placarding (flammable, corrosive, etc.)
            e. Traffic control information (streets blocked, etc.
            f. Weather report (rain, wind direction, etc.)
            g. Any other helpful information (nearby fuel farm, school, hospital, etc.)

II. First Arriving Units

    A. Position apparatus upwind, at a safe distance (300 feet) and headed in opposite direction from incident

    B. Information to Fire Alarm Office

        1. Location
        2. Disposition
        NOTE: When arriving on the scene, if information given initially is correct, then no amplification or clarification need be made. Simply report on the scene and give a short disposition. However, if location or disposition is different, go ahead and give up-dated information.

III. Approaching the Incident

    A. Determine if the incident can be safely approached through observation and any other available information

        1. If unsafe to approach—

            a. Notify the Fire Alarm Office of any unsafe condition (leakage of class "A" poisonous gas, trailer of explosives on fire, BLEVE, etc.).
            b. Move company and apparatus to a safer area.
            c. Designate staging area for next-arriving apparatus.
            d. Continually up-date information to Fire Alarm Office.
            e. First arriving chief officer will respond to location.
            f. The possibility of a large scale evacuation should be considered at this time; however, immediate area evacuation should be started at once.

        2. If an approach can be made—

            a. Do it with 2 men.
            b. Wear full turn-out (protective) clothing.
            c. Wear air mask.
            d. Use natural barriers (hills, gullies, etc.)

        3. Secure shipping papers, wheel report, or lay lines for heavy stream appliances. (For immediate self-protection, any available options may be used, as 1½ light water line, CO-2, dry powder, booster line).

Figure 3. Continued.

IV. Set Up Command Posts

    A. Primary Command Post (Key Personnel)

        1. Fire Department

            a. Director of Fire Services
            b. Deputy Director
            c. Staff Officers

        2. Mayor

        3. Law enforcement agencies

        4. Director of Civil Defense

        5. Shipper representatives

        6. Manufacturer representatives

    B. Secondary Command Post

        1. Support Personnel

            a. Public Works
            b. Light, gas, and water
            c. U.S. Coast Guard
            d. News Media
            e. Hospital representatives

        2. Support Equipment (Fire Department)

            a. A-6
            b. Air truck
            c. Hose tender
            d. Dump truck
            e. Front-end loader
            f. Helicopter
            g. Off-loading fuel tankers
            h. Wreckers

Figure 3. Memphis Fire Service Transportation Incident Guidelines Hazardous Material.

1. Identification of the Commodity.

First responders often arrive on-scene knowing only that "something" is causing a serious problem. Only after a commodity has been accurately identified can response personnel determine the proper tools, equipment, materials, methods and procedures to safely deal with that particular commodity. Because problems with identification arise most commonly during transportation, the U.S. Department of Transportation has placed stringent requirements on shippers and carriers relative to the use of hazard information labels, placards, identification numbers and shipping papers.

For emergency response personnel labels, placards, identification numbers and shipping papers are a primary source of identification of hazardous materials involved in transportation-related incidents. Labels are four-inch square diamonds that depict an internationally recognized symbol, a one-digit United Nations class number, the D.O.T. hazard class, and a distinctive coloring of the label background for each hazard class; labels must be provided for any regulated hazardous material contained within a package, overpack or freight container offered for transportation. (Figure 4 shows the symbol and United Nations class number used on labels for various D.O.T. hazard classes of regulated materials.) Labeling requirements are extensive; any person responsible for labeling a regulated hazardous material for transportation should refer to the *Code of Federal Regulations,* Title 49, Part 172, Sections 172.400 through 172.448.

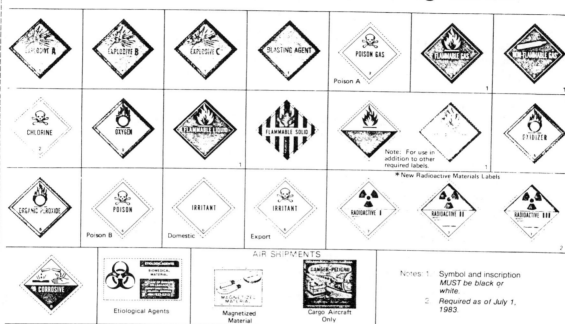

DOT Hazardous Materials Warning Labels

General Guidelines on Use of Labels

1. The Hazardous Materials Tables, Sec. 171.101 and 172.102, identify the proper label(s) for the hazardous materials listed.

2. Any person who offers a hazardous material for shipment *must label* the package, if required. [Sec. 172.400(a)]

3. Labels *may* be affixed to packages (even though not required by the regulations) provided *each* label represents a hazard of the material in the package. [Sec. 172.401]

4. Label(s), when required, *must* be printed on or affixed to the surface of the package near the proper shipping name. [Sec. 172.406(a)]

5. When two or more different labels are required, display them next to each other. [Sec. 172.406(c)]

6. When two or more packages containing compatible hazardous materials are packaged within the same overpack, the outside container *must* be labeled as required for each class of material contained therein. [Sec. 172.404(b)]

7. Material classed as an **Explosive A, Poison A,** or **Radioactive Material** also meeting the definition of another hazard class *must* be labeled for *each* class. [Sec. 172.402(a)]

8. Material classed as an **Oxidizer, Corrosive, Flammable Solid,** or **Flammable Liquid** that also meets the definition of a Poison B *must* be labeled POISON, in addition to the hazard class label. [Sec. 172.402(a)(3) and (5)]

9. Material classed as a **Flammable Solid** that also meets the definition of a water-reactive material *must* be labeled with FLAMMABLE SOLID and DANGEROUS WHEN WET labels. [Sec. 172.402(a)(4)]

10. Material classed as a **Poison B, Flammable Liquid, Flammable Solid,** or **Oxidizer** that also meets the definition of a Corrosive material *must* be labeled CORROSIVE in addition to the class label. [Sec. 172.402(a)(6) through (9)]

This Chart does not include all of the labeling requirements. For details, refer to the Code of Federal Regulations, Title 49, Part 172, Sec. 172.400 through 172.448.

U.S. Department of Transportation

Research and Special Programs Administration
Materials Transportation Bureau
Office of Operations and Enforcement
Washington, D.C. 20590

Chart 7   June 1981

Figure 4.

With some exceptions, each motor vehicle, rail car and freight container carrying hazardous materials must be placarded on each end and side with square-on-point (diamond-shaped) placards that measure 10¾ inches on each side, the outer one-half inch of which must be white. Although the requirements for labels and placards differ in a number of instances, placards also depict an internationally recognized symbol, a one-digit United Nations hazard class number, the D.O.T. hazard class, and a distinctive coloring of the placard background for the D.O.T. hazard class of a regulated hazardous material. For detailed placarding requirements, one should refer to the *Code of Federal Regulations,* Title 49, Part 172, Sections 172.500 through 172.558. However, depicted by Figure 5 are diagrams of 18 placards and basic requirements for their use excerpted from material distributed by the Materials Transportation Bureau of the U.S. Department of Transportation.

For a number of years now, labels and placards such as those described above have been required to be displayed on hazardous materials offered for transportation. More recently, the Department of Transportation has required that four-digit identification numbers also be used in certain circumstances. In selecting the specific identification system now in use, (a combination of four-digit identification numbers, labels and placards) nine factors were considered by the Department of Transportation.

- Capability of the general public to recognize the existence of the immediate dangers presented by a material.
- Presentation of information in a manner so that the general public would be able to accurately transmit basic information to response personnel.
- Compatibility, intermodally and internationally.
- Compatibility of application to both bulk and non-bulk shipment.
- Capability of functioning without use of a manual or other subsidiary documents.
- Capability to meet the needs of emergency response personnel, carriers, shippers and the general public.
- Capability of integration with documentation, packaging and vehicle identification requirements to help insure accuracy.
- Capability of implementation without undue economic burden on shippers and carriers.

In addition, the Department of Transportation sought a system that would have the following characteristics.

- It must be immediately recognizable as a system describing the existence of a hazard.
- It must be readily understandable by anyone who comes in contact with it.
- The information gained by anyone who comes in contact with it must be readily communicable to all others with a need to know.
- For those with a greater degree of ability, it must be indicative of a response pattern applicable to the particular incident.
- It must be capable of indicating degree of hazard.
- It must be uniform among all transportation modes.
- It must be capable of accomodating new hazardous commodities and, for higher levels of sophistication, it would be desirable that it be capable of easy computer implementation.

The Department of Transportation ultimately decided to require use of the United Nations four-digit numbering system to supplement the information provided by previously required placards and labels. A new column (Column 3A) was added to the *Hazardous Materials Shipping Table* contained in *Code of Federal Regualtions—49* (Part 172.101) that is a basic source of information for shippers when offering for transportation any of the approximately 1600 hazardous materials that are regulated by CFR-49. (See Figure 6 for a sample from the "Hazardous Materials Shipping Table.")

Presently, display of the United Nations four-digit identification number (listed for each regulated material in column 3A of Table 172.101) is required on all shipping papers and on the outside of packages of 110 gallons capacity or less. In addition, the U.N. number must be displayed on orange panels affixed to portable tanks, cargo tanks, and tank cars; that is, on bulk shipments of regulated materials. Requirements for use of U.N. numbers as described

DOT Hazardous Materials Warning Placards

*Numbers in each square (illustration numbers) refer to TABLES 1 and 2.

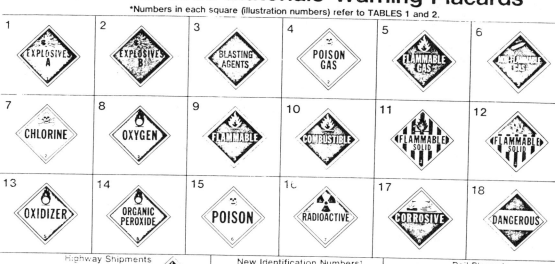

RADIOACTIVE MATERIAL PLACARD (Square Background): Use on "large quantity" shipments of radioactive materials requiring *special routing*. NOTE: Required for use Feb. 1, 1982.

IDENTIFICATION NUMBER—The four-digit ID number is found in the Hazardous Materials Table Sec. 172.101, Column 3A and the Optional Hazardous Materials Table Sec. 172.102 Column 4. They are used on orange panels, certain placards and on shipping papers.

PLACARD—When ID numbers are used on placards, the ORANGE PANEL is not required.

²**ORANGE PANEL**—When ID numbers are used on ORANGE PANELS, the appropriate placard is also required.

³NOTE: As of Nov. 1, 1981, ID numbers are required for use on cargo tanks, tank cars and portable tanks transporting hazardous materials. For details, see Sec. 172.332 through 172.338 and Sec. 172.519.

TABLE 1

Hazard Classes	*No.
Class A explosives	1
Class B explosives	2
Poison A	4
Flammable solid (DANGEROUS WHEN WET label only)	12
Radioactive material (YELLOW III label)	16
Radioactive material:	
Uranium hexafluoride, fissile (containing more than 0.7% U²³⁵	16 & 17
Uranium hexafluoride, low-specific activity (containing 0.7% or less U²³⁵	16 & 17

Guidelines

- Placard motor vehicles, freight containers, and rail cars containing *any quantity* of hazardous materials listed in TABLE 1.

- Placard motor vehicles and freight containers containing 1,000 pounds or more gross weight of hazardous materials classes listed in TABLE 2.

- Placard *any quantity* of hazardous materials classes listed in TABLES 1 and 2 when offered for transportation by air or water.

- Placard rail cars containing *any quantity* of hazardous materials classes listed in TABLE 2 except when less than 1,000 pounds gross weight of hazardous materials are transported in TOFC (Trailer on Flat Car) or COFC (Container on Flat Car) service.

TABLE 2

Hazard Classes	*No.
Class C explosives	18
Blasting agent	3
Nonflammable gas	6
Nonflammable gas (Chlorine)	7
Nonflammable gas (Fluorine)	15
Nonflammable gas (Oxygen, pressurized liquid)	8
Flammable gas	5
Combustible liquid	10
Flammable liquid	9
Flammable solid	11
Oxidizer	13
Organic peroxide	14
Poison B	15
Corrosive material	17
Irritating material	18

This Chart does not include all the placarding requirements. For details, refer to the Code of Federal Regulations, Title 49, Part 172, Sec. 172.500 through 172.558.

US Department of Transportation

Research and Special Programs Administration

Materials Transportation Bureau
Office of Operations and Enforcement
Washington D.C. 20590

Chart 7   June 1981

Figure 5.

(1) */E/W/W	(2) Hazardous materials descriptions and proper shipping names	(3) Hazard class	(3a) ID Number	(4) Label(s) required (if not expected)	(5) Packaging		(6) Maximum net quantity in one package		(7) Water shipments		
					(a) Exceptions	(b) Specific requirements	(a) Passenger carrying aircraft or railcar	(b) Cargo only aircraft	(a) Cargo vessel	(b) Passenger vessel	(c) Other requirements
	Accumulator, pressurized *(pneumatic or hydraulic),* containing nonflammable gas	Nonflammable gas	NA1956	Nonflammable gas	173.306	No limit	No limit		1,2	1,2	
	Acetal	Flammable liquid	UN1088	Flammable liquid	173.118	173.119	1 quart	10 gallons	1,3	4	
EA	Acetaldehyde ammonia *(RQ-1000/454)*	ORM-A	UN1841	None	173.505	173.510	No limit	No limit	1,3	5	
E	Acetaldehyde *(ethyl aldehyde)(RQ-1000/454)*	Flammable liquid	UN1089	Flammable liquid	None	173.119	Forbidden	10 gallons	1,2		
•E	Acetic acid *(aqueous solution)(RQ-1000/454)*	Corrosive material	UN2790	Corrosive	173.244	173.245	1 quart	10 gallons	1,2	1,2	Stow separate from nitric acid or oxidizing material.
E	Acetic acid, glacial *(RQ-1000/454)*	Corrosive	UN2789	Corrosive	173.244	173.245	1 quart	10 gallons	1,2	1,2	Stow separate from nitric acid or oxidizing materials. Segregation same as for flammable liquids
E	Acetic anhydride *(RQ-1000/454)*	Corrosive material	UN1715	Corrosive	173.244	173.245	1 quart	1 gallon	1,2	1,2	
	Acetone	Flammable liquid	UN1090	Flammable liquid	173.118	173.119	1 quart	10 gallons	1,3	4	
E	Acetone cyanohydrin *(RQ-10/4.54)*	Poison B	UN1541	Poison	None	173.346	Forbidden	55 gallons	1	5	Shade from radiant heat. Stow away from corrosive materials.01
	Acetone oil	Flammable liquid	UN1091	Flammable liquid	173.118	173.119	1 quart	10 gallons	1,2	1	
	Acetonitrile	Flammable liquid	UN1648	Flammable liquid	173.118	173.119	1 quart	10 gallons	1	4	Shade from radiant heat
	Acetyl benzoyl peroxide, solid	Forbidden									
	Acetyl benzoyl peroxide solution, *not over40% peroxide*	Organic peroxide	NA2081	Organic peroxide	None	173.222	Forbidden	1 quart	1,2	1	
E	Acetyl bromide *(RQ-5000/2270)* material	Corrosive	UN1716	Corrosive	173.244	173.247	1 quart	1 gallon	1	1	Keep dry. Glass carboys not permitted on passenger vessels.
E	Acetyl chloride *(RQ-5000/2270)*	Flammable liquid	UN1717	Flammable liquid	173.244	173.247	1 quart	1 gallon	1	1	Stow away from alcohols. Keep cool and dry. Separate longitudinally by an intervening complete compartment or hold from explosives.
	Acetyl iodide	Corrosive material	UN1898	Corrosive	173.244	173.247	1 quart	1 gallon	1	1	Keep dry. Glass carboys not permitted on passenger vessels.
	Acetyl peroxide solution, *not over 25% peroxide*	Organic peroxide	UN2084	Organic peroxide	173.153	173.222	Forbidden	1 quart	1,2	1	
	Acetylene	Flammable gas	UN1001	Flammable gas	None	173.303	Forbidden	300 pounds	1	1	Shade from radiant heat.
A	Acetylene tetrabromide	ORM-A	UN2504	None	173.505	173.510	10 gallons	55 gallons	1,2	1,2	Glass carboys in hampers not permitted under deck.
	Acid butyl phosphate	Corrosive material	UN1718	Corrosive	173.244	173.245	1 quart	5 gallons			
	Acid carboy empty: *See* Carboy empty.										
•	Acid, liquid, n.o.s.	Corrosive material	NA1760	Corrosive	173.244	173.245	1 quart	5 pints	1	4	Keep cool.
•	Acid, sludge	Corrosive material	UN1906	Corrosive	None	173.248	Fobidden	1 quart	1,2	1	
E	Acrolein, inhibited *(RQ-1/.454)*	Flammable liquid	UN1092	Flammable liquid and poison	None	173.122	Forbidden	1 quart	1,2	5	Keep cool. Stow away from living quarters.
	Acrylic acid	Corrosive material	UN2218	Corrosive	173.244	173.245	1 quart	5 pints	1	1	
E	Acrylonitrile *(RQ-100/45.4)*	Flammable liquid	UN1093	Flammable liquid and poison	None	173.119	Forbidden	1 quart	1,2	5	Keep cool.
	Actuating cartridge, explosive *(fire extinguisher or valve)*	Class C explosive	NA0276	Explosive C	173.114		50 pounds	150 pounds	1,2	1,2	Keep cool and dry
•	Adhesive, n.o.s. *See* Cement, liquid, n.o.s.		UN1133								
E	Adipic acid *(RQ-5000/2270)*	ORM-E	NA9077	None	None	173.510	No limit	No limit	2	2	
	Aerosol product, each aerosol container exceeding 50 cubic inches capacity. See Compressed gas, n.o.s.		UN1956								
•	Air, compressed	Nonflammable gas	UN1002	Nonflammable gas	173.306	173.302	150 pounds	300 pounds	1,2	1,2	
	Aircraft rocket engine *(Commercial)*	Flammable solid	NA2791	Flammable solid	None	173.238	Forbidden	550 pounds	1,3	5	
	Aircraft rocket engine igniter *(Commercial)*	Flammable solid	UN2792	Flammable solid	None	173.238	Forbidden	25 pounds	1,3	5	
	Airplane flare, See Fireworks, special										
•	Alcohol, n.o.s.	Flammable liquid	UN1987	Flammable liquid	173.118	173.125	1 quart	10 gallons	1,2	1	
•	Alcohol, n.o.s.	Combustive liquid	NA1987	None	173.118a	None	No limit	No limit	1,2	1,2	
EA	Aldrin, cast solid *(RQ-1/.454)*	ORM-A	NA2761	None	173.505	173.510	No limit	No limit			
EA	Aldrin mixture, dry, with 65% or less aldrin *(RQ-1/.454)*	ORM-A	NA2761	None	173.505	173.510	No limit	No limit			

Figure 6. Continued.

E	Aldrin mixture, dry *(with more than 65% aldrin)(RQ-1/.454)*	Poison B	NA2761 Poison	173.364	173.376	50 pounds	200 pounds	1,2	1,2	
E	Aldrin mixture, liquid *(with more than 60% aldrin)(RQ-1/.454)*	Poison B	NA2762 Poison	173.345	173.361	1 quart	55 gallons	1,2	1,2	If flash point less than 141 °F, segregation same as for flammable liquids.
EA	Aldrin mixture, liquid, with 60% or less aldrin *(RQ-1/.454)*	ORM-A	NA2762 None	173.505	173.510	No limit	No limit			
E	Aldrin *(RQ-1/.454)*	Poison B	NA2761 Poison	173.364	173.376	50 pounds	200 pounds	1,2	1,2	
	Alkaline corrosive battery fluid	Corrosive material	NA2797 Corrosive	173.244	173.249 173.257	1 quart	5 gallons	1,2	1,2	
	Alkaline corrosive battery fluid with empty storage battery	Corrosive material	NA2797 Corrosive	None	173.258	Forbidden	5 pints	1,2	1,2	
•	Alkaline corrosive liquid, n.o.s.	Corrosive material	NA1760 Corrosive	173.244	173.249	1 quart	5 gallons	1,2	1,2	
•	Alkaline liquid, n.o.s.	Corrosive material	NA1760 Corrosive	173.244	173.249	1 quart	5 gallons	1,2	1,2	
•	Alkanesulfonic acid	Corrosive material	NA2584 Corrosive	173.244	173.245	5 pints	1 gallon	1,2	1	
	Alkyl aluminum halides. See Pyrophoric liquid, n.o.s.		NA2221							
A	Allethrin	ORM-A	None	173.505	173.510	No limit	No limit			

Figure 6. Hazardous Materials Table

above are in addition to requirements for use of placards and/or labels. When U.N. numbers are used on placards, the orange panel is *not* required. For example, placarding/U.N. number requirements for acetone, a flammable liquid with a U.N. identification number of 1090, could be met in *either* of the following ways.

• An orange panel with the number 1090 *AND* the appropriate "Flammable" placard—

or

• A "Flammable" placard that contains the United Nations four-digit identification number for acetone. (1090)

The U.N. four-digit identification number (For example: 1090/acetone; 1165/dioxane; 1264/paraldehyde; 1466/ferric nitrate; etc.) should not be confused with the one-digit United Nations hazard class number which may be printed at the bottom of placards and labels. (For example: 1-Explosives; 2-Gases; 3-Flammable and Combustible Liquids; 4-Flammable Solids; 5-Oxidizers and Organic Peroxides; 6-Poisons; 7-Radioactive Materials; 8-Corrosives.) The four-digit United Nations identification number identifies specific types of commodities; the one-digit United Nations hazard class number identifies a broad class of hazardous materials. Response personnel can identify specific hazardous materials by the U.N. four-digit number, and obtain basic response information for a similar range of products, by using a handbook developed by the U.S. Department of Transportation, *Hazardous Materials—Emergency Response Guidebook*. (DOT—P 5000.2; Materials Transportation Bureau, Research and Special Programs Division, U.S. Department of Transportation, Washington, D.C. 20590. This handbook is also available for approximately $5.00 a single copy from a number of commercial suppliers.)

Any person shipping a hazardous material must properly prepare a shipping paper. A shipping paper may be a shipping order, bill of lading, manifest, or other shipping document containing the information described below.

• The proper shipping name of the material being offered for transport. (e.g. methyl ethyl ketone; thionyl chloride; etc.)
• The proper hazard class of the material being offered for transport. (e.g. Flammable Liquid; Corrosive Material; etc.)
• The four digit identification number of the material being offered for transport. (preceded by "U.N." or "N.A." as appropriate. The number is preceded by "U.N." if the description preceding it is exactly the same or sufficiently similar to the international description contained in the Intergovernmental Marine Consultive Organization (IMCO) code. If the description in Table 172.101 is significantly different but addresses the same material as a U.N. entry, it will be given the same number but will be preceded by "N.A." for "North American" in the United States and Canada. A four-digit number beginning with "9" (e.g.

NA 9194 for Oxidizer, Corrosive Solid, n.o.s.) indicates there is no corresponding U.N. description for the material assigned that number.)

- Except for empty packagings, the total quantity of the hazardous materials covered by the description.

For a tank car of acetone, the shipping paper entry might properly read—

1 Tank Car   Acetone   Flammable Liquid   UN 1090   85,000 lbs.

For ease of reference, on a shipping paper that lists both hazardous materials and non-hazardous materials, hazardous materials must be indicated by *one* of the following methods.

1. The hazardous materials must be entered first.
2. The hazardous materials must be entered in a contrasting color. (for a reproduction of shipping paper such as photocopy, the hazardous material(s) may be highlighted rather than printed in a contrasting color.)
3. The hazardous material(s) must be identified by an "X" entered before the proper shipping name in a special column captioned "HM."

Complete requirements for hazardous materials shipping papers are far more extensive than the selected requirements covered here. A person responsible for completing shipping papers must refer to the actual regulations contained in the *Code of Federal Regulations* Title 49, Part 100–177.

Labels, placards, four-digit identification numbers and shipping papers are primary sources of information about hazardous materials involved in transportation that are available to emergency resopnse personnel; but there are additional sources of information that responders must be aware of and know how to use. A few of these are reviewed below.

The railroads in the United States established and use a numerical, seven-digit "Standard Transportation Commodity Code" (STCC) to identify materials. STCC numbers appear on railroad waybills covering all shipments. An STCC number beginning with "49" indicates the commodity is a hazardous material. The third, fourth, and fifth digits identify potential hazards presented by the material while the final two digits identify the specific commodity. Thus, the middle three digits of a 49-series Standard Transportation Commodity Code number provide valuable information for railroad personnel and emergency responders. As an example, the middle three digits for Flammable Liquids run from 060 through 105 and deliniate potential secondary hazards for the broad hazard class of flammable liquids. That is, numbers beginning—

- 49–060 indicate pyrophoric flammable liquids;
- 49–062 indicate thermally unstable, poisonous flammable liquids;
- 49–064 indicate polymerizable, poisonous flammable liquids;
- 49–066 indicate thermally unstable, corrosive flammable liquids;
- 49–068 indicate polymerizable, corrosive flammable liquids;
- 49–070 indicate thermally unstable flammable liquids;
- 49–072 indicate polymerizable flammable liquids;
- 49–074 indicate poisonous flammable liquids;
- 49–076 indicate corrosive, acidic flammable liquids;
- 49–078 indicate corrosive, basic flammable liquids;
- 49–081 indicate flammable liquids with a flashpoint below 20 °F.

Thus, 49–060–60 is the STCC number for Pentaborane, a pyroforic flammable liquid. Anhydrous hydrazine; a thermally unstable, poisonous flammable liquid that is also corrosive, is identified by STCC number 49–062–25. As a final example, gasoline, a flammable liquid with a flashpoint below 20 °F is identified by STCC number 49–081–78. For other hazard classes as well, (Explosives, Nonflammable Compressed Gases, Flammable Solids, Oxidizers, Poisons, etc.) the middle three digits of the STCC number also deliniate potential secondary hazards.

To identify specific products by STCC 49-series number, and obtain basic information on proper emergency response procedures for each product, emergency response personnel utilize a 600 page handbook, *Emergency Handling Of Hazardous Materials In Surface Transportation,* published by the Bureau of Explosives of the Association of American Railroads and edited by Patrick J. Student. (Price: $12.50; available from Elizabeth P. Rabben Supervisor-Publication Services; Bureau of Explosives/Association of American Railroads, 1920 L Street, N.W.; Washington, D.C. 20036. Tel. 202-835-9500.) Alphabetical arrangement of this handbook by product name, plus numerical indexes for both United Nations four-digit codes and STCC seven-digit codes, allows the user to find response guidance for a particular commodity if any one of the following is known; the proper shipping name, the U.N. code, or the STCC 49-series code.

While the railroad waybills provide information on the contents of a specific car, another document known as a consist shows the position of each car relative to either the caboose or the power unit. A number of major railroads use computerized recordkeeping systems that in addition to other functions allow matching of waybill/consist/emergency response information. Thus, should the need arise, response guidelines pertaining to hazardous materials carried by specific cars in a specific train can be printed out and provided to railroad personnel or other emergency responders. Figures 7a, 7b, and 7c provide examples of emergency response outputs for specific hazardous materials cars printed in response to computer inquiries.

If shipping papers, placards, orange panels, United Nations four-digit numbers, Standard Transportation Commodity Code seven-digit numbers, waybills or consists are unavailable or nonproductive in identifying the hazardous contents of a specific rail car; the following could be potential sources of identifying information.

```
                         *** RESPONSE ***
CONRAIL EMERGENCY RESPONSE INQUIRY OUTPUT - TRAIN  ENTRY
  TRAIN -    PLA  18
  8622 FROM CABOOSE NATX 34253
  8621 FROM CABOOSE NATX 34343
  8622 FROM CABOOSE NATX 34313
 STCC CODE              - 4905731
 PROPER SHIPPING NAME - PROPANE
 HAZARD CLASS          - FLAMMABLE GAS
 PLACARD REQUIRED      - FLAMMABLE GAS
 PLACARD ENDORSEMENT   - DANGEROUS
     PROPANE IS A COLORLESS GAS WITH A FAINT PETROLEUM LIKE ODOR.  IT IS SHI
 GAS A LIQUEFIED GAS UNDER ITS VAPOR PRESSURE.  FOR TRANSPORTATION IT MAY BE
 ODTENCHED.  CONTACT WITH THE LIQUID CAN CAUSE FROSTBITE.  IT IS EASILY
 IGNITED.  ITS VAPORS ARE HEAVIER THAN AIR AND A FLAME CAN FLASH BACK TO THE
 SOURCE OF LEAK VERY EASILY.  THIS LEAK CAN BE EITHER A LIQUID OR VAPOR
 LEAK.  IT CAN ASPHYXIATE BY THE DISPLACEMENT OF AIR.  UNDER FIRE CONDITIONS
 THE CYLINDERS OR TANK CARS MAY VIOLENTLY RUPTURE AND ROCKET.
    IF MATERIAL ON FIRE OR INVOLVED IN FIRE
       DO NOT EXTINGUISH FIRE UNLESS FLOW CAN BE STOPPED
       USE WATER IN FLOODING QUANTITIES AS FOG
       COOL ALL AFFECTED CONTAINERS WITH FLOODING QUANTITIES OF WATER
       APPLY WATER FROM AS FAR A DISTANCE AS POSSIBLE
    IF MATERIAL NOT ON FIRE AND NOT INVOLVED IN FIRE
       KEEP SPARKS, FLAMES, AND OTHER SOURCES OF IGNITION AWAY
       KEEP MATERIAL OUT OF WATER SOURCES AND SEWERS
       ATTEMPT TO STOP LEAK IF WITHOUT HAZARD
       USE WATER SPRAY TO KNOCK-DOWN VAPORS
    PERSONNEL PROTECTION
       AVOID BREATHING VAPORS
       KEEP UPWIND
       WEAR PROTECTIVE GLOVES AND SAFETY GLASSES
       DO NOT HANDLE BROKEN PACKAGES WITHOUT PROTECTIVE EQUIPMENT
       APPROACH FIRE WITH CAUTION
    EVACUATION
       IF FIRE BECOMES UNCONTROLLABLE OR CONTAINER IS EXPOSED TO DIRECT FLAME -
          EVACUATE FOR A RADIUS OF 2500 FEET
       IF MATERIAL LEAKING (NOT ON FIRE), DOWNWIND EVACUATION MUST BE CONSIDERED
```

Figure 7a.

```
TRAIN 01063   09

HAZARDOUS MATERIAL HANDLING INSTRUCTIONS 07/10/78 1234 PST


  31 BEHIND POWER CCBX  2992
FLAMMABLE LIQUID, N.O.S.
    CLASSIFICATION:  FLAMMABLE LIQUID
    COMMODITY NUMBER IS:  4910185

    FLAMMABLE LIQUID, N.O.S. IS THE PROPER SHIPPING NAME FOR THOSE MATERIALS
HAVING A CLOSED CUP FLASH POINT OF LESS THAN 100 DEG. F. AND NOT SPECIFICAL-
LY MENTIONED IN THE HAZARDOUS MATERIALS REGULATIONS.
  IF MATERIAL ON FIRE OR INVOLVED IN FIRE
     DO NOT EXTINGUISH FIRE UNLESS FLOW CAN BE STOPPED
     USE WATER IN FLOODING QUANTITIES AS FOG
     SOLID STREAMS OF WATER MAY BE INEFFECTIVE
     COOL ALL AFFECTED CONTAINERS WITH FLOODING QUANTITIES OF WATER
     APPLY WATER FROM AS FAR A DISTANCE AS POSSIBLE
     USE "ALCOHOL" FOAM, CARBON DIOXIDE OR DRY CHEMICAL
  IF MATERIAL NOT ON FIRE AND NOT INVOLVED IN FIRE
     KEEP SPARKS, FLAMES, AND OTHER SOURCES OF IGNITION AWAY
     KEEP MATERIAL OUT OF WATER SOURCES AND SEWERS
     BUILD DIKES TO CONTAIN FLOW AS NECESSARY
     ATTEMPT TO STOP LEAK IF WITHOUT HAZARD
     USE WATER SPRAY TO KNOCK-DOWN VAPORS
  PERSONNEL PROTECTION
     AVOID BREATHING VAPORS
     KEEP UPWIND
     AVOID BODILY CONTACT WITH THE MATERIAL
     WEAR BOOTS, PROTECTIVE GLOVES, AND SAFETY GLASSES
     DO NOT HANDLE BROKEN PACKAGES WITHOUT PROTECTIVE EQUIPMENT
     WASH AWAY ANY MATERIAL WHICH MAY HAVE CONTACTED THE BODY WITH COPIOUS
        AMOUNTS OF WATER OR SOAP AND WATER
```

Figure 7b.

```
HAZARD INFORMATION EMERGENCY RESPONSE  0957 NOV 22/78


4918311

AMMONIUM NITRATE (NO ORGANIC COATING)                                    4?
OXIDIZER, THERMALLY UNSTABLE

** AMMONIUM NITRATE IS A WHITE CRYSTALLINE SOLID.  IT IS SOLUBLE IN WATER.
THE MATERIAL ITSELF DOES NOT READILY BURN BUT WILL DO SO IF CONTAMINATED
BY COMBUSTIBLE MATERIAL.  IT WILL ACCELERATE THE BURNING OF COMBUSTIBLE
MATERIAL.  TOXIC OXIDES OF NITROGEN ARE PRODUCED DURING COMBUSTION.

** IF MATERIAL ON FIRE OR INVOLVED IN FIRE
     FLOOD WITH WATER
     COOL ALL AFFECTED CONTAINERS WITH FLOODING QUANTITIES OF WATER
     APPLY WATER FROM AS FAR A DISTANCE AS POSSIBLE

** IF MATERIAL NOT ON FIRE AND NOT INVOLVED IN FIRE
     KEEP SPARKS, FLAMES, AND OTHER SOURCES OF IGNITION AWAY
     KEEP MATERIAL OUT OF WATER SOURCES AND SEWERS

** PERSONNEL PROTECTION
     WEAR BOOTS, PROTECTIVE GLOVES, AND SAFETY GLASSES
     DO NOT HANDLE BROKEN PACKAGES WITHOUT PROTECTIVE EQUIPMENT
     WASH AWAY ANY MATERIAL WHICH MAY HAVE CONTACTED THE BODY WITH COPIOUS
        AMOUNTS OF WATER OR SOAP AND WATER
     WEAR SELF-CONTAINED BREATHING APPARATUS
        WHEN FIGHTING FIRES INVOLVING THIS MATERIAL
     APPROACH FIRE WITH CAUTION

** EVACUATION
     IF FIRE BECOMES UNCONTROLLABLE - EVACUATE FOR A RADIUS OF 5000 FEET


END
```

Figure 7c.

For knowledgable responders, the shape of a tank car, the configuration of its dome, and whether or not a car has a certain type of bottom outlet valve arrangement provide clues to the types of products normally carried in a specific car. For this reason, response teams often obtain tank car manuals that provide basic information about the various types of tank cars used in the United States, including schematic diagrams of each type of car.

In excess of 40 hazardous materials transported in tank cars, including all flammable gases, require the name of the product be stenciled on the side of the car in letters four-inches high. (Anhydrous Ammonia, Chlorine, Formic Acid, Hydrogen, Liquefied Petroleum Gas, Nitric Acid, etc.) Color is not a reliable identifier of tank car contents; only cars dedicated to carriage of Hydrocyanic Acid are *required* to be painted with a standard, distinctive color scheme, a white background with a horizontal red stripe around the length of the car plus vertical red bands near each end of the car.

Tank car identification numbers, a combination of letters and numerals that identify each specific car and its owner, are required on the left end of every car. If such a number is the only identification available for a specific car and thus the product contained within the car, CHEMTREC (the Chemical Transportation Emergency Center in Washington, D.C. operated by the Chemical Manufacturers Association) may be able to identify the shipper or the owner and thus learn what commodity is currently contained within the car. Tank car identification numbers would look something like the following: NATX 3434, GATX 1406, etc.

The type and configuration of tank trucks can also provide clues to the type of commodity carried should all other identification be lacking. Motor carrier cargo tanks MC 306 and MC 307 are nonpressure vessels designed for products such as gasoline or fuel oil. MC 311 and MC 312 tanks are nonpressure vessels designed to carry corrosives; MC 330 and MC 331 tanks are pressure vessels built to haul commodities such as liquefied petroleum gas or anhydrous ammonia.

Once a hazardous material being transported or stored has been identified, the manufacturer or handler may be able to provide detailed information about the product's characteristics from *Material Safety Data Sheets* (MSDS) maintained for many hazardous materials due to Department of Labor/OSHA requirements. A Material Safety Data Sheet can provide one of the most complete descriptions of a chemical normally available to response personnel. When making preplanning visits or responding to incidents at facilities that manufacture, store, process, ship, carry, use or dispose of hazardous materials; responders should determine if MSDS forms are available for products present within the facility. (See Figure 8 for a sample Material Safety Data Sheet.) At least one commercial supplier (Information Handling Services, 15 Inverness Way East, Englewood, Colorado 80150. Tel. 303-779-0600) makes available on an annual subscription basis a comprehensive collection of current Material Safety Data Sheets (MSDS) acquired from the manufacturers of chemicals and chemical products. The MSDSs are produced on microfiche at a 24 to 1 reduction ratio yielding a maximum of 98 frames (pages) per microfiche card. Individual MSDSs are index-accessible to a precise location on a particular microfiche card if any of the following identifiers are known: United Nations or North American identification number, chemical name, synonym, trade name, brand name, Chemical Abstracts Registry (CAS) number, or the name of the supplier.

In attempts to identify hazardous materials involved in emergency situations, plant personnel, truck drivers, freight train conductors and other knowledgable persons can often supply crucial information. Also, CHEMTREC has been assisting emergency responders to identify unknown chemicals involved in transportation emergencies, and providing information on proper initial control actions, since 1971.

2. Dealing With Specific Hazard Class Materials.

Effective hazardous materials incident response requires that response personnel understand a variety of specific hazards and use a common language in identifying and defining such hazards. The paragraphs that follow provide basic information relative to

U.S. DEPARTMENT OF LABOR
Occupational Safety and Health Administration

Form Approved
OMB No. 44-R1387

MATERIAL SAFETY DATA SHEET

Required under USDL Safety and Health Regulations for Ship Repairing,
Shipbuilding, and Shipbreaking (29 CFR 1915, 1916, 1917)

SECTION I

MANUFACTURER'S NAME	EMERGENCY TELEPHONE NO.

ADDRESS (Number, Street, City, State, and ZIP Code)

CHEMICAL NAME AND SYNONYMS	TRADE NAME AND SYNONYMS

CHEMICAL FAMILY	FORMULA

SECTION II - HAZARDOUS INGREDIENTS

PAINTS, PRESERVATIVES, & SOLVENTS	%	TLV (Units)	ALLOYS AND METALLIC COATINGS	%	TLV (Units)
PIGMENTS			BASE METAL		
CATALYST			ALLOYS		
VEHICLE			METALLIC COATINGS		
SOLVENTS			FILLER METAL PLUS COATING OR CORE FLUX		
ADDITIVES			OTHERS		
OTHERS					

HAZARDOUS MIXTURES OF OTHER LIQUIDS, SOLIDS, OR GASES	%	TLV (Units)

SECTION III - PHYSICAL DATA

BOILING POINT (°F.)		SPECIFIC GRAVITY (H₂O=1)	
VAPOR PRESSURE (mm Hg.)		PERCENT, VOLATILE BY VOLUME (%)	
VAPOR DENSITY (AIR=1)		EVAPORATION RATE (_____ =1)	
SOLUBILITY IN WATER			
APPEARANCE AND ODOR			

SECTION IV - FIRE AND EXPLOSION HAZARD DATA

FLASH POINT (Method used)	FLAMMABLE LIMITS	Lel	Uel
EXTINGUISHING MEDIA			
SPECIAL FIRE FIGHTING PROCEDURES			
UNUSUAL FIRE AND EXPLOSION HAZARDS			

PAGE (1)   (Continued on reverse side)   Form OSHA-20
                                                                           Rev. May 72

Figure 8. Continued.

SECTION V · HEALTH HAZARD DATA

THRESHOLD LIMIT VALUE

EFFECTS OF OVEREXPOSURE

EMERGENCY AND FIRST AID PROCEDURES

SECTION VI · REACTIVITY DATA

STABILITY	UNSTABLE		CONDITIONS TO AVOID
	STABLE		
INCOMPATABILITY *(Materials to avoid)*			
HAZARDOUS DECOMPOSITION PRODUCTS			
HAZARDOUS POLYMERIZATION	MAY OCCUR		CONDITIONS TO AVOID
	WILL NOT OCCUR		

SECTION VII · SPILL OR LEAK PROCEDURES

STEPS TO BE TAKEN IN CASE MATERIAL IS RELEASED OR SPILLED

WASTE DISPOSAL METHOD

SECTION VIII · SPECIAL PROTECTION INFORMATION

RESPIRATORY PROTECTION *(Specify type)*

VENTILATION	LOCAL EXHAUST		SPECIAL
	MECHANICAL *(General)*		OTHER
PROTECTIVE GLOVES		EYE PROTECTION	
OTHER PROTECTIVE EQUIPMENT			

SECTION IX · SPECIAL PRECAUTIONS

PRECAUTIONS TO BE TAKEN IN HANDLING AND STORING

OTHER PRECAUTIONS

PAGE (2)
GPO 930-940

Form OSHA-20
Rev. May 72

Figure 8.

hazard classes, subclasses and types of dangers presented by specific classes of materials. Definitions commonly recognized by responders and used here are taken from the *Code of Federal Regulations,* Title 49, Parts 100–199.

Explosives refers to any chemical compound, mixture or device the primary or common purpose of which is to function by explosion, i.e., with substantially instantaneous release of gas or heat. (CFR-49, Section 173.50)

Class A Explosives are those that function by detonation or are otherwise of maximum hazard. There are nine subclasses of Class A Explosives (CFR-49, Section 173.53)

- Solid explosives such as black powder and low explosives.
- Solid explosives which contain a liquid explosive ingredient such as high explosives and commercial dynamite containing a liquid explosive ingredient.
- Solid explosives which contain no liquid explosive ingredient such as high explosives, commercial dynamite containing no liquid explosive ingredient, trinitrotoluene, amatol, tetryl, picric acid, etc.
- Solid explosives such as initiating and priming explosives, lead azide, fulminate of mercury, and certain high explosives.
- Desensitized liquid explosives such as desensitized nitroglycerine and certain high explosives.
- Liquid explosives such as nitroglycerine.
- Blasting caps. (Blasting caps in quantities of 1,000 or less are classified as Class C Explosives)
- Any solid or liquid compound, mixture or device not specifically included in any of the above types which may be designated a Class A Explosive. (shaped charges, ammunition for cannon, explosive projectiles, grenades, etc.)

Class B Explosives generally function by rapid combustion rather than by detonation and include some explosive devices such as special fireworks, flash powder, etc. The major hazard is flammability. (CFR-49, Section 173.88)

Class C Explosives are certain types of manufactured articles containing Class A or Class B explosives, or both, as components but in restricted quantities, and in certain types of fireworks; that present a minimum hazard. (CFR-49, Section 173.100)

Approximately two billion pounds of explosives are produced each year in the United States and widely distributed, and thus transported, throughout the country. On December 3, 1956 in Brooklyn, New York a pier fire caused the explosion of 37,000 pounds of explosive initiating cord killing ten and injuring 245. In June of 1964 at Marshall's Creek, Pennsylvania a tire fire on a truck carrying explosives resulted in the detonation of cargo that killed six and caused widespread property damage. At Richmond, Indiana on April 6, 1968 gunpowder stored in a sporting goods store ignited causing an explosion and fire that killed 41, injured more than 100, and destroyed approximately 15 buildings. A Western Pacific train carrying military bombs was wracked with explosions at Tobar, Nevada on June 29, 1969. On June 4, 1971 at Waco, Georgia an automobile crossed over the centerline and collided with a tractor-trailer carrying 25,414 pounds of explosives. Two firemen, a wrecker operator, and two bystanders died in the resultant explosion. At Roseville, California on April 28, 1973 a railcar containing military bombs exploded in the Southern Pacific railyards setting off explosions in approximately 18 additional cars loaded with military bombs, leveling the town of Antelope, injuring more than 50, and forcing the evacuation of all people within a two-mile radius. Near Benson, Arizona on May 24, 1973 a Southern Pacific freight hauling 12 boxcars of munitions exploded over a period of several hours. Three died in early 1978 when a gunpowder mixing plant exploded at DuPont's Carney's Point, New Jersey facility. In March of 1978 one died and seven were injured in an explosion at a commercial fireworks plant being operated in a house within a residential neighborhood. Six persons were injured September 14, 1978 when an explosion ripped through a plant manufacturing pyrotechnical equipment used in mining and construction. Nine workers were injured April 23, 1979 when a truckload of dynamite exploded at a quarry near Buffalo, New York. A number of homes were damaged and 16 persons injured at Keystone, West Virginia on April 5, 1979 when a six-wheel flatbed truck loaded with 1,000 pounds of dynamite and boxes of blasting caps ignited shortly after witnesses saw the truck stopped at a traffic light with smoke pouring from the rear end. Fireworks explosions on May 6, 1980 in Spangler, Pennsylvania; June 4, 1980 in Laramie, Wyoming; and July 29, 1980 in San Diego, California; killed a total of eight persons. In Shelbyville, Kentucky on January 29, 1981 a truck loaded with explosives overturned and exploded with a blast felt up to 25 miles away.

The potential destructive capacity of certain explosives, particularly Class A and Class B explosives, is so great that fires involving such cargos may present a "no fight" situation. Oftentimes the only action that can or should be taken is to evacuate all persons from within a 5,000 foot radius. Except in rare circumstances where major loss of life is a distinct possibility if no action is taken, efforts to extinguish fires involving larger quantities of Class A or Class B explosives are generally viewed as unwarranted. In some circumstances, water streams applied to explosives involved in fire can be the direct cause of an explosion. Industrial representatives tend to advise response personnel not to fight fires in cargos of explosives; rather, evacuate the area and let the cargo burn.

Explosive cargos not involved in fire can still be extremely dangerous. Various types of explosives can be ignited by impact, (shock sensitive) electrical initiation, (radio signals, lightening, static electricity) or heat ignition. (Common sources of heat that pose serious threats are exhaust systems and catalytic converters on vehicles.) Even grit or hobnails on shoes or boots can cause enough friction to set off certain explosives. Critical steps for incidents of spilled explosives not involving fire include every possible effort to prevent fire through control of ignition sources, initiation of severe site control measures to keep unauthorized persons out of the area, (only experts should be allowed anywhere near spilled explosives) immediate evacuation, and control of equipment. (extremely careful handling of shock-sensitive materials; control of metal-on-metal, or metal-on-stone friction by equipment; and avoidance of use of metal tools that can cause a spark and thus heat ignition.) In addition, spilled explosive materials not involved in fire should be kept moist.

Simple tire fires or carburator fires on motor vehicles carrying explosives, and journal box overheating on rail boxcars, can present extremely dangerous situations for emergency personnel responding to transportation incidents. "Only a tire fire" is wishful thinking until positive identification has been made of a transport vehicle's cargo.

Oftentimes, explosives do not *look* dangerous. Certain plastic explosives look like green or white putty; "Prima Cord" may look like simple plastic clothesline. There may be little or no smoke in the early stages of a fire involving explosives, yet a straight hose stream directed at such a fire could set-off a massive explosion due to shock or the action of steam on the product. Other explosives look like powder, sand, or even cornmeal. The variety of explosives used in construction, mining, farming and industry is great; explosives are manufactured, stored, transported or used in every part of the United States.

Flammable Liquids refer to any liquid with a flash point less than 100°F. (CFR-49, Section 173.115)

Combustible Liquids refer to any liquid with a flash point from 100 to 200°F., or any liquid mixture with 99 percent or more combustible components. (CFR-49, Section 173.115)

Pyroforic Liquids are any liquids that ignite spontaneously in dry or moist air at or below 130°F. (CFR-49, Section 173.115)

During the period 1976 through 1980 approximately 600 persons died and nearly 4,400 others were injured in tank truck incidents involving flammable and combustible liquids according to a recent report sponsored by the U.S. Department of Transportation. On September 4, 1977 a tank truck carrying 8,255 gallons of gasoline overturned in Beattyville, Kentucky. Escaping gasoline ignited killing seven and destroying six buildings and 16 vehicles. Massive quantities of flammable liquids are not required to cause extensive injuries and destruction. On June 20, 1978 an explosion caused by a 2½ gallon can of gasoline on an ice cream vending truck in New York City's financial district injured 106 persons, including 24 hospitalized with serious injuries. At Kenner, Louisiana at least seven died on November 25, 1980 when a gasoline truck failed to beat a train to a downtown rail crossing.

The term "flash point" (the flash point of a liquid is the minimum temperature at which it gives off sufficient vapor to form an ignitable mixture with the air near the surface of the liquid or within a vessel) is important in defining flammable/combustible liquids because the vapor burns rather than the liquid. For liquids with low flash points, the possibility of ignition is greater than for liquids with higher flash points. Also, the higher the temperature of the liquid on the surrounding air the more vapor will be formed thus increasing the danger of ignition.

Spills of flammable liquids are generally considered to be more hazardous if ignition does not occur immediately because both the unignited liquid and vapor can quite rapidly spread the hazard over a far greater area than if the material had been immediately ignited. In general, it is unwise to extinguish a flammable liquid fire until the flow of material can be stopped; otherwise, greater dispersion and reignition of flammable liquid/vapor is a dangerous possibility. It is crucial to recognize that successful response to a flammable liquid spill requires attention to both liquid and vapor. Earthen dikes can be used to control the spread of liquid. While liquids can travel only downhill, vapors are much more difficult to confine and can be spread quite quickly by air currents over wide areas where any ignition source can cause them to flash back to their point of origin. Water used in firefighting efforts will float nearly all flammable liquids spreading them about the area; thus application of water should be strictly controlled. Water fog can be used to disperse flammable liquid vapors, but the run-off must be strictly controlled and closely monitored since it may contain flammable levels of product. Although the vapors of gasoline and many other flammable liquids are heavier-than-air, such vapors can travel considerable distances with air currents and will tend to seek low-points in the terrain. Continuous monitoring with combustible gas indicators is necessary to determine combustible concentrations of vapor.

As with any fire involving hazardous materials, the first five minutes of a flammable liquid fire can be the most important. Every attempt should be made to shut off the source of the fuel, then contain the spread of liquid and vapor. Preplanning of flammable liquid facilities and transport vehicles can determine the location and operation of shut-off devices prior to an emergency. Utilize plant personnel who are familiar with a facility's valves, piping, and electrical and mechanical control systems. Firefighting foam can be used not only to extinguish flammable liquid fires but also to blanket vapors being released from a yet unignited flammable liquid. Often, control of a flammable liquid situation comes first; extinguishment of fire may be secondary to control of the commodity. Once the spread of a flammable liquid and its vapors has been controlled, response personnel tend to feel extinguishment of any resultant fire is comparitively easy.

Class A Poisons are extremely dangerous poisonous gases or liquids of such nature that a very small amount of gas, or vapor of the liquid, mixed with air is dangerous to life. Sample Poisons A are liquid bromoaceton, liquefied hydrocyanic acid, phosphine, parathion and compressed gas mixture, phosgene, etc. (CFR-49, Section 173.326)

Class B Poisons are less dangerous substances, liquids, or solids, (including pastes and semi-solids) other than Class A Poisons or Irritating Materials, which are known to be so toxic to man as to afford a hazard to health during transportation; or which in the absence of adequate data on health toxicity are presumed to be toxic to man. Class B Poisons include carbolic acid or phenol, a wide range of pesticides, liquid nicotine, arsenic trichloride, dry cyanide, and selenium oxide. (CFR-49, Section 173.343)

Irritating Materials (formerly called Class C Poisons) are liquid or solid substances which upon contact with fire or when exposed to air give off dangerous or intensely irritating fumes, but not including any poisonous material, Class A.

Although Poison A gases are handled and transported by both the military and by industry, potential exposure to Poisons A is somewhat more limited than potential exposure to Poisons

B simply due to the smaller volume and variety handled. Nevertheless, the volume of Poison A gases or liquids transported and handled in this country is substantial. Shipments of military poison gas, for example, may be transported in unescorted civilian vehicles, and any incident might initially be handled by local response personnel. Most military poison gases are transported in liquid form, although some are solids that are made poisonous by adding water. Military poison gases are of four main types. Nerve agents such as Sarin (a colorless liquid) are rapid acting but can be counteracted by the antidote, atropine. Blister agents such as distilled mustard (colorless to pale yellow liquid) or Lewisite (dark oily liquid) can kill by both inhalation and skin absorption and cause blisters inside the lungs as well as externally. Action may be either delayed (mustard) or rapid (Lewisite). Whereas the nerve agent, Sarin, mentioned above has no ordor, blister agents such as mustard or Lewisite have distinctive odors (Garlic and geranium respectively). Blood agents such as hydrogen cyanide (colorless gas or liquid) and cyanogen chloride (a colorless gas) follow the bloodstream once inhaled and work on interior tissue by starving cells to bring about the death of tissue. Hydrogen cyanide has an odor of bitter almonds while cyanogen chloride has a slight odor of bitter almonds. Choking agents such as phosgene (a colorless gas) or diphosgene (a colorless liquid) cause the lungs to fill with fluid, thus excluding oxygen, in a process popularly known as "dry land drowning." There are no antidotes for a choking agent. Both phosgene and diphosgene have a distinct odor of new mown hay or green corn.

Proper equipment, specialized training and prior planning will greatly affect the survival rate in a poison gas incident. Response to poison gas incidents is a job for experts, either military or industrial, who generally take the following steps.

- Always approach from upwind.
- Avoid visible concentrations of fumes, smoke or liquid.
- *Immediately* evacuate the area of unnecessary persons.
- Establish a "hot line" outside the evacuation area and upwind from it.
- Initiate rigid decontamination procedures for anything or anybody brought out of the evacuation area.
- Attempt to roll or upright a leaking container if liquid is leaking. Highly volatile agents such as hydrogen cyanide turn into vapors very quickly.
- Plug the leak if possible and safe to do so.
- Cool the container if possible to decrease escape rate of vapors.
- Dig a sump around the container.
- If possible, wrap container in plastic and seal, or seal container in overpack.

Industrial Poison A gases and liquids such as hydrocyanic acid or cyanogen chloride might be transported in single unit or multiunit tank cars or trucks or in cylinders. For example, cyanogen chloride can be shipped only in cylinders, while hydrocyanic acid can be shipped in either tank cars or cylinders. Tanks used to transport poison gas are not insulated and do not have safety release devices; therefore, lengthy exposure to fire or radient heat can cause cylinder or tank to violently rupture and rocket.

Many pesticides are Poisons B and may be toxic by inhalation, skin absorption or ingestion. The potential for exposure to pesticides is high. A number of pesticides are flammable as well as poisonous, and water can make many pesticides ultradangerous. Many pesticides can readily pass through the skin without causing any sensation or provide any warning; the area near the eyes is particularly susceptible. Solvents in pesticides solutions can be absorbed through clothing, even through rubber boots or protective garments, so repeated washdown of garments and footwear is a critical precaution during prolonged or repeated exposure to certain pesticides. Massive, prolonged exposure to a low toxicity pesticide can be as damaging as minor, short term exposure to a highly toxic pesticide. In addition, a number of pesticides are cholinestrerase inhibitors; they affect the ability of nerves to transmit impulses so that a pest, or a responder, will not realize he is being poisoned. The effect of pesticides is often cumulative so a small exposure one day may not present a serious hazard while prolonged or repeated exposure may produce a killing cumulative effect. Persons who routinely work around pesticides, or have prolonged or repeated exposure such as in fire situations, should obtain a cholinestrerase test to identify decreased levels of cholinestrerase.

Ingestion of pesticides may seem easy to avoid, yet response personnel at a pesticide fire or incident may remove their gloves to smoke a cigarette, or take a break for a cup of coffee and a doughnut—the end result is exactly the same. Smoke from burning pesticides provides a common route for poison inhalation. When exposed to pesticides, it is important to recognize that poisoning may have occurred. Effects are often delayed and cumulative. Common symptoms of pesticide poisoning are nausea, tearing, unusual sweating, stomach cramps, or trembling. *Any* unusual discomfort, illness or appearance can be a symptom of pesticide poisoning, and these symptoms can be delayed for up to 12 hours. A major difficulty in judging symptoms of pesticide poisoning is that many such symptoms are the same as for heat prostration, smoke inhalation and the flu. Pesticide poisoning, unless suspected and considered, can easily be misdiagnosed or overlooked. Anyone exposed to pesticides should routinely wash repeatedly with copius amounts of soap and water and not wear possibly contaminated clothing until it too has been laundered.

Labels on pesticide containers provide important information pertaining to safety and health. When seeking medical attention after pesticide exposure, take along an uncontaminated labelled container, if possible, to provide guidance for the examining physician. Typical pesticide label information of interest to physicians and emergency medical services personnel is provided below.

- *Carbaryl Insecticide:* "Carbaryl is a moderate, reversible, cholinestrerase inhibitor. Atropine is antidotal. Do not use 2-PAM, opiates, or cholinestrerase-inhibiting drugs."
- *Phosphamidon Liquid Insecticide:* "In case of skin contact, remove contaminated clothing and immediately wash skin with soap and water. For eye contact, wash eyes immediately for 15 minutes and see a physician. Wash hands, arms and face thoroughly with soap and water before eating and smoking. Wash all contaminated clothing with soap and hot water before reuse. Keep all unprotected persons out of operating areas or vicinity where there may be drift. ANTIDOTE: If Swallowed—Drink one to two glasses of water and induce vomiting by touching back of throat with finger or blunt object and repeat until vomit fluid is clear. Do not induce vomiting or give anything by mouth to an unconscious person. Have a victim lie down and keep quiet. CALL A PHYSICIAN IMMEDIATELY. Note to Physicians: Atropine is antidotal. 1-PAM is also antidotal and may be used in conjunction with atropine but should not be used alone."
- *Naled Insecticide:* "Note to Physician: Naled is a cholinestrerase inhibitor. Atropine is antidotal. 2-PAM is also antidotal and may be used in conjunction with atropine. Mucosal damage may occur, and if gastric lavage is indicated it should be performed cautiously . . ."

A number of pesticide manufacturers such as Chevron Chemical, Kerr-McGee, FMC Corporation, and Shell Chemical maintain emergency medical centers and personnel to provide toxicological data on their products through a 24-hour telephone number which is printed on labels and other documents. Such assistance is almost wholly dependent on proper identification of the product involved. The exact product name and brand name must be known. **GET THE LABEL.**

Pesticides are differentiated by toxicity levels; high, moderate, or low. The most toxic pesticides can be identified by a skull-and-crossbones insignia and the words, "DANGER-Poison." (Those exhibiting an acute oral lethal dose/50 percent of from zero to 50 mg/kg) Less toxic pesticides (Those exhibiting an acute oral lethal dose/50 percent of from 50 to 90 mg/kg) can be identified by the word, "WARNING" while the least toxic pesticides (Those with an acute oral lethal dose/50 percent of from 500 to 5,000 mg/kg) are identifiable by the word, "CAUTION." These signal words are set by law and provide a quick method of determining the relative toxicity of a particular pesticide.

Pesticide labels also contain an ingredients statement such as the sample for Endrin pesticide provided below.

ACTIVE INTREDIENTS
  Endrin (Hexachloroepoxyoctahydro-endo, endo-dimethanonaphthalene)........10.50%
  Xylene 76.00%
INERT INGREDIENTS 4.50%

contains 1.6 lbs. Endrin per gallon

The term "active" in an ingredients statement refers to function as a pesticide rather than to other characteristics such as flammability, although in the example of Endrin given above it should be noted that 76 percent of the solution is Xylene, a flammable liquid with a flash point of from 81 to 90°F. Thus, response personnel called in to control a spill of this particular product would be dealing with a Poison B that is toxic by inhalation, skin absorption, and/or ingestion; as well as a flammable liquid and an environmentally hazardous substance. Also, since only the active ingredients are required to be listed on a pesticide label, some solvents and other chemicals could be present without being named.

Response to a fire involving quantities of pesticides or other agricultural chemicals requires extreme care. Burning pesticides will likely produce toxic gases, fumes and smoke—all of which must be avoided. Common procedure is to stay upwind and wear full protective clothing. For incidents at stationary storage facilities a crucial first step is to locate the facility manager who knows the hazards of the various products stored as well as volumes and storage locations. Storage locations and volumes change often in a warehousing operation, so even if the facility has been preplanned recently it is wise to obtain current information from facility personnel. Nowadays, many companies maintain computerized up-to-date inventories of products on-hand.

Early on a decision has to be made whether to fight a pesticide fire or let it burn. Facility personnel can provide assistance in making this decision based on their knowledge of toxic properties of products on-hand, potential salvage considerations, and factors related to post-fire clean-up. As soon as a need is indentified, responders may also want to establish communication with the firm's principal supplier, particularly when highly technical toxicity data or poison control information may be needed.

Immediate consideration must be given to evacuating areas downwind and isolating the area close by. Access should be controlled to limit contamination and transfer of contaminents. If clothes become contaminated through contact with fumes, smoke, puddles, run-off or dust; responders should immediately leave the area, remove contaminated clothing, and shower even if the only shower facilities available are those created by fire hoses.

During firefighting operations, efforts should be made to remain upwind of possibly toxic smoke and to stay away from drums and other containers which may rupture violently, prevent fire from spreading by cooling nearby containers, and move portable tank or tank cars away from the fire site if possible. Minimum amounts of water should be used on pesticide fires. Most agricultural chemicals are sold in concentrated form; mixing such chemicals with water can create toxic solutions. Too much water spreads contamination and increases the amount of toxic material in the soil. Run-off should be diked or otherwise contained and prevented from entering streams and sewers. If toxic materials have already entered a waterway used for drinking or recreational purposes, or entered a sewer, officials should be notified at once.

Post-fire pesticide residues can be more toxic than the original chemicals. A hot fire may decompose pesticides to less toxic compounds and thus cause less air pollution. On the other hand, water will boil causing steam to rise and carry with it toxic material that may drift over a wide area and drop as toxic fallout. Too much water may merely leave additional toxic debris and residue as well as quantities of contaminated run-off that will have to be properly disposed of. For pesticide fires, there is general agreement that water should be used sparingly.

Once a pesticide fire is extinguished or allowed to burn itself out, the work has just begun. Personnel leaving the site should be required to remove contaminated clothing and gear and then shower using large amounts of soap and water before changing into clean clothing. Inner clothing should be washed with detergent and bleach using a separate washer load. Clothing should not be merely dropped in a pile with the expectation it will be washed; someone must insure that clothes are in fact washed, and individuals should refuse to wear contaminated clothing until laundering has been completed. Protective clothing, hoses, vehicles and tools should be isolated until they have been washed with strong detergent and thoroughly rinsed. A pesticide facility fire site should be secured and patrolled 24 hours a day. The facility manager, in consultation with appropriate authorities, is responsible for developing an acceptable restoration plan. Cleanup personnel must wear appropriate gloves, hats, boots and proper respiratory equipment. Work under such conditions can be demanding, and frequent rest periods or relief may be required. Material handling equipment should be utilized whenever possible to limit human contact with debris. Equipment should be of smooth metal construction for ease of decontamination once cleanup has been completed. Trucks used to haul debris to approved disposal sites should be lined with impervious sheeting and the loads covered. Covers and linings, as well as other materials used in cleanup may have to be properly discarded if decontamination proves too difficult or costly, so selection of materials and equipment to be used in a cleanup effort should be done carefully. During cleanup, debris may have to be lightly sprinkled with water to control dust. Contaminated earth may have to be excavated and structures and fixtures may require repeated, extensive scrubbing with a selected decontamination solution that is then picked up with sorbent materials.

Many if not all pesticide manufacturers widely distribute information sheets that provide emergency guidance relative to specific products. Figure 9 provides a sample of one such sheet created for Methyl Parathion, a Poison B.

Nonflammable Gases refer to any compressed gas other than a flammable compressed gas. More specifically, a nonflammable gas is any material or mixture having in the container pressure exceeding 40 psi at 70 °F., or having an absolute pressure exceeding 104 psi at 130 °F. (CFR-49, Section 173.300) Examples of nonflammable gases are anhydrous ammonia, chlorine, argon, helium, and nitrogen. Two of the hazardous materials transported most widely in commerce are nonflammable gases: anhydrous ammonia used as a fertilizer and refrigerent; and chlorine used to purify drinking water, treat sewage, and make paper. Sixteen million tons of ammonia and eight million tons of chlorine are produced each year in the United States. Nationally, railroads alone transport roughly 50,000 car loads of anhydrous ammonia and a like amount of chlorine. On February 18, 1969 at Crete, Nebraska nine died and 53 were injured due to exposure to a toxic cloud of anhydrous ammonia caused by a train derailment. At Houston, Texas on May 11, 1976 six died and 178 were injured by a vapor cloud of anhydrous ammonia resulting from a highway accident. On May 16, 1976 14 persons were injured by a release of anhydrous ammonia following a derailment and tank car rupture in Glyn Ellen, Illinois. Three died and 46 were injured in Pensacola, Florida after a train derailment on November 9, 1977 resulted in a toxic cloud of anhydrous ammonia that rapidly moved over a wide area. At Youngstown, Florida on February 26, 1978 eight died and 138 were injured by chlorine escaping from a punctured tank car. At Steubenville, Ohio on March 14, 1978 an estimated 263 persons, including 32 of the city's 64 firemen and eight police officers, were taken to area hospitals when a cloud of chlorine gas seeped into the downtown area during a fire at a chemical warehouse.

Chlorine and anhydrous ammonia are two of the most widely used nonflammable gases; therefore, exposure is common. Both are compressed gases that are stored, transported and

METHYL PARATHION SAFETY

NO SMOKING, EATING, OR CHEWING IN WORK OR STORAGE AREAS

DANGER! POISON! CAN KILL YOU!

Methyl parathion is extremely poisonous by skin and eye contact, inhalation, or swallowing. It is rapidly absorbed through the skin and eyes. The fatal dose is extremely small — as little as one drop of the concentrated material splashed into the eye may be fatal. Repeated exposure may increase susceptibility without giving rise to symptoms.

READ THE LABEL

Do not get methyl parathion insecticides in eyes, on skin, or on clothing. Do not handle until the label has been read carefully.

Additional information is contained in Technical Data Sheet available from Kerr-McGee Chemical Corporation.

POISONING SYMPTOMS

Symptoms most often appear in this order: headache, fatigue, giddiness, nausea, salivation, sweating, blurred vision, tightness in chest, abdominal cramps, vomiting, and diarrhea. In severe poisoning, difficult breathing, tremors, convulsions, collapse, coma, pulmonary edema, and respiratory failure may follow. The more advanced the poisoning, the more obvious are the danger signals of miosis or narrowed pupils, rapid asthmatic breathing, and marked weakness coupled with excessive sweat and bronchial fluids.

Symptoms may not appear until one to four hours after contact.

PERSONAL PROTECTION

To avoid contact with methyl parathion pesticides, wear protective clothing at all times. This includes protective eye goggles, washable work clothes, latex rubber gloves, rubber apron, rubber or impervious head covering, rubber work shoes or overshoes, and a NIOSH approved respirator. Work clothing should be tight-fitting around the wrists, ankles, and neck; no skin should be exposed. Destory and replace gloves frequently. Wash shoes, apron and gloves with soap and water before removing. Wash hands, face, and arms thoroughly with soap and water and clean under fingernails before eating, drinking, or smoking. Bathe immediately after work and change all clothing including underwear. Wash clothing, including headgear, thoroughly with soap and hot water before re-use.

HANDLING AND STORAGE

Store methyl parathion insecticides in a well-ventilated, fire-resistant building away from sparks and flame. Protect drums from overheating.

DO NOT store or transport with food or feed products, wearing apparel, or other sensitive materials where contamination could be hazardous.

Use adequate mechanical ventilation when handling.

SPILLS AND LEAKS

1. Use personal protective equipment.
2. Keep area roped off and posted.
3. Dike large spills.
4. Sprinkle with hydrated lime or soda ash (one pound per square foot).
5. Dampen slightly with water from a hose or sprinkling can.
6. Let the lime or soda ash stand for several hours or overnight, depending on the surface.
7. Remove the caked material; sweep or shovel it into a disposal drum. Bury the drum in an approved area or burn the contents safely.
8. Sprinkle spill area with more lime or soda ash.
9. If a yellow color appears (indicating that the methyl parathion has not been completely removed or hydrolyzed), repeat procedure until there is no yellow color.

Figure 9. Continued.

FIRE FIGHTING

CALL THE FIRE DEPARTMENT. Tell them what has happened, and that it involves methyl parathion. If they have any questions about hazards, tell them to call the emergency telephone number listed below. Fire fighters must keep a respectful distance from smoke, spills, and hot drums and keep upwind of fire.

Use PROTECTIVE CLOTHING and BREATHING APPARATUS. Foam, dry chemical, or carbon dioxide extinguishers should be used for fighting fires. A water spray may be used to reduce rate of burning and cool containers. Water streams must not be used as they may spread contamination.

Let large fires burn. Drums should be kept cool with a water spray during and after extinguishing the fire and until all embers are dissipated.

Smoke or fumes from parathion fires are extremely harmful by inhalation and skin or eye contact.

After a fire, the entire affected area must be decontaminated according to the procedures indicated under SPILLS AND LEAKS. Protective clothing must also be decontaminated and rinsed before it is removed.

FIRST AID

CALL A PHYSICIAN AT ONCE IN ALL CASES OF SUSPECTED ORGANOPHOSPHATE POISONING.

BEFORE GIVING FIRST AID, BE SURE TO PROTECT YOURSELF FROM CONTAMINATION.

Anyone suspected of having been exposed to methyl parathion should be kept under medical observation for at least 24 hours.

Immediately remove the victim to fresh air. Support respiration. Keep airway clear. Use artificial respiration if necessary. Remove clothing. Decontaminate skin, hair, and fingernails with soap and water. If in eyes, flush with water for at least 15 minutes. If swallowed, induce vomiting (finger down throat) until fluid is clear. Save fluid for physician's examination. Take label to physician with patient.

Don't send victim to the doctor or hospital — take him.

ATROPINE FOLLOWED BY 2-PAM IS THE ONLY KNOWN ANTIDOTE. THEY SHOULD BE ADMINISTERED ONLY BY A PHYSICIAN.

NEVER GIVE ANYTHING BY MOUTH TO AN UNCONSCIOUS PERSON.

Figure 9.

handled as liquids. Anhydrous ammonia is a liquid which at atmospheric pressure boils at minus 280 °F. It remains a liquid when the temperature is higher than the boiling point if kept under pressure in a container. Once pressure is removed as in the case of a tank rupture, and the ambient temperature is above the boiling point, the liquid will rapidly convert to gas at a ratio of 877 to one. That is, one unit of liquid will expand to 877 units of gas. When inhaled, anhydrous ammonia will result in damage to the respiratory tract ranging from minor irritation to severe pulmonary edema depending on concentration of gas in the atmosphere and duration of exposure. Contact with the liquid can cause immediate danger to exposed tissue (frostbite). Anhydrous ammonia is water soluble and mixes with water to form a corrosive liquid. It is lighter than air and can form a vapor cloud over a wide area. When two anhydrous ammonia tank cars were punctured in a derailment at Pensacola, Florida on November 9, 1977 a concentrated cloud of ammonia vapor was initially held near the ground by a combination of overhanging trees and a light rain. Eventually, the cloud expanded to cover an area one mile in diameter and approximately 125 feet high, and traveled about 15 miles on a light wind blowing from the southeast. Had there been an easterly wind, the lethal cloud would have moved through a heavily populated area of Pensacola. Anhydrous ammonia is not generally considered a fire hazard due to its narrow flammable range; however, it is a toxic gas and very caustic. Contact between anhdrous ammonia liquid/vapor and the eyes is to be avoided at all costs.

Chlorine is a nonflammable gas shipped as a liquid under pressure. When released from a containerized environment, liquid chlorine expands 470 times to gas. Chlorine reacts violently and may form explosive mixtures with certain common chemicals. (Ammonia, acetylene, butadiene, benzene, hydrogen, sodium carbides, turpentine, and finely divided powdered metals for example.) Contact with liquid chlorine can cause frostbite; its vapors are much heavier than air and will tend to settle in low areas. When a single tank car of chlorine was punctured in a derailment at Youngstown, Florida on February 26, 1978 eight persons died when trapped in a low lying vapor cloud that seeped across a highway in the early morning hours causing automobile engines to stall for lack of oxygen. Although the hazard class of which chlorine is a member, Nonflammable Gas, may sound far less threatening to the ear than certain other hazard class designations; chlorine inhalation is not a pretty way to die. At Youngstown victims' faces, upper body skin, and lungs had turned purple; arms were clasped to chests; mouths were covered with foam-like discharge. Some casualties had their eyeballs eaten away, while others had broken bones suggesting they had run into obstacles in a frantic effort to escape the cloud.

Chlorine is widely transported in both tank cars and cylinders. Cylinders are equipped with fusible plugs designed to melt at 156–164 °F and release the chlorine. That is, chlorine cylinder safety release devices are temperature-triggered rather than pressure-triggered. Chlorine is 2½ times as heavy as air hence basic advice for escaping from a chlorine release is to head for the high ground. Water should not be applied to the point of leak in a chlorine container as it will form an acid and tend to enlarge the leak. In addition, chlorine is a powerful oxidizer and should never be stored near organic materials or hydrocarbons.

Chlorine can be detected by humans at a concentration level of one part-per-million (PPM). Four PPM is the maximum concentration that can be inhaled for an hour or more without serious problems. Fifteen PPM will cause throat irritation; 30 PPM induces coughing. A concentration of 40–60 PPM would be dangerous to life over a period of 30 minutes, and 1,000 PPM would kill most animals in a short period of time. Chlorine leaks are said to always get worse and never get better because chlorine and moisture form a corrosive mixture. Small leaks in a container can be located by wetting a rag with aqua ammonia and passing around the container until white smoke appears. Household ammonia can also be used to locate small chlorine leaks, but the effect will not be as dramatic.

In any response to nonflammable gas incidents, particularly incidents involving chlorine or anhydrous ammonia, contact with either the liquid or the vapor should be avoided as chemical burns or frostbite is a real danger. Water spray can be used to absorb vapors, but resultant run-off is toxic or corrosive and should be diked for containment and disposal. Full protective clothing and selfcontained breathing apparatus is required, and evacuation downwind must be an initial consideration. Any material that may have contacted the body must be washed away with soap and water. Short exposure to either chlorine or anhydrous ammonia can seriously injure even if prompt medical attention is provided.

Oxidizers are substances such as chlorate, permanganate, inorganic peroxide, nitro carbo nitrate, or a nitrate that yields oxygen readily to stimulate the combustion of organic matter. (CFR-49, Section 173.151) Oxidizers can react violently and support combustion in organic materials and hydrocarbons. A typical demonstration of an oxidizer's potential familiar to most firefighters is the addition of brake fluid or another hydrocarbon to dry chlorine which is a strong oxidizer. Within seconds sufficient heat will be generated to ignite the hydrocarbon. Technically, oxidizers do not themselves burn but can present a severe fire and explosion hazard through contamination or contact with organic materials, acids or oxidizable substances. A basic problem in handling, transporting and storing oxidizers is that so many extremely common materials in wide usage may react violently when placed in contact with an oxidizer. (Examples: wood, lubricating grease, various petroleum-based products, paper, clothing, solvents, paint, cleaning products—the list is long and varied.) Oxidizers may also react violently with certain oxidizable substances such as ammonium and sulfur compounds, powdered metals and metal salts. When contaminated by an oxidizer, many combustible materials will burn fiercely due to release of oxygen.

A hazardous material that does not itself burn but merely releases oxygen to support combustion is in danger of being perceived as "only semihazardous." For that reason, it is necessary to remember that the April 16, 1947 explosion of the French freighter, Grandcamp, at Texas City, Texas that killed at least 468 and seriously injured 1,000 involved an oxidizer, ammonium nitrate fertilizer.

Because of potential reactivity with a wide variety of other materials, storage and handling techniques are extremely important when working with oxidizers. Separation is the key. Oxidizers should never be allowed to come in contact with organic materials or hydrocarbons. Common safety practices applied in storage situations involve use of metal containers as differentiated from wood/paper/fibre containers; extreme care in the reuse of containers so possible residue from prior contents does not contact an oxidizer; stringent segregation of oxidizers in storage situations; continuous attention to dust control so building materials do not become impregnated; use of metal conveyor systems as differentiated from fabric or rubber systems; and restrictions of the use of common petroleum-based lubricating products.

Response personnel and others must take care that personal and protective clothing does not become impregnated with oxidizing material. Whether apocryphal or not, stories are legion of individuals who after working around spilled oxidizers returned home or to the stationhouse for a cup of coffee—and a cigarette, and received a classic lesson in the combustion enhancement powers of oxidizers. Clothing impregnated with an oxidizer must be washed thoroughly, and may have to be discarded.

Spilled oxidizers in both fire and nonfire situations can present difficult problems for responders. There does not appear to be universal acceptance of the "best" method for dealing with a fire involving oxidizers. Use of water for extinguishment is viewed as basically acceptable if there is recognition of potential problems that may occur, while dry chemical agents or CO-2 are seen as ineffective. Water in flooding quantities applied as fog or spray rather than in straight streams is normally used on fires involving either large or small quantities of liquid oxidizers with the understanding that use of water on oxidizers that are also acids (nitric acid is an example) may well cause spattering and must be done from a safe distance with due caution. Water spray/fog is also used on small quantities of solid oxidizers. Use of water on fires involving large quantities of solid oxidizers is viewed as ineffective by some and "the only method normally available" by others.

Flammable Compressed Gases are any material or mixture having in the container a pressure exceeding 40 psi at 100°F. (CFR-49, Section 173.300) (Examples: hydrogen sulfide, acetylene, butane, butadiene, ethylene, liquefied hydrogen, hydrogen isobutane, liquefied petroleum gas, vinyl chloride, etc.)

On January 25, 1969 two died and 33 were hospitalized at Laurel, Mississippi when one broken wheel on a 144 car train caused a derailment and violent rupture of tank cars of liquefied petroleum gas that destroyed 54 homes; and damaged 1,350 residences, six schools and several local businesses. Two tank cars rocketed over long distances spewing burning propane. One tank car was propelled through the air striking the ground three times before coming to rest on a dwelling 1600 feet from the point of take-off. A second tank car rocketed 1,100 feet striking the ground five times enroute.

Sixty-six persons were injured and many buildings destroyed at Crescent City, Illinois on June 21, 1970 when up to nine tank cars of liquefied petroleum gas ruptured, beginning about one hour after a derailment. Seven were injured at Sound View, Connecticut on October 8, 1970 when a passenger train struck a freight train that had derailed earlier and punctured a liquefied petroleum gas tank car designated as empty. On October 19, 1971 at Houston, Texas one died and 50 were injured by a violent rupture of vinyl chloride tank car that released 50,000,000 BTUs of energy in one second approximately 35 minutes after derailing. Eight newsmen and photographers were among those injured as well as numerous firefighters and bystanders. The ruptured tank car rocketed 300 feet trailing liquefied vinyl chloride.

At East St. Louis, Illinois on January 22, 1972 more than 230 persons were injured when a tank car loaded with propylene collided with a standing hopper car in a "hump yard" accident at a rail classification yard. Two died and five were injured March 9, 1972 on U.S. Route 501 in Lynchburg, Virginia when a tractor-semitrailer carrying propane overturned and ruptured. Two died and 28 persons were injured at Exit 8 of the New Jersey Turnpike on September 21,

1972 when a tank car load of propylene BLEVEd. (experienced a Boiling Liquid Expanding Vapor Explosion.) Forty workmen were killed in New York City on February 10, 1973 while repairing an "empty" liquefied natural gas (LNG) storage tank. Thirteen died, including 12 firefighters, and 95 were injured at Kingman, Arizona on July 5, 1973 when propane tank cars BLEVEd and rocketed. At Oneonta, New York on February 12, 1974 five tank cars of liquefied petroleum gas ruptured after a derailment and injured 54 firefighters and members of the press. In a rail "hump yard" incident at Decatur, Illinois on July 19, 1974 a tank car carrying liquefied isobutane ruptured killing seven and injuring 52. Felt as far as 45 miles away, the explosion damaged 2,000 homes and businesses, damaged 14 schools and blasted debris over a 20 block area. On April 29, 1975 a tank truck struck a highway abutment near Eagle Pass, Texas. The resultant LPG explosion killed 16 and injured 51. At Belt, Montana on November 26, 1976 a derailment and rupture of propane tank cars killed two, injured 23, and destroyed 40 homes and 19 automobiles plus a number of businesses including a bulk storage plant. Sixteen died at Waverly, Tennessee on February 22, 1978 when a tank car of liquefied petroleum gas ruptured 40 hours after a derailment. Approximately 6,500 persons were evacuated from Muldraugh, Kentucky on July 26, 1980 when two tank cars of vinyl chloride ruptured after a derailment.

Many flammable gases have immense destructive capacity when released from their containerized environment. Temperature, pressure and volume play critical roles in the handling of flammable gases; a relationship that was recognized long ago by two scientific laws. Boyle's Law says in effect that volume is inversely proportional to pressure; that is, as pressure increases volume decreases. Thus, by applying pressure to flammable gases it is possible to significantly decrease their volume and make possible large savings in transportation and storage costs. In discussing compressed gases it is necessary to differentiate between pressurized gases and liquefied gases. Pressurized gases are those that do not liquefy when compressed within a cylinder because their boiling points are comparitively low. (less than minus 150°F.) Liquefied gases, on the other hand, have boiling points in the range of 32°F to minus 150°F; compressed within a pressure vessel, they assume a liquid/vapor form. Although both flammable and nonflammable gases can be liquefied, in this section the focus will be on flammable gases.

Charles Law holds that volume is directly proportional to temperature. As temperature increases, volume increases. When a compressed gas container is heated, the product within will expand. Pressure vessels used to transport compressed flammable gases are therefore equipped with safety relief devices designed to respond to and control increased pressure.

Liquefied petroleum gas, (LPG) a compressed flammable gas, has a liquid-to-vapor expansion ratio of 270 to one. Thus a tank car of LPG contains product that through pressurization has been reduced in volume 270 times. LPG tank cars are never completely filled in order to allow for expansion of the liquid due to increases in temperature. The major portion of the tank car contains liquid while the upper portion contains vapor. Normal atmospheric temperature changes will cause the liquid to expand and contract thus increasing and decreasing pressure. Under normal conditions, as temperature rises and the liquid expands increased pressure is vented through a safety relief device. As temperature decreases and the liquid contracts, decreased pressure allows the safety relief device to close.

However, abnormal temperature increases such as those resulting from fire, particularly direct flame impingement on the tank as differentiated from radiant heat, can overcome the capacity of a safety relief device to adequately compensate for increased pressure and may possibly lead to drastic tank failure popularly known as a Boiling Liquid Expanding Vapor Explosion (BLEVE). The BLEVE of a compressed flammable gas container the size of a pressurized tank car can be extremely destructive and involve shock waves, flying shrapnel, an immense fireball, and the rocketing of tank car sections trailing flaming product over considerable distances.

For a BLEVE to occur, the liquid contained within a pressure vessel must be at a temperature well above its boiling point at normal atmospheric pressure. Boiling point refers to the temperature at which the vapor pressure of a liquid equals the atmospheric pressure. (In simpler terms, boiling point refers to the temperature at which a specific liquid will give off vapors.) A liquid with a boiling point of less than 100°F has a high vapor pressure and thus an increased probability of becoming involved in a BLEVE.

Using the example of LPG begun previously, it may be useful to develop a scenario to illustrate how a railroad derailment can set the stage for a BLEVE. As the train derails, physical forces may cause one or more cars of LPG to rupture or be punctured, releasing liquid to the atmosphere which immediately expands 270 times to vapor. Being heavier than air, the vapor remains close to the ground, finds an ignition source, and flashes back. The resultant fire is massive and extremely hot. Other compressed flammable gas tank cars in the consist that survived the derailment without rupture or puncture are now subjected to greatly elevated temperatures from surrounding fires. Radiant heat increases the temperature on remaining tank cars, but perhaps the safety relief devices are able to compensate for increased heat by venting to relieve pressure buildup. Direct flame impingement on a liquefied gas car, however, creates a far more hazardous condition. Steady flame impingement on the tank car exterior causes near-continuous boiling of the liquid within and expansion of the vapors. The safety relief device may be unable to compensate for increased expansion, and enormously increased pressure begins to strain the limits of tank construction. Where flame impingement strikes the tank may make the difference between BLEVE/No-BLEVE. Like a teapot filled with water, the portion of the tank containing liquid is able to absorb heat to a significant degree. However, like a teapot empty of water, the vapor filled portion of the tank is unable to absorb heat as effectively as the liquid filled portion and container failure is much more likely to occur. Flame impingement of the liquid filled portion is serious but not necessarily critical; flame impingement on the vapor space is indeed critical. The problem and probabilities are compounded by the fact that as the liquid continues to boil the liquid level decreases while the portion of the tank filled with vapor increases. As the liquid level is lowered, additional vapor space area may be exposed to flame impingement. At some point in time, and the time period may vary widely due to multiple factors, the metal tank portion above the liquid level (in the vapor area) is weakened to such an extent it ruptures. The boiling LPG suddenly experiences a vast drop in pressure, expands instantaneously 270 times to vapor and is ignited.

Perhaps the scenario just created was too severe. Perhaps the first tank car did indeed rupture and release a vapor cloud, but responders arrived immediately and were able to control ignition sources, (sparks, open flame, vehicle exhaust systems and catalytic converters, etc.) use straight hose streams to speed vaporization of spilled liquid, and apply water as spray to dissipate the vapor cloud. However, ignition sources are normally so widely available that the above variation would require optimum conditions rather than normal conditions.

More likely, the vapor cloud did ignite and response personnel were faced with a fire situation. If tank shells are subjected to radiant heat alone, chances are good that combined operation of the safety release device to relieve pressure and applications of large quantities of water to cool the tank surface can keep pressure within the tank to a level that will not result in tank failure.

Even when there is direct flame impingement on the tank surface, but such flame impingement is below the liquid level; the capacity of liquid to absorb heat, operation of the safety relief device, and application of large volumes of cooling water to the exterior tank surface to reduce temperature may prevent a BLEVE.

Response options available to prevent a BLEVE when flame impingement strikes the vapor space are more limited and possibly less effective. It may be possible to move the tank away from the impinging flame. It may be possible to insert a barrier between the impinging flame and the endangered tank car. These two possibilities may be remote ones given the jumbled, jackstraw nature of some derailment scenes where cars are piled atop and around each other. It may be possible to sufficiently cool the exterior of the car with large volumes of water to the extent metal will not become dangerously weakened and the safety relief device will not be overpowered by vastly increased pressure. Insulated tank cars are not as readily subject to failure through external application of heat, yet by the same token the cooling effect of water applied to an insulated tank car surface is minimal. Still, insulation may have been torn away in sections during the actual derailment, and in such instances application of water to the exposed metal surface would be called for. In any event, the amount of water that would have to be applied to an uninsulated car subject to direct flame impingement on the vapor area in order to diminish the possibility of a BLEVE, although large, is open to question. Water application should always be done using unmanned nozzles to lessen the very real danger to response personnel. It is only partly a joke when an experienced firefighter says—"Whenever we have to set up unmanned nozzles to bracket a tank car subject to direct flame impingement in the vapor area, we always send in one of the newer men."

If there is no serious threat to civilian life posed by a potential BLEVE, response personnel should withdraw to a safe distance. When a BLEVE threatens, potential property loss is not sufficient justification for endangering emergency responders.

3. Maintenance of Incident Vigilance and Discipline.

Successful response to hazardous materials incidents requires an organized, disciplined effort. Perhaps the hazards involved cannot be eliminated, but they can be minimized. Approaching a transportation or stationary facility incident with full knowledge of the product(s) and container(s) involved is a rare occasion. More often there is only limited information available, perhaps only a report that "Something" is creating a Hell of a problem. Professional responders learn by experience to develop a special awareness of clues that suggest proper response and control activities, and they *never* underestimate the incident. A simple truck tire fire can turn into a disaster in seconds if the vehicle is loaded with unrecognized hazardous materials.

Effective response presupposes prior planning. Do you have a plan? Are your personnel trained, equipped and knowledgable? Have they made preplanning and inspection visits to facilities that manufacture, process, transport, store and dispose of hazardous materials? Have they established working relationships with expert individuals and organizations able to provide specialized advice, assistance and equipment when needed? Is it understood within your community or organization who is in-charge when an incident occurs? Is there agreement? If not, iron out any difficulties or misunderstandings. Once you arrive on-scene, it is too late to start getting organized.

A number of response organizations employ a "Safety Officer", a knowledgable individual whose overriding responsibility on-scene is personal and public safety related to placement of apparatus, evacuation distances and areas, adequacy of protective equipment, elimination of potential ignition sources, and maintenance of vigilance and discipline throughout the duration of any incident. This person is kept free of other duties on-scene so that he/she can devote full attention to insuring utilization of proper methods and procedures.

Correct identification of the material(s) is a critical area of concern. Too many times, a failure to first identify the commodity and its reactive characteristics before taking action has resulted in unnecessary injury. Even when a material has been correctly identified, care must be taken that transmission of this information to others does not result in error or misunderstanding. A commercial response group in Massachusetts recently received a call to respond to a truck terminal where "liquid vinyl chloride was leaking from a trailer." While enroute, the response team asked for a double-check on the commodity involved since they recognized that vinyl chloride is a flammable gas. Although it is shipped as a liquefied gas under its vapor pressure, liquid escaping from a container would be expected to turn to vapor almost immediately. The product proved to be thionyl chloride, a corrosive/acidic/poisonous liquid. "Vinyl chloride" and "thionyl chloride" sound the same to a person relaying a message by radio, but required response and control measures, and personal protective clothing are entirely different. Persons unfamiliar with pronounciation of chemical names can easily make "propionyl chloride" sound like "propyl chloride;" "phenol" may sound like "phenyl;" "chloride" may be confused with "flouride;" "hydride" may be repeated as "hydrate;" and "benzyl" may come across as "benzol" or "benzoyl." Did the voice on the radio say Phosphorous-tribromide, trichloride, trifluoride, trioxide or trisulfide? Was that key word "methyl" or "ethyl?" Did he say "ethane" or methane?" Was that term "isopropanol" or "isopropenyl?" Is the product "hexane" or "hexene?" It sounded like "alkyl" or was it "allyl?" Did he mean "acetyl iodide" or "acetaldehyde?" Extreme caution is required. Spell it out when transmitting, relaying or receiving chemical names. One misplaced letter can make a world of difference. Cite the United Nations or STCC (49-series) number to make certain everyone is talking about the same chemical. Such stringent procedures may seem severe to the uninitiated but mistakes occur all the time.

Experienced responders recognize they may be presented with "no-fight" situations. When faced with flame impingement on the vapor space of a compressed flammable gas tank, insufficient supplies of water or foam, fires involving explosives, radiation incidents—ultrahazardous situations where no life and minimal property is endangered—they are not too proud to pull back and regroup. They recognize that eventually they may find themselves in a situation where no control and containment efforts are practical, that a chemical reaction is about to run its course and evacuation is the only action they can take. They are willing to consider alternatives, to ask the question: "What happens if we do nothing?"

Experienced responders attempt to utilize facility personnel such as gas company technicians, truck drivers, warehouse supervisors, traincrews, petroleum storage area managers, container specialists and others who know the equipment, containers, vehicles, valves, piping, ventilation systems, and mechanical shut-off devices at a specific facility. They also know where to locate and how to use specialists such as explosive ordnance disposal teams, radiological response teams, chemists, industrial teams, and product response groups such as CHLOREP and members of the Pesticide Safety Team Network. They establish and maintain strong relationships with agencies, industries, commercial contractors and others who may be able to supply equipment, training or expertise when needed.

4. Control of the Response Effort.

Experienced responders do not rush in. They stop as soon as they have visual sighting and perform a detailed size-up, using binoculars to evaluate the situation if necessary, before commiting personnel and equipment. They identify fire/explosion/reactive/health/environmental hazards and evaluate potential impacts. They consider rescue needs and estimate danger areas, identify exposures, consider wind direction and velocity if a plume or toxic cloud is present, consider potential secondary emergencies, and decide if evacuation may be necessary.

They initiate a staging area outside the danger zone where all personnel and equipment report in and receive their orders. Those not needed immediately are placed on remote standby until needed. Personnel and equipment are commited to the danger zone only as needed and sufficient reserves are obtained and held in readiness. A command post is established immediately to serve as a focal point for information gathering and decision making. Approach of committed personnel and equipment is from upwind and upgrade using natural barriers for protection. Specific individuals are assigned to protect equipment and tools from run-off, vapors, sprays or residues. Ignition sources are identified and controlled, and unmanned equipment is used when possible. Standby personnel are used to lug and carry equipment so the attack team will not expend valuable energy before arriving at the point of attack. Long lays of hose are utilized rather than driving apparatus close to the danger zone. Responders recognize that some people think they are immune, so attention is paid to insuring that protective equipment is not only available but used, that face shields are down and selfcontained breathing apparatus is worn. Operations in the danger zone are performed with the minimum number of personnel to limit casualties if the situation gets out of hand.

5. Control of Access to the Scene.

Only those persons who need to be there, and only while they are needed, are allowed access to the danger zone. Roadblocks, barriers, barrier tapes, and traffic and crowd control are used to seal off the area. Silent pagers may be used so only specifically trained and equipped response personnel absolutely necessary to achieve control and containment are alerted and called to the scene. Radio discipline is imposed so that idle chatter and "radio checks" are discouraged. There are so many radio scanners in use that a telephone may have to be used to obtain information from CHEMTREC, the manufacturer or the shipper. Patching through by radio means everyone with a scanner may hear the conversation. Rumors or incorrectly repeated information can cause unnecessary public concern. Radio and television broadcasts are monitored so that if incorrect information is broadcast it can be corrected at once. A specific location away from the danger zone is provided for media representatives, and one person is assigned to provide them with correct information. An anxious public must be provided with accurate information as quickly and completely as possible; but heresay, subjectivity and rumor must be guarded against.

6. Evacuation.

Although largescale evacuations are becoming relatively common actions in response to hazardous materials incidents, few jurisdictional agencies have had to implement the process of emptying an entire city. Early on the morning of Sunday, November 11, 1979 the Peel Regional Police Force in Mississauga, Ontario initiated the largest peacetime evacuation in North American history. Between 1:47 A.M. and 10:55 P.M. that day in excess of 217,000 persons were evacuated from an area that ultimately covered 45 square miles. Just before midnight, a Canadian Pacific freight train hauling 106 cars derailed 24 cars at the Marvis Road crossing in Mississauga. Derailed cars included 11 of propane, four of caustic soda, three of toluene, three of styrene, and a single car of chlorine. Within one minute of the derailment there began a series of tank car BLEVEs. (Boiling Liquid Expanding Vapor Explosions.) Within 21 minutes of the derailment the Peel Regional Police Force implemented a formal disaster and emergency plan. Lessons learned during the next six days have been identified and widely disseminated by an informative report prepared for the Canadian Police College with the assistance of the Peel Regional Police Force.[1]

Noncommissioned and senior officers of the Peel Regional Police Force are required to know the agency's "Major Emergency and Disaster Manual" in order to pass examination to higher rank. The plan requires initiation of the following basic steps during a major emergency.

- Control at the scene is to be under the direction of a designated and trained On-Scene Commander responsible for all police operations.
- To assure adequate command and control, the On-Scene Commander is to work from a command post and maintain direct, open-line communications with the Peel Police dispatch center.
- Senior officials are to be notified, and other personnel are either to be held on duty or called in.
- A special location is to be established as a media relations center, and a senior officer is to be assigned as media spokesman.

The emergency plan also provides guidance as to Communications, Duties of the On-Scene Commander, Duties of a Disaster Control Group, (operations at staging area, police lines, command post, and the emergency scene) Maintenance of a Command Post Log, Operation

[1]*The Peel Regional Police Force and The Mississauga Evacuation* ("How a police force reacted to a major chemical emergency.") by Joseph Scanlon and Massey Padgham. Minister of Supply and Services, Canada, 1980. (Catalogue No. J 566-1/1980 E)

of a Staging Area, Establishment of Police Lines, Initiation of Crowd Control, Evacuation Considerations, and Dangerous Gases. Few municipal or regional police agencies are as well prepared for a major emergency as was the Peel Regional Police Force.

While the Mississauga Fire Department sought to control the actual incident involving fire, BLEVEs, and release of a sizable quantity of chlorine gas; Peel Regional Police sealed off the surrounding area. Within two hours of the derailment, police began a stage-by-stage evacuation of the entire city. The logistics of such an exodus are incredible. Initial efforts were directed toward establishing a command post, communications systems, and attempting to learn the exact contents of the burning rail cars.

As a decision to evacuate was being made, a staff sergeant began collecting police officers as they arrived at the staging area and formed them into work groups, each led by a sergeant. Maps were prepared to show each group exactly which streets they were to cover door-to-door to get the people out. Evacuation assignments were given out in small, controlled areas to keep errors to a minimum. Officers systematically called personally at each door; police cars with public address systems cruised the streets to alert residents; and the media was kept informed of all actions. Evacuees were instructed to leave the area immediately and find accomodations on their own if possible. If transportation or accomodations were needed, Mississauga Transit buses were standing by to take people to a major shopping plaza being used as a collection point while additional facilities were being activated as emergency shelters. In addition, ambulances were standing by to transport those unable to relocate without assistance. As each building was cleared, yellow crayon normally used to mark taffic accident scenes and readily available to all police officers was used to mark doors and sidewalks indicating a particular building had been covered. Although police used persuasion to motivate people to leave the area, no one was forced to leave although identifying information was recorded on anyone remaining in the evacuated area. As each area was cleared, police assigned patrols to guard against criminal activity. As the evacuation area increased in size; the manpower requirements for conducting additional evacuations, maintaining the overall perimeter, and patroling the evacuated areas became enormous, requiring large numbers of police personnel from agencies outside Mississauga.

Police officials arranged for continuous weather reports from various agencies as a knowledge of wind direction and velocity was critical in determining specific areas requiring evacuation. The command post was relocated three times, both to obtain better equipped facilities and to escape chlorine fumes that were readily apparent in certain areas. The Ontario Ministry of Environment provided mobile ''Trace Atmospheric Gas Analyzer'' units to monitor the presence of gases in areas around the derailment site. An Ontario Provincial Police helicopter was used regularly to monitor the area, and Queen Elizabeth Way—Canada's busiest highway—was closed down as were the computer ''Go'' trains that service Mississauga enroute between Toronto and Hamilton.

Peel Regional Police arranged for and relied heavily on Mississauga Transit to have buses and drivers available as needed, the Canadian Red Cross to implement emergency reception centers, and various ambulance services to relocate handicapped residents as well as patients of three hospitals and a number of nursing homes that ultimately had to be evacuated. In addition, it was arranged for Humane Society officers to enter the evacuated area accompanied by a police officer to care for pets. Residents were instructed to leave a key at city hall and sign an authorization for entry to their premises. Ultimately, 1,861 homes were entered and 2,500 pets cared for.

Peel Regional Police established a media relations center in order to keep the public informed of developments. (190 media representatives eventually covered the Mississauga incident.) A senior officer was asssigned as Media Relations Officer as required by the agency's emergency plan, and a French-speaking officer was assigned to assist representatives of the French language media. In addition, Peel Regional Police and Mississauga Fire Department officers escorted media personnel by bus to the fire safety line around the derailment.

Certain factors allowed Peel Regional Police to concentrate their efforts on the evacuation.

Because of the emergency, normal police activity was greatly reduced. Also, outside police agencies immediately provided massive support both to effect the evacuation and to cover duties normally handled by Peel constables. In addition, Peel Regional officials had great confidence in Mississauga Transit, the Red Cross, and various ambulance services—due in large part to assistance provided by these organizations at previous emergencies—and unhesitently called on them to handle extensive functions. There was a recognized faith that these organizations could in fact do what they were asked to do. As but one example, the decision to evacuate three hospitals and a number of nursing homes was a major one. While 1,023 hospital patients were evacuated, 431 (42 percent) could be released to relatives; on the other hand, of 929 nursing home patients evacuated, only 63 (less than 10 percent) could be discharged to relatives.

Every time the municipal transit authority was requested to have buses and drivers at a certain point, they were there ready and waiting when needed by police. One indication of the effectiveness of the Red Cross service officer in initiating and operating emergency reception centers was her later selection of "Citizen of the Year." Overall, an estimated 100 agencies and companies contributed services, goods or personnel in various ways. As noted in the report previously cited, "The presence of extra persons anxious to help is common in disasters; Peel's experience, however, was that those who did standby were those with both experience and special skills."

Initial evacuations were felt to be easier than subsequent efforts, and perimeter control became more difficult as time passed. At first residents had heard explosions and could see flames. As the evacuation spread over time and distance from the derailment, problems increased and crowd control became more difficult. It was noted that persons who repeatedly declined to move back when requested by a uniformed officer did so immediately as soon as two canine officers and their charges arrived. Throughout the six days of evacuation considerable numbers of people, many of them essential to the overall public safety effort, moved through police lines and it was necessary to separate the essential traffic from the nonessential. Bus drivers, firefighters, ambulance personnel, Red Cross workers, telephone company employees, food and equipment providers, Humane Society officers, police officers and other essential personnel had to be passed through the lines and some times escorted back and forth. Firefighters in uniform may be easily able to establish their credentials but when reporting for duty in civilian clothes they are more difficult to distinguish from nonessential personnel also seeking entry. Badly needed bus drivers initially had difficulty passing perimeter points until bus company officials identified perimeter points staffed by receptive police officers and told drivers to enter there. Eventually, persons with a recognized need to enter the evacuation zone were instructed to go to a specific entry point, and officers on perimeter control duty there were informed exactly who would be coming through and at what time. Lack of standard identification for persons needed within the perimeter was a problem for police throughout the six day evacuation.

A number of general problems were identified during the week. Fifteen or so hours into the incident police officers were becoming exhausted. A number had reported for emergency duty after working a full regular shift. Radio broadcasts requested individuals who had been on duty for 24 hours to report in. Peel Police noted ". . . one of the difficulties in all disasters is reacting to the first hours as if the incident would be brief."

Police found they had to mediate a number of disputes, both at perimeter points and at emergency reception centers, as the duration of the evacuation became extended. Officers had to provide crowd control at a shopping plaza used as a gathering point for evacuees, had to mediate arguments between welfare groups seeking to provide services to "their" evacuees, and sooth irritated evacuees at reception centers. Many people wanted to gain access to their property within the evacuated area for one-thousand-and-one different reasons, some legitimate, others not. The longer the evacuation, the greater the irritation and the longer the lines of people appearing at perimeter access points. Peel Regional Police organized six conciliation teams to man access points and evaluate requests for re-entry or temporary access. If a request for access was judged valid, a police officer usually escorted the person to his home or business.

Police were hesitant to transmit certain information, such as gas concentrations or changes in evacuation zones, in plain English over the radio system as it could easily be monitored. Officers calling in boundry changes would use the telephone. For temporary gas problems, a code was developed and used in radio transmittals to avoid making this information available to other than emergency personnel.

Early on the command post was innundated by a wide variety of officials, politicians and representatives from every level of government. Everyone wanted to jam into the mobile command trailer. It was necessary to create a list of persons who would be allowed to attend future disaster planning sessions. Also, it appeared that many "visitors" provided the command post telephone numbers to their offices so the numbers quickly became widely known and lines were jammed with traffic. (A direct line from the mobile command trailer to police communications was secure and unaffected.) When the command post was later shifted to a telephone company building, numbers were changed every night to solve the problem. The police communications center also experienced an immense amount of both radio and telephone traffic. Although the communications center had 25 telephone lines, (plus four special crises lines the numbers of which are sealed until a major emergency occurs) existing facilities were inadequate to handle the traffic.

Persons with medical problems often could not locate their prescriptions or make contact with their normal physicians or druggists. Personnel from an evacuated hospital opened four temporary clinics in areas surrounding the evacuation zone and made arrangements for other hospitals to provide special services through their emergency rooms. A number of police officers, firefighters, chemical specialists and others were required to work nonstop for many hours under extremely stressful conditions. A health department representative who is a physician arranged through the Ontario Medical Association for volunteer physicians to stand by at the command post on a 24 hour basis.

Planning the ultimate re-entry was perhaps more difficult than the original evacuation. Thirteen separate areas had been evacuated by stages, but conditions made it impossible for areas to be considered for re-entry in the same order they had been evacuated. Perhaps residents who had been evacuated first felt they should be allowed to re-enter first, but the determination of re-entry order was far more complex and confrontations arose. Also, police had to insure that when one perimeter was done away with, a new, inner perimeter had already been established. Designations and changes of perimeters were sometimes viewed by evacuees as entirely arbitrary. Considering that some evacuees were out nearly six full days, irritation was not unexpected.

Briefings were held for media personnel to explain planned boundry changes so the public would be informed. However, sometimes the media was broadcasting such proposed changes even before the briefing had been completed or police at perimeter points had been given the word. Immediately people began arriving at control points seeking re-entry from officers who had not yet been informed re-entry was to be allowed. Confusion, irritation and radio traffic rose considerably. In a limited number of instances, information put out by the media was incorrect and added to the confusion. Peel Police monitored commercial radio and television broadcasts and immediately called to correct any erroneous information. As might be expected, police officers on duty at control points took to listening to commercial radio in order to remain as quickly and fully informed as the citizenry.

Because of the size, duration and effective administration of the Mississauga evacuation; the Peel Regional Police Force has been besieged with queries, requests, and questions from officials throughout Canada, the United States and a number of foreign countries seeking to learn how better to organize response to potential similar emergencies in their own communities. Peel Police have unstintingly responded to such interest through reports, presentations and correspondence. They caution however—". . . that those studying the Mississauga incident must be cautious about drawing conclusions relative to their own concerns about emergency planning."

7. Rescue

If certain types of casualty-producing hazardous materials incidents, such as those involving radiation contamination or pesticide poisoning, are treated as routine situations by emergency medical services and hospital personnel; a serious secondary emergency can easily be created. Rescue and care of contaminated casualties present special problems requiring special methods. There is presently a particularly high level of interest among response personnel and others regarding proper handling of victims contaminated by radiation.

The potential for exposure to radiation has increased substantially during the past 40 years. Stationary nuclear power generating facilities, radioactive wastes and waste sites, nuclear source materials widely used in industry and medicine, civilian and military special nuclear processing and storage facilities—and related transportation—have created a situation where nearly any community in the United States could become the scene of a casualty-producing radiation emergency. The incident at Three Mile Island at Harrisburg, Pennsylvania on March 28, 1979 led a number of communities to reassess their readiness to handle and care for persons contaminated during a hazardous materials incident involving radiation.

By the time emergency medical service (EMS) personnel arrive at the scene of a radiation incident police officers and/or firefighters should have secured the scene by establishing perimeter lines both to keep unnecessary persons out and possibly contaminated victims in. Normally, site control involves establishment of three concentric circles that are clearly identified by ropes, tape, barriers or other deliniation methods. The innermost circle is "dirty;" all persons and materials within are *presumed* to be contaminated. A buffer zone around the inner circle may be entered only by properly protected workers and may become contaminated as rescue operations progress. An outermost third circle deliniates a "clean" zone; every possible effort is made to keep the clean zone from becoming contaminated. Stringent control of zones is maintained to keep radiation contamination confined to the smallest possible area.

Any personnel or equipment entering the innermost or dirty zone must be monitored for contamination before being allowed to exit to the buffer zone, and decontaminated if necessary. Such monitoring and decontamination-if-necessary is repeated for personnel and equipment moving from the buffer zone to the outermost clean zone. If sufficient trained personnel are available, two separate teams are normally used to minimize transfer of contamination. One team works only in the contaminated/possibly contaminated zones; the other only at the edge of the clean zone. Decontaminated casualties and equipment are passed from one team to the other but the "clean" team never actually enters the buffer zone.

Within the contaminated or possibly contaminated zones experienced radiation monitoring personnel are used to identify the types and levels of radiation present, monitor both injured and noninjured persons, and assist with decontamination. The first rescue (EMS) crew clad in filter-type gas masks, disposable surgical gloves, clothlined rubber gloves, coveralls with hoods, shoe covers or boots, and impervious rain suits; and carrying a supply of sheets and plastic bags, crosses the outermost perimeter line from upwind (particulary if fire is present) making use of natural protective barriers such as terrain and buildings. This crew locates victims, carefully strips them of all clothing to remove gross contamination, places them on a stretcher atop a disposable sheet and covers the victim with another disposable sheet before passing the victim to a second EMS crew over the perimeter line to the clean area. Monitor readings are taken and recorded at each step of the process. As time is available the first crew will secure victims' clothing in plastic bags tagged with appropriate information. When this "inside" crew eventually exits the site they remove and bag their outer protective clothing, are monitored for contamination, and decontaminated if necessary before being allowed to cross to the clean zone. Once allowed into the clean zone, this first crew will shower and change clothing, bag and tag inner clothing previously worn, and report for medical evaluation.

Meanwhile, the second rescue (EMS) crew, dressed in protective clothing identical to that worn by the inside crew; remains in the clean zone to receive patients from the inside crew, record monitor readings a third time, and transport patients to a hospital. Once all patients have been delivered to the hospital, members of the second crew also remove and bag their outer protective clothing, are monitored by hospital staff for radiation contamination before being released and sent to shower with copius amounts of soap and water.

Stringent controls initiated by rescue (EMS) crews are maintained by hospital personnel who have already been informed by rescue crews of the number and condition of uncontaminated and contaminated casualties, the type of radiation source involved in the incident, and the form of radiation damage suspected or known. "Contaminated" persons are those to whom radioactive dust, particles or liquids had become or are physically attached to clothing or skin. "Incorporated" persons are those more serious cases involving inhalation or ingestion of radioactive matter, or open wounds contaminated by radioactive matter. An "irradiated" person has been penetrated by high levels of gamma radiation likely to cause tissue damage but is himself not radioactive; no radiation may be detectable on his person or clothing.

By the time ambulances begin arriving at the hospital emergency entrance, all nearby patients and unnecessary personnel will have been moved to other areas of the hospital, the entire area from the entranceway to the treatment area will have been roped off and the floor covered with plastic sheeting. Pregnant or possibly-pregnant personnel are assigned duties in other areas of the hospital. Within the treatment room loose equipment has been moved to other areas and immovable equipment covered with plastic sheeting. Here again plastic is used to cover the floor. Additional surfaces likely to be touched such as light switches and door handles are covered with tape, and any ventilation systems are turned off. A designated person remains outside the treatment room to obtain any materials needed by staff within, and passes materials across a line without ever actually entering the treatment room.

As patients arrive at the hospital they are examined in the ambulance by a physician and a radiation control person. Contaminated persons are placed on stretchers covered with plastic sheeting; persons who are both critically injured and contaminated are moved first. Ambulance attendants must remain with their vehicle at all times, and are not released until checked by the radiation control person. In the treatment room, medical personnel wear full surgical dress including surgical trousers and shirt, surgical mask, shoe covers taped to trousers, surgical gown, inner surgical gloves taped to sleeves and outer surgical gloves that can be changed as needed; dosimeters are worn, and readings are periodically reported to the radiation control person. During decontamination of a patient in the treatment room cotton swab swipe samples taken from ears, nose and mouth are labeled as to patient, site and time and stored in a closed glass container that is then placed in a lead shielded container. A radiation control person does a complete head-to-toe monitoring of the patient, and any areas and amounts of contamination are recorded. Lengthy decontamination and treatment is then undertaken. Since much decontamination involves washing the patient with water and other liquids, the treatment table has first been covered with plastic sheeting to form a trough to the bottom end of the table where run-off drains into a plastic bucket and is contained for later proper disposal. All previously contaminated areas on the patient are again swabbed, swabs are then marked and stored for later evaluation. Once again, the radiation control person monitors the patient from top to bottom. Additional plastic is laid on the floor before a clean stretcher is brought in, the patient is placed on the clean stretcher by persons not involved in the decontamination process, and removed from the treatment room only after the radiation control person has again monitored the patient as well as the clean stretcher.

Once the last contaminated patient has been handled, decontamination personnel remove all surgical clothing following a specified manner of undressing to minimize spread of contaminated materials and place their clothing in a sealed plastic container. They turn over their dosimeters to the radiation control person, are monitored for contamination one final time and then sent to the showers. The treatment area is then thoroughly cleaned and monitored. Disposable materials are collected in sealed plastic bags for proper disposal. Other instruments and materials are cleaned with soap and water; and all specimens collected, identified and evaluated. All hospital areas that might have become contaminated are remonitored. If readings indicate contamination; floors, walls and ceilings are washed until all contamination is removed. The stringent controls imposed on-scene, in transit, and at the hospital may seem like overkill to inexperienced persons, but each step of the process is absolutely necessary to minimize transfer of contaminants.

In any mass casualty producing hazardous materials incident, rescue and medical personnel can be presented with an extremely difficult task requiring repeated decision-making, instant and repeated communication of basic information previously determined, a method for insuring coordination and reducing confusion, avoidance of duplication of effort, and adequate recordkeeping—all tasks that must be performed within extremely tight time constraints. An international field triage tag called METTAG (Medical Emergency Triage Tag) developed by and available from the *Journal of Civil Defense* (P.O. Box 910, Starke, Florida 32091. Tel. 909-964-5397) provides a systematic approach to such tasks. Whether triage is performed on-scene by EMS personnel or physicians, or later at a medical facility, METTAG and similar tagging systems promote effectiveness by helping to coordinate and expedite processing of disaster casualties for appropriate medical care. (See Figure 10 for a black-and-white facsimile METTAG triage tag.)

Multicolored triage tags are made of heavy cardstock nearly impervious to weather or rough handling and are designed to provide maximum visibility, clarity and ease of handling under a wide range of emergency conditions. Individual serial numbers printed repeatedly on MET-TAG and its multiple tear-offs provide temporary data on a casualty even when basic data cannot be obtained from the victim due to death or lack of consciousness or when the emergency situation precludes initially obtaining such information. Perforated, colored, serial-numbered tear-offs permit immediate visible priority injury designations, avoid time-consuming repetitive casualty checks, and may be used as temporary ambulance records, to mark positions where victims were located at the scene, or to identify personal effects and

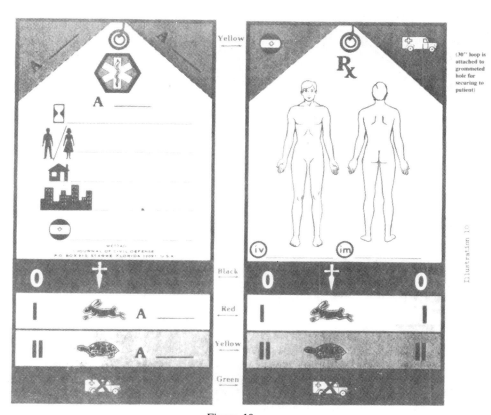

Figure 10.

clothing of each casualty. Triage tags use colors, numbers and symbols for communication so usage is not dependent on any one language.

When a decision is made that triage should be performed at an incident scene involving mass casualties; trained, experienced medical personnel move in to locate victims and provide an initial evaluation of medical treatment required. A triage tag with any one of four priority markers exposed is attached to each victim. Priority I (Red) indicates immediate care is required. Priority II (Yellow) indicates an urgent need for treatment although time is not a critical factor. Priority III (Green) indicates treatment is needed but the need is not urgent. Priority 0 (Black) indicates the victim is either dead or mortally injured and attention should be witheld until all others are cared for. Using METTAG triage tags, all casualties with Red markers exposed would be treated/transported first, those with Yellow markers exposed second, and those with Green markers exposed third. Those with Black markers would be dealt with last. Serialized tear-offs are used to mark a victim's clothing or personal effects, to identify the location on-site where the victim was located, and for other purposes.

13 BUILDING A LIBRARY

A number of specific publications commonly appear in reference libraries maintained by hazardous materials incident response personnel. On the pages that follow is a sampling of publications often used by responders for reference and training. Where given, prices are approximate and should be checked with the publisher or supplier before ordering.

1. *Acute Chemical Disasters: A Perspective For And Guide To Improving Community Response,* The Disaster Research Center, 128 Derby Hall, 154 North Oval Mall, The Ohio State University, Columbus, Ohio 43210 ($5.00).

2. *Air Freight Guide for Hazardous Materials,* Flying Tigers Line, 7401 World Way West, P.O. Box 92935, Los Angeles, California 90009 ($5.00).

3. *American National Standard for the Storage and Handling of Anhydrous Ammonia,* (Standard #K61.1), American National Standards Institute, Inc., 1430 Broadway, New York, New York 10018.

4. *Basic Principles· of Radiation Protection,* Training Resource Center (HFX-70), DTMA, BRH, FDA, 5600 Fishers Lane, Rockville, Maryland 20857 (Free).

5. *Basic Requirements for Personnel Monitoring,* Unipub, 1180 Avenue of the Americas, New York, New York 10036 ($8.75).

6. *Biological Affects of Neutrons TR38,* Training Resource Center (HFX-70), DTMA, BRH, FDA, 5600 Fishers Lane, Rockville, Maryland 20857 (Free).

7. *Chemical Engineers Handbook,* Robert H. Perry and Cecil H. Chilton, McGraw-Hill Book Company, Princeton Road, Hightstown, New Jersey 08520 ($65.50).

8. *Chemical Hazards Response Information System,* (U.S. Coast Guard Manual)
   Vol. 1: *Condensed Guide to Chemical Hazards.*
   Vol. 2: *Hazardous Chemical Data.*
   Vol. 3: *Hazard Assessment Handbook.*
   Vol. 4: *Response Methods Handbook.*
   Superintendent of Documents, U.S. Government Printing Office, Washington, D.C. 20402.

9. *Chemical Safety Slide Rule,* (SN #129.91–9), National Safety Council, 444 N. Michigan Ave., Chicago, Illinois 60611 ($6.00).

10. *Chemistry of Hazardous Materials,* Eugene Meyer, Prentice Hall, Inc., 200 Old Tappan Road, Old Tappan, New Jersey 07675 ($25.00).

11. *Chlorine Manual* (Publ. #1), The Chlorine Institute, 342 Madison Avenue, New York, New York 10017 ($5.00).

   Also Available from the Chlorine Institute

   Instruction Booklet: Chlorine Institute Emergency Kit "A" for 100 lb. and 150 lb. Chlorine Cylinders, (Publ. #1B/A) ($6.00).

   Instruction Booklet: Chlorine Institute Emergency Kit "B" for Chlorine Ton Containers. (Publ. #1B/B) ($6.00).

   Instruction Booklet: Chlorine Institute Emergency Kit "C" for Chlorine Tank Cars and Tank Trucks, (Publ. #1B/C) ($6.00).

   Location of Chlorine Emergency Kits. (Publ. #35) ($5.00).

12. *Clinical Toxicology of Commercial Products,* Gosselih, Hodge, Smith and Gleason, Williams & Lilkins Publishing, Baltimore, Maryland ($80.00).

13. *Code of Federal Regulations Title 49 Transportation* (Parts 100–177), *Code of Federal Regulations Title 49 Transportation* (Parts 178–199), Superintendent of Documents, U.S. Printing Office, Washington, D.C. 20402.

14. *Condensed Chemical Dictionary,* Gessner G. Hawley, Van Nostrand Reinhold Company, 135 West 50th Street, New York, New York 10020 ($42.50).

15. *Control of Internal Radiation Hazards,* TP51 Training Resource Center (HFX-70), DTMA, BRH, FDA, 5600 Fishers Lane, Rockville, Marlyand 20857 (Free).

16. *Correlation of Exposure Dose and Absorbed Dose,* TP52 Training Resource Center (HFX-70), DTMA, BRH, FDA, 5600 Fishers Lane, Rockville, Maryland 20857 (Free).

17. *Correlation of Units of Activity and Exposure,* TP53 Training Resource Center (HFX-70), DTMA, BRH, FDA, 5600 Fishers Lane, Rockville, Maryland 20857 (Free).

18. *Critical Reviews in Toxicology (Vol. XI),* CRC Press, Inc., 2000 Corporate Blvd., N.W. Boca Raton, Florida 33431 ($96.00).

19. *Dangerous Properties of Industrial Materials,* N. Irving Sax, Van Nostrand Reinhold Company, 135 West 50th Street, New York, New York 10020 ($57.50).

20. *Dealing With Chlorine Emergencies-Fire,* The Chlorine Institute, Inc., 342 Madison Avenue, New York, New York 10017 ($1.50).

21. *Determination of Halflife,* TP265 Training Resource Center (HFX-70), DTM, BRH, FDA, 5600 Fishers Lane, Rockville, Maryland 20857 (Free).

22. *Diagnostic X-Ray Equipment,* TP65 Training Resource Center (HFX-70), DTMA, BRH, FDA, 5600 Fishers Lane, Rockville, Maryland 20857 (Free).

23. *Directory of Chemical Producers,* Stanford Research Institute International, 333 Ravenswood Avenue, Menlo Park, California 94025 ($595.00).

24. *Directory of Personnel Responsible for Radiological Health Programs,* U.S. Public Health Service Food & Drug Administration, Bureau of Radiological Health, 5600 Fishers Lane, Rockville, Maryland 20857.

25. *Do's and Dont's—Explosives,* (Publication #4), Institute of Makers of Explosives, 1575 Eye Street, N.W., Suite 550, Washington, D.C. 20005 (Free).

26. *Effects of Exposure to Toxic Gases,* ("First Aid & Medical Treatment"), Braker, Mossman and Siegel, Matheson, Lyndhurst, New Jersey 07071 ($17.00).

27. *Effects of Radiation on Living Tissue & Chemical Structure,* TP67 Training Resource Center (HFX-70), DTMA, BRH, FDA, 5600 Fishers Lane, Rockville, Maryland 20857 (Free.)

28. *Emergency Handling of Hazardous Materials In Surface Transportation,* (Manual of the Bureau of Explosives of the Association of American Railroads), Bureau of Explosives/AAR, 1920 L Street, N.W., Washington, D.C. 20036 ($12.50).

29. *Emergency Handling of Radiation Accident Cases,* U.S. Department of Energy, Assistant Secretary for Environment, Washington, D.C. 20545 (Free).

30. *Emergency Repair of Pressure Tank Car Leaks,* Phillips Petroleum, Bartlesville, Oklahoma 74033 (Free).

31. *Exploxives & Toxic Hazardous Materials,* James Meidl, Glencoe Publishing Company, Inc., Front and Brown Streets, Riverside, New Jersey 08370 ($21.95).

32. *Farm Chemicals Handbook,* Meister Publishing Company, 37841 Euclid Avenue, Willoughby, Ohio 44094 ($35.00).

33. *Firefighters Handbook of Hazardous Materials,* Charles J. Baker, Maltese Enterprises, Inc., P.O. Box 34048, Indianapolis, Indiana 46234 ($10.00).

34. *Fire Officer's Guide to Dangerous Chemicals,* (Publication #FSP-36A), Charles W. Bahme, National Fire Protection Association, Batterymarch Park, Quincy, Massachusetts 02269 ($12.00).

35. *Fire Officer's Guide to Disaster Control,* (Publication #FSP-48), Charles W. Bahme, National Fire Protection Association, Batterymarch Park, Quincy, Massachusetts 02269 ($23.50).

36. *Fire Protection Guide on Hazardous Materials,* (Manual of the NFPA), National Fire Protection Association, Batterymarch Park, Quincy, Massachusetts 02269 ($17.50).

37. *Flammable Hazardous Materials,* James H. Meidl, Glencoe Publishing Company, Inc., Front and Brown Streets, Riverside, New Jersey 08370 ($21.95).

38. *Flash Point Index of Trade Name Liquids,* (Publication #SPP-51), National Fire Protection Association, Batterymarch Park, Quincy, Massachusetts 02269 ($8.50).

39. *Fruehauf Tank & Commodity Bulletins,* Fruehauf Corporation, Omaha, Nebraska 68101.

40. *GATX Tank Car Manual,* General American Transportation Corporation, 120 South Riverside Plaza, Chicago, Illinois 60606.

41. *Guide for Safety and The Chemical Laboratory,* Van Nostrand Reinhold Company, 135 West 50th Street, New York, New York 10020.

42. *Handbook for Chemical Technicians,* Howard J. Strauss, Ph.D., McGraw-Hill Book Company, Princeton Road, Hightstown, New Jersey 08520 ($32.50).

43. *Handbook of Analytical Toxicology General Data,* (Vol. 1, Section A; Vol. 2, Section B), CRC Press, 2000 Corporate Blvd., N.W., Boca Raton, Florida 33431 ($66.00).

44. *Handbook of Compressed Gases,* (Compressed Gas Association), Van Nostrand Reinhold Company, 135 West 50th Street, New York, New York 10020 ($44.50).

45. *Handbook of Emergency Toxicology,* Sidney Kay, Ph.D., Charles C. Thomas Publishing, 301–327 East Lawrence Avenue, Springfield, Illinois.

46. *Handbook of Laboratory Safety,* CRC Press Inc., 2000 Corporate Blvd., N.W., Boca Raton, Florida 33431 ($59.95).

47. *Handbook of Poisoning, Diagnosis and Treatment,* Robert H. Dreisbach, Lange Medical Publications, Inc., Drawer L, Los Altos, California, 94022 ($19.00).

48. *Handling Guide for Potentially Hazardous Materials,* The Richard B. Cross Company, 103 South Howard Street, P.O. Box 405, Oxford, Indiana 47971.

49. *Handling Radiation Emergencies,* Robert G. Purington & H. Wade Patterson, National Fire Protection Association, Batterymarch Park, Quincy, Massachusetts 02269 ($14.50).

50. *Hazardous Chemical Spill Clean-Up,* Edited by J. S. Robinson, Noyes Data Corporation, Park Ridge, New Jersey 07656.

51. *Hazardous Commodity Handbook,* National Tank Truck Carriers, Inc., 1616 P Street, N.W., Washington, D.C. 20036.

52. *Hazardous Materials,* Warren E. Isman & Gene P. Carlson, Glencoe Publishing Company, Front and Brown Street, Riverside, New Jersey 08370 ($21.95).

53. *Hazardous Materials,* Leroy Schieler & Denis Pauze, Delmar Publications, Albany, New York.

54. *Hazardous Materials—A Guide For Firefighters,* Myron P. Gerton, Colorado Div. of Highway Safety, 4201 E. Arkansas Avenue, Denver, Colorado 80222.

55. *Hazardous Materials: A Guide For State and Local Officials,* (Document #050–012–00187–1), Superintendent of Documents, U.S. Government Printing Office, Washington, D.C. 20402 ($4.50).

56. *Hazardous Materials Emergency Response Guidebook,* (The D.O.T. Manual), Department of Transportation, Materials Transportation Bureau, Research & Special Programs Administration, Washington, D.C. 20590 (Free).

57. *Hazardous Materials Guide,* United Parcel Service, 51 Weaver Street, Greenwich Office Park, Greenwich, Connecticut 06830.

58. *Hazardous Materials Guidebook,* American Trucking Association, Traffic Department, 1616 P Street, N.W., Washington, D.C. 20036.

59. *Hazardous Materials Handbook,* James H. Meidl, Glencoe Publishing Company, Front and Brown Streets, Riverside, New Jersey 08370 ($14.95).

60. *Hazardous Materials Handbook,* American Trucking Association, 1616 P Street, N.W., Washington, D.C. 20036.

61. *Hazardous Materials Injuries,* ("A Handbook for Pre-Hospital Care"), Douglas R. Stutz, Ph.D., Robert C. Ricks, Ph.D., and Michael F. Olsen, Bradford Communications Corporation, 7500 Greenway Center Drive, Greenbelt, Maryland 20770 ($16.95).

62. *Hazardous Materials Reference Manual,* ("Complete Pocket Digest"), Labelmaster, 7525, N. Wolcott Avenue, Chicago, Illinois 60626 ($5.00).

63. *Hazardous Materials Transportation Accidents,* (Publication #SPP-49) (Collection of 38 *Fire Journal* and *Fire Command* Articles), National Fire Protection Association, Batterymarch Park, Quincy, Massachusetts 02269 ($7.50).

64. *Hazardous Materials Transportation Accidents, Illustrated,* Explosives Research Institute, Inc., P.O. Box 2103, Scottsdale, Arizona 85252 ($32.00).

65. *Highly Hazardous Materials Spills and Emergency Planning,* J. E. Zajic and W. A. Himmelman, Marcel Dekker, Inc., 270 Madison Avenue, New York, New York 10016 ($35.00).

66. *Index to the Hazardous Materials Regulations,* (Title 49, Code of Federal Regulations, Parts 100–199), U.S. Department of Transportation, Materials Transportation Bureau, Office of Hazardous Materials Operations, Washington, D.C. 20590.

67. *Industrial Gases Data Book,* Airco Welding Products, Clermont Terrace, Union, New Jersey 07083.

68. *Industrial Applications of Radiation,* TP-84 Training Resource Center (HFX-70), DTMA, BRH, FDA, 5600 Fishers Lane, Rockville, Maryland 20857 (Free).

69. *International Air Transport Association Dangerous Goods Regulations,* UNZ & Co., (Catalog #65–660), 190 Baldwin Avenue, P.O. Box 308, Jersey City, New Jersey 07303 ($40.00).

70. *International Maritime Consultive Organization Dangerous Goods Code,* UNZ & Co., 190 Baldwin Avenue, P.O. Box 308, Jersey City, New Jersey 07303 ($135.00 per five-volume set).

71. *Lange's Handbook of Chemistry,* John A. Dean, McGraw-Hill Book Company, Princeton Road, Hightstown, New Jersey 08520 ($41.50).

72. *Linde Specialty Gases, Safety Precautions and Emergency Procedures,* Linde Division, Union Carbide Corp., 51 Cragwood Road, P.O. Box 372, South Plainfield, New Jersey 07080.

73. *Managing Hazardous Substances Accidents,* Al J. Smith, Jr., McGraw-Hill Book Company, Princeton Road, Hightstown, New Jersey 08520 ($21.95).

74. *Medical First Aid Guide For Use in Accidents Involving Dangerous Goods,* Unipub, 1180 Avenue of the Americas, New York, New York 10036 ($16.50).

75. *Medical X-Ray Protection,* TP105, Training Resource Center (HFX-70), DTMA, BRH, FDA, 5600 Fishers Lane, Rockville, Maryland 20857 (Free).

76. *Merck Index,* ("An Encyclopedia of Chemicals & Drugs"), Merck and Company, Inc., P.O. Box 2000, Rahway, New Jersey 07065 ($25.00).

77. *Mutual Aid Information Handbook,* (A listing of carrier safety representatives by state; includes business address and telephone, residence telephone), National Tank Truck Carriers, 1616 P Street, N.W., Washington, D.C. 20036.

78. National Fire Protection Association Codes and Standards (Selected).

NFPA 30: Flammable & Combustible Liquids Code ($9.00).

NFPA 43C: Code for Storage of Gaseous Oxidizing Materials ($6.00).

NFPA 43D: Storage of Pesticides In Portable Containers ($6.00).

NFPA 44A: Manufacture, Storage, Transportation of Fireworks ($7.50)

NFPA 45: Fire Protection for Labratories Using Chemicals ($7.50).

NFPA 49: Hazardous Chemicals Data ($10.50).

NFPA 56F: Nonflammable Medical Gas Systems ($7.00).

NFPA 58: Liquefied Petroleum Gases, Storage & Handling ($9.50).

NFPA 59: Liquefied Petroleum Gases at Utility Plants ($7.50).

NFPA 59A: Liquefied Natural Gas, Production, Storage and Handling ($8.50).

NFPA 325M: Fire Hazard Properties of Flammable Liquids, Gases, Volatile Solids ($7.50).

NFPA 329: Underground Leakage of Flammable & Combustible Liquids ($7.00).

NFPA 491: Hazardous Chemical Reactions, Manual of ($10.50).

National Fire Protection Association, Batterymarch Park, Quincy, Massachusetts 02269.

79. *National Tank Truck Carriers Directory,* National Tank Truck Carriers, 1616 P Street, N.W., Washington, D.C. 20036.

80. *NIOSH/OSHA Occupational Health Guidelines for Chemical Hazards,* NIOSH, 5600 Fishers Lane, Rockville, Maryland 20857.

81. *NIOSH/OSHA Pocket Guide to Chemical Hazards,* (Publications #78–210; Stock #017–033–00342–4), Superintendent of Documents, U.S. Government Printing Office, Washington, D.C. 20402 ($7.50).

82. *NIOSH Registry of Toxic Effects of Chemical Substances,* (HE 20.7112:980/vl. 2. Stock #017–033–00399–8) Superintendent of Documents, U.S. Government Printing Office, Washington, D.C. 20402 ($27.00).

83. *Oil Spills and Spilled Hazardous Substances,* U.S. Environmental Protection Agency, Oil & Special Materials Control Division, 401 M Street, S.W., Washington, D.C. 20460 (Free).

84. *Peel Regional Police Force and The Mississauga Evacuation (The),* ("How a police force reacted to a major chemical emergency."), Minister of Supply and Services Canada (Cat. #J–566–1/1980E).

85. *Personal Instruments,* TP129, Training Resource Center (HFX-70), DTMA, BRH, FDA, 5600 Fishers Lane, Rockville, Maryland 20857 (Free).

86. *Pesticide Handbook—Entoma,* Entomological Society of America, 4603 Calvert Road, Box AJ, College Park, Maryland 20740.

87. *Pesticide Products Safety Manual,* Shell Chemical Company, 2401 Crow Canyon Road, San Ramon, California 94583.

88. *Planning for Handling of Radiation Incidents,* (Safety Series #32–STI/PUB/227), Unipub, 1180 Avenue of the Americas, New York, New York 10036 ($9.75).

89. *Proceedings of National Conference on Control of Hazardous Materials Spills*
    1980: (494 pages) Order #HM50. ($30.00).
    1978: (458 pages) Order #HM56. ($15.00).
    1978: (456 pages) Order #MM56. ($10.00).
    Publications Department, Hazardous Materials Control Research Institute, 9300 Columbia Blvd., Silver Springs, Maryland 20910.

90. *Proceedings of National Conference on Management of Uncontrolled Hazardous Waste Sites*
    1981: (426 pages) Order #UHWL. ($35.00).
    1980: (285 pages) Order #UHWO. ($30.00).
    Publications Department, Hazardous Materials Control Research Institute, 9300 Columbia Blvd., Silver Spring, Maryland 20910.

91. *Properties and Production of X-Rays,* TP483, Training Resource Center (HFX-70), DTMA, BRH, FDA, 5600 Fishers Lane, Rockville, Maryland 20957 (Free).

92. *Protective Clothing for Chlorine,* (Publication 65), The Chlorine Institute, 342 Madison Avenue, New York, New York 10017 ($4.00).

93. *Quantity and Units of Radiation,* TP481, Training Resource Center (HFX-70), DTMA, BRH, FDA, 5600 Fishers Lane, Rockville, Maryland 20857 (Free).

94. *Radiological Emergencies—A Handbook for Emergency Responders,* N. A. Klimenki & J. F. Redington, Bradford Communications Corporation, 7500 Greenway Center Drive, Greenbelt, Maryland 20770 ($12.95).

95. *Radiation Protection Guide,* TP145, Training Resource Center (HFX-70), DTMA, BRH, FDA, 5600 Fishers Lane, Rockville, Maryland 20857 (Free).

96. *Radioactivity,* TP152, Training Resource Center (HFX-70), DTMA, BRH, FDA, 5600 Fishers Lane, Rockville, Maryland 20857 (Free).

97. *Recognition and Management of Pesticide Poisoning,* U.S. Environmental Protection Agency, Office of Pesticide Poisoning, Washington, D.C. 20460.

98. *Red Book on Transportation of Hazardous Materials,* Lawrence Bierlein, Cahners Publishing Company, 221 Columbus Avenue, Boston, Massachusetts 02116 ($100.00).

99. *Response Method Handbook,* (Document CG 446-4), Superintendent of Documents, U.S. Government Printing Office, Washington, D.C. 20402 ($6.50).

100. *Risk Evaluation for Protection of the Public in Radiation Accidents,* (Safety Series #21–STI/PUB/124), Unipub, 1180 Avenue of the Americas, New York, New York 10036 ($7.00).

101. *Safety in Transportation, Storage, Handling and Use of Explosives,* Publication #17), Institute of Makers of Explosives, 1575 Eye Street, N.W., Suite 550, Washington, D.C. 20005 (Free).

102. *Service Bulletins,* Shippers Car Line Division, ACF Industries, Inc., 620 North 2nd Street, St. Charles, Missouri 63301.

103. *Shell Chemical Safety Guide,* Shell Chemical Company, 2410 Crow Canyon Road, San Ramon, California 94583.

104. *Shielding Properties of Common Building Materials for X-Ray and Gamma Rays,* Training Resource Center (HFX-70), DTMA, BRH, FDA, 5600 Fishers Lane, Rockville, Maryland 20857 (Free).

105. *Sources of Radiation Exposure,* TP465, Training Resource Center (HFX-70), DTMA, BRH, FDA, 5600 Fishers Lane, Rockville, Maryland 20857 (Free).

106. *Threshold Limit Values (TLVs) for Chemical Substances and Physical Agents in the Work Room,* American Conference of Government Industrial Hygenists, Publications Office, 6500 Glenway Avenue, Cincinnati, Ohio 45211 ($4.00).

107. *Toxicology of Drugs & Chemicals,* W. Deichmann & H. W. Gerade, Academic Press, Harcourt Brace Jovanovich Building, 1001 Polk Street, San Francisco, California 94109.